Human–Computer Interaction Series

More information about this series at http://www.springer.com/series/6033

Gerrit Meixner · Christian Müller
Editors

Automotive User Interfaces

Creating Interactive Experiences in the Car

Springer

Editors
Gerrit Meixner
UniTyLab
Heilbronn University
Heilbronn
Germany

Christian Müller
Automotive IUI
German Research Centre for Artificial Intelligence (DFKI)
Saarbrücken
Germany

ISSN 1571-5035
Human–Computer Interaction Series
ISBN 978-3-319-49447-0 ISBN 978-3-319-49448-7 (eBook)
DOI 10.1007/978-3-319-49448-7

Library of Congress Control Number: 2016959404

Printed on acid-free paper

This Springer imprint is published by Springer Nature
The registered company is Springer International Publishing AG
The registered company address is: Gewerbestrasse 11, 6330 Cham, Switzerland

Preface

This book is about automotive user interfaces. In the last years the importance of user interfaces for in-vehicle usage has increased strongly. Different studies show that over 80% of today's innovations in the automotive industry are based on car electronics and its software. These innovations can be categorized into hidden technologies (e.g., ASP, ESP), comfort functions (e.g., navigation, communication, entertainment) or driver assistance (e.g., distance checking). Especially the last two categories have to be configurable by the driver and therefore require a certain amount of driver interaction. This results in a need for a modern and consistent automotive user interface which on the one hand allows the configuration of these systems and on the other hand conforms to the specialized requirements of the automotive industry. Some of these requirements are: the interaction devices have to be integrated into a limited space; the automotive user interface has to be intuitively usable and adaptable, since drivers generally do not get an extensive explanation and the automotive user interface has to be very easy to use and should distract the driver as little as possible from his main task of driving. The increased complexity of automotive user interfaces, the importance of using consumer electronic devices like smartphones in the car as well as autonomous driving has induced a lot of research at universities and industrial companies.

The specific chapters in this book cover a relatively broad spectrum of detailed research topics in the area of automotive user interfaces concerning, e.g. usability and user experience, interaction techniques and technologies, applications, etc. This book provides an outstanding overview as well as deep insights into the area of automotive user interfaces, which is an important topic in the field of human–computer interaction. Besides aiming to be a reference in its area, this book is intended as a very significant and valuable source for professional practitioners, researchers as well as senior and postgraduate computer science and engineering students.

This book could not be completed without the help of many people. We would like to thank all the authors for their contribution to the book. Finally, we would like to thank Beverley Ford and James Robinson at Springer (London, UK) for their support and assistance in publishing this book in a timely fashion.

Heilbronn, Germany Gerrit Meixner
Saarbrücken, Germany Christian Müller
September 2016

Contents

Part III Interaction Techniques and Technologies

Part IV Tools, Methods and Processes

Part I
Introduction and Basics

Chapter 1
Retrospective and Future Automotive Infotainment Systems—100 Years of User Interface Evolution

Gerrit Meixner, Carina Häcker, Björn Decker, Simon Gerlach, Anne Hess, Konstantin Holl, Alexander Klaus, Daniel Lüddecke, Daniel Mauser, Marius Orfgen, Mark Poguntke, Nadine Walter and Ran Zhang

Abstract The history of automotive HMI development reveals that the development of new interactive in-car functionalities (such as infotainment systems) has often been influenced by upcoming new technologies that customers got used to in their daily lives. Examples of such technologies include the first in-car radio, which was introduced around 1922, or the first in-car phone, which was introduced around 1952. Today, a car without such functionalities is hard to imagine and the automotive industry is aiming to develop and integrate more and more innovative functionality to stay competitive on the market. The development of such functions is motivated by the construction of safer, more efficient, and more comfortable

G. Meixner (✉) · C. Häcker
UniTyLab, Heilbronn University, Max-Planck-Strasse 39, 74081 Heilbronn, Germany
e-mail: gerrit.meixner@hs-heilbronn.de

B. Decker
Comlet Verteilte Systeme GmbH, Amerikastrasse 27, 66482 Zweibrücken, Germany

S. Gerlach
Vorentwicklung Connected Car and Infotainment, Volkswagen AG, 38440 Wolfsburg, Germany

A. Hess · K. Holl · A. Klaus
Fraunhofer IESE, Trippstadter Strasse 121, 67663 Kaiserslautern, Germany

D. Lüddecke
Group Research/Human Factors and HMI, Volkswagen AG, P.O. Box 1777, 38436 Wolfsburg, Germany

D. Mauser
Communications (INST-ICM/BSV-BS), Bosch Software Innovations GmbH, Stuttgarterstrasse 130, 71332 Waiblingen, Germany

M. Orfgen
Innovative Factory Systems (IFS), DFKI, Trippstadter Strasse 122, 67663 Kaiserslautern, Germany

G. Meixner and C. Müller (eds.), *Automotive User Interfaces*, Human–Computer Interaction Series, DOI 10.1007/978-3-319-49448-7_1

vehicle systems. Current trends in the area of in-car infotainment applications include, for instance, Internet-based applications or social network applications, whereas extendable, hybrid, adaptive, or even personalized HMI are emerging as future trends. Not only technologies have evolved throughout history, the development processes themselves have also been adapted continuously due to the challenges the automotive industry had to face with new technologies. Thus, the authors have summarized their experiences, their knowledge, and the results of literature studies in this article which covers the history of automotive HMI development from the past in 1922 to the present with an outlook on upcoming trends for future automotive user interfaces.

Whoever wishes to foresee the future must consult the past—Machiavelli

1.1 Introduction

The development of Human Machine Interfaces (HMI) is a complex, interdisciplinary challenge (Bader and Fallast 2012). Besides the technical aspects, the development is also challenged by the need to adhere to cognitive principles manifested in the need to choose interaction patterns that fit the mental model of the user. For example, the efficient usage of electric windows by pressing a flip switch is possible if the window is lowered when the bottom of the flip switch is pressed. The other way around would not be intuitive. The technical realization is responsible for the adequate implementation of the concept. The simplicity of developing automotive HMIs, as in the example above, was common many decades ago. Comparing present and future developments, the main differences to past developments are the aspects of information processing and entertainment. Hence, this article focuses primarily on the HMI of automotive infotainment systems using the representative term "automotive HMI" or simply "HMI."

The types and the complexity of automotive HMIs have rapidly changed in the last decades proportional to the development of computer systems: from

---

M. Poguntke
Advanced User Experience Engineering, Mercedes-Benz Research and Development North America, Inc., 309 North Pastoria Ave, Sunnyvale, CA 94085, USA

N. Walter
Advanced Development Car Multimedia, Robert Bosch GmbH, Daimlerstrasse 6, 71226 Leonberg, Germany

R. Zhang
Diesel Gasoline Systems (DGS-EC/ECD2), Robert Bosch GmbH, Wernerstrasse 51, 70469 Stuttgart, Germany

rudimentary command line interfaces to a wide variety of graphical user interfaces, speech dialog systems, and gesture-based systems like touch interfaces.

The first automotive HMIs were primarily mechanical. Their main purpose and implemented functionalities aimed at providing the driver with relevant information about the car or about driving, such as speed, gas level, or rev counter. Later on, displaying only this information was not sufficient anymore. The drivers also wanted to be entertained while driving.

Therefore, entertainment functions like radios were progressively integrated into the car, leading to an increase in automotive HMI complexity. The HMI as well as the different functions together became an infotainment system, i.e., a system that combines the provision of information with entertainment functionalities (Bosshart and Hellmueller 2009).

In Fig. 1.1, an example of an early infotainment system is depicted. The picture shows the car dashboard of a Ford Taunus in 1958. Moreover, the type of information provided to the driver has also evolved and been enhanced over time. Besides status information about the car, information about the traffic or navigation has been integrated as well. Today, standard functionalities of HMIs encompass the display of vehicle-related information, advanced driver assistance functionalities, and entertainment components like radio, media player, etc. An example of today's developments is shown in Fig. 1.2. It illustrates the driver's view in a cockpit of the Ford S-Max, which was initially delivered in 2016.

Fig. 1.1 FORD Taunus 17M P2 (TL) deluxe two door 1958 steering wheel (Wikimedia Commons; User: Yeti.bigfoot 2009)

Fig. 1.2 Ford S-Max 2015 Interior (Wikimedia Commons; User: Ranger 1 2016)

Due to the increased complexity of the HMI, which consists of a variety of different input and output interfaces, its usability has become a very important quality factor (Ariza et al. 2009). Modern HMIs consist of a graphical user interface and a control unit as well as speech dialog systems and gesture-based systems like touch interfaces. The application of up-to-date hardware and software components enables a steadily rising number of use cases.

Modern automobiles provide complex functionalities and can be connected to different mobile devices. The complexity in functionality has a direct impact on the complexity of the HMIs because the driver has to manage the provided functions. The established HMIs of the past have to be improved and adapted to those requirements to make them more modern and innovative and to reduce the complexity. Therefore, designs from everyday interfaces of the users could be taken as role model for the HMI in the automotive field. For example, users know the graphical user interface of their smartphones and are used to their utilization.

Electrical interfaces have emerged and evolved rapidly and are continually replacing their mechanical counterparts due to many advantages. For instance, the replacement of mechanical mirrors with cameras allows expanding the field of vision and eliminates blind angles. The range of expectations is also widening, i.e., simple driving support versus high-quality entertainment. This variety of functionalities is always a competitive attribute for automobile manufacturers but this requires also an optimal handling of a large number of different user qualifications. A big challenge is for example to design infotainment systems in a way that also people with minor technical background can easily use them. In this domain, the ability of configuration with respect to target groups and individuals is also an important issue when it comes to increasing usability.

Varying preconditions directly affect the development of HMI-based software. New tools and methods are necessary to handle the development of more and more complex functionalities. For example, in the future, tools have to support the design of graphical user interfaces based on 3D graphics, or manage the interface of multimodal systems as well.

In addition, the significance of certain development process phases that might have been neglected until now is increasing, such as the process of software testing. With growing complexity, the cost of testing rises, too. More complex HMIs result in more complex testing methods and automated testing becomes even more essential due to the numerous test cases. These facts finally lead to the challenge that within the entire HMI developing process, new concepts and tools are necessary to handle the steadily growing complexity. The process requires new roles and responsibilities to be assigned, such as requirements and usability engineers. Besides that, education in new concepts is necessary. The concept of extendable HMIs for example could reduce the problems regarding the up-to-dateness of the HMI. Functionalities can be added after roll-out, which decouples the development cycle from the life cycle of a vehicle.

In the literature, articles can be found reporting on the histories of user interfaces in general like Myers (1998), Jørgensen and Myers (2008), or Myers et al. (2000). However, information can also be found on facts and challenges in the history of HMI development within the automotive domain. Thus, the authors have summarized their experiences, their knowledge, and the results of literature studies in this article, which covers the history of HMI development from the past in 1922, to the present and also provides an outlook on upcoming trends for future automotive user interfaces.

The remainder of this paper is structured as follows: Sect. 1.2 describes the history of automotive HMI development. The representative infotainment systems are developed by Audi and Daimler. Section 1.3 describes early technology trends and development processes. Section 1.4 contains the state of the art. Besides the driving factors and the input/output capabilities, the development process is one of the main topics. The HMI development of the future is part of Sect. 1.5, which tries to predict tomorrow's HMI, e.g., by considering upcoming trends and user expectations. Past, present, and future HMI development is summarized in Sect. 1.6.

1.2 The Past of Automotive HMI

1.2.1 From 1915 till 1993

As outlined in the introduction, the first HMIs in vehicles were primarily mechanical. Their main purpose and implemented functionalities aimed to provide relevant information about car functions in the form of simple diagrams and digits

Fig. 1.3 Mercedes-Benz 22/50 Open Tourer (1915) (Meixner 2013b)

that were required for vehicle handling to the driver. The HMI of the *1915 Mercedes-Benz 22/50 Open* Tourer shown in Fig. 1.3 is an example for such a simple HMI. There are only a few knobs and mechanical displays. But in addition to its simplicity the instruments were placed in the footwell which seems untypical from a today's perspective due to higher driver distraction.

Besides the development of motor and chassis technologies, the improvement of comfort in cars became another important topic as well. One aspect of such improvements was the addition of entertainment, information, and telematics systems to vehicles.

1.2.1.1 Music

In 1922, the first car radio was introduced experimentally (Gesellschaft für Unterhaltungs- und Kommunikationselektronik 2010) on a Ford Model T (see Fig. 1.4). The American automobile manufacturer Packard introduced the Packard 120 in 1935. Figure 1.5 shows the cockpit of a 1936 built Packard 120 and also indicates that comfort and quality became more and more important in these years. All instruments are placed next to the driver and are designed in a consistent more superior look.

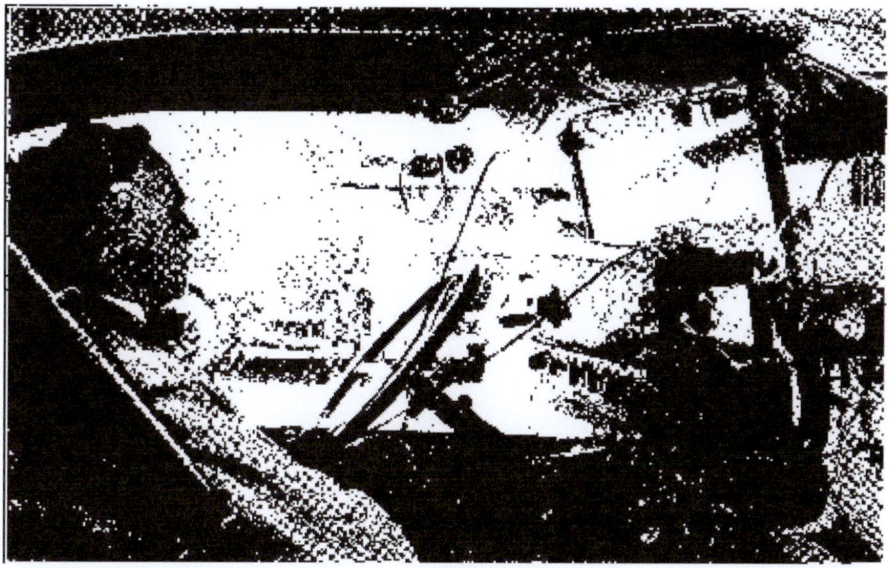

Fig. 1.4 First car radio (General Photographic Service 1922)

Fig. 1.5 Packard 120 (1936) (Meixner 2013d)

Fig. 1.6 Mercedes-Benz 300 SL (1955) (Meixner 2013c)

In 1954 radios were still accessories available only for some cars. Even in high-quality cars like the *Mercedes-Benz 300 SL* radios were no standard. Figure 1.6 shows a 1955 Mercedes-Benz 300 SL without a radio. But the significance of in-car entertainment grew steadily. In the 1960s in-car radios became more and more popular probably from the fact that there was not yet any possibility to listen to other media than radio broadcasting.

In 1956 an in-car record player called "Highway Hi-Fi" which was built by CBS/Columbia was offered in vehicles from Chrysler, Dodge, DeSoto, and Plymouth. With this custom records could be listened to in the vehicle. In 1968 *Philips* released the first in-car cassette player which quickly enabled users to listen to tapes with their favorite music. An in-car cassette player by *Blaupunkt* can be found in the cockpit of the *1972 Maserati Indy America 4700* shown in Fig. 1.7. Shortly after inventing the compact disc, Philips also developed an in-car CD player in 1983. But only in the late 1990s in-car CD players take over due to the ability to read CD-RW disks and MP3 files. Compared with the use of cassette tapes it was now possible to skip forward or back which leads to less driver distraction.

Fig. 1.7 Maserati Indy America 4700 (1972) (Meixner 2013a)

1.2.1.2 Navigation

With increasing number of vehicles on the road, traffic information became more and more important. In the 1970s, the first traffic reports were broadcasted via radio. For example, in Germany the *Autofahrer-Rundfunk-Information* was developed by Blaupunkt and provided on many German radio channels (Gesellschaft für Unterhaltungs- und Kommunikationselektronik 2010). In the 1990s traveling by car was further facilitated by GPS navigation, at first using simple visualization on large LCD displays (Bellis 2011). In 1990s navigation systems (GPS) were introduced as well and within this the era for automotive infotainment started. Data processing, technical components for managing the GPS signals, and different sensors for allocating the car were now required. So, the user interface had to be expanded with displays which offer higher information density and resolution as well as with new input possibilities such as rotary push buttons.

1.2.1.3 Telephone

In 1910, Lars Magnus Ericsson had installed the first telephone in his car, which could be connected with electrical wires to telephone poles installed along the road (Wheen 2011). In the 1940s and 1950s, the development of cell towers enabled the

Fig. 1.8 A trucker rolls with one of the first in-car phones, used in Chicago in 1946 (AT&T 2016)

further development of car telephones. An example for this technology is shown in Fig. 1.8. In the 1970s, a car phone service, which used the Autoradiopuhelin, a Car Radiophone service network, became popular.

After the first 1G systems were created in 1982 by Nordic Mobile Telephone, mobile telephone service became mainstream for automotive phone services. In the 1990s, car phones lost their popularity because personal cell phones became affordable to the public. As a result, the first hands-free car kit for mobile phones as well as an enhanced version featuring speech recognition were introduced by the Bluetooth Special Interest Group in 2001. From the year 2000 the wireless signal technology Bluetooth was used in cars for hands-free calling while the first mobile phone with Bluetooth was available on the market. Since 2002 there are advanced voice integration features available, thanks to Bluetooth.

Constantly adding comfort functionality like entertainment or telematic systems to cars soon led to increasingly complex systems. For this the overall operation complexity raises, since every new functionality brought its own dials, switches, and displays.

Over time, in-vehicle HMIs became one of the most important components in the automotive industry. For the early automotive HMIs, there was an exact mapping between control unit and function. Examples of these are steering wheels, pedals, switches for turn signals, and wipers. Driven by the rapid progress of microchip technology and computer science, mechanical devices were replaced by electronic counterparts. At the end of the 1990s, the large number of functions across a wide range of electronic devices required the development of a new system architecture concept for automotive HMIs. Automobile manufacturers started to aggregate functions within a single device in order to reduce complexity (Bellis 2011). The complete system could then be accessed via one graphical user interface with a hierarchically structured menu. At that point, premium car manufacturers like Audi, BMW, and Mercedes-Benz presented their first in-car infotainment systems combining informative and entertaining functionalities. Until today, the main subjects are multimedia (e.g., radio, mp3, and television), car information (e.g., trip length, temperature), navigation, and telecommunication.

1.2.2 *Infotainment Systems of the Last Decades*

In the last 20 years, various technologies such as the Internet, computers, and smartphones have become more and more important in everyday life and have influenced the development of in-vehicle infotainment systems. The users expect more than simple entertainment and information functions in their car. They want to enjoy using these infotainment systems. Therefore, the HMIs for in-vehicle infotainment systems were further developed especially regarding a better usability. In the following chapter infotainment examples from Mercedes-Benz and Audi of the last decades are explained to figure out how the design changed. Figure 1.9 shows the history of HMI development between 1998 and 2009.

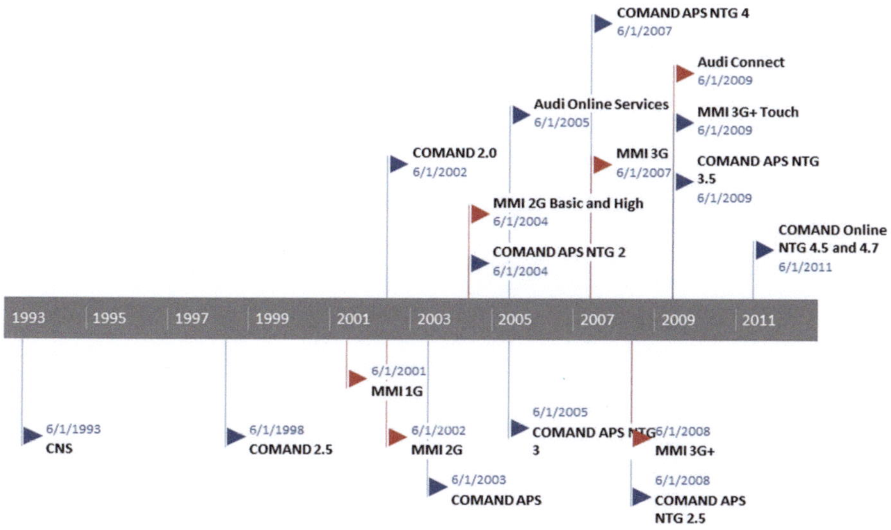

Fig. 1.9 History of HMIs from Mercedes and Audi (1993–2016) (Häcker 2016)

1.2.2.1 Mercedes-Benz

1993—CNS: In 1993, Mercedes-Benz introduced its first infotainment system named Communication and Navigation System (CNS), which was one of the first fully integrated telematics system in the automotive domain.

1998—COMAND 2.5: Since 1998, the infotainment system has been sold under the brand name COMAND, which is an acronym of Cockpit Management and Data System. The functionality of COMAND has been extended over the years based on the current state of the practice in terms of development and vehicles. The main components of COMAND in version 2.5 comprised a simple radio and tape deck (see Fig. 1.10). Extras included CD changer, phone, TV, and navigation system. The latter consisted of a 4:3 color screen located in the center console and displaying the current route, whereas the instrument panel showed arrows indicating the direction. The number 2.5 comes from the height of the display which measures 2.5 Din units.

Optionally, the system provided dynamic route guidance based on current traffic information received by a mobile phone. On the faceplate of COMAND 2.5, the hard keys were located in the area of the display. Besides the classic push buttons, COMAND 2.5 had an assembly button group, which consisted of push and rotary buttons for several frequently used functions such as setting the volume and switching between different radio stations. The design concept of this assembly

Fig. 1.10 COMAND 2.5 (Wikimedia Commons; User: Guido Gybels 2008)

button group enabled the user to access frequently used functions more quickly and easily, leading to reduced distraction by menu operation.

2002/2003—COMAND 2.0 and COMAND APS: The two versions released in 2002/2003 comprised the components radio and CD drive. In contrast to the previous version, the navigation system consisted of a 16:9 color screen including the Auto Pilot System (APS), which calculated the routes based on information from Traffic Message Channel (TMC) services.

The navigation maps provided with COMAND APS NTG (new telematics generation) released in 2003 were displayed via a 6.5-in. color screen with an aspect ratio of 16:9. Maps on DVD covered Europe as well as points of interests like hotels, restaurants, gas stations, etc. Furthermore, the availability of an "Aux-In"-port enabled the connection of external devices. In 2003, the Mercedes-Benz Portal provided services targeted at business people, such as calendar, emails, text messages on portable computers (personal digital assistant PDA), and PC, respectively, from COMAND.

2004 COMAND APS NTG 2: In comparison with the first COMMAND APS NTG the main change here was that the navigation processor was attached near to the display in the center console.

2005 COMAND APS NTG 3: With COMAND APS NTG 3, a radically new system was developed that was completely integrated into the interior of the vehicle for the first time. The 8-in. color display (16:9) was not located on the center console anymore but higher, on the right-hand side of the instrument panel, putting it in the driver's field of view. Interaction with the system's functionalities was possible via a central control element (CCE) positioned in the central armrest, which enabled single-handed interaction. Additionally, seven keys were available: three of them dedicated to the quick selection of menu functions, the ON key, the key for individually selecting a favorite function, the return key (which enabled quick return to the previous menu level), as well as the mute key. Furthermore, volume control was integrated. Navigation data was stored on a 2.5-in. hard disk, which enabled faster route calculation. Passengers in the back were provided with a monitor integrated into the headrests of the front seats that enabled independent usage of the entertainment program. To interact with this system, an additional CCE had been integrated into the central armrest in the back.

2007—COMAND APS NTG 4: Since 2007 the COMAND APS NTG 4 is offered as a special equipment package for the c-class. In comparison to the previews COMAND systems there were changes within the design of the navigation map. With a so-called *birdview* variant the map is not still 2d but has a kind of possibility to look inside the map "from the side." For this also attractions on the map were now 3D. To optimize the route guidance there were information about the traffic via *TMC Pro*. Concerning the way of displaying there was an innovation. Because there was not a display in the middle console anymore, a 7-in. display could be flipped out for use. The system is controlled with a rotary push button near to the gear lever.

2008—COMAND APS NTG 2.5: COMAND in version APS NTG 2.5 was introduced in April 2008. There were changes in the design of the keys and the layout of the menus, which were adapted from NTG 3. Furthermore, the CCE was not located in the central armrest anymore but on the device itself. Innovations in the navigation system included a top-down view that enabled the driver to look sideways on a map. Additionally, some places/points of interests could be viewed in 3D. Traffic jams could be detected automatically via TMC Pro. In some series, the mounted display was replaced by a 7-in. color display that could be folded out electronically.

2009—COMAND APS NTG 3.5: With the launch of the COMAND APS NTG 3.5 the capacity of the disk was expanded up to 7.2 GB. Also the speech communication system "Linguatronic" is a basic element of the car so the user could control the telephone and navigation. Within this system Splitview (compare to Sect. 4.2.2), a splitted view for the driver and the front-seat passenger is used for the first time.

2011—COMAND Online NTG 4.5: Since January 2011 the COMAND Online NTG 4.5 is available in the c-class of Mercedes under the abbreviation COMAND Online because it has a connection to Mercedes-Benz online internet services with any Bluetooth mobile phone via VPN. The list of available services includes weather forecast and Google search to find points of interest and send them to the navigation system. Also, these services reside physically on a remote server. They look as if they were implemented in the local infotainment system using the same HMI design. The services are designed to avoid driver distraction and can therefore be used while the car is moving. Furthermore, the infotainment system can be used to browse the Internet which is only possible when the car is stopped. The connection to the Internet is implemented by pairing a mobile phone via the Bluetooth Dial-up Networking Profile (DUN).

There is no flip display anymore but a fixed mounted TFT color display on the right side of the instrument cluster with 7-in. and with a resolution of 800 × 400 pixels. There is still a hard disk with the information of the navigation maps. Additionally, the system supports a Media Interface where an iPod kit is fully integrated, a digital radio, and also Digital TV. Furthermore, Facebook and other similar social media web sites could be used with this generation of COMAND.

2011—COMAND Online NTG 4.7: This COMAND Online system is the second generation of the 4.5. There are changes in form of different hardware elements and now a Bluetooth PAN is installed to allow online access with an iPhone.

1.2.2.2 Audi

Before 2001: Audi cars offered an in-car radio named Radio Chorus with basic radio functions such as FM and AM receivers. Radio Chorus was soon extended with a cassette or a CD player. The input modality of Radio Chorus consisted of physical control elements, such as rotary knobs for volume and tuning as well as some functional hard keys. It provided six hot keys that could be assigned to radio stations by the user. As the output modality, an eight-segment digital display was used to show the information from the radio, such as the station and the frequency.

2001—First-generation MMI: At the IAA in Frankfurt am Main, Germany, Audi presented its concept study Audi Avantissimo, the Avant version of the A8. It provided the first Audi MMI (Multimedia Interface), and its usability concept established the conceptual basis for all subsequent Audi infotainment systems. It consisted of a control unit and a graphical user interface displayed on a color screen (Audi 2001).

2002—Second-generation MMI 2G: The MMI 2G (see Fig. 1.11) came built-in in the Audi A8. Its main components were a 7-in. color display, a radio, and a CD player. Extras were a CD changer, a navigation system based on DVD, a simple speech dialog system with command input, Internet connection, television reception, and a satellite radio for the USA. The control unit consisted of a *central control element* (CCE) with four control keys and eight functional keys for accessing the four main menus media/entertainment, phone/communication, navigation/information, and car functions shown on the display. The ordering of the function keys was reflected in the graphical user interface (Elektronik Automotive 2002). Figure 1.11 also shows the dialing screen available in the telephone menu.

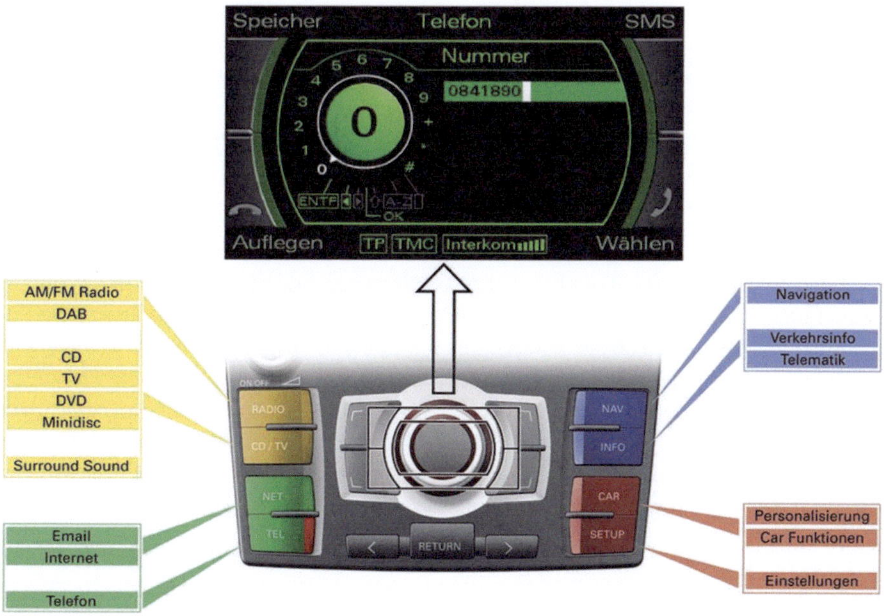

Fig. 1.11 Control unit and graphical user interface of the first MMI 2G, which already shows the typical control logic (Elektronik Automotive 2002)

A green color theme immediately indicated to the user that the phone/communication menu was active. Other menus had different colorings, namely orange for media/entertainment, blue for navigation/information, and red for car functions.

2004—MMI 2G variants Basic and High: The new A6 offered two variants of the MMI 2G. The variant High corresponded to the MMI of the A8; the variant Basic had a smaller 6.5 in. monochromatic display (Wikipedia 2015). Introducing this kind of variants was an important step in infotainment system development. At first glance, these variants were only distinguished by different display sizes and slightly different hardware control elements. However, behind the scenes these variants were completely independent systems, even developed by different suppliers. Whereas they shared a common look and feel, they were based on different hardware platforms of the main unit as well as the attached control units. As a consequence, the software also had to be developed separately. This fact had a major impact on the complexity of the development. First, OEMs had to design a look and feel that suited both the High and the Basic variant regardless of functional differences. Then it had to be ensured that the common look and feel was correctly implemented by the suppliers.

2005—Audi Online Services: Audi and Google formed a partnership to provide online services within the infotainment system (Audi 2001). At that time, Google was already able to provide web-based services for route planning that made use of a variety of additional information. Whereas in-car systems relied on built-in map data and restricted traffic information through TMC, there was more detailed data available online, e.g., traffic flow information. In a web service it was also much easier to solve the problem of keeping the map data up-to-date or providing user-specific points of interest. By integrating a navigation component based on an online service, rich and up-to-date information could be made available in the automotive HMI. Other services such as weather forecast or Internet radio were following soon.

Insertion: 2006—Touchpad: In 2006, a new input device was presented: the touchpad (White 2010). Touchpads are used more and more widely in in-vehicle infotainment in premium cars such as Mercedes-Benz S-Class and Audi A8. The touchpad we are used to from laptops has a touch-sensitive surface that is built with a capacitive, resistive, or infrared array and may be integrated into a multifunctional steering wheel or center stack. The Input/Output interface of touchpads supports key events, moving pointers, or pictures, as well as handwriting recognition software. Additional touchpads may have backlight to display numbers or soft keys that are predefined.

2007—Third-generation MMI (3G): The MMI 3G offered many new functionalities. Additionally, the number of different models and country variants grew. For the first time the new A4 and A5 provided an MMI with a navigation system and a 7-in. display. Bluetooth phone connectivity is an option for the A4 and the A5 with navigation system, as well as for the A6, Q7, and A8.

The MMI 3G was the first system providing advanced driver assistance systems like cruise control and parking sensors in cars made by Audi (ATZ/MTZ 2007). The first Audi lane deviation warning system was presented in the new Q7 and later also put into the new A4 (HELLA 2007). Besides the development of the MMI, the basic Radio Chorus was also replaced by Radio Concert and Radio Symphony. Radio Concert allowed playing mp3 files and Radio Symphony provided a CD changer with six slots and the possibility to store traffic information (TP memory).

2008—Third-generation plus MMI (3G+): Audi Q5 provided the first MMI 3G +. It had a 7-in. display with a resolution of 800 × 480 pixels and a speech dialog system with full-word input. The map of the navigation system is 3D and its data is stored offline on an integrated hard disk.[1]

2009—MMI 3G+ Touch: The Audi A8 provided a touchpad with handwriting recognition to input phone numbers, addresses, and navigation destinations. It supported Latin as well as non-Latin characters like Chinese and Japanese. Additionally, for the first time in an Audi, rear-seat entertainment became available with two 10.2-in. displays (Audi 2009).

1.3 Early Technology Trends and Development Processes

OEMs always aimed at augmenting the comfort available in their cars. The requirement of personal customization and the diversity of the functionality to be realized enforced a modular design and thus an increasing number of electronic control units. Initially these devices were invented independently and there was no need for mutual interaction. Later it was recognized that even better results could be achieved if the devices shared some of their information that otherwise would be unavailable for certain devices. For example, information from wheel sensors can significantly improve position calculations in navigation systems, especially in situations where GPS is unavailable. Technical solutions were developed that enabled inter-device communication. Early, direct, proprietary device-to-device connections were soon replaced by standardized communication systems. The CAN (Controller Area Network) technology, a serial bus standard for distributed control systems, was introduced by Bosch in 1986 (Kurfess 2011). Each device connected to the bus can read relevant messages and use the contained information as well as put its own data on the bus. The bus features a communication protocol preventing message collisions. For interpreting messages and contained data there exists a system-wide database that is centrally maintained. Modern systems typically feature complex inter-device communication. To a certain extent, this is caused by the goal of system designers to provide a single-user interface rather than allowing each device to provide its own. This allows uniformly controlling all connected devices as

[1]http://www.tomshardware.de/MMI-3G-Audi,news-241334.html, accessed 07/2016.

though all the functionality were realized within a single device. The true complexity of the underlying distributed system remains hidden from the user.

Automotive HMIs provide a variety of different kinds of input and output interfaces (I/O interfaces). There are three kinds of feedback available: visual, auditory, and haptic (Kern and Schmidt 2009). The steady change of controller types over time, from simple switches via graphical user interfaces to speech dialog systems, touchpads, and touchscreens aimed at simplifying access, decreasing distraction, and thereby increasing safety. Nowadays, some major functions can be controlled by switches near the steering wheel for faster access, e.g., radio or telephony.

Due to the increasing complexity of the HMI, the usability of the interfaces has become a very important quality factor. Since the 1980s, standards have been defined to develop user interfaces with high usability. One of the first general models was the so-called *IFIP* (user interface reference model). With IFIP, a user interface is structured into the four parts input/output, dialogue, functional, and communication. Also in the 1980s with the growing impact of software engineering, a lot of software architecture models, such as *MVC* (Model–View–Controller), were invented. Soon it became apparent that for the automotive domain special user interface standards had to be established because automotive HMIs differ in major points from HMIs in other domains. One big difference is the focus on user attention. Whereas in many domains the main task of the user is to interact with the application, with automotive HMIs driving must remain the highest priority. When the functionalities of infotainment systems increase, the causes for driver distraction increase, too. In addition, the cognitive load for performing a task can grow immensely (Kern and Schmidt 2009). It has become more and more important over the years to ensure safety when developing automotive HMIs.

The European Statement of Principles on HMI was issued in 1998. It gives advice for developing the automotive HMIs in such a way that the provided functions do not distract the driver from safely driving the car. The statement is updated from time to time due to the steady increase of functionalities; the last update was done in 2013 (Commission of the European Communities 2013). Stevens et al. (2002) also discuss guidelines for ensuring more safety in cars in the face of a complex HMI. Another difference to other user interfaces is that the devices in cars are normally at fixed positions and the user can only interact with them within a limited radius. In this context, (Kern and Schmidt 2009) discuss the proper use of the so-called design space, i.e., the proper ordering of the devices with respect to their functions within the interaction space.

These varying preconditions directly affect the development of HMI-based software. New tools and methods are necessary to handle the development of more and more complex systems. The importance of certain development process phases that might have been neglected until now also increases, such as the process of software testing. Increasing complexity raises the cost of testing. HMIs of higher complexity require new testing methods (e.g., automated testing) to handle the increasing number of test cases. In the early days of software testing, everything was tested manually. Unfortunately, manual test procedures were not feasible for broad verification of complex systems, which led to the invention of automated testing methods. In the first step, scripted test procedures were introduced. Current trends in testing point to more complex methods, e.g., model-based testing. At first, the test models were relatively simple flow charts. Later the Unified Modeling Language (UML) was established as a standard for specifying the behavior of HMI in state chart diagrams (Reich 2005). There are many challenges to be solved in automated model-based HMI testing, e.g., where the machine-readable test models are coming from (Stolle et al. 2005). For automatically deriving test models from specifications, the existence of formal machine-readable specifications is essential. Various specification languages for different purposes are known in literature (Hess et al. 2012a, b). For the automotive domain, one approach of a specification for the formal description of the HMI of infotainment systems is presented in Fleischmann (2007).

1.4 The Present of Automotive HMI Development

As discussed in the previous chapter, automotive HMI has rapidly changed since the invention of the automobile and continues to grow in complexity. Combined with more and more new technologies from other domains, infotainment systems increase mobility and comfort in modern cars (Amditis et al. 2010). Therefore, developing infotainment HMI in the automotive context is intensely interdependent with the experiences and expectations of car customers with regards to other technologies.

Early computer systems and their interaction concepts used to be highly efficient but not easy to use or learn. Systems were primarily operated by technophiles using command line interfaces. In contrast, today it is possible to just use computers without knowing many details about the system architecture or the technical background. The potential of computers in daily life is clear. For almost every job, computer skills are required and knowing how to handle digital data with computers is a fundamental skill. In personal lives the importance of computers is also growing steadily. Today's youth grows up with digital multimedia and social networking services. The upside for the development of automotive HMI is that users are not afraid of interacting with computers nor do they have to be motivated to do so. They know the basics of data processing and have acquired strategies for learning how to operate new computer systems. On the other hand, since people know about the potential of computers in other contexts, they have high expectations.

Especially the rise of so-called nomadic devices has increased this trend. Today, most people are used to being on call anywhere and anytime. Anderson (2015a, b) shows that in 2015, 68% of all adults in the USA had a smartphone. Smartphones provide similar functionality as mobile computers, especially regarding multimedia content or contact management. Additionally, basic functionality can easily be enhanced by apps.

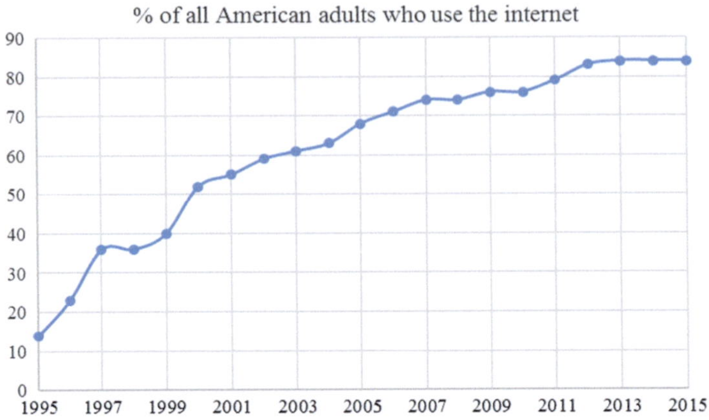

Fig. 1.12 Internet adoption by adults in the USA from 2000 to 2015 (Perrin and Duggan 2015; Pew Research Center 2014)

Another driving factor for HMI development is the rise of the World Wide Web, as Internet provision has become faster, more stable, and more comfortable to use. Technologies such as VDSL or fiber optic link have overcome earlier restrictions and allow transferring several hundred Mbits per second. Internet has become mobile as well (Gow and Smith 2006). Fourth-generation mobile communication, such as the Long Term Evolution (LTE), and the increasing spread of wireless LAN hotspots have made it possible to be online anywhere and anytime with a reasonable transfer rate. As one example, the Internet adoption by adults in the USA rose from about 14% in 1995 to 84% in 2015 (Fig. 1.12) (Perrin and Duggan 2015; Pew Research Center 2014). According to Smith (2011), 59% of all adults in the USA went online wirelessly in 2011. In 2013 this number grows up to 63% (Pew Research Center 2016). With the perspective of reaching a growing number of users via the Internet, more and more online services are provided. According to (Kellar 2007), information retrieval and information exchange via the Internet are part of everyday life. Especially in combination with smartphones, Internet-originated communication (e.g., instant messaging and email) is integrated with cell phone communication (e.g., calling and text messaging). Much personal information, such as contact information (telephone numbers, mail/email addresses) or birth dates, are stored in mobile devices. People tend to be highly dependent on the availability of this data and not being able to access this in their cars would not be acceptable.

1.4.1 Fields of Application

Applications of in-vehicle infotainment systems include navigation, media, TV, car configuration, data interfaces, telephone, and so on. The combination of previously independent functions is typical for today's automotive HMI. The *navigation system* helps the driver to arrive at his destination. Via TMC (traffic message channel), traffic information is visually presented on the navigation map. Also, higher safety is achieved through so-called *driver assistance systems*, e.g., collision avoidance system. For relaxation during driving, the infotainment system provides several types of *audio and/or visual entertainment*, such as radio, music (mp3, music CD), and TV. The *configuration functions*, e.g., climate control, seat function, and in-vehicle lighting, enable the driver to configure his car easily and comfortably. Additionally, the driver can access data on *connected devices* such as MP3 players, smartphones, USB devices, and flash drives. The driver can also make hand-free calls via his mobile phone that is connected to the car.

Nowadays, in more and more cars a WLAN router is available at an additional cost. Audi, Mercedes, Peugeot, and Citroën offer their users the possibility to, e.g., check emails, log into Facebook and use Google via an Internet browser in their car. Because of the wide usage of smartphones, a lot of users expect to have flexible access to more useful applications for infotainment, e.g., via smartphones or directly from app marketplaces via the Internet. To enable the use of *external applications*, several approaches have been developed such as the use of technologies like, e.g., Android Auto or Apple CarPlay.

Moreover, besides smartphones, the personalization of in-car infotainment is becoming increasingly important for the automotive industry. Some car manufactures allow the user to customize his infotainment system, e.g., by installing applications and storing personalized configurations. In this context, user identification is a current field of research. For example, in (Feld and Müller 2011) a speaker classification system is presented to personalize in-car services automatically via speech input.

Providing an automotive HMI with lots of features is no longer a distinguishing criterion for premium cars. Nowadays, infotainment systems are installed in almost every car and are part of the standard equipment. So, car manufacturers have to look after a solution to enhance the user experience. Keywords are user experience, connectivity, and multimodal solution by understanding the user and his expectations (IQPC 2016).

1.4.2 Input/Output Devices

In many non-automotive domains, interaction with computers or smartphones is the primary task. Users can spend their cognitive capacity completely on human–computer interaction and concentrate less on their surroundings. In a typical automotive setup, users drive their car at the same time and additionally have to stay in the driving lane, watch their speed or react to the current traffic situations (cf. Sect. 1.3). Less attention can be given to selecting destinations or changing the radio station.

Examining the influence of human–computer interaction as a secondary task is a very active field of research as summarized by Ersal et al. (2010) and von

Zezschwitz et al. (2014). Tasks are termed secondary when they engage the driver voluntarily or involuntarily and "do not directly pertain to the primary task of safe vehicle operation" (Ersal et al. 2010). (Governors Highway Safety Association 2011) adds that this task "uses the driver's eyes, ears, or hands."

Results show that things such as talking on a cell phone or writing a text message on a mobile device while driving are distracting (Drews et al. 2008). According to National Safety Council (2015), cell phone use (talking and texting) is estimated to be associated with a minimum of 27% of all accidents. As a consequence, in the automotive context it is crucial to design systems aims at reducing distraction: Interaction concepts have to be obvious, plausible, and consistent; modalities have to be chosen appropriate to tasks and users; constraints and affordances must clarify valid and invalid input. The user must not have any doubts or questions about how to interact with the system.

With the rising complexity and diversity of functionality these aspects are becoming even more critical. A modern in-vehicle infotainment system is an integrated set of Input and Output devices (I/O devices) which (a) communicate with each other using bus systems, such as CAN, and (b) enable interaction between the driver and the vehicle. I/O devices are a major hardware component of in-vehicle infotainment and offer the largest number of physical HMIs. Automotive suppliers do not develop the physical user interfaces as single hard keys, such as three push buttons or two rotary knobs. They are usually supplied as complete component assemblies of a set of hard keys. Additionally, these I/O components are controlled by separate *Electronic Control Units (ECU)*, which are connected to each other by the vehicle bus. These separate responsibilities are also reflected in the organizational structure of car manufacturer's and supplier's development departments.

A vast number of solutions are available on the market that differs in appearance and functionality. Although their design and their integration into the vehicle's interior is dependent on the car manufacturer, they can be divided into a set of groups. This allows different combinations, which can be located in different areas of a car to build variants that can be easily distinguished by the users. For the above reasons, in the following sections the I/O devices, including the physical and speech, HMI will be introduced based on current industrial categories.

1.4.2.1 Input Devices

Most of the conventional integrated infotainment interaction devices described in Sect. 1.2 are still common in the latest infotainment systems. Many of the input devices discussed in the following are shown in Fig. 1.13. Compared with the faceplate of previous head units, the *Infotainment faceplate* nowadays uses displays that come with touchscreens or even force feedback. The number and layout of the buttons usually depends on functional considerations and/or the philosophy of the car manufacturers. The infotainment faceplate enables some input functions for navigation and other systems.

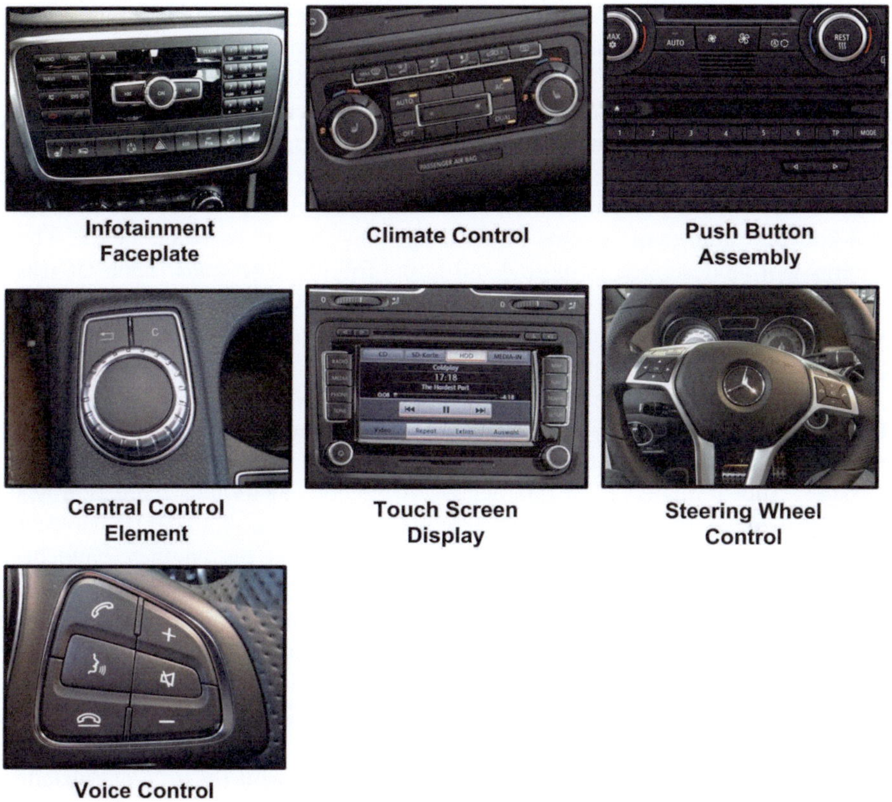

Fig. 1.13 Overview of haptic input elements of automotive infotainment systems (Zhang 2015)

In addition to the infotainment faceplate, the *climate control panel* is used for setting air condition parameters, such as the fan speed and the temperature by means of turning wheels, push buttons, and sliders (Zhang 2015). There may be separate panels for setting the air condition parameters for front and rear passengers. In order to lower prices, some climate control panels are implemented without electronics. More complex climate control panels use electronic components, such as displays (Zhang 2015). Although the infotainment faceplate and the climate control panel are often located next to each other, they are controlled by separate ECUs and developed in different departments.

Some cars are equipped with a *push button assembly* that enables direct access for major infotainment functions or functions that are frequently used (Zhang 2015). Examples include driver assistance functions, Electronic Stability Control, door lock, or voice control. Some of these functions, such as the latter ones, are also provided by duplicate buttons, which can also be commonly found on the steering wheel. Such buttons may use indicator lights to show the current status of the operated system or can even provide miniaturized displays. They can also be implemented as capacitive surfaces or approximation sensors.

However, due to the limited space, the maximum number of such buttons in a car is limited, although the number of functions to be controlled is constantly increasing. For that reason, some manufactures like BMW allow the user to assign functions to the available push buttons. Some manufacturers apply usage concepts that are based on center control elements (CCEs). These devices are multi-purpose controllers used to navigate complex menu structures (Zhang 2015). As described in Sect. 1.2.2, MMI (Audi), iDrive (BMW), and COMAND (Mercedes-Benz) are well-known examples of systems whose usage concepts are based on a CCE. These systems will be described in more detail in the following sections. Their CCE is a rotary controller with force feedback technology. Around that controller they provide buttons for switching between the major infotainment contexts or provide quick access to common functions.

However, rotation and selection by means of a rotary controller is not convenient in some usage scenarios, such as navigation on a map. That is why, from the beginning, the rotary controllers applied by BMW allowed for pushing and pressing. Audi refrained from this degree of freedom in its first-generation MMI system but added a coolie hat to the top of their rotary controller later on. However, these systems are restricted to simple inputs. For this reason they have been replaced in modern systems by *touchpads*, which also allow for convenient character input (see Sect. 1.2.2). These touchpads are mounted next to or even as part of the CCE. They allow users to provide input such as pointing, clicking, gestures, and characters. By activating a backlight, some touchpads can display predefined symbols like numbers. They are used to mark that clicking a region is currently associated with a specific function.

Touchscreen displays in the head unit are equipped with resistive or capacitive surfaces enabling direct manipulation of interactive objects by means of touch. Just like the touchpads in the CCE, current touchscreens can also provide gesture operations such as scrolling and zooming, as well as handwriting recognition.

Common steering wheel controls are buttons, scroll wheels, or little touchpads. Their main advantage is reduced distraction of the driver compared to touchscreens in the middle of the car because the driver does not need to take his hands off the steering wheel. For the same reason, *control levers* were used in the past for controlling the turn signals and the windshield wipers. Modern infotainment systems also use these levers to control infotainment functions or driver assistance functions such as adaptive cruise control (ACC). The levers as well as the buttons and scroll wheels on the steering wheel can be assigned to a single function, such as to accept incoming calls, activate voice recognition, or change the audio volume. They can also be used as multi-purpose control for navigating in lists displayed in the cluster or head-up displays like in the digital cockpit of Audi.

Voice control enables the user to input commands in natural language without taking his hands off the wheel and his eyes off the road. An array of microphones, which may need to be activated explicitly by, e.g., pressing a button located on the steering wheel, records the user's commands as acoustic signals, which are then processed by the Speech Dialog System (SDS) (Lamel et al. 2000). In SDS, these recorded acoustic signals are transformed by a speech recognizer, which is often based on a probabilistic approach for modeling the production of speech using the technique of Hidden Markov Models (HMMs) (Schuller et al. 2006), into the most probable word sequence.

Voice control was first built into a regular series model in 1996, when Daimler integrated Linguatronic into its S-Class Mercedes-Benz cars (Heisterkamp 2001). This system supported the telephone application.

Nowadays, the driver can access functions like music selection, destination input for the navigation, or even climate control changes with the vehicle's embedded voice recognition system. But beyond the in-vehicle solution also smartphone functions which let the user interact with music, social media or phone contacts via voice control are common like Apple's Siri, Google's Google Now, or Microsoft's Cortana. These personal assistants are familiar to the driver and always up-to-date.

Especially, Ford is one of leading car manufacturers that pushes the smartphone integration combined with voice recognition. The so-called Ford SYNC is a voice-based communication system that is connectable via Bluetooth. Ford SYNC with AppLink goes one step further. With this refined system the user has access to his apps which he can control via voice, steering wheel buttons, or the middle console (Ford 2016).

A recent addition to the list of input devices is *camera controls*. They are used for monitoring the driver and for gesture recognition. It is possible to combine this with other functions that would require a camera, such as video telephony. The camera can be located on the dashboard or in the instrument cluster. Currently, camera controls are rarely utilized, except for some premium cars (e.g., the Lexus LS 460).

1.4.2.2 Output Devices

The optical channel is still the predominant output device. *Displays* are used to provide the user with information about the current system state. Today, the common locations of the major displays are the head unit and the instrument cluster (IC). Depending on the car model, different display sizes, resolutions, and color as well as monochromatic displays are used.

The primary display in modern cars is located in the head unit and shows the graphical user interface of the infotainment system. Some manufacturers apply usage concepts based on touchscreens, while the majority of premium cars combine a conventional display with a CCE. In addition, this central display can be enhanced with 3D ability. Because the display of the head unit is used by the driver as well as the front passenger, some recent systems use "Split View" displays. Depending on the viewing angle, they can show two different screens. This is done by applying a special optical filter on top of the display. It splits the image on the screen into two separate ones that are visible from different angles. Whereas one of these images consists of all odd pixel columns, the other one consists of all even pixel columns. Thus, the horizontal display resolution is halved compared to the nominal resolution of the display (Robert Bosch GmbH 2013).

Fig. 1.14 Audi TT Digital Cockpit (Häcker 2015)

Conventional instrument clusters consist of electro-mechanical tachometers, speedometers, odometers, oil gauges, etc. These are often complemented by a display used for showing information such as the on-board computer or ACC warning. Whereas in luxury class vehicles, full-color and high-resolution displays are common, simple seven-segment LC displays are still in use in lower price vehicles. However, there is a clear trend toward replacing the electro-mechanical components with virtual instruments shown in the displays, which in turn are becoming larger. This trend leads to free programmable instrument clusters (FPK) without the classical mechanical components. Figure 1.14 shows the Digital Cockpit of the Audi which also has a representation of the tachometer and speedometer but is reconfigurable in size. Depending on which information is important in a special situation, the user can change the graphical representation. For example, the car rounded elements will become smaller when the user needs a bigger few of the navigation system.

Head-up displays (HUD), which were used in primitive versions in airplanes, are a recent innovation in cars. A HUD consists of projector, optical lens, information source, reflecting element, as well as combiner (Wood 1988) and shows good readability of displayed information in daylight and darkness. Just like the FPK, the HUD can also show virtual instruments and navigation information such as maneuvers. Their image is shown on the windshield in the driver's line of sight. Using optical means, it is possible to have the HUDs image appear to be located in some distance to the driver, avoiding the need for the eyes to adapt when glancing at the HUD while driving. This is why using HUDs to display driving-related information to the driver promises to reduce distraction and thus increase safety.

Complementing to the optical channel, the acoustic channel is also used in cars. There are two major fields of application. On the one hand, nonverbal sounds are played to signal changed vehicle states, and confirm keystrokes or the activation of a specific system. On the other hand, speech output is used to request follow-up information from the user, e.g., say a name from the telephone book, or explain which commands can be used in a specific situation. With the help of Response Generator and Speech Synthesis (also called Text-to-Speech) (Lamel et al. 2000) in SDS, the output information is converted into natural language and played to the user. Jensen et al. found out that there are more advantages of speech output compared with visual output. They tested the driving behavior of participants in a real traffic driving situation. Doing this they became the result that an audio instruction for the navigation system was better than a visual instruction on a screen. But this depends on the quality of the auditory output and how the system is implemented in the whole infotainment array.

To increase accuracy and usability when operating menus, force feedback has been widely used in CCE to help the user get some kind of haptic feedback. Force feedback can also be used in the steering wheel to provide some driver assistance systems, such as the lane departure warning system (LDW).

1.4.3 Current Infotainment Systems

COMAND Online NTG 5: The latest version of Mercedes' COMAND (COMAND Online NTG 5) is available in the new series 222, also known as the S-class. It consists of two 12.3-in. displays placed in the head unit and the instrument cluster. One of them replaces the classic instrument elements in front of the driver, whereas the other one is for the infotainment and navigation. Although a small bridge separates the two 12.3-in. displays they look like one unit. Each of them has a resolution of 1.440 × 540 pixels within a pixel density of 125 ppi. The central input device is the rotary push button in the center console. Within this button the user can navigate through lists and menus. Around that button there are six hard keys for quick entrance to the main points of the infotainment system: Seats, navigation, radio, multimedia, telephony, and vehicle functions. Furthermore, there are buttons for back, On/Off, and volume. On the steering wheel the user can also interact with the system while driving. As a special, there is a touch element in the center console for entering letters or for using the mouse cursor. Additionally, speech input is available for the driver. Compared to further COMAND systems the complexity and number of infotainment elements increase. There are much more interaction possibilities and therefore also buttons.

Currently, depending on the car model and the country, Audi offers different MMI infotainment systems on the market. The MMI is offered in four different variants, which vary in display size and the infotainment features they support. The basic variant is called *MMI Radio*. It is equipped with a 6.5-in. display with 400 × 240 pixels and supports features such as phone connectivity, address book, CD player, and TP memory. In addition to that, the *MMI Radio plus* is equipped with two SD card readers, Bluetooth phone, a speech dialog system, and the ability play mp3 files. *MMI Navigation* extends the variant *MMI Radio plus* with a DVD-based navigation system, TMC traffic information, and a speech dialog system that allows controlling the address book, and the phone and entering navigation destinations via voice commands. The variant *MMI Navigation plus*, which is standard in the A8 and A6 Avant, provides an 8-in. display with 800 × 480 pixels, a hard-disk-based navigation system with 3D map and Google Earth satellite view, a DVD drive, USB port, iPod interface, advanced driver assistance systems, and a speech dialog system with full-word input to control the navigation system, the phone, and the address book.

The latest Audi infotainment systems also extend the range of available mobile online services, which are called *Audi Connect*. These services include Google Search, Google Street View, traffic information, news, and a weather forecast. In the Audi A8 and A7, it is possible to connect up to eight mobile devices at the same time to a WLAN hotspot provided by the infotainment system.

In addition to those MMI variants, Audi offers a completely new display called *virtual cockpit*. It is a fully digital instrument cluster focused on the driver. In the 12.3-in. TFT display all functions of a standard instrument cluster and the middle MMI monitor are combined. Here, the driver is able to configure the information representation form. There are different view modes where the speedometer and the rev counter are more or less dominant (Audi 2015).

1.4.4 Development Process

After having a closer look at the current infotainment systems of Daimler and Audi it is important to know that the development process of automotive systems is mainly characterized by an intense interchange between OEMs and suppliers (Bock 2007). In many cases, the OEM specifies the requirements and hands them over to a supplier responsible for the development. After the development, the product is handed over to the OEM again, who tests the product. In the following, the development process will be explained for the steps specification and design, implementation, quality assurance, and post-implementation.

The OEM builds the specification containing requirements, functions, design-related requirements, writes and translates texts for different languages, and perhaps also creates a model of the specification for model-based development. In rare cases, a formal specification is also created, but often this is done by the supplier. The documents are then given to the supplier, who can be put in charge of a special field of devices or software (see Sect. 1.4.1).

The supplier sometimes has to do additional work on the specification as refinement. However, as a first step, there is a technical review, which results in an offer for the OEM. Seldom, reviews for quality are made concerning the specifications, because these documents can be very large, up to several thousand pages in extreme cases (Bock 2007). However, this can cause problems later on in development, when inconsistencies or similar are detected. After negotiations for price, the development starts. The analysis of requirements is often done using Microsoft

Office applications. Some OEMs also already develop prototypes, Flash animations, Photoshop files, etc.

Sometimes, there is a need to clarify specification-related questions. This is often done using ticket management or bug tracking systems, such as JIRA, which are used for communication between OEMs and suppliers. At some point in time, a relatively stable version of the requirements documents exists. Then a feature freeze is set, which means that changes to the requirements can only be done via a change request (at extra cost). For the OEM, this means that testing can start to develop test cases based on the requirements. For suppliers, this implements that development can now be based on a defined and stable set of requirements. For development, a feature roll-out plan is developed and agreed on between the OEM and the supplier. Some OEMs also want to have prototypes during development, e.g., when defined quality gates are reached.

During development, the suppliers often use V-model-like processes, but aligned to the feature roll-out plan, thus with an iterative component (Ganzhorn et al. 2011; Amditis et al. 2010). The whole process, from start to delivery of the final product, may take between 1.5 and 3 years. The development process is filled with the typical elements: architecture, design, development, static, and dynamic quality assurance.

This brief explanation serves to demonstrate the development process. However, some elements are not as easy as it seems, and thus require a deeper discussion in order to shed light into the situation as it is found today. For example, the traditional V-model, although adjusted to include iterative cycles, can often not be processed in that way due to the large number of prototypes, which has a significant impact on further development.

Once the product has been received, the quality assurance at the OEM starts. Using the test specification created on the basis of the stable requirements, the test cases can be executed. If deviations from the specification are found, a ticket is entered into the ticket management system and transferred to the supplier for clearance and correction. Both the OEM and the supplier use quality assurance, but with a different scope. Suppliers normally use unit, integration, and software (system) tests. OEMs use testing on the system level. Model-based testing in particular is currently a trend. According to Duan (2012), a concept for model-based testing of HMIs (e.g., using UML state charts) ensures the quality and reduces testing costs.

Apart from the traditional roles in development, requirements engineer, architect, designer, developer, quality assurance-related experts, and professionals from other disciplines are included, often due to the focus on the human actors, the users, and the market. The variety of different roles which take part in HMI development leads to several problems.

One of the main problems is communication. All of these specialists have their own vocabulary with special terms. Clear communication always requires definitions in terms of a glossary. Such a glossary, though important, cannot be found in all projects, which leads to the need for additional communication, for questions and explanations due to the different languages (i.e., vocabularies) being used.

Even if there is a unified language, the contents of the documents are often not unambiguous. This leaves room for interpretations, and since the different domains have different concepts, misunderstandings can occur. It is possible that these misunderstandings may be detected during quality assurance. Correcting these results in high costs, depending on when they are detected (Shull et al. 2002).

The broad range of people and disciplines involved in specification and development also leads to a variety of tools being used in this domain. Besides the technical tools traditionally used in software engineering, such as IDEs, compilers, testing tools, bug tracking systems, build environments, or tools for configuration and version management, non-technical roles involved in the development use their own tools. Overall, the variety of tools ranges from general tools suitable for nearly all users up to very complex and special tools usable only by domain experts. The tools themselves often consist of a mix of commercial and open-source tools. In many companies, the tool suites are complemented with self-developed tools.

1.4.4.1 Specification and Design

Today, requirement specification is done with the intensive inclusion of stakeholders, users, and external test persons for evaluations. Drivers for the requirements specification are often workshops with stakeholders, interviews, subject studies, or car clinics. Thus, a very wide range of topics is covered in the requirements. In recent years, the focus of development has shifted to the customers or users, which leads up to a new type of requirements, together with specially designed interaction concepts. These concepts are needed to make interaction with the new features easy and safe. Safety is as issue, because driver distraction is something all companies are engaged in mitigating, as it is also demanded by various international standards and laws.

The parties involved in the specification include not only requirements experts (i.e., technical personnel), but also marketing and sales people, end users, and others. This makes the process of defining the requirements on the part of the OEMs a difficult task. The involvement of end users and the need to make complex interactions and usage of the HMI feel easy leads to a kind of cycle: Concepts are specified, developed, refined, assessed by users, and so on. This human-centered design approach is defined in the ISO 9241-210. The intension in here is to enhance human–system interaction through both, hardware and software.

Before development starts, during preproduction, concepts and features to be included have to be selected and specified. Here, various different roles come into play. Psychologists and market researchers often conduct experiments or user studies, such as car clinics, and even management may have special requirements, for example due to the market situation.

The set of requirements finally given to the supplier is then prioritized according to criteria such as cost, attractiveness, match with the brand and the corporate identity, competitors, relevance for the end users, time needed, and so on. Finally, a plan for the development of the features is defined. This leads to a feature roll-out plan, which may be aligned not only with quality goals, but also with real prototypes delivered to the OEM.

The elements in the specification include golden rules, state charts, use cases, GUI widget catalogs, style guides, graphical files (example screens), formal models, and often simulations. However, most of the documents are not formal and not readable by machines. Thus, they can also not be processed by computers, which may lead to problems due to media breaks and the manual (and thus error-prone) effort needed to transform these data in the design phase.

Notations used for the specification include semi-formal notations, UML or similar notations, pseudo-programming languages, textual descriptions, schematic representations, or even flash prototypes. Even at this early point in time, various different tools are used, such as Office applications, DOORS, Flash, or Photoshop. This is due to the involvement of various types of actors, not only with purely technical skills, but also with "designing skills." Once the concepts have been agreed on, the ways users interact with the system, its different features and its look have to be designed. During this stage, interaction designers and display/UI designers are involved. Graphic designers and speech dialog designers are responsible for optical and acoustic feedback and general appearance of the HMI. Depending on the requirements, this also needs to include specialists for haptic feedback.

In this phase, software programs such as Adobe Flash or Photoshop are used for designing prototypes or the visual design of the display. Finally, tools for animation and 3D content are also employed. Again, there is no tool chain in terms of a seamless transfer from one tool to another. This results not only in manual and thus error-prone work, but also in a probable loss of traceability from one phase to another.

In general, the tools used are aimed at the requirements and goals of the different roles/groups involved. They offer the best functionality for the respective target group, but one problem arising at this point is that it might not always be clear how tools from different disciplines and their contents relate to each other. The specification is spread across several different instruments with different foci, and the connections between the different parts captured in different notations have to be managed. It might not be easy to exchange content between these documents, as these tools often do not maintain interfaces for each other. For example, while content can be interchanged between Office documents, it might be difficult to perform imports or exports between these documents and a requirements database.

Nowadays, development is no longer concentrated on one specific spot; there are projects which are developed by teams around the world. The tools need to be usable across the web, and often more than one user wants to use a certain tool at the same time. However, not all tools are currently multi-user enabled. Besides that, some of the tools have to be able to maintain variants since such variants play a crucial role in the automotive domain.

So, the complexity of managing the interface between the different actors increases the more different tools is used. However, the complexity of modern HMI development can also be seen in the way requirements are managed: Each requirement is annotated with attributes for the series, the market (regions), line, release, and so on. The market, for example, does not only define the language used, but can result in different interactions and even changed features. The features may also change according to the series: The top model has the largest range of features, the cheapest model has only a few of them (e.g., no smartphone integration). Also, different versions may exist according to the equipment in the car (e.g., some HMIs contain a TV, some do not). Additionally, local habits (e.g., the way addresses are entered in the USA is different from that in Europe) as well as laws and standards (e.g., regional standards concerning driver distraction are ESoP (European Statement of Principles) in Europe, JAMA (Japan Automobile Manufacturers Association) in Japan, and AAM (Alliance of Automobile Manufacturers) in North America) have to be followed (Blessing et al. 2010). This means that there is not "one specification." Whenever we talk about a specification for an HMI, we have to keep in mind that there are numerous variants in different versions.

Accordingly, there are different challenges and even problems: Informal specifications are not always clearly interpretable and possibly not machine-readable, which results in media breaks. This may result in a loss of traceability. Additionally, transfers of data between different tools are often only possible manually.

1.4.4.2 Implementation

As mentioned above, the software is being developed by the supplier, sometimes with the help of subcontractors. The development is processed in an iterative way, oriented on the feature roll-out plan. Additionally, the hardware to be filled with the software is often developed by other parties. Both have to be integrated after development. Therefore, constraints have to be taken into account.

The software itself has to be developed so that the corporate branding of the OEM is integrated. However, sometimes the development of different HMIs for different vehicles of one company is split across several suppliers. Nevertheless, they have to adhere to the corporate branding. For this, style guides are developed, which have to be used by the suppliers.

Model-based development is being increasingly used by the developing companies. That way, the specification is modeled and brought to a formal level. The code is then generated from that. This also shows some of the problems in the domain. There is no tool chain, and thus the model has to be built on the basis of the specification documents. It is not possible to transform these documents, as they are from different applications and lack sufficient means for exporting them and for allowing easy import into other tools. The results of model building can be used during quality assurance; however, this is not possible in a fully automated way, either.

1.4.4.3 Quality Assurance

Quality assurance is used throughout the development. The suppliers use static and dynamic quality assurance. In particular, reviews, code analyses, and testing on the unit, integration, and (software) system level are used. Reviews are used for checking requirements specifications, test-related documents, and code.

The OEMs also employ testing, on the system level. They also face the largest number of challenges; several of the recent advancements in the domain pose a challenge and require special methods for testing. An emerging discipline in the field is model-based testing. However, even with model-based development, there are still some challenges.

One of the main challenges is *multi-modality* (see Sect. 1.4.3). As we have seen, modern HMIs allow several different types of interaction, ranging from traditional keystrokes to gestures, speech, and touch enabled for drawing, e.g., numbers or interacting directly on the screen. This means that all test cases have to be modified and executed several times in order to include the various possibilities for interaction. However, this is not a simple change in the test cases, as it requires special environments to enable the use of the respective interaction devices. Since this multiplies the testing effort needed, many OEMs have tried to automate testing. For this reason, robots are employed, which use the respective devices and capture the output of the system for test evaluation.

There is not one single product to be tested; since there are many different *variants* of the same product (see above), tests have to be tailored to the correct version of the program and the correct variant. Again, this increases the effort needed for testing, as well as the complexity. There have to be annotations for specifying which functions are present in which variant, and which are not. The test cases have to be adapted to these, so that for each variant, there is a distinct set of test cases. Otherwise, all the tests targeted at features not existent in the actual variant will fail. Tracking all the differences of the variants and the different versions requires thorough configuration and variation management.

In the past, there was often a very simple interface, but now we have *digital screens with graphical widgets*, overlaying text, and others. Testing the correct positioning and size of the elements on the screens is very cumbersome. A taxonomy of failures in graphical user interfaces of modern In-Vehicle Infotainment Systems is published in (Mauser et al. 2013b). Some companies have started using cameras to capture the screens of the HMI in different states. Afterwards, the elements have to be compared to the specification. This does not require only a very detailed specification, but capturing the screens is also error-prone. For example, it is not advisable to just film the displays of the head units because there are too many sources of error, for example changing or extreme lighting conditions, problems with the lens of the camera, or others. Finally, position, size, and color have to be compared to the specification, which itself has to be machine readable. However, the length of the text is subject to the language used. Again, different tests have to be employed for different languages, as the results change not only in terms of the length, but also in terms of the position of the texts, as some languages are written from right to left, or from top to bottom.

Lately, increased attention has been given to the verification of *animations*. Modern systems no longer perform hard, visual transitions between screens. Rather, they are filled with smooth animations, even within single widgets, providing a comfortable user experience. On the other hand, it is a challenge for testers to cope with the increased complexity, and currently there is no fully automated solution.

To overcome the difficulties of these modern developments, some companies have therefore started to educate their testers and demand a certificate of skills, e.g., *certified tester* by the International Software Testing Qualifications Board (ISTQB) (2013). In addition, ongoing research discusses, whether particular testing goals can be covered by partially or fully automatic testing process (Mauser et al. 2013a; Duan 2012).

When the product is ready for release, management is responsible for the final approval. Even at this late stage, it is possible that changes to the product features may be requested. If approval is granted, the HMI may go to production.

1.4.4.4 Post-Implementation

Even after the start of production, many companies continue to work on the HMIs, to eliminate defects found in the field, or to include new functions. For example, some companies have added functionalities for accessing external mobile devices through the HMI. These updates cannot be applied automatically (e.g., via download) but have to be deployed when the car is in the garage. Since there is no possibility to easily perform updates, (i.e., by the customer) this has to be done very carefully. This also means that the software (and the hardware) has to be tested extensively in order to prevent the introduction of new errors.

1.5 HMI Development Tomorrow

The current trends and developments in automotive HMI development are still in the process of establishing themselves, while new, future trends are already emerging. Some are a consequent progression from current trends, such as connected technologies. Future HMIs could be permanently connected to mobile devices or to the Internet, and make use of server- or cloud-based applications, which will not only foster the ability to change software easily even after roll-out, but also to reduce the constantly growing hardware demands. However, HMIs of the future may also be dynamically adaptive, depending on the habits and demands of the drivers and the driving situation, and could also be personalizable. One very important domain will continue to be safety functions included in the car, and also technologies concerning autonomous driving. Other developments will introduce new concepts in automobiles. As already known from TVs, 3D technologies will be examined for displays, and sooner or later, small touch displays will replace the traditional buttons. Hardware could be modularized and then be replaced in small parts, as development progresses, instead of having to change a whole HMI system.

1.5.1 Upcoming Trends and Changing User Expectations

In 2015, the average age of new car customers in Germany was nearly 53 years.[2] One of three new car customers is currently older than 60 years. For the younger generation, i.e., adults between 18 and 25 years, the Internet and mobile phones are more important than a car, as the results of the study "The young generation and the connected car" by the Center for Automotive Management indicate (Bratzel 2011). The study also identified high expectations within the targeted younger generation concerning a connected car. However, there is also a willingness to pay for the respective value-added services in such a connected car. Thus, there is a good chance for automotive manufacturers to inspire the young generation with new products and features. Focusing on the needs of the older generation can also be reasonable. Holl et al. (2011) used a digital pen as an enabler for the effective interaction between modern cars and elderly drivers.

These trends are a paradox for the automotive industry. On the one hand, the younger generation expects highly innovative infotainment in cars. They are interested in new technologies like Internet access, extensibility with application updates, and installation of new applications (apps) as well as the seamless integration of mobile devices in the car. On the other hand, this generation of car customers is not able to pay for premium segment cars, which used to be the car models where technical innovations are introduced first. A change of strategy is required from automotive manufacturers to stimulate the customers' inducement to buy. It can be expected that innovations in the area of connected technologies will increasingly be introduced in lower priced segments (Dick 2011).

New challenges furthermore arise from the growing market of consumer electronics, such as smartphones and tablet PCs. The number of smartphone users in the United States increases from approx. 62.6 millions in 2010 to 171 million in 2014 and it is expected to increase to 236.8 million in 2019 (eMarketer 2016). In their cars, users now expect options similar to those they know from mobile contexts. 46% of the Americans are even dependent on their phones, as they are saying that "they couldn't live without" their smartphone (Anderson 2015a, b). Especially so-called *digital natives* (Selwyn 2009) have grown up with computers and often strongly rely on them. The influence of smartphones and integration of personal information is summarized in Bratzel (2011): Young adults aged between 18–25 years were asked what they would rather relinquish for one month: their car or their mobile phone. The results show that they would rather give up their car than

---

[2]Study by the CAR—Center for Automotive Research, 2015.

be without their mobile phones. A global survey from frog design confirms this decision (Giacchetto and Gregorio 2015). One-third of the car owners of this survey would give up their vehicle rather than their smartphones. This indicates that the concept of mobility is being redefined and refers no longer only to spatial but also to virtual dimensions, including communication or information retrieval anytime, anywhere.

Modern smartphones provide powerful hardware with high-definition touch-screens and sensory input- and output-like compass and GPS. Relying on permanent Internet availability, manifold functions and applications are possible at low costs. Furthermore, new apps can be installed easily. This leads to increased customer expectations, which carry over to in-car infotainment systems because customers compare their in-car infotainment systems with other devices of their everyday life, such as mobile phones (Meroth and Tolg 2008). Examples are spoken dialog systems, which were first introduced in cars to allow operation while driving without affecting visual attention and without the need to take the hands off the steering wheel (Tièschky et al. 2010). Modern smartphones also provide speech operation, although the context of use is usually not as safety critical as in the car. Since Apple uses speech operation as the most important selling point for the iPhone-4S mostly every smartphone has the possibility to interact with it via speech. This will lead to ever-increasing expectations concerning in-car speech operation.

The number of infotainment functions that can be better performed by in-car systems than by consumer electronic devices will decrease. However, automotive manufacturers can profit from some determining factors such as the possibility to communicate with other components in the car and to design the operational concept and appearance of the Human Machine Interface in a way that perfectly fits a car's interior design. Automotive manufacturers have to be aware of this advantage and need to make strategic use of it. Rather than implementing more and more functions that the customers would expect anyway, such functions can be brought into the car via consumer electronic devices or Internet services. Furthermore, automotive manufacturers should adopt certain concepts and functions from these areas and provide solutions for easy integration of such devices into the in-car environment. Some of the major challenges with regard to achieving this goal are the differences in the development and product life cycles for automotive products and consumer electronics.

1.5.2 Extendable HMIs

A survey by IHS Inc. with 4000 people from US, UK, Germany, and China in 2015 showed that nearly 45% of respondents would use in-car apps in course of a growing driving experience, and also 75% of those surveyed would be willing to pay for updates of an app. This increased from 25% compared to a study by IMS Research with 2250 people from the US, the UK, and Germany in 2012. Because of the rapidly evolving consumer electronics, time to market for infotainment functions is becoming more important than before. At the same time, the complexity of infotainment functions is increasing. Closed, proprietary automotive infotainment systems cannot keep pace with such short innovation cycles. So, one possibility to improve this situation is the development of extendable systems where functions can be added after the roll-out (Infotainment app), thus separating the development cycle from the life cycle of a car.

Another possibility to integrate new functions into an infotainment system after its roll-out would be server-based applications where no new software needs to be installed on the in-car system. This enables easy deployment of new functions and allows for simple billing concepts (e.g., pay-per-use). Furthermore, OEMs could easily prevent the installation of unwanted applications (Schönfeld et al. 2011). The in-car infotainment system running these apps provides them with input and output devices, whereas the application logic is executed on a server. Thus, the HMIs have to be described in a way that they suit different input and output technologies as well as operation concepts found in different car models. In contrast to conventional telematics services, where the web service only provides machine–machine interfaces and no form of presentation, these apps require new forms of realization for HMIs similar to web technologies.

1.5.3 Hybrid HMIs with Mobile Devices

Traditionally, integration of mobile devices was limited to external data such as address book entries or music data, or to the use of certain functions like the actual phone call function where a respective HMI was already present in the in-car system. In the future, this will change, so that the mobile device will not only provide functions and data for the in-car system, but also use functions and data from the car. One example of such a function that is already implemented and used by mobile devices is the charging status of electric vehicles, which can be read by a corresponding smartphone app. These apps extend the in-car HMI in a certain manner. Thus, OEMs may want such apps to be designed to suit the respective brand and meet the respective quality targets.

Future technologies will also provide possibilities to extend the functional range of current infotainment systems by integrating new functions from external sources. There are different setups possible in which infotainment systems, external devices, and web services take over different roles. This requires technologies such as Mirror Link, Apple Car Play, or Android Auto, which enable remote operation of mobile phone applications (Bose et al. 2010). This is achieved by transferring the display content from the mobile phone to the in-car infotainment system and passing input signals from the infotainment system back to the phone. Another possibility for using mobile phone applications in the car is to run a web server on the phone that provides HTML pages. These pages are displayed in the in-car infotainment system, making it possible to operate the applications in the mobile phone. This requires deep browser integration into the HMI software (Müller 2011).

The most important challenge for the future will be to create added value for the customer by enabling continuous data and information flow between different domains (Sauerzapf et al. 2010). No matter what sources the data and functions may come from, the HMI has to provide a consistent look and feel, giving the impression that the data is coming from one single source. In order to bridge the life cycle gap between mobile devices and in-car infotainment systems, automotive and mobile device manufacturers have to cooperate. Exchange formats and interfaces have to be defined and flexible software architectures should be developed.

1.5.4 New Operation Concepts

With the increasing number and complexity of infotainment applications, new operation concepts have to be designed and improved to avoid driver distraction. New forms of speech operation allow more natural dialogs, similar to those currently promoted by Apple's Siri. New technologies such as gesture recognition are being pushed continuously by the games industry and can be found in current products, e.g., Nintendo Wii or Microsoft Kinect. This can give rise to customer expectations regarding similar technologies in new cars. However, how such technologies can be applied to the automotive context has yet to be investigated in detail.

Another trend is the continuous increase in the size and number of displays in the car cabin replacing former buttons or knobs. Future display technologies such as 3D displays will enable new HMI concepts for communicating certain data to the driver or passenger. More powerful hardware will also make augmented reality applications possible at affordable costs.

1.5.5 Flexible Presentation Concepts

The increasing number and size of in-car displays lead to new possibilities for presenting information to the driver or passenger. In some cars, for example, freely configurable displays are used to replace former analog instrument cluster elements with virtual designed counterparts (e.g., speedo-/tachometer, see also Sect. 1.2.2 AUDI). The first car containing such a freely configurable display was the Toyota Crown Hybrid from 2008 (Burghardt 2009). This concept greatly simplifies the construction of displays, as only a (strong) GPU and a display are necessary; CPUs are already included in the ECUs for the standard displays (i.e., the analog instrument clusters) (Burghardt 2009). Such innovations are established in the premium car segment and in the future will trickle down to the medium and finally to the lower car segments to achieve more flexible presentation concepts. This will also enable the OEMs to achieve significant economies of scale because the same hardware platforms can be used to achieve a look and feel adapted for different brands, segments, and car models.

1.5.6 Adaptive HMIs and Personalization

The availability of different kinds of displays enables the creation of situation-dependent presentation sets or individually adaptable presentation forms the driver can choose from. For example, there might be a route guidance mode, an audio mode, and a night driving mode (Burghardt 2009). It might even be possible to create one's own personal presentation profiles. This may include choosing preferred sounds or background images or adapting the layout of the presentation elements in the available display areas. The Cadillac Cue, for instance, allows the driver to choose between different arrangements for the digital instruments and to define the set of values that is displayed. However, there is also a drawback, in the form of possible driver distraction concerning some presentation concepts (Burghardt 2009), which has to be minimized as much as possible. In the future, there will be more possibilities to personalize and adapt the in-car infotainment system. The users will be allowed to download, install, and update software for their infotainment system, ranging from simple stand-alone apps to new design styles that adjust the look and feel of the in-car environment. With the possibility to install third-party applications, the OEM has to ensure that these apps are presented in an appropriate manner within the existing input and output devices in the car. Furthermore, new apps have to comply with certain standards assuring minimal driver distraction, and they have to be seamlessly included into a dynamic HMI adaptation process. With variably equipped car models, the quantity and functional range of available input and output devices may differ from car to car, thus leading to an increased need for flexibly designed HMIs that are able to adapt to the respective context of use. The behavior of the current HMI systems is statically predefined. One example is the prioritization of warning messages and the definition of when and how these warnings are presented to the user. With the increased connection of different car components in modern cars and improved sensor systems, it is also possible to add more dynamically adaptive HMIs based on knowledge of the current driving situation or the current driver. Volvo presented such a system called Intelligent Driver Information System (IDIS) with adaptive HMI technology already in 2003 (Brostroem et al. 2006) and has thereby contributed to future developments in this area. The trend of HMI systems goes toward a personal assistant, which means that the driver will enter into a personal relationship with the system. The HMI system will learn the driver's needs and preferences in order to offer the relevant information and functions at the right time. A driver can be supported better when the system has some knowledge about him. First steps are being made in this direction: BMW is working on their system called BMW ConnectedDrive to be more personal by including an emotional browser that presents information depending on person, position, and mood. The system learns which information in what kind of situation is the right one (BMW 2011). A project of the USC Mobile and Environmental Media Lab MEML also funded by BMW explores how a relationship between the driver and an overall vehicle system could be realized. A user profile and system character parameters are held in cloud storage to allow access from everywhere. This enables the use of user-specific data in different vehicles. An interactive timeline represents the relationship between the driver and the system.

1.5.7 HMIs for New Mobility Concepts

The increasing popularity of car sharing communities like *car2go*[3] leads to new requirements for HMIs in cars. Customers use a car only temporarily and share the same car with hundreds or thousands of other possible users. Referred to the consulting company Frost & Sullivan the amount of car sharing users worldwide increased from 0,35 million in 2006 to 4,94 million in 2014 (Frost and Sullivan 2016). Thus, one customer may use many different car models and would have to adapt to the respective in-car environment each time. This implies the goal of creating HMI concepts not only for one type of car, but for a whole brand or model range of vehicles the driver may use. This also includes corresponding concepts for smartphone apps, web pages, or portals that belong to the car sharing solution.

1.5.8 Future Challenges for Upcoming Infotainment Systems

The previously described innovations in the field of HMI systems will highly influence the hardware and software architectures of upcoming infotainment generations as well as the underlying development processes. In the following sections future challenges for upcoming infotainment systems will be explained split in hardware and software.

1.5.8.1 Hardware

Integrating the latest infotainment functions and presenting them in an appealing manner in complex HMIs leads to an increased demand for processor performance and memory capacity while cost pressure remains intense. At the same time, new and improved forms of interaction require incorporating new hardware elements

---

[3]http://www.car2go.com, accessed 12/2013.

such as approximation sensors, control elements including display capabilities, multi-touch displays, or touchpads providing haptic feedback.

Further expansion of mobile broadband networks such as LTE raises both the bandwidth and availability of the mobile Internet. The global average broadband speed continues to grow and will more than double from 2014 to 2019, from 20.3 to 42.5 Mbps (Cisco 2015). In the future, increased bandwidth and availability of the mobile Internet will enable permanent Internet connection for infotainment systems. Functions that nowadays are locally realized within the systems could then be implemented as cloud services. As a result, the steady rise in demand for hardware resources for infotainment systems could be mitigated, which in return would help to keep the respective hardware costs per unit within limits. Furthermore, such cloud-based functionalities allow easy maintenance and modifications even after a car's roll-out to the customer. Thus, car manufactures may achieve further benefits by applying different development and deployment processes.

Another strategy is the separation of short-lived hardware components, such as graphics processing unit, main processor, or memory from long-lived parts, such as audio amplifier or CAN transceiver. Audi has already developed and adopted such a hardware architecture for its new infotainment generations, where the hardware, originating from the consumer world, is placed on a replaceable MMX board (multimedia extension) that is detachable from the hard-mounted RCC module (Radio Car Control) (Hudi 2010). The RCC module contains those functions that are stable during the whole lifecycle of a specific car, e.g., power management, tuner, and diagnostics. The exchangeable MMX module contains those functions that change over the lifecycle of a specific car, e.g., media, navigation, phone, or even the user interface. Breaking up the system into distinct modules enables OEMs to combine modules with different features and performance depending on the vehicle configuration. In doing so, model upgrading becomes much less complex because inventions affect only single modules rather than the whole system, and can be done only a short time after new and improved hardware is available from the manufacturers. The development cycles for such systems could be reduced from 4 years to 2 years. Customers may also benefit from variable hardware architectures as they will have the opportunity to upgrade their existing infotainment system by exchanging the MMX board. Such a concept was realized in the A3 from Audi in 2012. In 2015 the second generation of the so-called MIB ("Modularer Infotainment Baukasten") was delivered from the VW concern. The Audi TT and the product value build-up of the A6 and A7 are equipped with the MIB2 for the first time. With a T-30-processor from Tegra (Nvidia) it has double the storage capacity, double the processing power, double the graphics performance, and its flexibility is a great advantage to the intensive competition field (Hudi 2014).

The implementation of a growing number of connectivity services leads not only to a wider functional diversity of future infotainment systems, but also to increased multiplicity of the necessary hardware variants. To cope with the latter, it is likely that the car manufactures will adopt such platform- or module-based development strategies for their infotainment systems, as these are well-known and approved solutions for similar problems at the vehicle level (e.g., large growing number of different car models). Hence, decoupling short- and long-lived hardware components can be seen as an initial step in an ongoing large-scale change process. In future development processes, the early identification of similarities and diverse fragments between system variants will become a key task.

Facing high cost pressure in combination with increased complexity and condensed development cycles, automotive hardware will continue to align itself with consumer world devices. When adopting technology from different sectors, it must be considered that various automotive-specific regulations have to be satisfied, e.g., undervoltage, range of temperature, or crash safety. In most cases, redesigning the hardware or at least parts of it becomes necessary in order to meet the higher requirements. Apart from hardware elements for direct user interaction such as switches or displays, there will be growing competition in the field of software engineering. With regard to the overall development costs, there will be a shift toward software elements, which will be responsible for a substantial percentage of the total time and effort. This trend has already begun and will continue in the future.

1.5.8.2 Software

The modularization and separation of MMX board and RCC modules sketched in the section above has a counterpart in the software, which is also modularized and detached from the hardware (Hudi 2014). This flexible development allows the integration of several newly developed parts of the software, such as navigation or telephone modules, into the software system. Furthermore, there is an interface for MirrorLink, Android Auto, and CarPlay called *App Connect* within the user is able to have access to different smartphone functions like SMS, speech recognition like Siri, or music like Spotify or Audible.

At present it can be observed that partnerships between OEMs and suppliers are changing in the field of software engineering. In the past, software was developed almost exclusively by suppliers, whereas OEMs concentrated on conceptual design, specification, integration, and acceptance test. Nowadays, more and more large car manufacturers undertake strategic in sourcing of activities dedicated to the development of brand identity, forming elements of a car such as the HMI system (Hüttenrauch and Baum 2008). In practice, this is realized by shareholdings in supplier companies or by an OEM forming its own subsidiary companies that fulfill special-purpose tasks.[4] In addition, OEMs, suppliers, and service providers have recently begun launching common businesses that perform their software development activities.[5] The common strategy pursued by OEMs in all these approaches is to build up and keep software engineering expertise under their control. Such

[4]Volkswagen has shareholding in IAV, which is the biggest service provider for electronics development. Other examples are subsidiary IT companies such as BMW Car IT or Carmeq.

[5]E-solutions is a common business of Audi and Elektrobit.

activities require large-scale investments that will only amortize if the OEM is willing to take responsibility for software development in the long term. Thus, it will become possible to individually customize and divide the assignment of development tasks between the OEM and its suppliers based on the particularities of the respective HMI project. As a consequence, it is likely that OEMs will realize a higher added value than they did before in development partnerships.

Industry standards like AUTOSAR and GENIVI allow development costs to be kept under control for both OEMs and suppliers, as there is no further need to develop new adaptation layers in each new partnership. At the same time, they make it easier for an OEM to change to another supplier. Suppliers, on the other hand, benefit by being less dependent on single OEMs. Since it becomes easier for both sides to create new partnerships as well as to end one, it is likely that future development partnerships will become more dynamic. This means that both a trend toward closer cooperation and a trend toward more dynamic partnerships are emerging.

The rapid evolution of consumer electronic devices forces OEMs to operate in increasingly shorter time-to-market cycles in developing their infotainment functions and HMIs. In order to keep this pace up while possibly varying partnerships and task assignments emerge, there is an urgent need for optimized development processes in HMI design and implementation. Being able to exchange data efficiently and without the commonly occurring format mismatch caused by the use of different tools on both sides are critical success factors of such a method. This can be achieved by standardizing exchange formats and appropriate integrated processes that are properly supported by development tools (Consulting et al. 2004). Current tools are used in isolation from each other, which causes issues in terms of the budget and quality of the developed systems (Ougier and Terrier 2010). A necessity to abandon established but company-specific special-purpose tools in software engineering is evolving. They have to be replaced by tool chains enabling the use of free software packages or available solutions originating from different industrial application domains such as mobile devices, web services, or desktop applications.

Both, software and hardware will be changed due to the upcoming trend of (semi-)autonomous driving. More and more assistant functions or pilots are available in modern cars, such as adaptive cruise control, parking systems, or lane change assistants. Hence, the vehicle is going to be rather a cooperative partner for the driver than only a mobility device.

In this case not only original automotive branches are going to be competitors of automobile companies of today. For example, Google is working on a self-driving car since 2009. Google's electro vehicles are equipped with several sensors to drive autonomous (more than 1.5 million miles until today). In contrast to other cars their interior is not designed for driving but for riding. So, the HMI of upcoming cars will have a different focus. It is more about cooperation between the driver and the car, to enhance trust in the new technic. For example, there will be time to do other things while the car is driving autonomous and for this there is not a need for the steering wheel in this period. Perhaps, the interior of the vehicle is transformed more and more into a living room or working station.

1.5.9 Ongoing Research

The issues mentioned above are addressed by the research project automotiveHMI (AutomotiveHMI 2014), which is funded by the German Federal Ministry of Economics and Technology. The project aims to improve the process of the development of Human Machine Interfaces in the automotive sector. An integrating approach based on standardized languages (Meixner et al. 2013), models, and interfaces leads to an improvement of efficiency for all companies involved, from car manufacturers and component suppliers to the producers of the used tools (Hess et al. 2012a, b). The description based on abstract models also enables the convergence of new multimodal forms of interaction in completely new operating systems (AutomotiveHMI 2014). Designing and realizing HMIs in the automotive sector involves a multitude of highly diverging and concurrent development processes, each of them focusing on different aspects of the system or different development phases. Hence, one major objective is the creation of a standardized description language that can be used across workflow boundaries. This requires the language to comprehensively model every aspect relevant to one of the stakeholders and to provide views on the system from various perspectives, e.g., from the viewpoint of designers, engineers, or testing people. With proper tool support, this technique will facilitate communication between the involved parties, resulting in faster overall development while at the same time reducing error-proneness. For a historical overview of model-based user interface development outside the automotive industry, we refer to (Meixner et al. 2011).

1.6 Summary

Today's infotainment systems are rather closed systems that come with a statically defined set of OEM-defined functions. Their interaction with external devices is limited at the moment to selected functionalities or even selected connectivity solutions like MirrorLink, Android Auto, or CarPlay. However, the rapid evolution of the consumer electronics sector and the broad acceptance and spread of its inventions in the general public drives customers' expectations regarding automotive infotainment systems. This challenge can be mastered by bringing both worlds closer together, using the vitality of the consumer electronic market as a driving force and ramping up in new inventions for automotive infotainment systems. Therefore, in future HMI systems one can expect better connectivity, which will enable new functionality and customer value with less borders. So, for example, the smartphone integration will be seamless and much easier in the next decades.

A further challenge the OEMs have to deal with is the growing desire of consumers to personalize their vehicles. Consumers want to configure their vehicles according to their personal preferences and requirements, which may change from time to time. Taking into account the extended lifetime of vehicles and installed HMI systems—compared to the average lifetime of smartphones and other consumer trends—it is quite obvious that after-sale solutions for system upgrades and modifications are urgently needed and represent a non-negligible market for the industry. In terms of the HMI system, individually configurable and skinnable digital instrument clusters are first steps toward a higher degree of infotainment personalization.

However, personalization is not limited to pure appearance modifications but rather involves many other parts and aspects of an infotainment system. It is likely that downloadable content and functionalities (infotainment apps) will enable customers to individually extend the functional range of their HMI systems in the future. While the concept of app-based individualization has become a major factor in the market for consumer electronics, the automotive industry is still hesitant to adopt similar approaches. Nevertheless, it is obvious that OEMs will not have the resources to develop and offer a similar range of applications and add-ons on their own, thus paving the way for more and more third-party providers entering the market. In the future, one might even doubt the ability of a single OEM to develop on its own infotainment systems that meet the rising customer expectations in this area. This might force the OEMs to leave the concept of closed proprietary systems behind and move toward the disclosure of assorted signals and vehicle interfaces for third development parties.

Implementing the technological foundations for downloadable content and appearance updates open up ways for focusing on the primary function of the HMI: direct user interaction. For example, dialog design and control modalities may be kept up-to-date. Modern techniques in natural language recognition enable system designers to create more naturally feeling user–machine conversations and to mix different modalities. The human–machine interaction in a car could be pushed to a next level by successfully combining natural language speech commands and touchpad gestures with coherent audio-visual feedback of the system.

Once technical obstacles are overcome and the automotive industry succeeds in providing user-centered solutions for the individual configuration of in-car infotainment systems, one can expect that customers will start identifying themselves with their self-designed and personalized systems to a certain extent (e.g., similar to communities of users that exchange individually created desktop themes for their personal computers). This may lead to a new customer desire to transcend the physical boundaries of their cars and to extend and spread the interaction with "their" infotainment system to occasions that might go beyond the time that is actually spent in the car. Depending on the functional range of future infotainment systems, users may have the wish to share their configurations and to interact with the systems anywhere at any time. One example could be a user on vacation checking via his smartphone if everything is alright with his car, which is waiting for him on the public parking lot at the airport. Faced with such future scenarios, it is likely that OEMs will want to leave behind the idea of isolated in-car infotainment systems and move toward the development of a whole infotainment universe. In addition to the in-car hardware, such an OEM-branded HMI universe would comprise corresponding smartphone applications, websites, and driving portals, thus offering future customers a holistic brand experience.

From a technical perspective, the future scenarios presented in this article already seem to be conceivable but one major issue has not been addressed yet. Whereas continuous research efforts are being undertaken to cope with technical and infrastructural hurdles, questions about legality and liability aspects and responsibilities concerning the interaction with infotainment systems are still pending issues. As long as clarification of these legal basics remains an issue and as long as vehicles cannot drive fully autonomously, the driver and his distraction remain the bottleneck to the integration of new functionalities into infotainment systems. Neither will OEMs disclose car interfaces to third-party developers in a legal gray area nor will customers spend time and money on HMI systems without knowing the actuarial implications of their usage. Hence, one can say that in addition to the necessary technical solutions, clear legal directives need to be established to define exactly which function of an infotainment system is made available to which vehicle occupant under which specific driving conditions.

1.7 Conclusion

From the history of automotive infotainment, it can be seen that user interaction in the car has always been influenced by upcoming new technologies customers first get used to in the world outside the car. This already began in the early twentieth century with the first car radio and with the installation of the first in-car phone. A car without these functions is hard to imagine today. Currently, Internet-based applications and social network applications are finding their way into in-car infotainment applications. The car usually is not the first device utilizing new technologies like these. However, there have always been applications uniquely designed for in-car use, such as navigation applications. With the increasing market of portable devices and smartphones, these are not restricted to built-in navigation devices anymore. Consumer electronics providers are big competitors for the automotive infotainment market. Last, but not least, there are some distinctive applications that are hard to replace with external applications and devices. Driver assistance applications and all sorts of in-car settings and comfort applications are major examples. Looking into the future, driver assistance applications may not be needed anymore for self-driving cars. The car will turn into a mobile living room, a mobile office, a mobile child's room, and maybe also a mobile dining room. This means that even more external devices and applications will influence in-car user interaction. Office workstations or home-entertainment devices may find their way into the car when the actual driving task is more and more reduced.

For a seamless experience, interfaces between different types of external applications and devices have to be developed and maintained, and perfect integration into the in-car infotainment world has to be achieved. This requires accurate design, development, and testing. Perfect engineering tools and the development of standard engineering approaches adjusted to the automotive industry will help keep development cycles shorter and ensure future-proof automotive infotainment development and deployment.

Acknowledgementss First and foremost, the help of all the people involved in this project and, more specifically, the authors and reviewers who contributed their time and expertise to this book chapter should be acknowledged. Furthermore, a special thanks goes to Mister Horst Koch from Koch KLASSIK Automobil GmbH for providing his old timers for illustrations in this book chapter.

Parts of the research described in this paper were conducted within the project automotiveHMI. automotiveHMI was funded by the German Federal Ministry of Economics and Technology.

References

Amditis, A. et al. (2010). Towards the automotive HMI of the future: Overview of the AIDE-integrated project results. *IEEE Transactions on Intelligent Transportation Systems, 11* (3) s.l.:IEEE, 567–578.

Anderson, M. (2015a). *6 facts about Americans and their smartphones.* http://www.pewresearch.org/fact-tank/2015/04/01/6-facts-about-americans-and-their-smartphones/.

Anderson, M. (2015b). *Technology device ownership: 2015.* http://www.pewinternet.org/2015/10/29/technology-device-ownership-2015/.

Ariza, M., Zato Recellado, J. G., & Naranjo Hernández, J. E. (2009). *HMI Design in Vehicles based in Usability and Accessibility Concepts.* s.l., s.n.

AT&T. (2016). *AT&T.* Retrieved June 07, 2016 from http://www.corp.att.com/attlabs/reputation/timeline/46mobile.html.

ATZ/MTZ. (2007). Audi A4. *Sonderausgabe von ATZ und MTZ.*

Audi, A. G. (2001) *Konzeptstudie Audi Avantissimo.* Retrieved April 04, 2016 from http://www.audi.ch/ch/brand/de/erlebniswelt/design___technologie/design-studien/avantissimo.html.

Audi, A. G. (2009). Der neue Audi A8—Es werde Licht. *Audi mobil Extra.*

Audi, A. G. (2015). *Audi virtual cockpit.* Retrieved June 20, 2016 from http://www.audi.com/com/brand/en/vorsprung_durch_technik/content/2014/03/audi-virtual-cockpit.html.

AutomotiveHMI. (2014). *The automotiveHMI Project.* Retrieved May 11, 2016 from http://www.automotive-hmi.org/index.php?id=5&L=3&cHash=1ba93b4caed0f38635b514d9bff8ad4f.

Bader, M., & Fallast, M. (2012). The development process as a complex and interdisciplinary team based challenge. *New Technologies—Trends, Innovations and Research* s.l.:InTech, 327–347.

Bellis, M. (2011). *Automobile History.* http://inventors.about.com/od/cstartinventions/a/Car_History_2.htm.

Blessing, J., Feltgen, C., & Green, M. (2010). Vom Bediensystem zu HMI-Signaturen. *ATZelektronik* 21–25.

BMW. (2011). *BMW Vision ConnectedDrive.* Retrieved May 10, 2016 from https://www.press.bmwgroup.com/pressclub/p/de/pressDetail.html?title=bmw-vision-connecteddrive&outputChannelId=.

Bock, C. (2007). Model-driven HMI development: Can meta-CASE tools do the job? In *HICSS 2007. 40th Annual Hawaii International Conference on System Sciences, 2007* (Vol. 01, p. 287b).

Bose, R., Brakensiek, J., & Park, K.-Y. (2010). Terminal mode—transforming mobile devices into automotive application platforms. In *The Second International Conference on Automotive User Interfaces and Interactive Vehicular Applications* (pp. 148–155).

Bosshart, L., & Hellmueller, L. (2009). Pervasive entertainment, ubiquitous entertainment. *Communication Research Trends, 28*, 3–19.

Bratzel, S. (2011). *„i-Car": Die junge Generation und das vernetzte Auto. Eine empirische Studie zu den Einstellungen und Verhaltensmustern der 18–25 Jährigen in Deutschland.* Bergisch Gladbach: Center of Automotive Management.

Brostroem, R., Engstroem, J., Agnvall, A., & Markkula, G. (2006). Towards the next generation intelligent driver information system (IDIS): The Volvo car interaction manager concept. In *Proceedings of the 13th ITS World Congress.*

Burghardt, S. (2009). *Anzeigekonzept für ein frei program-mierbares Kombinationsinstrument.* Fachbereich Informatik der Universität Koblenz-Landau: s.n.

Cisco. (2015). *The Zettabyte Era: Trends and Analysis.* Retrieved May 10, 2016 from http://www.cisco.com/c/en/us/solutions/collateral/service-provider/visual-networking-index-vni/VNI_Hyperconnectivity_WP.pdf.

Commission of the European Communities. (2013). *Commission recommendation of 22 December 2006 on safe and efficient in-vehicle information and communication systems: Update of the European Statement of Principles on human machine interface.* s.l.:s.n.

Consulting, M. M., Fraunhofer-Institut für Produktionstechnik und Automatisierung -IPA-, S., & Fraunhofer-Institut für Materialfluss und Logistik -IML-, D. (2004). *Future Automotive Industry Structure (FAST) 2015.* Frankfurt: Henrich Druck + Medien GmbH.

Dick, M. (2011). *Im Interview: Audi-Entwicklungsvorstand Michael Dick* [Interview] (April 03, 2011).

Drews, F. A., Pasupathi, M., & Strayer, D. L. (2008). Passenger and cell phone conversations in simulated driving. *Journal of Experimental Psychology, 18*(11), 392–400.

Duan, L. (2012). *Model-based testing of automotive HMIs with consideration for product variability.* München: s.n.

Elektronik Automotive. (2002). Elektronik Automotive. *Sonderausgabe Audi A8.*

eMarketer. (2016). *Number of smartphone users in the United States from 2010 to 2019 (in millions).* http://www.statista.com/statistics/201182/forecast-of-smartphone-users-in-the-us/.

Ersal, T., et al. (2010). *IEEE Transactions on Intelligent Transportation Systems, 27*(05), 692–701.

Feld, M., & Müller, C. (2011). The automotive ontology: managing knowledge inside the vehicle and sharing it between cars. In *Proceedings of the 3rd International Conference on Automotive User Interfaces and Interactive Vehicular Applications* (pp. 79–86).

Fleischmann, T. (2007). *HCI, 8*, 31–39.

Ford. (2016). *Ford.de.* Retrieved April 26, 2016 from http://www.ford.de/Rund-um-den-Service/Ford-SYNC/AppLink.

Frost & Sullivan. (2016). *Anzahl der weltweiten Carsharing-Nutzer von 2006 bis 2014 (in Millionen).* Retrieved May 10, 2016 from http://de.statista.com/statistik/daten/studie/388022/umfrage/anzahl-der-weltweiten-carsharing-nutzer/.

Ganzhorn, M., Diederichs, F. J., & Widlroither, H. (2011). Enhancing the reproduction of road signs in automotive HMIs to protect children. In *11th International Conference on Intelligent Transport Systems Telecommunications, ITST 2011* (pp. 338–344).

Gesellschaft für Unterhaltungs- und Kommunikationselektronik (Hrsg.). (2006). Ce im Auto—Vom ersten Autoradio zum mobilen Multimedia-Center. p. http://www.gfu.de/home/historie/autoradio.xhtml.

Gesellschaft für Unterhaltungs- und Kommunikationselektronik. (2010). *CE im Auto.* http://www.gfu.de/ifa-berlin/historie/.

Giacchetto, A., & Gregorio, A. (2015). *How to Respond to Declining Car Ownership*. Retrieved April 19, 2016 from http://designmind.frogdesign.com/2015/09/how-to-respond-to-declining-car-ownership/.

Governors Highway Safety Association. (2011). *Distracted driving: What research shows and what states can do*. Washington, D.C.: Governors Highway Safety Association (GHSA).

Gow, G. A., & Smith, R. K. (2006). *Mobile and wireless communications: An introduction*. UK: McGraw-Hill Education.

Häcker, C. (2015). *Audi TT digital cockpit*. s.n.: Heilbronn.

Häcker, C. (2016). *History of HMIs from Mercedes and Audi (1993–2016)*. Heilbronn: s.n.

Heisterkamp, P. (2001). Linguatronic product-level speech system for Mercedes-Benz cars. In: *Proceedings of the First International Conference on Human Language Technology Research* (pp. 1–2). Stroudsburg, PA, USA: Association for Computational Linguistics.

HELLA. (2007). *prnewswire.com*. Retrieved April 21, 2016 from http://www.prnewswire.com/news-releases/hella-adds-lane-change-warning-to-audi-q7-52780752.html.

Hess, S. et al. (2012a). *Standardizing model-based IVI development in the German automotive industry*. New Hampshire, USA: s.n.

Hess, S. et al. (2012b). Standardizing model-based IVI development in the German automotive industry. In *Proceedings of the 4th International Conference on Automotive User Interfaces and Interactive Vehicular Applications (AutomotiveUI-12), 4th, September 17–19, Portsmouth, New Hampshire, USA*, 17–19 09.

Holl, K., Hambsch, D., & Bröckl, U. (2011). Effektive und effiziente Fahrzeuginteraktion durch die Verwendung des digitalen Stifts. *i-com Zeitschrift für interaktive und kooperative Medien, 10*(1), 74–79.

Hudi, R. (2010). Wir werden noch effizienter. *VDI-FVT-Jahrbuch, 2010*, 76–77.

Hudi, R. (2014). *Audi-Connect-Manager Hudi: "Es muss das Beste auf dem Markt sein"* [Interview] (July 11 2014).

Hüttenrauch, M., & Baum, M. (2008). *Effiziente Vielfalt: Die dritte Revolution in der Automobilindustrie*. s.l.: Springer.

IQPC. (2016). *3rd International Conference Automotive Cockpit HMI USA 2016*. Retrieved April 25, 2016 from http://www.automotive-hmi-usa.com/.

ISTQB. (2013). Retrieved May 09, 2016 from http://istqb.org/display/ISTQB/Home.

Jørgensen, A. H., & Myers, B. A. (2008). *User interface history*. New York: ACM.

Kellar, M. (2007). *An examination of user behaviour during web information tasks*. Canada: Ph.D. thesis Dalhousie Univ.

Kern, D., & Schmidt, A. (2009). Design space for driver-based automotive user interfaces. In *AutomotiveUI '09: Proceedings of the 1st International Conference on Automotive User Interfaces and Interactive Vehicular Applications* (pp. 3–10).

Kurfess, T. R. (2011). *The Growing Role of Electronics in Automobiles: A Timeline of Electronics in Cars*. http://www.chicagofed.org/digital_assets/others/events/2011/automotive_outlook_symposium/kurfess_060211.pdf.

Lamel, L., Minker, W., & Paroubek, P. (2000). Towards best practice in the development and evaluation of speech recognition components of a spoken language dialog system. *Natural Language Engineering* (pp. 305–322). New York: Cambridge University Press.

Mauser, D., Klaus, A., & Holl, K. (2013a). Towards a GUI test model using state charts and programming code. *Testing Software and Systems* 271–276.

Mauser, D., Klaus, A., Holl, K., & Zhang, R. (2013b). GUI failures of in-vehicle infotainment: analysis, classification, challenges, and capabilities. *International Journal on Advances in Software* 142–154.

Meixner, G. (2013a). *Maserati Indy America 4700 (1972)*. Heilbronn: s.n.

Meixner, G. (2013b). *Mercedes-Benz 22/50 Open Tourer*. Heilbronn: s.n.

Meixner, G. (2013c). *Mercedes-Benz 300 SL (1955)*. Heilbronn: s.n.

Meixner, G. (2013d). *Packard 120 (1936)*. Heilbronn: s.n.

Meixner, G., Orfgen, M., & Kümmerling, M. (2013). Evaluation of user interface description languages for model-based user interface development in the German automotive industry. *HCI, 1,* 411–420.
Meixner, G., Paternò, F., & Vanderdonckt, J. (2011). Past, present, and future of model-based user interface development. *i-com* 2–11.
Meroth, A., & Tolg, B. (2008). Infotainmentsysteme im Kraftfahrzeug (p. 364). Wiesbaden: Vieweg + Teubner Verlag.
Müller, T. (2011). Die Revolution des Kundenerlebnisses. *Elektronik im Kraftfahrzeug/Electronic Systems for Vehicles* 150–168.
Myers, B. A. (1998). A brief history of human computer interaction technology. In: *Interactions* (pp. 44–54). Pittsburgh: ACM.
Myers, B., Hudson, S. E., & Pausch, R. (2000). Past, present, and future of user interface software tools. *ACM Transactions on Computer-Human Interaction (TOCHI)—Special Issue on Human-Computer Interaction in the New Millennium, 03*, 3–28.
National Safety Council. (2015). *Annual Estimate of Cell Phone Crashes 2013*. Retrieved April 25, 2016 from http://www.nsc.org/DistractedDrivingDocuments/Cell-Phone-Estimate-Summary-2013.pdf.
Ougier, F., & Terrier, F. (2010). *Eclipse based architecture of the EDONA platform for automotive system development.* Toulouse, Conference ERTS 2010.
Perrin, A., & Duggan, M. (2015). *Pew Research Center.* http://www.pewinternet.org/2015/06/26/americans-internet-access-2000–2015/.
Pew Research Center. (2014). *Pew Research Center.* http://www.pewinternet.org/data-trend/internet-use/internet-use-over-time/.
Pew Research Center. (2016). *Pew Research Center.* http://www.pewinternet.org/fact-sheets/mobile-technology-fact-sheet/.
Reich, B. (2005). *Abstrakte Beschreibung automobiler HMI-Systeme und deren Erweiterung für neue Dienste.* s.l.: Institut für Medieninformatik, Universität Ulm.
Robert Bosch GmbH. (2013). *Bosch-presse.de.* Retrieved April 04, 2016 from http://www.bosch-presse.de/presseforum/details.htm?txtID=6094&locale=en.
Sauerzapf, S., Mutschler, R., & Krstic, Z. (2010). Innovative Bedienkonzepte—Mobile Endgeräte im Fahrzeug. *ATZ elektronik* 18–23.
Schönfeld, A., Stümpfle, M., & Freitag, R. (2011). Mobiles Internet und automotive APPs. *Elektronik im Kraftfahrzeug/Electronic Systems for Vehicles* 255–263.
Schuller, B., et al. (2006). Speech communication and multimodal interfaces. *Advanced man-machine interaction* (pp. 141–190). Berlin: Springer.
Selwyn, N. (2009). The digital native: myth and reality. In: *Publication Cover* (pp. 364–379). London: s.n.
Shull, F. et al. (2002). What we have learned about fighting defects. In *Proceedings. Eighth IEEE Symposium on Software Metrics, 2002* (pp. 249–258).
Smith, A. (2011). *Technical Report Pew Internet & American Life Project.* http://www.pewinternet.org/Reports/2011/Cell-Phones.aspx.
Stevens, A. et al. (2002). *Design guidelines for safety of in-vehicle information systems.* s.l.: s.n.
Stolle, R., Benedek, T., Knuechel, C., & Heinecke, H. (2005). Challenges in automated model-based HMI testing. *35. Jahrestagung der Gesellschaft für Informatik e.V. (GI)* 186–190.
Stueber, G. L. (2011). *Principles of mobile communication* (3rd ed.). New York: Springer Publishing Company.
Tièschky, S., Hamberger, W., Schröder, J., & Mauter, G. (2010). Audi MMI—von der Idee zum Produkt. *Elektronik Automotive (Sonderausgabe Audi A8)* 10–18.
von Zezschwitz, E., et al. (2014). *Secondary tasks.* Munich: LMU.
Wheen, A. (2011). *Dot-Dash to Dot.Com: how modern telecommunications evolved from the telegraph to the internet.* New York: Springer.
White, J. B. (2010). *The Wall Street Journal.* Retrieved April 21, 2016 from http://online.wsj.com/article/SB10001424052748704561004575013152608105546.html.

Wikimedia Commons; User: Guido Gybels. (2008). *Wikimedia Commons.* Retrieved June 20, 2016 from https://commons.wikimedia.org/wiki/File:Mercedes_W220_interior_view.JPG.

Wikimedia Commons; User: Ranger 1. (2016). *Wikimedia Commons.* Retrieved May 23, 2016 from https://commons.wikimedia.org/wiki/File:Ford_S-Max_2015_Interior.jpg.

Wikimedia Commons; User: Yeti.bigfoot. (2009). *Wikimedia Commons.* Retrieved May 23, 2016 from https://commons.wikimedia.org/wiki/File:FORD_Taunus_17M_P2_deLuxe_Steering_wheel.jpg.

Wikipedia. (2015). *Infotainmentsystem.* Retrieved April 20, 2016 from https://de.wikipedia.org/wiki/Infotainmentsystem.

Wood, R. (1988). *Head up displays system.* U.S. Patent No. 4,763,900.

Zhang, R. (2015). *Referenzarchitektur zur Erhöhung der Flexibilität in der GUI-Applikation-Anbindung für automotive Infotainmentsysteme.* Kaiserslautern: Techn. Univ.

Chapter 2
Engaged Drivers–Safe Drivers: Gathering Real-Time Data from Mobile and Wearable Devices for Safe-Driving Apps

Fabius Steinberger, Ronald Schroeter and Diana Babiac

Abstract Regardless of punitive strategies such as fines and demerit points, drivers continue to bring their own devices into cars and use them while driving. In this chapter, we explore the opportunities for gamified safe-driving apps provided by real-time data gathered from mobile and wearable devices. The study is grounded in our interest in providing engaging experiences for drives that otherwise lack engagement, both in manual and semi-automated vehicles. We developed *Brake-Master*, a smartphone app built around vehicle and road data, and evaluated it in a simulator study looking at system performance, usability, and affect. We found the app to perform responsively and accurately, and self-reported data indicate good usability and increased pleasure. Besides exploring *vehicle* and *road* data, we investigated wearable activity monitors for gathering *driver* data such as arousal. Consumer wearables are more cost and size effective than advanced biofeedback systems and are capable of revealing heart rate patterns and trends across drives. We conclude that road and particularly vehicle data can be leveraged to develop novel driving experiences, whereas driver data is more challenging to exploit in this unique design context.

2.1 Introduction

Mobile and wearable devices are prevalent in our everyday lives, including our cars. As a result, there has been an increase in people accessing social media and apps while driving as reported in Germany (Vollrath et al. 2016) and the US (NHTSA 2016). Not only are drivers calling and texting; they do not stop short of

---

F. Steinberger (✉) · R. Schroeter · D. Babiac
Queensland University of Technology (QUT), Brisbane, Australia
e-mail: fabius.steinberger@qut.edu.au

R. Schroeter
e-mail: r.schroeter@qut.edu.au

G. Meixner and C. Müller (eds.), *Automotive User Interfaces*,
Human–Computer Interaction Series, DOI 10.1007/978-3-319-49448-7_2

browsing Facebook, taking selfies, and video calling (AT&T 2015). Indeed, the *"car is merging with consumer electronics"* (Normark 2015) as gadgets such as smartphones, fitness trackers and dashboard cameras enter the driving space. These devices have essentially become automotive user interfaces, even though they were never designed as such.

Statistics show the average age of all vehicles in Australia (Australian Bureau of Statistics 2016) and the United States (United States Bureau of Transportation Statistics 2014) is approximately ten years. Since this time span vastly exceeds the product lifetime of consumer electronics, smartphones and other gadgets are likely to be used in the car for another decade without dedicated vehicle integration (Android Auto, Apple CarPlay, etc.). This presents an opportunity for new driving experiences to take advantage of.

Mobile phones are often portrayed as unwanted distractions, which is reflected in a vast array of research on this topic. Distractions such as unsafe phone use can be caused by a lack of engagement in the driving task (Steinberger et al. 2016). Our previous studies indicate that driver boredom is most likely to occur in low-stimulation conditions such as routine drives, speed maintenance, cruise control, or low traffic. Semi-automated driving further amplifies the significance of this issue. A cutback in manual control causes a lack of engagement in the driving task more often, yet requires drivers to remain vigilant and take over control at any time (Casner et al. 2016). Vehicles are gradually evolving through several levels of partial (L2), conditional (L3), and high (L4) automation (SAE International 2014). One can expect some form of manual control for at least another two decades to come (Litman 2015), which emphasises the ongoing need for task engagement.

Drivers perform best and safest if their arousal levels are neither too high nor too low. The relationship between arousal and performance is referred to as the Yerkes–Dodson Law (Yerkes and Dodson 1908). It suggests that during periods of low arousal, added workload may improve performance, while during heightened arousal, higher workload may reduce performance. Added workload and distraction have received extensive attention in the driving context, whereas lack of engagement and low arousal have not.

The ability to add engaging, yet safe stimuli when needed, e.g., through gamified driving apps, can therefore, have direct impact on road safety and user experience (Schroeter et al. 2016; Heslop 2014). Capitalising on real-time driving data for gamified safe driving provides novel driving experiences. These have not been explored to date. In the near future, capabilities of connected cars and semi-automated vehicles, which can be even less engaging than manual driving in terms of the primary driving task (Casner et al. 2016), further broaden the design space. Biometrics, e.g., related to arousal, can contribute as an indicator for engagement in the driving task. This information may be one way to help determine appropriate points in time to present gamified interventions. For example, vigilance or stress experienced while driving in a big city for the first time would be reflected in high arousal levels and would therefore suggest that adding stimuli is inappropriate. Biometrics may furthermore feed into dynamic difficulty adjustment (DDA,

Tijs et al. 2008), e.g., to adapt challenges presented in gamified driving apps to the current driver state so as to ensure optimal levels of engagement.

The *research aim* of this study is to explore how real-time driving data gathered from present day mobile and wearable devices can facilitate novel experiences such as gamified driving. To address this research aim, we sought to answer the following research questions.

RQ1: How can we develop driving apps built around real-time vehicle and road data gathered from mobile devices?

RQ2: To what extent can wearable devices be used as a data source in the design of engaging driving apps?

2.2 Contribution Statement

The contribution of our explorative work is threefold. First, we report on the design and implementation of the smartphone app *BrakeMaster* that makes use of vehicle and road data to provide an engaging driving experience. Second, we present an evaluation of the app in a driving simulator study. Third, we discuss to which extent data from consumer wearables, e.g., related to driver arousal, can be used to develop novel driving apps. We believe our work to be useful for both researchers and practitioners who aim to enhance driver engagement without compromising safety. The relevance of our contribution applies to manual as well as semi-automated driving as here the driving task is even less engaging and requires further research attention.

2.3 Related Work

Couben and Zhu (2013) suggested that technological capabilities should be used to render phones inoperable while cars are in motion to mitigate the risks of driver distraction. We argue that drivers are likely to continue bringing more consumer electronics into the car and use them, and it is our responsibility as human–computer interaction (HCI) researchers to address this scenario. We are not alone in this way of thinking. Insurance companies and software developers offer commercial apps that employ quantified-self and gamification approaches to improve driving behaviour. Axa Drive[1], e.g., reward good driving behaviour with points and allow users to share their accomplishments with their social networks. These apps, however, will only offer insights *after* the drive has been completed. Unlike our approach, they do not enhance safe-driving in-situ and *in real-time*. Other related

[1]https://www.axa.ie/car-insurance/young-drivers-insurance/products/drivesave/.

applications do provide feedback during the drive but without taking into account when it is safe to do so (McCall and Koenig 2012; Prokhorov et al. 2011). As a consequence, they become distractions themselves.

In order to implement new driving apps, several enabling data streams are needed: access to vehicle information, context awareness for detecting road and traffic conditions, and, ideally, a means of detecting driver arousal in order to activate or mute the app when appropriate. The following subsections provide a review of three types of data relevant for providing feedback to drivers on their on-road behaviour:

1. Vehicle data (location, direction, speed, acceleration, etc.)
2. Road data (following distance, road signs, traffic situation, etc.)
3. Driver data (driver state, mental and physiological conditions, etc.)

2.3.1 Sensing Vehicle Data

Smartphones have previously been utilised to gather vehicle data relevant to safe driving, e.g., detecting speeding violations (Eren et al. 2012). Bluetooth connectors for on-board diagnostics (OBD), which are available at a cheap price (less than US $50), complement this set of information. Paired with a smartphone, they allow anybody to display accurate data such as current speed or fuel intake. OBD dongles such as Automatic[2] and Wayray[3] track and visualise data in their respective smartphone apps, offering location information to family members and coaching features to improve driving skills. A study (Meng et al. 2014) shows that the various sensors found in smartphones can orchestrate information to achieve near complete similarity to OBD. Eren et al. (2012) utilised accelerometer, gyroscope and compass data to detect fatigue, inattention and speeding violations. Similarly, Dai et al. (2010) compared accelerometer and orientation data of Android phones with existing drunk driving patterns as part of accident prevention. All of the above examples illustrate the richness of vehicle data available from consumer electronics.

2.3.2 Sensing Road Data

Data about the driving environment can be derived in several ways. Camera imagery has previously been investigated to identify collision danger or traffic signals

[2]https://www.automatic.com.

[3]https://www.wayray.com/element.

(Koukoumidis et al. 2012). Crowdsourcing data (e.g., traffic light schedules) can facilitate novel applications and benefit drivers by saving fuel or recognising changed road conditions. Another possible solution would be using OpenStreetMap[4] and its APIs to gather road data such as upcoming intersections or speed signs. Similarly, open data and smart city initiatives increasingly enable vehicles to receive real-time information from traffic management systems and other infrastructure elements. A combination of map data and smartphone sensors could furthermore detect the orientation of the car and whether the road network will lead the vehicle to the upcoming location in question. Furthermore, studies have shown how an array of GPS, accelerometer and microphone data can help in the detection of road bumps or vehicle braking (Eriksson et al. 2008; Mohan et al. 2008). More broadly, Andreone et al. (2005) envisioned an information manager that collects data about the vehicle and environment to estimate safety risks at any given moment to present to the driver.

2.3.3 Sensing Driver Data

Highly accurate and reliable physiology platforms are heavyweight and expensive (e.g., Liang et al. 2007), and therefore neither suitable nor affordable for deployment in vehicles. In the near future, however, advanced physiological measures are likely to become part of mainstream wearables such as fitness trackers and smart watches. Therefore, they provide real-time data streams and an ideal platform that we intend to capitalise on towards driver state detection. For example, driver states may serve as an indicator for the appropriateness of activating driving apps or for adjusting the difficulty in gamified driving challenges. Studies suggest that daily stress levels (Bogomolov et al. 2014) or daily moods (LiKamWa et al. 2013) can be accurately detected using smartphones. A problem with many of the methods employed is that they result in low granularity data (e.g., daily). In-vehicle biometrics need much higher granularity, ideally in real time. Therefore, researchers have looked into live imagery from smartphones, e.g., as a means to detect fatigue using blink detection algorithms (You et al. 2012; Schroeter et al. 2013). Dashboard or action cameras could potentially extend these capabilities. Hong et al. (2014) constructed a platform consisting of a smartphone and cheap sensors that assesses aggressive driving style. While this is not a heavyweight, expensive setup, it still requires additional components that are neither part of the car itself nor the driver's everyday belongings. The automotive community has started to explore the potential of wearable technologies for in-car usage. Fitness trackers provide information such as heart rate activity and whether the user is seated or not.

[4]https://www.openstreetmap.org.

Smart watches provide biometrics that have been trialled as an indicator for driver drowsiness (Aguilar et al. 2015). Lastly, pattern recognition has been increasingly applied, e.g., to characterise driver skills (Zhang et al. 2010).

2.3.4 Research Gap

In summary, consumer electronics such as smartphones or fitness trackers have increasingly sophisticated sensing capabilities. Therefore, they present a largely unexplored potential to develop driving apps and to assess driver engagement. Rather than restricting their usage, we investigate how mobile and wearable devices can be considered allies in the quest towards safer driving. We argue that real-time driving data gathered from these devices can be taken advantage of for engaging safe-driving apps. Furthermore, the capabilities of such lightweight devices may create cheaper and increasingly accurate methods for assessing driver engagement, thereby laying the groundwork for affordable future safety interventions.

2.3.5 **BrakeMaster:** *An App Built Around Vehicle and Road Data*

Addressing *RQ1,* this step serves to explore the vehicle and road data streams provided by mobile devices. We developed *BrakeMaster*, a smartphone application that gamifies approaches to red lights. The app shows a black screen while the car is in motion to avoid unnecessary distractions. Upon approaching a red light, an audio cue signals the beginning of a new challenge. The driver is invited to match a deceleration curve instead of breaking abruptly. When the car has come to a halt, the app will display an assessment of the braking performance (Fig. 2.1).

2.3.6 User Experience Design

Designing engaging experiences in the safety critical space of the car requires a careful balance between fun and safety. *BrakeMaster* represents a first attempt at striking that balance and is an outcome from our design approach for driving gamification, where we defined conceptual layers for designing driving gamification such as *verbs, mechanics, core, theme* and *concept* (Steinberger et al. 2015). For this initial game, we picked the routine driving scenario of approaching a red light. At the *verb* layer of the game, which describes user input and simple actions associated with it, users control the app by applying force to the brake pedal and therefore slowing down the vehicle. This aspect is novel; existing driving apps do not read user input from pedals or the steering wheel.

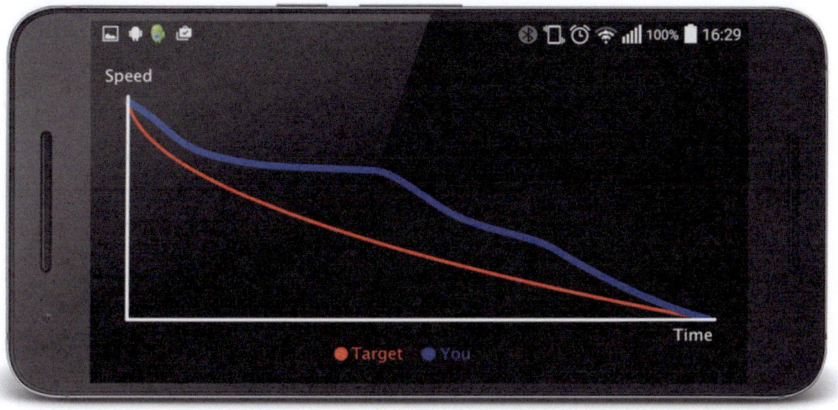

Fig. 2.1 Sample target curve (*red*) and braking performance (*blue*) shown in the prototype implementation

BrakeMaster is a result of design activities such as storyboarding and sketching, which allowed us to explore combinations of existing videogame concepts and driving situations (Steinberger et al. 2015). Figure 2.2 shows an early version of *BrakeMaster* that was designed around an Angry Birds[5] theme. The target in this version is a bull's eye. The bird's position on it reflects the smoothness of braking (y-axis) and steering (x-axis). It was later refined based on participant feedback gathered from an extensive qualitative user study. In that study, potential users expressed their interest in accessing more raw data about their driving performance and skills. As a result, *BrakeMaster* now offers a technical representation of the performance (Fig. 2.1) rather than an elaborate theme.

Auditory instructions were chosen to avoid additional visual load, given driving is mostly a visual task (Sivak 1996). An assessment of the performance is conveyed visually as a graph (Fig. 2.1) and via audio feedback based on a matching score. The app distinguishes between 0–33% matches (e.g., "*Try harder next time!*"), 33–66% matches (e.g., "*Close but not close enough!*"), and 66–100% match (e.g., "*Excellent*"). Both the visual and auditory feedback allow users to quickly comprehend their performance before the traffic light turns green and the vehicle is set into motion again.

2.3.7 Prototype Implementation

BrakeMaster was prototyped as an Android application. In terms of gathering vehicle data, the app connects to on-board diagnostics (OBD). OBD is a prevalent interface for monitoring a wide range of vehicle parameters such as speed, pedal

[5]https://www.angrybirds.com.

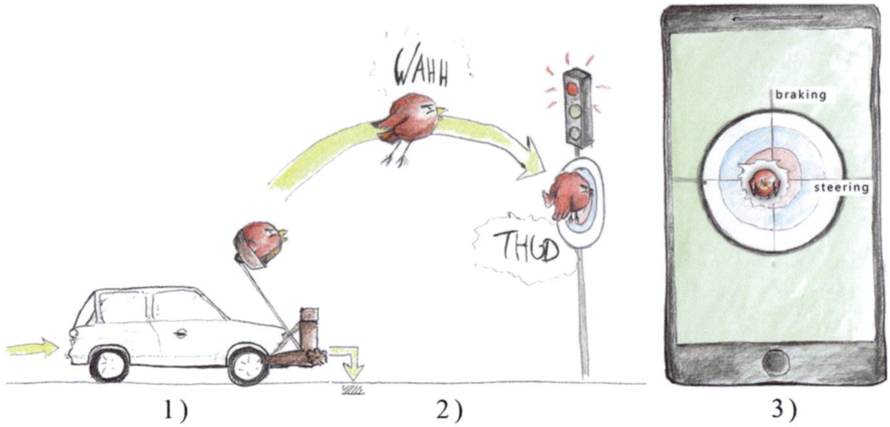

Fig. 2.2 Excerpt of a storyboard illustrating an early version of *BrakeMaster* inspired by an Angry Birds theme

use, steering wheel angle, or fuel intake. There are many cheap adapters from different manufacturers offering USB, COM, or Bluetooth access to OBD data. We used the ELM327 OBD Bluetooth Interpreter,[6] which allows for a wireless connection, and an open source library[7] based on the OBD-II Java API to process vehicle data within the Android app. This implementation was used for initial test drives around a car park.

For the purpose of the driving simulator study (cf. next section), however, vehicle data had to be gathered in a different way. Since the OBD socket was already occupied as a result of the laboratory setup, speed data needed to be acquired through the simulation software. We therefore added functionality to the *BrakeMaster* code to allow connections from either Bluetooth (for the OBD adapter) or WiFi (to read network data transmitted by an Intempora RTMaps[8] middleware application). Figure 2.3 illustrates the hardware software mapping used in the implementation.

Relevant road data for *BrakeMaster* include locations of traffic lights as well as their status (green or red). Detecting traffic lights can be achieved in several ways. For example, vehicles are able to receive signal information from advanced traffic management systems in smart cities. This approach has recently been pursued by Audi for their traffic light information system, which indicates on the instrument cluster the time remaining until the signal changes to green.[9] Aside from vehicle-to-infrastructure integration, OpenStreetMap APIs could be utilised to

[6]https://www.elmelectronics.com/obdic.html#ELM327.

[7]https://www.github.com/pires/android-obd-reader.

[8]https://www.intempora.com/products/rtmaps.html.

[9]https://www.audiusa.com/newsroom/news/press-releases/2016/08/audi-announces-first-vehicle-to-infrastructure-service.

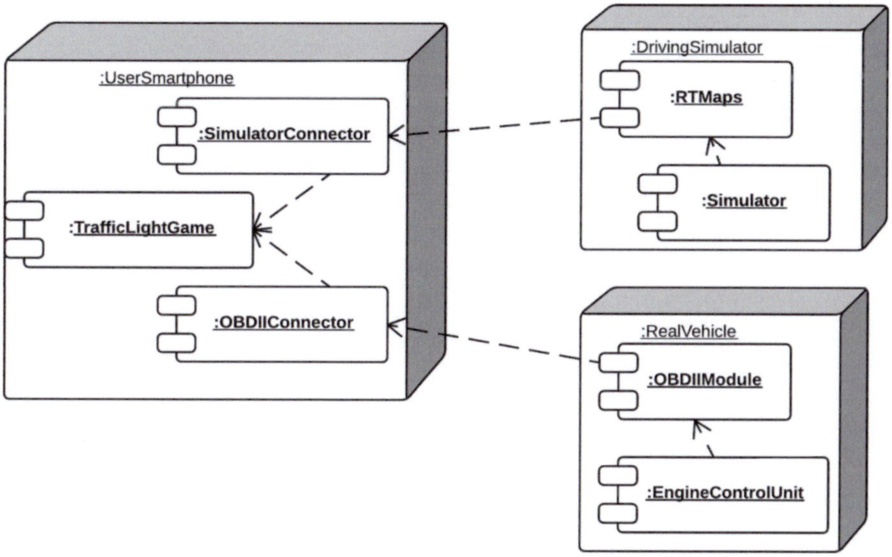

Fig. 2.3 Interface connection allows input from two data sources: simulation software (WiFi) or OBD (Bluetooth)

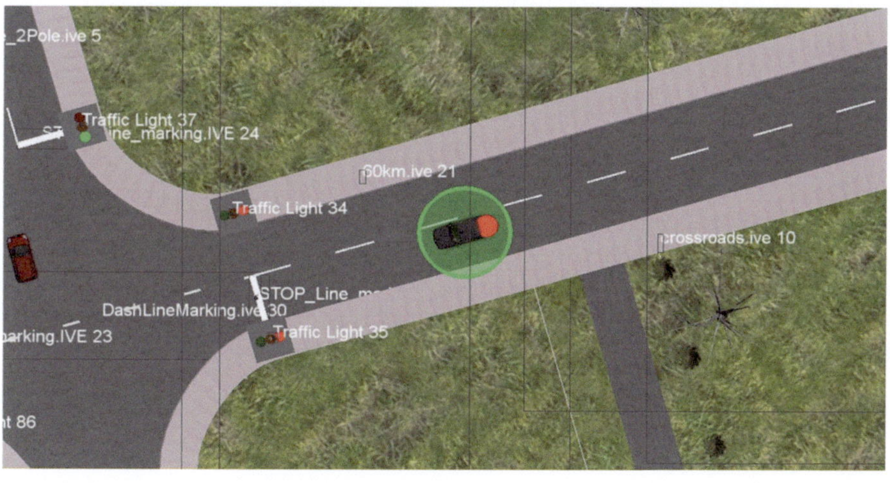

Fig. 2.4 Red lights trigger *BrakeMaster* challenges

identify nearby signalled intersections, in combination with camera image recognition as outlined in the related work section. The *BrakeMaster* implementation for use in the driving simulator gathered traffic light information from the simulation software (Fig. 2.4).

Between the beginning of a braking challenge and stopping, the app records the speed of the vehicle. This data is used to display the braking curve and to calculate the performance score. Once the car has come to a stop, the user is presented a graph illustrating the desired pattern versus the user's braking performance (Fig. 2.1).

The exact nature of desired braking patterns as well as the comparison algorithms could be based on transportation literature and/or be context-dependent. For example, for smooth braking aimed at optimising passenger comfort, a pattern matching algorithm could be used before comparing the two braking curves, while for economical braking, a dynamic time warping algorithm could be better suited. We did not take those into consideration at this point, but applied a software pattern that decouples relevant components and allows for future implementations of different target curves. Every target curve can present a different challenge or level in the game design.

2.4 User Study

We conducted a driving simulator study using the *BrakeMaster* prototype, which served two purposes: (a) identify performance or usability challenges in the prototype; and (b) learn about the user experience of the app's gamification concept.

2.4.1 Participants

Overall, 10 people (5 female) aged between 21 and 37 ($M = 29$, $SD = 4.42$) participated in the study. Given the explorative nature of our approach, the only selection criterion was the possession of a valid driver's license. Before commencing data collection, we obtained approval from the university's ethics committee (approval number 1500000046 in accordance with the *Australian Code for the Responsible Conduct of Research*) and written consent from participants.

2.4.2 Procedure

The study took place in a motion driving simulator with six degrees of freedom, which presents a safe, yet immersive way to conduct controlled experiments. Each session lasted approx. ninety minutes. To begin with, two five-minute familiarisation drives ensured that participants were acquainted with the driving simulation and the app. Afterwards, participants were asked to complete a motion sickness

Fig. 2.5 Smartphone placement in the driving simulator study

survey to ensure they were feeling well and able to continue. We designed the evaluation as a within-subjects, repeated measures experiment with two counter-balanced conditions across participants, *control* and *game*, and one drive per condition. The two research drives (*control* and *game*) consisted of approximately 8 min suburban driving each. The speed limit was 60 km/h throughout the road scenario, which included trailing, oncoming and cross-traffic. Participants encountered nine signalled intersections and five red lights, which triggered challenges through the *BrakeMaster* app in the *game* condition (Fig. 2.4). A smartphone running *BrakeMaster* was placed behind the steering wheel where dashboard displays are usually positioned (Fig. 2.5).

2.4.3 Data Collection

We acquired a combination of objective and self-reported data. The selected methods were chosen to gain insights related to app performance, usability, user experience and driver arousal.

To monitor app performance, we looked into three different metrics. First, we were interested in response time to see whether a change in vehicle status was detected and signalled to the user in a responsive manner. The vehicle status can vary between *driving* (>100 m from next red light), *approaching* (<100 m from the next red light), or *stopped* (stopped at red light). A fast response time is essential, because the transition from one state to another often takes only a few seconds and represents an indirect user input. Regarding throughput, the system should process data at a minimum rate of 10 Hz to project the braking curve and to support a response time of at least 0.1 s. Therefore, the third performance metric is accuracy. Invalid speed readings, noise and rounding errors should be avoided by checking the data for consistency upon aggregation.

We asked participants to fill in the system usability scale (SUS) questionnaire (Brooke 1996), which provides a metric for overall product usability and can be used on small sample sizes with reliable results. We were particularly interested in the interaction technique that takes user input from the brake pedal, which is a novel aspect compared to existing driving apps.

In order to explore the subjective experience with the app, participants were asked to self-report their affective state using a paper-based version of the Circumplex Model of Affect (Russell 1980). This was done after the *control* drive and after the *game* drive.

Lastly, semistructured interviews were conducted with five of the participants to further explore the experience with the app.

2.4.4 Results

2.4.4.1 Objective Data

In terms of app performance, response time and throughput was assessed post-experiment by comparing the application log with the driving simulator log. Based on these observations, the vehicle data was processed at a frequency of 20 Hz, exceeding our desired prerequisite. In terms of accuracy, the vehicle speed log revealed that the majority of erroneous data was received while the vehicle was stopped, in the form of negative values close to zero. These kinds of imprecise readings can be prevented in the user interface by, e.g., rounding up values.

2.4.4.2 Subjective Data

The mean SUS score across all participants is $M = 78.5$ ($SD = 16.55$), on a scale from 0 (worst) 100 (best). All individual scores can be seen in Fig. 2.6 as a boxplot diagram, as proposed (Young and Wessnitzer 2016). According to literature (Bangor et al. 2008; Lewis and Sauro 2009), scores above 68 are considered above average and systems scoring in the high 70s to upper 80s are considered to have

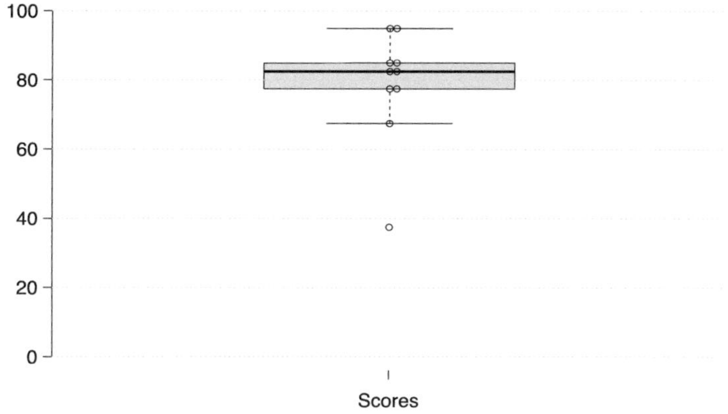

Fig. 2.6 Boxplot diagram showing the SUS scores distribution

above passable usability and below superior usability. Our results show superior usability for two participants (P2, P9), below average usability for another two participants (P3, P6), and scores in between for the remaining six participants. These results suggest good overall usability and indicate that the game objective was well understood and desirable enough to be pursued. P3's score of 37.5 is particularly low. In the post-experiment interview, she revealed that she does not enjoy games and driving tests, and therefore she had a generally negative attitude towards her study participation.

The Circumplex Model of Affect data indicate that, in general, participants perceived an increase in arousal and a more positive valence in the *game* drive. The aggregated data can be found in Fig. 2.7a, b. Looking back at the motivation behind the app, it can be argued that this result is ideal. That is, the game creates arousal and increases driver engagement, without over-catering to hedonistic needs, thus not creating too much distraction from the driving task.

Finally, the interview data reveal participants' subjective experience using *BrakeMaster* and simulated driving. P1 felt *"happy to engage with the challenge and to get positive results."* His game performance was good overall, and he reported feeling *"a small sense of accomplishment"* when receiving positive feedback.

Similarly, P5 was deeply immersed in the driving simulation and enjoyed *BrakeMaster*. He said, *"I felt like playing the game was raising the sense of awareness, I was committed to complete the task that I was assigned, and in doing that I was alert. So I was more aware of the obstacles and the traffic signs and very engaged in the idea of braking in a proper way."* Furthermore, the participant reported after-effects of the study and said, *"days after this test, actually every time now I am more aware of my braking. I think it keeps on doing something good for me."* He felt it was owed to a *"sense of competition with myself."*

P6 described herself as an easily bored person and felt that *BrakeMaster* targets an issue that she is familiar with. She said, *"highway driving or typically*

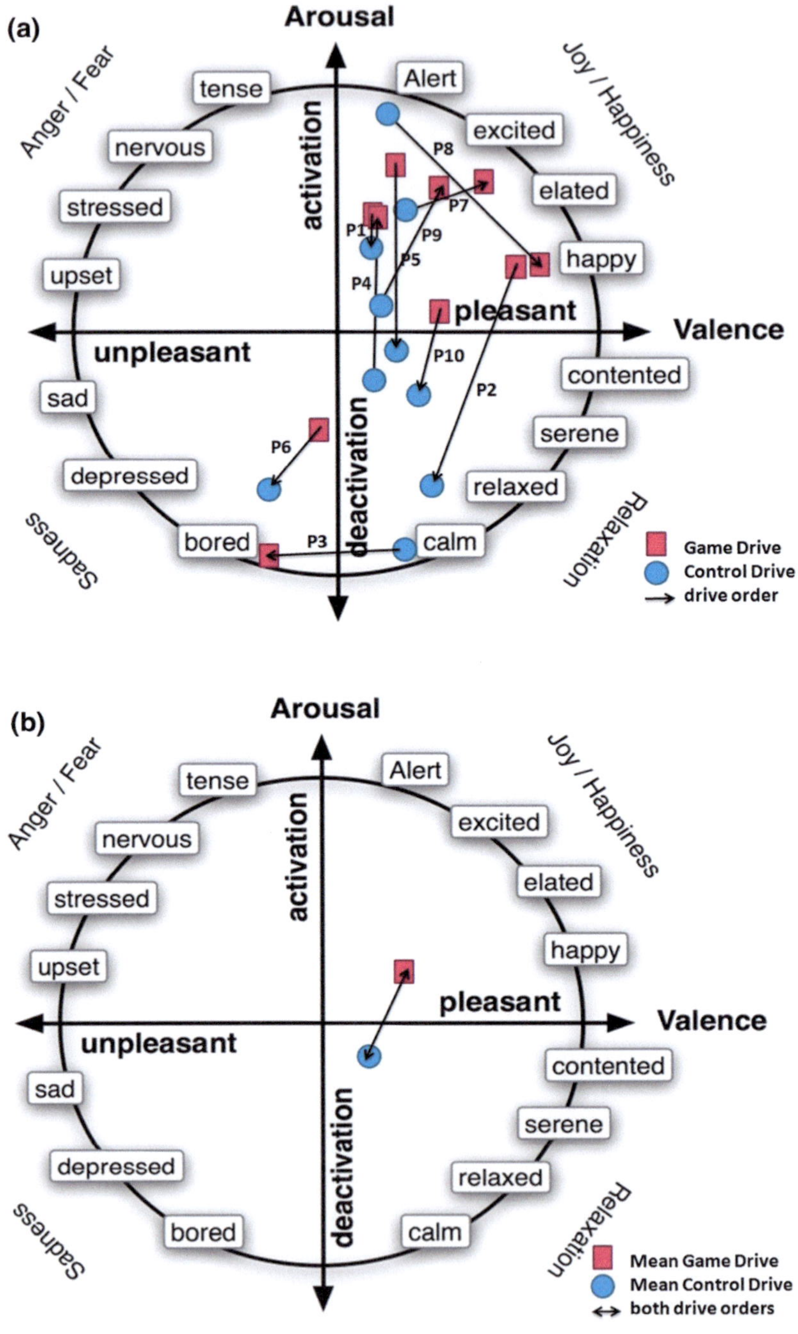

Fig. 2.7 **a** Individual data on the Circumplex Model of Affect. **b** Mean data on the Circumplex Model of Affect

monotonous driving is just repetitive, but you have to focus on it." Furthermore, P6 said she is *"fairly competitive"* and *"likes the idea of playing the game,"* although in the long run, the game should change from time to time to remain interesting.

For P7, playing the game was a generally positive experience, *"I think it improved my mood and my arousal."* He said that the visual feedback helped him improve his driving performance. However, he felt that the audio feedback was *"a bit mean"* at times and suggested using more positive audio messages.

P3 agreed as she expressed dissatisfaction with the audio feedback and stated that it felt discouraging. P3, as pointed out previously, began the interview by stating, *"I don't enjoy playing games, I don't like driving tests."* This led to an overall frustrating experience. Lastly, some participants reported that the simulator brakes did not perform as expected based on their real-world driving experience, which emphasises the need for longer practice drives in future studies.

2.4.4.3 Discussion

Through developing *BrakeMaster* as a prototype and studying user interactions with it, we have seen promising results. Real-time vehicle and road data can be obtained and collected in smartphones and built upon for safe-driving apps as we have demonstrated with *BrakeMaster*. Vehicle data in particular can be gathered in a feasible and accurate way through the prevalent OBD interface. It is technologically more challenging to gather road data. Although much information can be pulled from OpenStreetMap, e.g., upcoming traffic lights or speed signs, a combination of this information with other data, e.g., the direction in which the vehicle is moving, is often required to make sense of the information in a useful way.

Furthermore, gamifying just one aspect of the drive may not be enough. A combination of several challenges might provide more stimulation and pleasure. *BrakeMaster* is just one example to illustrate the trajectory of our ongoing programme of research, but provides a platform for extensions. Gamifying driving can be spun further to incorporate more driving scenarios, other drivers, and challenges. For example, highway driving could gamify the keeping of accurate following distances relative to current speed to prevent tailgating, driving in start-stop peak hour traffic could turn into a game that facilitates the least amount of acceleration, braking and lane changes with the view to create a smoother traffic flow. In the future, the increased contextual awareness of connected and semi-automated cars will provide greater scope for exploring more alternatives.

2.5 Sensing Driver Arousal from Wearable Devices

Through the development and evaluation of *BrakeMaster,* we explored the challenges and opportunities in gathering *road* and *vehicle* data (addressing *RQ1*). We now explore to what extent wearables can be used to gather *driver* data (addressing

RQ2). In particular, we are interested in the acquisition of *driver arousal* data, which is an indicator of task engagement (Yerkes and Dodson 1908), as a data stream in the design of engaging driving apps. Such biometrics may be useful to determine appropriate points in time for presenting driving games such as *Brake-Master* or to dynamically adjust their difficulty. In the future, this type of driver data may be acquired through sensors integrated into steering wheels or seats, but in this study we are exploring if and what kind of information we may be able to gather from present day consumer wearables such as fitness trackers and smart watches.

We compare two consumer wearables to an advanced biofeedback system. The aim of this step is to understand in what ways they differ in terms of data accuracy as well as procedure and handling.

2.5.1 Requirements for Driver Arousal Detection Systems

The driving context is a unique and challenging space for acquiring physiological data. Unlike lab settings, the driving context is more constrained in a number of ways. For example, vehicles move during drives, just like participants move their arms and legs whilst steering and accelerating. Furthermore, there is limited space in a vehicle for instrumentation. Lastly, driving contains frequent situational changes caused by traffic conditions and driver behaviour, which underlines the necessity for highly granular data.

Based on the above constraints, we identified the following device requirements:

1. Portable and wireless
2. Robust (not affected by movement)
3. Real time and patterns

Regarding the first requirement, portability is necessary because of the nature of the driving environment. Data collection devices would have to be located in the vehicle. Wireless devices are portable and also provide convenience in regards to set up and positioning in the vehicle. Wired devices would be difficult to hook up, both to the driver and to the collection device, and having wires run through the vehicle is additionally troublesome, as any movement of the leads and the sensors would contribute to noise in the data.

Addressing the second requirement, it is vital that the data collection systems are robust and not easily susceptible to movement artefacts, such as those associated with driving, as this would also increase noise in the data (Stern et al. 2001; Baguley and Andrews 2016).

In terms of the third requirement, it would be valuable to collect physiological data in real time or to detect physiological patterns across drives. This allows apps to determine arousal at a particular time during the drive or for entire drives, e.g., if an app is meant to reflect or factor in driver states.

2.5.2 Devices Compared

We selected three devices to compare to represent the variety of biofeedback systems available today across different cost and portability factors: Polar H7,[10] Empatica E4[11] and Biopac MP150.[12]

The Polar H7 is an affordable (approx. US$80) heart rate monitor mounted on a chest strap. It uses an electrocardiogram (ECG) heart rate sensor to deliver continuous and resting heart rate data.

The Empatica E4, which is more costly (approx. US$1700) but still available off-the-shelf, is a wireless wristband. It has four embedded sensors: photoplethysmograph (PPG), electrodermal activity (EDA), three-axis accelerometer and temperature.

Lastly, the Biopac MP150 is an advanced biofeedback acquisition system. It offers up to sixteen channels for data collection including ECG, HRV, electroencephalogram (EEG), electromyography (EMG) and electrogastrogram (EGG) and is an established system used in research laboratories. Unlike the Polar H7 and the Empatica E4, it is far less affordable (>US$10,000) and neither wearable nor lightweight.

2.5.3 Testing Procedure

In light of our interest in driver arousal, we focus on EDA and HR/ECG, which are established indicators of arousal (Stern et al. 2001). One of our team members (male, aged 27) wore the Empatica E4 on the right wrist and the Polar H7 was attached around the chest as directed. In terms of the Biopac MP150, two EDA electrodes were placed on the inner arch and sole of the participant's left foot. For ECG, the ground electrode was placed on the forehead, the first electrode approximately 4 cm below the right clavicle, and the second electrode on the left side, just below the last rib.

Since the Biopac is a stationary device, we conducted the comparison in a lab setting. The participant completed two 4 min sessions of Need for Speed,[13] a driving video game that simulates typical movements while driving. The set up consisted of a desktop steering wheel and a two pedal attachment on the floor connected to an Xbox 360.[14] Note that automatic transmission was used to reflect the configuration in our driving simulator. Consequently, the participant's left foot

---

[10]https://www.polar.com/au-en/products/accessories/H7_heart_rate_sensor.

[11]https://www.empatica.com/e4-wristband.

[12]https://www.biopac.com/product-category/research/systems/mp150-starter-systems/.

[13]https://www.needforspeed.com.

[14]https://www.xbox.com/Xbox360.

did not have to move for shifting gears, which we exploited in the placement of electrodes to avoid movement artefacts in the EDA data. We would expect more noise in the data in a moving vehicle.

2.5.4 Results

Unsurprisingly, the Polar H7 and Empatica E4 were substantially easier to use and less intrusive than the Biopac, whereas the Biopac provided the most accurate data for both EDA and ECG. A comparison of the data showed that the Polar H7 heart rate and heart rate variability data very closely resemble the Biopac data. The data collected from the Empatica E4 was not similar for both EDA and heart rate and sometimes was even in the opposite direction (see Fig. 2.8).

Table 2.1 summarises the key characteristics of the three biofeedback systems and presents the main insights from the comparison study.

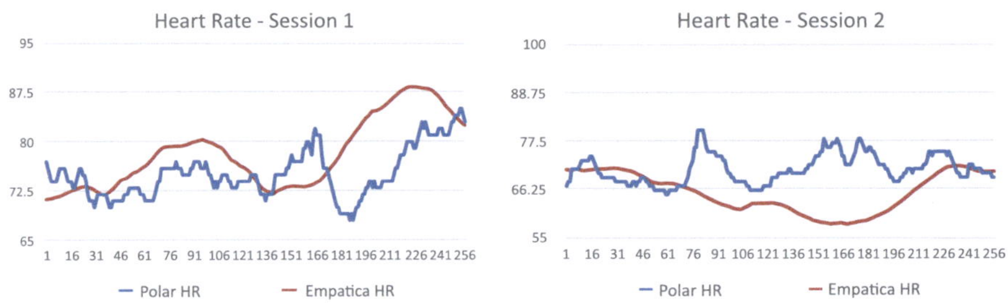

Fig. 2.8 In our experimental setup, heart rate data from the Empatica E4 (*red*) often does not resemble Polar H7 data (*blue*)

Table 2.1 Main characteristics of the three biofeedback systems compared

	Polar H7	Empatica E4	Biopac MP150
Cost	USD 80	USD 1700	>USD 10 000
Biofeedback	ECG	PPG, EDA	Up to 16 channels, e.g., EDA, ECG
Placement	Chest	Wrist	E.g., hands, feet
Portable	Yes (wearable)	Yes (wearable)	No
Wireless	Yes	Yes	Option available
Robust	Yes	No	Yes
Real-time	Yes (on smartphone)	Yes	Yes (in software)
Accuracy	Patterns overlap with Biopac	Patterns do not overlap with Biopac and Polar data	Yes (proven system for lab research)

2.5.5 Discussion

Acquiring biometrics in the driving context poses various technological and logistical challenges. We particularly faced difficulties in separating arousal levels associated with game engagement from those associated with bodily movements. In the driving context, people move their arms and legs to manoeuvre the car. This makes particularly capturing EDA challenging, where electrodes that are susceptible to noise caused by movement artefacts need to be attached to participants' hands or feet. Capturing HR is more feasible, and consumer wearables will do for certain use cases as trends and patterns are comparable with data acquired with the Biopac.

Our observations regarding varying HR data echo findings from a recent study by Nelson et al. (2016) who examined the accuracy of various FitBit[15] and Jawbone[16] devices for estimating energy expenditure (EE) and step counts. They found that consumer-based physical activity monitors should be used cautiously for estimating EE, although they provide accurate measures of steps for structured ambulatory activity.

We propose that consumer wearables such as smart watches or fitness trackers may be used as a cost-effective and non-invasive means to derive arousal scores or patterns across entire drives. As a design implication, gamified driving apps cannot yet rely on biometrics as a means to determine appropriate points in time to activate driving games or to dynamically adjust their difficulty. However, apps could present post-drive feedback for reflection, leaderboards, or to unlock levels. More accurate values would have to come from advanced measures, which may be more integrated into future vehicles.

2.6 Conclusion

In this chapter, we explored the opportunities for gamified safe-driving experiences provided by real-time *vehicle, road*, and *driver* data gathered from mobile and wearable devices.

We developed *BrakeMaster*, a smartphone app for drivers based on *vehicle* and *road* data, and evaluated it in a simulator study. We laid out how these data can be obtained through a combination of OBD, map data and sensors built into smartphones and similar devices. Participants self-reported that arousal and pleasure increased when using *BrakeMaster* while driving.

Based on our comparison of different biofeedback systems, we found that capturing *driver* data reliably and accurately is challenging. In particular, biosignals such as EDA are susceptible to noise caused by movements that naturally occur

[15]https://www.fitbit.com.

[16]https://www.jawbone.com/up.

while driving. However, there may be opportunities to use heart rate monitoring capabilities in fitness trackers and smart watches to indicate arousal patterns across drives. A major benefit is the small size and portability of such consumer wearables. More accurate data, e.g., to determine when to activate or dynamically adjust gamified driving challenges, would have to come from advanced sensors which may be integrated into future vehicles.

Taking these findings together, we conclude that *road* and especially *vehicle* data are most promising for developing novel driving experiences, whereas *driver* data is more challenging to gather in this unique design context. Smartphones in particular present an opportunity to develop driving apps for researchers and practitioners that aim to enhance safe-driving experiences. Future work should investigate how such experiences can be designed without causing distraction.

Acknowledgements This work is supported under the Australian Research Council's Discovery Early Career Researcher Award (ARC DECRA) funding scheme (project number DE140101542). We would like to acknowledge Sébastien Demmel, Madison Klarkowski, April Moeller, and Patrick Proppe who provided assistance with this study and Marcus Foth for providing valuable comments on this manuscript.

References

Aguilar, S. R., Merino, J. L. M., Sánchez, A. M., & Valdivieso, Á. S. (2015). Variation of the heartbeat and activity as an indicator of drowsiness at the wheel using a smartwatch. *International Journal of Artificial Intelligence and Interactive Multimedia 3*.

Andreone, L., Amditis, A., Deregibus, E., & Damiani, S. (2005). Beyond context-awareness: Driver-vehicle-environment adaptivity. From the COMUNICAR Project to the AIDE concept. In *16th IFAC World*.

Australian Bureau of Statistics. (2016). Motor Vehicle Census, Australia. Australian Government. http://www.abs.gov.au/AUSSTATS/abs@.nsf/mf/9309.0.

Average Age of Light Vehicles in the U.S. Rises Slightly in 2015 to 11.5 years, IHS Reports|IHS Online Newsroom. http://press.ihs.com/press-release/automotive/average-age-light-vehicles-us-rises-slightly-2015-115-years-ihs-reports.

AT&T: Smartphone Use While Driving Grows Beyond Texting. (2015). http://about.att.com/story/smartphone_use_while_driving_grows_beyond_texting.html.

Baguley, T., & Andrews, M. (2016). Handling missing data. In J. Robertson & M. Kaptein (Eds.), *Modern statistical methods for HCI* (pp. 57–82). Springer International Publishing.

Bangor, A., Kortum, P. T., & Miller, J. T. (2008). An empirical evaluation of the system usability scale. *International Journal of Human Computer Interaction, 24*, 574–594.

Bogomolov, A., Lepri, B., Ferron, M., Pianesi, F., & Pentland, A. (sandy). (2014). Daily stress recognition from mobile phone data, weather conditions and individual traits. In *Proceedings of the 22Nd ACM International Conference on Multimedia* (pp. 477–486). New York, NY, USA: ACM.

Brooke, J. (1996). Others: SUS-A quick and dirty usability scale. *Usability Evaluation in Industry, 189*, 4–7.

Casner, S. M., Hutchins, E. L., & Norman, D. (2016). The challenges of partially automated driving. *Communications of the ACM, 59*, 70–77.

Chohan, N. Young drivers ignore dangerous phone warning. http://www.heraldsun.com.au/news/victoria/young-drivers-ignore-dangerous-phone-warning/story-fni0fit3-1227339646687.

Coben, J. H., & Zhu, M. (2013). Keeping an eye on distracted driving. *JAMA, 309*, 877–878.
Dai, J., Teng, J., Bai, X., Shen, Z., & Xuan, D. (2010). Mobile phone based drunk driving detection. *Pervasive Health, 2010*, 1–8.
Eren, H., Makinist, S., Akin, E., & Yilmaz, A. (2012). Estimating driving behavior by a smartphone. In *Intelligent Vehicles Symposium (IV)*, 2012 IEEE (pp. 234–239).
Eriksson, J., Girod, L., Hull, B., Newton, R., Madden, S., & Balakrishnan, H. (2008). The pothole patrol: Using a mobile sensor network for road surface monitoring. In *Proceedings of the 6th International Conference on Mobile Systems, Applications, and Services* (pp. 29–39). New York, NY, USA: ACM.
Heslop, S. (2014). Driver boredom: Its individual difference predictors and behavioural effects. *Transportation Research Part F: Traffic Psychology and Behaviour, 22*, 159–169.
Hong, J.-H., Margines, B., & Dey, A. K. (2014). A smartphone-based sensing platform to model aggressive driving behaviors. In *Proceedings of the 32nd Annual ACM Conference on Human Factors in Computing Systems* (pp. 4047–4056). New York, NY, USA: ACM.
Koukoumidis, E., Martonosi, M., & Peh, L.-S. (2012). Leveraging smartphone cameras for collaborative road advisories. *IEEE Transactions on Mobile Computing, 11*, 707–723.
Lewis, J. R., & Sauro, J. (2009). The factor structure of the system usability scale. In M. Kurosu (Ed.), *Human centered design* (pp. 94–103). Berlin, Heidelberg: Springer.
Liang, Y., Reyes, M. L., & Lee, J. D. (2007). Real-time detection of driver cognitive distraction using support vector machines. *IEEE Transactions on Intelligent Transportation Systems, 8*, 340–350.
LiKamWa, R., Liu, Y., Lane, N. D., & Zhong, L. (2013). MoodScope: Building a mood sensor from smartphone usage patterns. In *Proceeding of the 11th Annual International Conference on Mobile Systems, Applications, and Services* (pp. 389–402). New York, NY, USA: ACM.
Litman, T. (2015). Autonomous vehicle implementation predictions: Implications for transport planning. In *Transportation Research Board 94th Annual Meeting*.
McCall, R., & Koenig, V. (2012). Gaming concepts and incentives to change driver behaviour. In *Ad Hoc Networking Workshop (Med-Hoc-Net), 2012 The 11th Annual Mediterranean* (pp. 146–151).
Mohan, P., Padmanabhan, V. N., & Ramjee, R. (2008). Nericell: Rich monitoring of road and traffic conditions using mobile smartphones. In *Proceedings of the 6th ACM Conference on Embedded Network Sensor Systems* (pp. 323–336). New York, NY, USA: ACM.
Meng, R., Mao, C., & Choudhury, R. R. (2014). Driving analytics: Will it be OBDs or smartphones? Zendrive Inc. Retrieved from http://blog.zendrive.com/103460130120/.
Nelson, M. B., Kaminsky, L. A., Dickin, D. C., & Montoye, A. H. K. (2016). Validity of consumer-based physical activity monitors for specific activity types. *Medicine and Science in Sports and Exercise, 48*, 1619–1628.
NHTSA. (2016). Distracted Driving 2014.
Normark, C. J. (2015). Vehicle interaction tailored to you. *Interactions, 22*, 32–36.
Prokhorov, D. V., Kalik, S. F., & Varri, C. K. R. (2011). System and method for reducing boredom while driving. http://www.google.com/patents/US7982620.
Russell, J. A. (1980). A circumplex model of affect. *Journal of Personality and Social Psychology, 39*, 1161–1178.
SAE International. (2014). J3016: Taxonomy and Definitions for Terms Related to On-Road Motor Vehicle Automated Driving Systems.
Schroeter, R., Soro, A., & Rakotonirainy, A. (2013). Social cars: sensing, gathering, sharing and conveying social cues to road users. In B. Guo, D. Riboni & P. Hu (Eds.) *Creating personal, social, and urban awareness through pervasive computing* (pp. 176–200). IGI Global.
Sivak, M. (1996). The information that drivers use: Is it indeed 90% visual? *Perception, 25*, 1081–1089.
Steinberger, F., Moeller, A., & Schroeter, R. (2016). The antecedents, experience, and coping strategies of driver boredom in young adult males. *Journal of Safety Research, 59*(December), 69–82. ISSN 0022-4375. http://dx.doi.org/10.1016/j.jsr.2016.10.007.
Steinberger, F., Schroeter, R., Lindner, V., Fitz-Walter, Z., Hall, J., & Johnson, D. (2015). Zombies on the road: A holistic design approach to balancing gamification and safe driving. In

Proceedings of the 7th International Conference on Automotive User Interfaces and Interactive Vehicular Applications (pp. 320–327). New York, NY, USA: ACM.

Steinberger, F., Schroeter, R., & Lindner, V. (2015). From gearstick to joystick—challenges in designing new interventions for the safety-critical driving context. *Presented at the OzCHI 2015 Workshop*, Melbourne, VIC December 7 (2015).

Stern, R. M., Ray, W. J., & Quigley, K. S. (2001). *Psychophysiological recording*. Oxford University Press.

Tijs, T. J. W., Brokken, D., & IJsselsteijn, W. A. (2008). Dynamic game balancing by recognizing affect. In: P. Markopoulos, B. de Ruyter, W. IJsselsteijn & D. Rowland (Eds.), *Fun and games* (pp. 88–93). Berlin, Heidelberg: Springer.

United States Bureau of Transportation Statistics. (2014). http://www.rita.dot.gov/bts/sites/rita.dot.gov.bts/files/publications/national_transportation_statistics/html/table_01_26.html_mfd.

Vollrath, M., Huemer, A. K., Teller, C., Likhacheva, A., & Fricke, J. (2016). Do German drivers use their smartphones safely?—Not really! *Accident Analysis and Prevention, 96*, 29–38.

Yerkes, R. M., & Dodson, J. D. (1908). The relation of strength of stimulus to rapidity of habit-formation. *The Journal of Comparative Neurology*.

You, C.-W., Montes-de-Oca, M., Bao, T. J., Lane, N. D., Lu, H., Cardone, G., Torresani, L., & Campbell, A. T. (2012). CarSafe: A driver safety app that detects dangerous driving behavior using dual-cameras on smartphones. In *Proceedings of the 2012 ACM Conference on Ubiquitous Computing* (pp. 671–672). New York, NY, USA: ACM.

Young, J., Wessnitzer, J. (2016). Descriptive statistics, graphs, and visualisation. In J. Robertson & M. Kaptein (eds.), *Modern statistical methods for HCI* (pp. 37–56). Springer International Publishing.

Zhang, Y., Lin, W. C., & Chin, Y. K. S. (2010). A pattern-recognition approach for driving skill characterization. *IEEE Transactions on Intelligent Transportation Systems, 11*, 905–916.

Part II
Usability and User Experience

Chapter 3
Driver and Driving Experience in Cars

Klaus Bengler

Abstract In connection with the automobile, user experience and emotion have always been contributing to a unique selling proposition and thus an important basis for the development of the product. In addition, requirements of traffic safety and usability have of course be taken into account. The corresponding trade-offs are not easy to solve; yet, existing premium products show that this is possible. While road safety provides a clear framework through basic requirements and regulations, the usability considered the interaction of a person with a technical system for a specific task in a given context, for example a navigation device in a motor vehicle. In this context, efficiency and satisfaction are optimized effectiveness. In addition, the emotional experience of users, as joy of use or user experience gain increasing importance. How these experiences can be translated into customer experiences in combination with current technology trends, for example in the area of perception of acceleration, electric mobility or automated driving is described in this chapter.

3.1 Introduction

In connection with the automobile, user experience and emotion have always been a unique selling point and thus an important basis for the development of the product beyond technical features. If we talk about the automobile we have to consider that, it was and is one of the most expensive and complex consumer goods that a heterogeneous user community experiences under again very heterogeneous circumstances.

Looking at the history of the automobile, we can see that from the beginning, it enabled its user to be mobile on an individual level at affordable cost. This individual mobility is not only rationally used to go to work or doing errands but also

K. Bengler (✉)
Chair of Ergonomics, Technical University of Munich (TUM), Garching, Germany
e-mail: bengler@lfe.mw.tum.de

G. Meixner and C. Müller (eds.), *Automotive User Interfaces*,
Human–Computer Interaction Series, DOI 10.1007/978-3-319-49448-7_3

for leisure activities and in extreme for driving per se doing sightseeing trips. Bubb et al. (2015) give an overview over different trip types. Already the selection of a given vehicle type influences seating position, possible driving styles and experience of the traffic environment like the sketches of Otl Aicher visualize (Fig. 3.1).

A look at historic BMW advertisement banners shows that already in the early 1900s, the sporty character of this brand was a central point of communication besides aspects of economy and affordable mobility (Fig. 3.2).

The first trip in the Motorwagen done by Bertha Benz in the year 1888 is a good example to understand the potential for experience by individual mobility at a novel speed and range, which was impossible before. This shows that a discussion of user experience of automobiles should not be led on a too simplistic level if we want to understand the different influencing factors (Fig. 3.3).

In general, many different definitions of user experience exist, that cover in many cases aspects of human computer interaction or interaction with mobile devices also UX is highly related to the construct of acceptance but not identical with it. Körber et al. (2013a, b) give a good overview over different definitory approaches and the related measurement problems.

Frequently the UX definitions highly overlap with the term usability following ISO 9241 (Hassenzahl 2008; Nielsen 1999; Tullis and Albert 2013). Several definitions stem from human computer interaction and describe the emotional change that is initiated by an interaction between a user and a product or the expectations that initiate these emotional state changes. Only some of these definitory aspects can be easily be transferred to the automotive domain.

First, it makes sense to differentiate between driver experience and driving experience. Where driver experience defines the general perception and emotional states of the driver and driving experience is especially focused on the experiences effected exactly by and during the negotiation of the driving task. In addition, it has to be noted that the automobile is a one of most complex and expensive consumer goods that enables its user to produce enormous energy in a variant public environment. Therefore, beyond user experience requirements of safety and usability have to be taken into account. Car concepts and interaction solutions have to meet user expectations simultaneously they have to stay with the framework defined by official guidelines, standards and laws. Such the corresponding trade-offs to user experience are not easy to solve.

Moreover, the success and tradition of existing premium brands and their products show that this it is possible arrange the very differing requirements. Such the automotive industry possesses a long-term knowledge in the definition, design, engineering, manufacturing and distribution of user experience via the automobile. (see also Akamatsu et al. 2013).

Moreover, the success and tradition of existing premium brands and their products show that this it is possible arrange the very differing requirements.

Compared the experiences with information systems and other artifacts that users are interacting with the automobile definitely provides an instant and ultimate multimodal feedback having the driver in dynamic environment.

Fig. 3.1 Seating positions in different concepts for individual mobility (Aicher 1984)

Fig. 3.2 Advertising for BMW automobiles in the early years of the 20th century

Fig. 3.3 Daimler Motorwagen and concept vehicle. Picture taken at Frankfurt Motorshow IAA 2009

Looking closer at the related control cycle of this human machine system shows that very simple input activities at pedals and steering wheel lead to very instant kinesthetic, auditory, and visual feedback. This feedback is unique in the combination of sensory modalities, purely physical and therefore highly synchronized (Fig. 3.4).

The evolutionary development of the automobile consists of a continuous improvement of existing functionality and the integration of additional functionality. This definitely led to an increase of safety and usability. A closer look shows that this process was also driven by the need to provide additional experiences corresponding to the user needs of the current period.

Examples are the introduction of communication facilities like the car phone and information systems like navigation. It is remarkable that most of these integrations are discussed under aspects of safety, efficiency and comfort. Allowing the driver to stay connected while being mobile or to travel time using exact route guidance. On the other hand, especially the integration of more and more information systems led to the discussion whether this could lead to safety critical driver distraction.

A current challenge gets obvious if we consider user expectations that are generated by the experience with mobile devices that offer a magnitude of functionalities and access to huge contents. On the other hand these concepts cannot be transferred 1:1 to in-vehicle interaction as they would be in conflict with existing

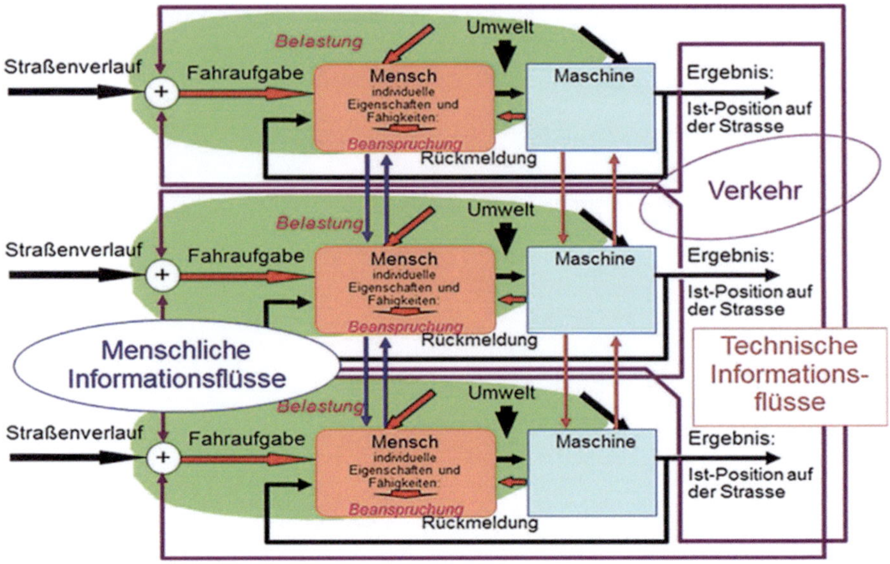

Fig. 3.4 Flow of information in the human-vehicle-traffic (Bubb et al. 2015)

Fig. 3.5 Several requirements relevant for human factors in automobiles

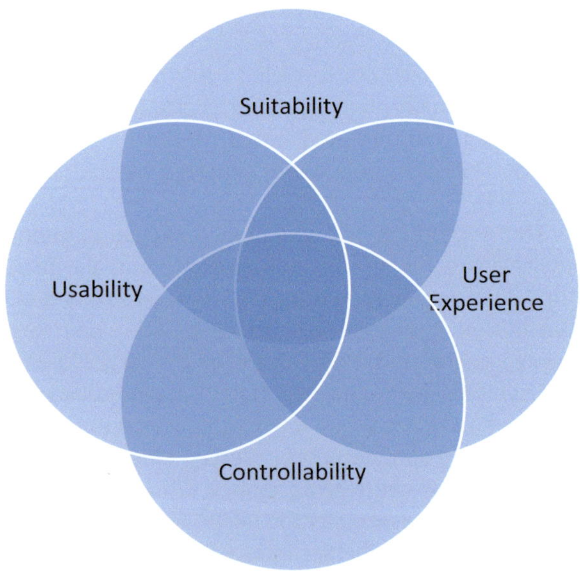

guidelines and recommendations (European Commission 2008) that focus on driver distraction under the aspect of suitability (Fig. 3.5).

If advanced driver assistance systems are considered a further requirement comes into play. The driver has to be able to gain the control over the car and driving situation also in situations where the technical system fails or breaks down.

In addition to these interaction-related requirements, it has to be considered that a positive driver experience requires that the basic product requirements of quality, functional reliability, and mobility have to be fulfilled.

Following Bhise (2012) also the following product features contribute to acceptance and user perception:

Visual quality
Tactile quality
Acoustic quality
Harmony
Olfactoric quality
see also Bubb et al. (2015)

Additionally design, driving performance, and acceleration characteristic play a dominant role and a majority of drivers likes to experience the acceleration potential of their car. (Atzwanger and Negele 2008). Especially, longitudinal dynamic (acceleration) is an important factor for drivers and studies show that a majority of drivers like to experiment with it (Tischler 2013; Müller et al. 2012). Especially, this interaction possibility via the accelerator pedal makes the automobile a specific field of user experience. By a simple pedal activity, a very intense synesthetic experience can be produced by the driver him/herself.

3.2 Trade-Offs and Examples for Solutions

While road safety provides a clear framework in form of basic requirements and regulations, the usability considering the interaction of a person with a technical system for a specific task in a given context is an additional requirement. For example, the design of a navigation device in a motor vehicle needs to fulfil safety requirements considering the suitability of this device for use while driving.

The device also has to be usable to support with dedicated information the negotiation of the navigation task and the solution of potential traffic problems by the driver.

Compared to desktop systems implementations must be excluded that could for example lead to increased driver distraction by animation effects, an overflow of information or graphical effects that harm readability (Heinrich 2012).

For example, navigation information can be provided in different ways and it can be shown that design elements that contribute to user experience like animations do not necessarily suffer from suitability and usability requirements. Bengler and Broy (2008) show that animations in the GUI can increase the likeability of a system, be very helpful for the learnability of a menu system on one hand and non-distracting if the stay with given in-vehicle design rules for duration (below 300 ms) and dynamic of the graphical effect.

Several technological developments are influencing driver experience and driving experience. Three selected examples will be discussed in the following.

3.2.1 UX Und Advanced Driver Assistance Systems (ADAS)

The development and introduction of ADAS follows the logic that the human driver could benefit from dedicated support for subtasks of the driving tasks (Bengler et al. 2014). This approach leads to a new role model between driver and car. Due to technical feasibility this can lead to the situation that parts of the driving tasks are done by the ADAS being monitored by the driver and that the remaining parts have actively be conducted. Automatic cruise control manages longitudinal control and stays in a safe distance to preceding cars, the driver stays with lateral control. Parking assistance systems automate the lateral control (i.e., steering) the driver accelerates and decelerates the car.

Mainly drivers select ADAS mainly for comfort reasons but less following safety considerations. Planing (2014) reports that if users want to experience control they show a reduced motivation to use ADAS. However, the new role-play leads to a driving experience that can vary from an intense experience of comfort and support to an experience that is dominated by the permanent monitoring task. Because most ADAS are disabled if the driver takes over control, the interaction between driver and ADAS is limited to monitoring and parametrization. In case of ACC systems, the driver is able to change the set speed and distance to the preceding car. If the driver brakes, the system is switched off. In overtaking scenarios, the ACC system would try to always keep a safe distance to the preceding car and the driver has actively to accelerate by a dedicated pedal activity. For experience this means that the driver has to interrupt the ACC monitoring by active driving activities and turn back to monitoring again.

Totzke et al. (2008) show that by a minor change in the system logic it is possible to intensify the interaction with the ACC. In their setup, the input element for speed adjustments allows the driver to accelerate or decelerate temporarily but keep the set speed after this intervention. Some drivers describe the interaction as a kind of hand gas/brake. This adaption of the interaction concept establishes a driver experience active driving together with ACC that goes beyond monitoring the ACC versus driving without ACC. Interestingly drivers characterized this interaction concept as less demanding and more motivating although they had shown more motoric interactions but did not have to check the system state (enabled/disabled).

Popiv et al. (2010) and Rommerskirchen et al. (2014) give further examples that also use the information of driver assistance systems but tries to increase driver experience. Their concept focus on the fact that meanwhile several systems try to support the driver in longitudinal control: ACC, traffic sign recognition, navigation, congestion warning. The authors describe a system, which motivates an anticipative driving style by an integrated visualization of the upcoming events that are relevant for longitudinal control (Fig. 3.6).

In addition, the Kolibri app described by Krause et al. (2014) follows this approach with an app running on a mobile device (i.e. smartphone).

The presentation of additional information with a horizon of about 10 s motivates a more efficient driving behavior that is characterized by less braking and

Fig. 3.6 Information for information integration in the cluster instrument to support driver anticipation (Popiv et al. 2010)

more sailing. The interesting effect is that drivers report very positive emotions and experience of competence and control compared to the usage of an automated system. It is evident that by the presentation of information instead of introducing an automated control system driving experience can be influenced sustainably. The studies show that although drivers received less automation support by control mechanisms they describe their driving behavior with the information system as more active and with positive emotions of control experience and system acceptance instead of being patronized by the system. Additionally by this self-initiated change of driving behavior efficiency gains in form of fuel savings go up to 20% (Fig. 3.7).

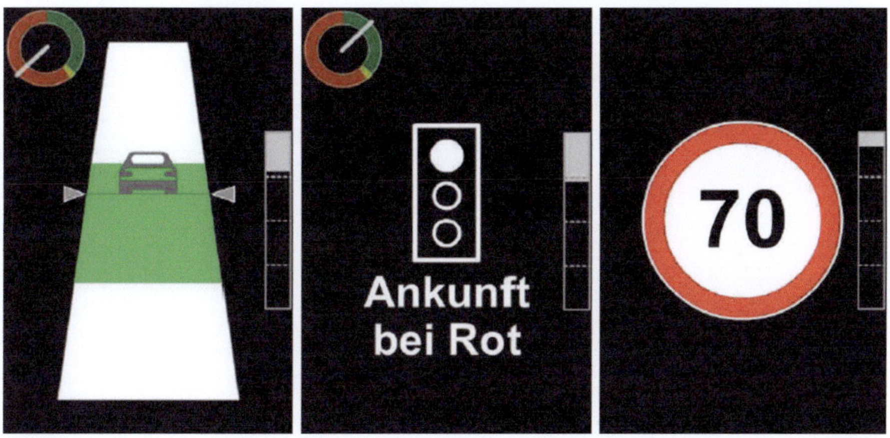

Fig. 3.7 Information concept of the KOLIBRI app to be displayed on an in-vehicle smartphone (Krause et al. 2014)

In subjective ratings in an ergonomic engineering process using the example of an in-vehicle information system.

Krause et al. (2014) compare different questionnaires (SUS, AttrakDiff, to investigate user acceptance and user experience related to above-mentioned information system.

This clearly shows that there is a remarkable potential to influence driving style in a positive way by information and connect it to positive user experiences instead of taking the driver out of the loop by automating the function.

3.2.2 UX und Automation

With increasing levels of automation the question arises, which driver experience can emerge from automated driving that goes clearly beyond assisted driving.

This question is highly related to the interaction concept that is implemented for an automated car. Different interaction paradigms have been realized. H-mode (Flemisch et al. 2014) and Conduct by Wire (Winner et al. 2006) represent two very different concepts how drivers could interact with an automobile that is capable of automated driving. Whereby H-Mode focuses on a permanent shared control interaction between driver and vehicle via driving elements like steering wheel or drivestick, Conduct by Wire allows the driver to instruct driving maneuvers to the car using a menu like concept. (see Kauer et al. 2010 for more details) (Figs. 3.8 and 3.9).

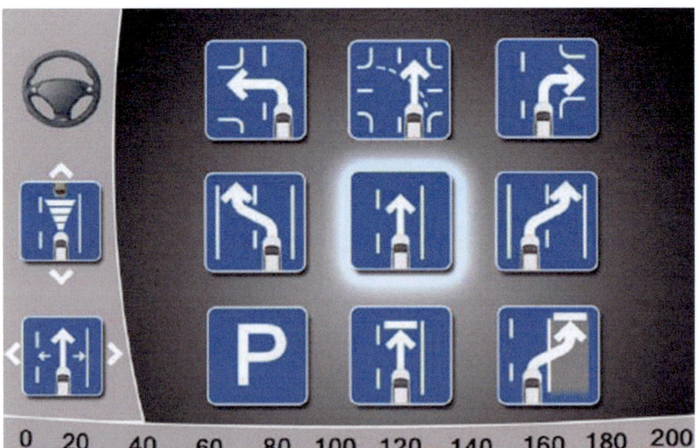

Fig. 3.8 Prototype of the CbW maneuver-interface (Kauer et al. 2010)

Fig. 3.9 H-Mode with trajectories displayed in the contact analogue head up. Display and active steering wheel/pedal. Application of H-Mode in the dynamic simulator with active sidestick. Flemisch et al. (2014)

Both concepts show that the integration of automation will lead to remarkable changes in driver vehicle interaction and especially driving experience.

Investigations by Albert et al. (2015) show that in addition driver and driving experience could move to the background and the interaction with non-driving related activities might get the main experience.

For automation levels 3 and 4 following the SAE taxonomy the driver will have to monitor the level 3 automated cars driving performance to intervene in critical situations. In case of level 4 to take over after dedicated take over requests given by the car without continuous monitoring.

Compared to active driving completely different experiences come to relevance that no longer benefit from direct interaction and continuous feedback but more from trust in automation (Gold et al. 2015) and mode awareness. Otherwise, the driver resources that are set free by an automation of the driving task would not be experienced with positive emotions.

For the design of the automated driving style, the experiments of Lange et al. (2015) show that the vehicle dynamics that play a central role for driver experience have to be treated very carefully in the automation case. First, an exact coordination of longitudinal and lateral acceleration is necessary to inform the driver about the maneuver that the automated car is going to execute. Therefore, the driver can differentiate a beginning overtaking scenario from an erroneous acceleration or a lane change with overtaking intention. This communication aspect between automobile and "driver" must not be underestimated in its value for the experience of trust into an intelligent machine. Furthermore, the experiments show that there is a remarkable difference in perception of driving dynamics dependent on whether the maneuver is conducted by the driver him/herself or by the automobile. The pure replication of driver patterns by an automated driving machine does not lead to the same experience like the "natural" human pattern.

3.3 UX and E-Mobility

Another technological development that heavily influences driver and driving experience of future automobiles is the introduction of electric mobility. The automobile offers in comparison to other products especially human computer interaction the potential to let the user experience dynamic via acceleration and deceleration in combination with different sound characteristics dependent on the drivetrain (i.e. combustion vs. electric engine).

First, vibration characteristics and sound level of an electric car are much lower. Electric vehicles are extremely silent in combination with a remarkable longitudinal dynamic. A second fact in combination with electric mobility is that many authors report the effect of "range anxiety" (Franke et al. 2011). Furthermore, users of electric vehicles positively experience the functionality of recuperation that allows them to gain control of their range if they replace brake pedal actions by releasing the accelerator pedal and thus fighting range anxiety. (see also Cocron et al. 2015)

Such due to the functional mechanisms of an electric drivetrain relevant perceptions and experiences differ significantly from a combustion engine vehicle.

On the other hand, these characteristics are combined with very specific longitudinal dynamics. Buyers and users of electric vehicles report that they prefer these cars, as they are innovative, green and economic. On the other hand, the experience of above-mentioned qualities might be the more convincing aspects.

Now the question is whether the absence of sound is perceived as a deficit and decreases the driver experience or it can lead to a new and specific experience. The driver and lead in combination with the low sound level to an experience of luxury can clearly perceive the remarkable dynamic qualities. The perception of dynamic is based upon the remarkable psychophysical sensitivity of humans to discriminate even small dynamic variations (Benson et al. 1986; Kingma 2005; Müller et al. 2013).

These effects show that not only amplification of sound can lead to more intense experiences. Especially the direct and instant feedback that the driver can achieve via the accelerator pedal contributes to an instant experience of control that can be combined with the luxury of a silent interior. Müller et al. (2012) gives very specific indications, which acceleration profiles are to be preferred. The studies of Helmbrecht show that drivers can easily and very quickly adapt to the different acceleration characteristic of an electric vehicle and the recuperation functionality. This adaptation and the related learning process is perceived with mainly positive emotions and shows no negative carry over effect if the drivers use intermittently a vehicle with combustion engine (Helmbrecht et al. 2014a, b).

3.4 Conclusion

The consideration of user experience in automobiles shows that there is an enormous potential and also need to take the challenge of an active design and engineering in this area. Important aspects of driver and driving behavior can positively be influenced via a successful experience design.

A closer view to different studies shows that there are not enough Methods to support especially automotive user experience engineering and evaluation. Still many approaches are transferred from the HCI domain or mobile device interaction importiert, but cannot cover all aspects of this complex product. The discussion mainly focusses on the integration von infotainment functions, while relations to existing knowledge around driving dynamic aspects is frequently addressed via acceptance and acceptability theory and methods.

The discussion in this chapter focusses on the driving task and here mainly on longitudinal control.

Important aspects like designing of shapes, selection, and combination of materials and interaction with infotainment systems have not been discussed but are of course in an important interplay with the driving task and the resulting experiences.

References

Aicher, O. (1984). Kritik am Auto: Schwierige Verteidigung des Autos gegen seine Anbeter. Eine Analyse von Otl Aicher. München: Callwey Verlag.

Akamatsu, M., Green, P., & Bengler, K. (2013). Advances of human factors research for future vehicular technology (editorial). In *International Journal of Vehicular Technology. Advances of Human Factors Research for Future Vehicular Technology* (Vol. 2013). Hindawi Publishing Corporation.

Albert, M., Lange, A., Schmidt, A., Wimmer, M., & Bengler, K. (2015). Automated driving—assessment of interaction concepts under real driving conditions. In T. Ahram, W. Karwowski, & D. Schmorrow (Eds.), *Proceedings of the 6th International Conference on Applied Human Factors and Ergonomics 2015 and the Affiliated Conferences (AHFE)* (Vol. 3, pp. 2832–2839). AHFE. Elsevier.

Atzwanger, K., & Negele, H.-J. (2008). Fahrdynamikstudie 2008. Kundenorientierung in der Fahrwerkstechnik. Automobiltechnische Zeitschrift (ATZ) (Vol. 108, Issue 6, pp. 510–514).

Bengler, K., & Broy, V. (2008). Animationen im Fahrzeug-GUI—Randbedingungen für deren ergonomische Gestaltung. Vortrag auf dem 54. Frühjahrskongress der Gesellschaft für Arbeitswissenschaft e.V. Produkt- und Produktions-Ergonomie—Aufgabe für Entwickler und Planer, 9.-11. April München.

Bengler, K., Dietmayer, K., Färber, B., Maurer, M., Stiller, C., & Winner, H. (2013). Three decades of driver assistance systems. *Intelligent Transportation Systems Magazin*, IEEE, (Vol. 6/4), 6–22. doi:10.1109/MITS.2014.2336271.

Benson, A. J., Spencer, M. B., & Stott, J. R. R. (1986). Thresholds for the Detection of the Direction Whole- Body, Linear Movement in the Horizontal Plane: Aviation, Space and Environmental Medicine, S. 1088-96.

Bhise, V. D. (2012). *Ergonomics in the automotive design process*. Boca Raton, London, New York: CRC Press Taylor & Francis Group.

Bubb, H., Bengler, K., Grünen, R. E., & Vollrath, M. (2015). *Automobilergonomie*. Springer Fachmedien Wiesbaden.

Cocron, P., Bühler, F., Neumann, I., Franke, T., Krems, J. F., Schwalm, M., et al. (2011). Methods of evaluating electric vehicles from a user's perspective—the MINI E field trial in Berlin. *IET Intelligent Transport Systems, 5*, 127–133.

European Commission Recommendation of 26 May 2008 on safe and efficient in-vehicle information and communication systems: Update of the European Statement of Principles on human-machine interface (2008/653/EC).

Flemisch, F. O., Bengler, K., Bubb, H., Winner, H., & Bruder, R. (2014). Towards a cooperative guidance and control of highly automated vehicles: H-Mode and conduct-by-wire. *Ergonomics, 57*(3), 343–360.

Franke, T., Neumann, I., Bühler, F., Cocron, P., & Krems, J. F. (2011). Experiencing range in an electric vehicle: Understanding psychological barriers. *Applied Psychology*. doi:10.1111/j.1464-0597.2011.00474.x.

Gold, C., Körber, M., Hohenberger, C., Lechner, D., & Bengler, K. (2015). Trust in automation—before and after the experience of take-over scenarios in a highly automated vehicle. In T. Ahram, W. Karwowski, & D. Schmorrow (Eds.), *Proceedings of the 6th International Conference on Applied Human Factors and Ergonomics 2015 and the Affiliated Conferences (AHFE)*. AHFE (Vol. 3, pp. 3025–3032). Elsevier. Retrieved from http://www.sciencedirect.com/science/article/pii/S2351978915008483.

Hassenzahl, M. (2008). User experience (UX): Towards an experiential perspective on product quality. In *Proceedings of the IHM'13* (pp. 11–15). New York: ACM.

Heinrich, C. (2012). Automotive HMI—international standards. In *Proceedings of the 4th International Conference on Applied Human Factors and Ergonomics*, San Francisco, July 21–25.

Helmbrecht, M., Olaverri-Monreal, C., Bengler, K., Vilimek, R., & Keinath, A. (2014a). How electric vehicles AFFECT driving behavioral patterns. *IEEE Intelligent Transportation Systems Magazine. Special Issue on Electro-Mobility*, 22–32. doi:10.1109/MITS.2014.2315758.
Helmbrecht, M., Rolnik, R., Bengler, K., Vilimek, R., & Keinath, A. (2014b). Identifying types of driver in electric and conventional vehicles. In *Proceedings of the FISITA 2014 World Automotive Congress*.
ISO DIS 9241–210:2008. Ergonomics of human system interaction-Part 210: Human-centered design for interactive systems (formerly known as 13407). International Organization for Standardization (ISO). Switzerland.
Kauer, M., Schreiber, M., & Bruder, Ralph. (2010). How to conduct a car? A design example for maneuver based driver-vehicle interaction. In *Paper presented at the IEEE Intelligent Vehicle Symposium*, San Diego, USA.
Kingma, H. (2005). Thresholds for perception of direction of linear acceleration as a possible evaluation of the otolith function. *BMC Ear, Nose, and Throat Disorders, 5*(1), 5.
Körber, M., & Bengler, K. (2013a). Measurement of momentary user experience in an automotive context. In *Proceedings of the 5th International Conference on Automotive UI 2013* (pp. 194–201).
Körber, M., Eichinger, A., Bengler, K., & Olaverri Monreal, C. (2013b). User experience evaluation in an automotive context: Workshop. In *Proceedings of the 2013 IEEE Intelligent Vehicles Symposium (IV)* (pp. 13–18). 23–26 June 2013. Gold Coast, Australia.
Krause, M., Rissel, A., & Bengler, K. (2014). Traffic light assistant -what the users want. In *The Seventh International Conference on Advances in Computer-Human Interactions ACHI 2014* (pp. 235–241), 23rd–27th March Barcelona.
Lange, A., Albert, M., Siedersberger, K.-H., & Bengler, K. (2015). Ergonomic design of the vehicle motion in an automated driving car. In T. Ahram, W. Karwowski, & D. Schmorrow (Eds.), *Proceedings of the 6th International Conference on Applied Human Factors and Ergonomics 2015 and the Affiliated Conferences (AHFE)*. AHFE (pp. 2761–2768). Elsevier. Retrieved from http://www.sciencedirect.com/science/article/pii/S2351978915007118.
Müller, T., Gold, C., Eichinger, A., & Bengler, K. (2012). Identifying customeroriented key aspects of perception with focus on longitudinal vehicle dynamics. In *Paper Presented at 4th International Conference on Human Factors and Ergonomics*, San Francisco, USA.
Müller, T., Hajek, H., Radic-Weissenfeld, L., & Bengler, K. (2013). Can you feel the difference? the just noticeable difference of longitudinal acceleration. In *Proceedings of the Human Factors and Ergonomics Society Annual Meeting* (Vol. 57 (1), pp. 1219–1223).
Nielsen, J. (1999). Designing web usability. In *The practice of simplicity*. Indianapolis: New Riders.
Planing, P. (2014). Innovation acceptance. In *The case of advanced driver-assistance systems*. Wiesbaden: Springer Gabler.
Popiv, D., Rakic, M., Laquai, F., Duschl, M., & Bengler, K. (2010). Reduction of fuel consumption by early anticipation and assistance of deceleration phases. In *World Automotive Congress of International Federation of Automotive Engineering Societies (FISITA)*, Budapest, Hungary, 30 May–4 June 2010.
Rommerskirchen, C., Helmbrecht, M., & Bengler, K. (2014). The impact of an anticipatory eco-driver assistant system in different complex driving situations on the driver behavior. *Intelligent Transportation Systems Magazin*, IEEE, (6(2)), 45–56. doi:10.1109/MITS.2014. 2307078.
SAE, J3016 (2014). Taxonomy and Definitions for Terms Related to On-Road Motor Vehicle Automated Driving Systems.
Tischler, M. A. (2013). Entwicklungsziel Fahrspaß. Gestaltungsempfehlungen für die Auslegung von Kraftfahrzeugen zur Optimierung des subjektiven Erlebens der Fahrzeugführung. Dissertation, Humbold-Universität. Berlin.

Totzke, I., Huth, V., Krüger, H.-P., & Bengler, K. (2008). Overriding the ACC by keys at the steering wheel: Positive effects on driving and drivers' acceptance in spite of a more complex ergonomic solution. In D. de Waard, F. O. Flemisch, B. Lorenz, H. Oberheid, & K. A. Brookhuis (Eds.), *Human factors for assistance and automation* (pp. 153–164). Maastricht: Shaker.

Tullis, T., & Albert, B. (2013). Measuring the user experience. In *Collecting, analyzing, and presenting usability metrics*. Amsterdam: Elsevier.

Winner, H., Hakuli, S., Bruder, R. Konigorski, U., & Schiele, B. (2006). "Conduct-by-Wire - ein neues Paradigma f¨ur die Weiterentwicklung der Fahrerassistenz," in 4. Workshop Fahrerassistenzsysteme 2006.

Chapter 4
"It's More Fun to Commute"—An Example of Using Automotive Interaction Design to Promote Well-Being in Cars

Marc Hassenzahl, Matthias Laschke, Kai Eckoldt, Eva Lenz and Josef Schumann

Abstract Automotive interaction design (AID) becomes more and more important. From advanced driver assistance systems to social media—the number of interactive systems available in cars increased substantially over the recent years. However, AID is still mostly concerned with making interaction either easy or exciting. In this chapter, we argue that AID should focus more on creating and shaping enjoyable and meaningful activities through sensible arrangements of interactive technologies. To advance this argument, we provide an overview of a well-being-oriented experiential approach to AID and discuss the *Perfect Commute* as an example of a well-being-oriented experiential interactive system.

4.1 Happiness in a Car!? The Role of Automotive Interaction Design

Life is too short to be anything but happy.

We are sure you'll agree: There is nothing worse than a *Facebook* news feed cluttered with inspirational quotes. However, the "inspiration" above contains a

M. Hassenzahl (✉) · M. Laschke · K. Eckoldt
Ubiquitous Design/Experience and Interaction, University of Siegen,
Kohlbettstraße 15, 57072 Siegen, Germany
e-mail: marc.hassenzahl@uni-siegen.de

M. Laschke
e-mail: matthias.laschke@uni-siegen.de

K. Eckoldt
e-mail: kai.eckoldt@uni-siegen.de

E. Lenz
Experience and Interaction, Folkwang University of the Arts,
Universitätsstraße 12, 45141 Essen, Germany

J. Schumann
BMW Group, 80788 München, Germany
e-mail: josef.schumann@bmw.de

G. Meixner and C. Müller (eds.), *Automotive User Interfaces*,
Human–Computer Interaction Series, DOI 10.1007/978-3-319-49448-7_4

spark of truth. It points at the intimate relationship between feelings of subjective well-being (i.e., happiness) and the way we spend our time. Time is the prime resource necessary to engage in enjoyable and meaningful activities (Kasser and Sheldon 2009). Since nowadays people spend a substantial part of their time in cars, the car as a potential site for well-being is important.

Enjoyment and meaning can be understood as a consequence of fulfilling psychological needs, such as autonomy, competence, relatedness, stimulation or popularity, through an activity (e.g., Ryan and Deci 2000; Sheldon et al. 2001). For example, a dinner with friends is certainly not primarily about "getting some food in." People engage in dinners to feel close and related to others. The wine and food just provides a pleasant chance for talking, bantering or serious debating. The dinner as an activity is enjoyable and meaningful because it fulfills—among others—a need for relatedness. Of course, we do not have dinners with friends every day. In fact, everyday life obligates us to many, not entirely voluntary activities. Take parking the car as an example. For some parking is simply a nuisance; others enjoy their ability to scout lesser known parking spots and to maneuver the car swiftly and elegantly into a lot just fitting. But even mundane everyday activities, such as parking, can create moments of joy and meaning. Studies support this. Howell and colleagues (Howell et al. 2011), for example, revealed substantial positive correlations between satisfaction as well as positive affect at the end of the day and the intensity of need fulfillment throughout the day. Moreover, daily need fulfillment was related to general feelings of life satisfaction and happiness. In other words, a day full of moments of fulfilled psychological needs is not only a happy day, but also contributes to more general feelings of life satisfaction (see also Reis et al. 2000). Some researchers estimate that about 40% of the variability in happiness among individuals may be attributed to differences in the activities they engage in throughout the day (Lyubomirsky et al. 2005).

As already hinted at above, we do not always have the freedom of choice with regard to our activities. Kahneman and colleagues (2004) used the *Day Reconstruction Method* to get an idea about how people spend their days and how they feel while doing so. For more than 900 US-American participants the three most positive activities were "intimate relationships", "socializing", and "relaxing;" the three least positive were "housework", "working", and "commuting." However, while people spent an average of 2.7 h a day on the activities they found especially positive, they spent 9.6 h a day on the three least positive activities, 1.6 h alone on commuting. One strategy to counteract this is to shift time spent on the least positive activities to the more positive. Efforts to reduce working hours, for example, to regain work-life balance and to become more "time affluent" are an example of this strategy. Based on Kahneman and colleagues' work, Kroll and Pokutta (2013) developed a number of happiness-optimized day schedules. For instance, they recommend to spend no more than 36 min per day on work. While an instantly appealing recommendation, it seems impossible to attain. An alternative strategy, though, is to pay more attention to whether activities, such as work or the daily commute, can be made more enjoyable and meaningful.

This is where design comes in. Any activity—or better practice—almost always relies on things. Things "mold" practices or as Reckwitz (2002) puts it: "[…] they enable and limit certain bodily and mental activities, certain knowledge and understanding" (p. 253). Designers deliberately or accidentally "inscribe" (to borrow Latour's 1992 term) certain ways of using things, which in turn creates new activities or re-structures well-known activities in particular ways. Things mold our daily experiences with the potential consequence of more or less happiness. Take parking as an example again. Obviously, a driver of a car equipped with hard-wearing bumpers may take a more risky stance to parking. She may evolve her skills of maneuvering into ever smaller lots, experiencing competence and pride. Parking assist systems, such as the ubiquitous proximity warning ("beep-beep-beep"), may have a similar effect. Nevertheless, to experience competence, the driver needs to believe that it was her and not the car, who managed the parking. A longitudinal study of the experience of a park assist system (Trösterer et al. 2014) provided some interesting insights of how park assist can corrupt feelings of competence. One participant, for example, was annoyed that the park assist was always on, even for the "easiest dork parking spot" (p. 5). To this driver, it is almost an insult that the system offers assistance no matter how challenging the parking lot. Another participant reported: "My parents laughed a little bit because I stopped much too early in the parking slot and had plenty of space backwards. […] It was a little embarrassing for me" (p. 5). In this anecdote, the result of the parking was deemed unprofessional *because* of the park assist, which in turn was an embarrassment for the driver. A car equipped with an autonomous parking system is yet another story. On one hand, autonomous parking makes own parking skills obsolete. Consequently, a driver can either be pleased with getting rid of an annoying inconvenience or be a little sorry about a lost opportunity to feel competent. On the other hand, the driver may let the car park while creating the impression of doing it all by himself. This would boost feelings of popularity, but only if the way the car parks is impressive enough. Another variant of this, would be—legal problems aside—to step out of her car and to watch the car doing the maneuvering. Hopefully a pedestrian stops to watch the show and to admire the driver for the cool gear he possesses. Quite obviously, even if this is experienced as enjoyable and meaningful, it is a different type of enjoyment compared to parking all by oneself. While the first satisfies popularity, the latter satisfies competence. To sum, through its very use the car creates the requirement to be parked. Parking as an activity, however mundane, can be a source of daily pleasure and meaning. However, any technology employed, any change in design, will inevitably change the activity, its meaning, and its potential to make people happy.

Studies show how subtle the molding of actives through technology and the link to well-being can be. In the domain of kitchen appliances, Hassenzahl and Klapperich (2014) compared two ways of preparing a cup of coffee: fully automated with a *Senseo* Pad Coffee Maker and more manually with grinding the coffee beans and preparing the coffee with an Italian stove-top coffee maker. The difference was striking. While the *Senseo* provided a fast and clean service, it focused their users entirely on the outcome—the cup of coffee. Positive feelings derived from the process itself were removed and replaced by impatience. This was different for the more manual preparation. Here participants enjoyed the process, because it induced stimulation and feelings of competence. While preparing a coffee with a *Senseo* was not experienced as negative, it became experientially "flat" and lost its potential to create enjoyment or meaning. Of course, we are aware that there are many potential reasons why a cup of coffee out of a pad coffee maker may seem sensible. The point is that the technology "inscribes" a certain meaning into the activity. The pad maker assumes preparing coffee to be an inconvenience. The more manual variant focuses on potential experiential gains.

In the automotive domain, studies of the experience induced by different variants of technology are rare. An exception is Eckoldt and colleagues (Eckoldt et al. 2012) study of the experiential consequences of using *Adaptive Cruise Control* (ACC). He interviewed five drivers (and owners) of a BMW series 5 car equipped with ACC. Note that all interviewees used ACC regularly and indicated a positive attitude toward ACC. The interview was mainly about how the experience and the meaning of the car changes, when ACC becomes active. All drivers agreed to especially enjoy driving as a control and competence experience. To them the car is a powerful beast, and they claim to become one with it while driving. In contrast, ACC creates a substantial distance between driver and car. It "tames" the beast and turns the driver into a spectator. While this is not necessarily experienced as negative, there is the feeling of losing the sportive, active, competence-inducing experience. Similar to the coffee makers, ACC as a technology subtly changes the activity of driving, making it more convenient, but at the same time experientially "flat".

Automotive Interaction Design (AID) should care about all this. The car is especially interesting for at least two reasons: First,—as already mentioned—people pass a substantial amount of time in cars. Time should be spend in a meaningful way. Second, the car as an interactive technology is quite exceptional. It is one of the few, which envelops its users completely. The car creates a secluded, highly structured space. Other than in more open situations, interactions with and in a car, and thus activities, are strongly determined by the car. Fortunately, Interaction Design and Human-Computer Interaction are disciplines, which possess a profound knowledge about how to structure activities through the design of technical artifacts. However, AID still seems to focus mainly on the operational level of design (i.e., the concrete motor actions, such as pushing, sliding, and turning) instead of focusing on the broader question of how to create meaningful moments in and through a car. AID is in many cases technology-driven. New types of displays (e.g., Head-Up, Glasses) and new, presumably "futuristic" interaction technologies (e.g., gestures, touch, gaze) seem in the fore. The result is at best steering wheels with touch controls or dashboards to point at. This is complemented by a primarily problem-driven approach, where either driving itself or operating the various functionalities of the car are understood as a problem to get rid of rather than an opportunity for a fulfilling activity. This stance may be the reason for automotive industry's rather inexplicable obsession with working on the disappearance of driving—in the form of autonomous cars. Obviously, driving is regarded as a "problem" (e.g., boring, demanding, dangerous), which can be solved by getting rid of it through automation. This is certainly at odds with what people actually say to enjoy about cars, which is primarily the driving. People who experience driving as a problem, tend to use a different mode of transport (e.g., a taxi, the railways) if possible. But even if driving is to vanish, one should expect as much investment of time and money into answering the question of how people will actually spend their time in the car, when *not* driving anymore. But far from it. Answering emails, reading newspaper, watching *YouTube*—"What people like to do, when they relax!" is the rather lame answer. On top of this, it is likely that truly autonomous driving will remain a fantasy for a long time. The future might be more about watching your car driving—which might be even more boring, demanding, and dangerous than driving (Casner et al. 2016).

All in all, we believe that car manufacturers must care about how people spend their precious time in cars and make sure that technologies are arranged in a way that enjoyable and meaningful moments become more likely. The discipline that has the most potential here is AID, however, only if the understanding of interaction design is broadened to the level of experiences. In the remainder of the chapter, we first discuss a well-being-oriented and experiential approach to AID. We then present an example of the type of design and product, one can expect from this approach. Finally, we discuss the challenges of a well-being-oriented AID.

4.2 An Experiential Approach to Automotive Interaction Design

Over the last 15 years, Human-Computer Interaction embraced the idea that pleasure, fun, positive emotions, or "experience" matters in interaction design (e.g., Diefenbach et al. 2014 for an overview of the hedonic in Human-Computer Interaction). Of course, what experience actually is, remains fuzzy. In many cases, "experiential" became associated with "cool" or "innovative" technologies, exciting interaction styles, and beauty. While the mantra of "usability" and "practicality" was somewhat defined down by acknowledging that quality comprises of more than just performing flawlessly, the object of all efforts pretty much remained the technical artifact itself. Automotive Interaction Design (AID) was and is no exception. On the one hand, the car is a consumer product in need for styling with interaction design having its part in this. That is why we are toying around with touch interfaces providing haptic feedback in a context where "good ol' mechanical pushbuttons" already did the trick years ago. On the other hand, AID often applies a "workplace"-metaphor to the car, emphasizing the complexity of driving and the need to streamline the demanding control task of driving. Obviously, the wish to impress and to inspire drivers through novel functions and interaction designs and the wish to increase road safety can create substantial contradictions and tensions in design.

In this chapter, we embrace a different understanding of experience. While providing moments of enjoyment and meaning should become the leitmotiv of

AID, we do not share the notion that this can be achieved solely on the level of the functionality or interaction itself. Becoming able to switch on the entertainment system with a wave of the hand (through free-form gesture recognition) may admittedly create a moment of awe, but at the end of the day it remains just this: switching on the radio. The primary question is not how to operate something, but how to use functionality and interaction to create and structure activities and resulting experiences in an enjoyable and meaningful way. Note that this does not imply that the particular interaction on a sensorimotor level is not important. Designed interaction has its own beauty, and variations in how we do things will surely make a difference. However, the particular aesthetic of interaction should not be independent of the experience it supports, but rather be derived from it (Diefenbach et al. 2017; Hassenzahl et al. 2015).

Fortunately, there are already some examples at least from a more academic background exploring a particular well-being-oriented, experiential approach to AID. Early work of Kesson and Nilsson (2002), for example, focused on the commute, its structure and the question of how information systems could help people making proper use of the time in the car. Esbjörnsson and colleagues (Esbjörnsson et al. 2004) created and tested *Hogman*, a system for sharing information among motorcyclists. In the spirit of the present chapter, they start their article with the observation that "Brief encounters between acquainted and unacquainted motorcyclists are enjoyable moments" (p. 92). *Hogman* itself is thoughtfully developed as a way to subtly enhance the practice of brief social interactions among motorcyclist, by for example, automatically sharing certain personal information. This "prolongs" the encounter, strengthens opportunities for later social interaction through other channels and provides identity to motorcyclists, who because of their clothing and the helmet appear rather anonymous on the street. This system is experiential, because it embodies a good understanding of social interaction in traffic. A more functionally oriented design may have proposed to just share information among all traffic members. *Hogman*, however, acknowledges and understands that social interaction among strangers needs common ground. Sharing personal information with an unknown, but fellow *Yamaha* rider is something completely different from sharing information with an unknown driver of a SUV, who just overtook without even keeping the safety distance. Another example of designed social interaction in traffic is Knobel and colleagues' *CliqueTrip* (2012). Through interviews they identified driving in a motorcade as an interesting situation for social interaction across cars. On the one hand, driving in a motorcade can be stressful since drivers need to make sure that they do not lose each other in traffic. On the other hand, finding each other again, overtaking the other car, waving and greeting, finding a nice place for a rest are enjoyable moments. People having some practice with motorcades bring walkie talkies to allow for an auditory connection between cars. Based on this, *CliqueTrip* offered only two functions. The navigation of the following car always pointed to the leading car. Even if both cars were not within sight of each other, the cars would remain in contact. On the navigation display, the link between both cars in the sense of the route for the following car was emphasized to create the impression of being tied together. If the cars were in proximity an auditory channel opened up, connecting both cars. This could neither be switched on or off, but was solely controlled by distance. This allowed for the subtle negotiation between closeness. A study with groups of people using *CliqueTrip* showed that indeed feelings of relatedness emerged from this set up. As one participant put it: "It feels as if the friends in the other car are sitting on the backseat" (Driver 5). Another remarked: "Thanks to the communication and the navigation we all drove together. It was a conjoint drive" (Driver 2) (Knobel et al. 2012, p. 36).

More examples exist. Eckoldt and Schulz (2009) explored making music together in the car as an enjoyable moment, building on the practice of singing together in a car. This is an example for social interaction among driver and passengers, which is an upcoming topic as well. Passengers are in focus (Perterer et al. 2013), with concepts such as the sharing of car-related information with passenger (Inbar and Tractinsky 2011) or even the collaborative modeling of tasks, such as wayfinding (Forlizzi et al. 2010).

In a conceptual paper, Eckoldt and colleagues (2013) explored how psychological needs, such as autonomy, competence, relatedness, popularity, stimulation, and security, could be addressed through experiential interactive systems in a car. One example is a minimal navigation that, while providing some guidance, at the same time attempts to promote own wayfinding skills and to keep the feeling of agency intact (competence). Another example is a vague navigation that ensures that the driver keeps on going into the right direction, but feels free to explore new routes and places on the way (autonomy) (see also Knobel et al. 2013a). Recently, we explore the notion of social assistance or designing for the prosocial driving (Eckoldt et al. 2015, 2016; Knobel et al. 2013b) with the goal of supporting the driver in becoming more considerate in traffic and to make cooperation with other traffic members more enjoyable and meaningful.

To summarize: The challenge of a well-being-oriented, experiential approach to AID is to shape everyday practices involving the car in such a way that they are experienced as enjoyable and meaningful. This goes way beyond driving and touches upon the multiple roles cars play in peoples' lives. Think of the daily commute, the weekly shopping, driving around the children, pleasure rides to the countryside, driving in a motorcade, parking, and long family journeys to places far far away. From the car as a mere transport mode, to the car as an explorer's vessel, a mobile office, a perfect place to talk or to have sex, or a cloister to contemplate life. Instead of leaving these different meanings and uses of the car to the ingenuity of the driver alone, a well-being-oriented design will ask the question of how to structure and support these practices to better fulfill relevant psychological needs. As in the example of *CliqueTrip* this is about understanding a practice and creating a vision of how an experience should ideally be like. When is driving in a motorcade fun? What are the details that distinguish an enjoyable motorcade-experience from a less enjoyable? This is about removing impairing elements, for example, the worry of losing sight of the leading car and being lost. More importantly, however, it is about modeling and amplifying the positive elements of the practice. People, who enjoy driving in a motorcade emphasize the moment, when the contact to the other car is reestablished. They overtake each other, send messages, or pick especially nice places for a rest and lead their fellow travelers there. All this are defining moments, able to distinguish a "good" motorcade from a "bad" one.

However, creating a clear vision of the experience and a rationale for how it supposedly contributes to well-being is only half of the challenge. The experience has then to be materialized through concrete functionalities and interaction. These have to be arranged in a way that the envisioned experience unfolds through using the interactive system. AID is, thus, more than information architecture or input/output devices. It is the deliberate design of an everyday activity involving the car through the sensible arrangement of interactive technology.

Much can go wrong on the way from the general notion of designing for well-being to concrete interactive systems. For example, not everything that seems "fun" or "practical" may be appropriate from a well-being perspective, even if there are some customers, who love the idea. The focus on speed suggested by modern cars, for example, may be the major source of stress. Nevertheless, mandatory and automatic adaption to the given speed limits seems out of questions. Verbeek (2009, p. 239 citing Steven Dorrestijn and the Transport Research Center) summarized the surprising findings from intensive field trials with mandatory and automatic speed limiting:

> [... C]ontrary to the great resistance that might have been expected, the system ultimately won a lot of praise. Users developed a quieter driving style that they enjoyed. Hectic driving behaviour was simply no longer an option and, in the end, this turned out to be a comfortable situation rather than a hindrance for most of the people involved.

Of course, simply restricting a certain behavior seems quite inelegant. The point is that from a well-being perspective "enforced" speed adherence can be a good feature, which becomes meaningful to drivers since it decreases stress while driving noticeably. At the same time, it seems unlikely that drivers would find such functionality exciting or would be willing to pay for it. Well-being-oriented AID needs to develop strategies to convince drivers of the positive, long-term impact of an interactive system on their daily well-being.

Moreover, the transformation of the target experience into an arrangement of interactive technologies can be tricky. Take the example of the enforced speed limit. In some trials the system automatically kept the speed limit, no matter what the driver did. There was an override button available, which was rarely used. Other trials used an accelerator pedal with force feedback. If the driver exceeded the speed limit, she or he needed more force to keep the pedal down. What appears to be quite similar materializations, may still make a substantial difference. Assume that the crucial point that makes the whole experience enjoyable is that the speed is externally forced. Consequently, the driver neither has to worry nor to explain himself to the passengers. There is simply no choice. The pedal is a different story. Here it remains in the responsibility of the driver, whether to stick to or to exceed the speed limit. The freedom of choice inscribed into the pedal may in fact corrupt the peace of mind the system could potentially create.

These intricacies of getting the experience as well as the materialization right calls for a detailed example. In the following, we present the concept of an app to turn the daily commute into a more enjoyable and meaningful experience. (The impatient or already convinced reader may skip the example and continue with the "Discussion and Conclusion").

4.3 "It's More Fun to Commute"—A Case

In Germany of 2012 (https://www.destatis.de/DE/ZahlenFakten/GesamtwirtschaftUmwelt/Arbeitsmarkt/Erwerbstaetigkeit/TabellenArbeitskraefteerhebung/Berufspendler.html), 66% of the working population used the car for their daily commute. Even in metropolitan areas with a good public transport infrastructure, 50% used the car as the main mode of transportation. All in all, many people spend at least an hour per weekday in the car. This makes practices of commuting with the car important, since they occupy a considerable amount of working population's time.

We start designing by better understand commuting—or to be more precise—better understand potential *joys* of commuting. In other words, we deliberately shift our perspective from the notion of "commuting as a problem" to "commuting as an opportunity or possibility" (Desmet and Hassenzahl 2012). Based on this, we develop an application in two steps. First, we structure and design an ideal commute. Second, we provide a potential materialization (a "system" with an "interface") able to shape the commuters experience in the desired way.

4.3.1 Understanding Potential Joys of Commuting

In their review of the literature, Lyons and Chatterjee (2008) summarize the ambiguous nature of the daily commute (see also Novaco and Gonzales 2009). On the one hand, the commute is a burden, a significant sacrifice of time, related to the experience of stress, fatigue and decreased well-being. This is well in line with the low positivity ratings commuting received in Kahneman and colleagues (2004) study mentioned above. On the other hand, even a commute can have attributes of a more leisurely, undirected journey. Mokhtarian and Salomon (2001) suggest that commuters may enjoy the exposure to the environment, the demanding and skillful movement through this environment, the experience of scenic beauty, or attractions along the way. The opportunity to engage in skillful movement and to savor resulting competence feelings, for example, may be a reason to commute by car even if public transport is faster and cheaper. Besides the joys of being *en route*, a commute can be a "pocket of time" providing opportunities for a smoother transition between life roles, for a time out, or being productive (Jain and Lyons 2008). Blumen (2000), for example, studied a small group of married, fully employed mothers in Israel with long daily commutes. The women had positive attitudes to commuting and used the trip for a "mental shift, contemplation and relaxation." The commute afforded a "pause" otherwise denied to them in their daily routine.

Jain and Lyons (2008) emphasize that whether commuting is experienced as positive is a question of how it is "crafted." First of all, the commute must be considered as a potential time-space for enjoyable and meaningful activities. In fact, in a survey (Mokhtarian and Salomon 2001) almost a half of the respondents disagreed with the statement that "travel time is generally wasted time," about 40% agreed that their "commute trip is a useful transition between home and work" (only about 35% disagreed), and about 35% found that they use their "commute time productively" (about 30% disagreed). All in all, there is a substantial proportion of people who find commuting meaningful. But, obviously, it must be structured appropriately. Commuting must provide travel related experiences or other ways to help people to benefit from the available "pocket of time."

Whether commuting is enjoyable and meaningful is shaped by the "materials" involved. Gattersleben and Uzzel (2007), for example, compared different modes of transportation for the daily commute (walking, by public transport, by bike, by car) and their affective consequences. First, the commutes differed in terms of how relaxing or stressful they were. Walkers and cyclists were more likely to feel relaxed, while car users were more likely to feel stressed. Second, commutes differed in terms of how depressing/boring or exciting they were. Public transport users were bored, while walkers, cars users and cyclists found their commute rather exciting. Asked about enjoyable experiences, beautiful scenery was mentioned by all commuters. However, walkers and cyclists were more likely to enjoy the activity itself, while public transport and car user were more likely to enjoy reading or listening to music. Not surprisingly, the mode of transport and the technology available shaped affective experience and the specific activities people engaged in.

Admittedly, the difference in experience between commuting by car and a bike is quite obvious. In the present case our goal is to improve the car commute itself. Kesson and Nilsson (2002) studied a small sample of car commuters. They found that people either spend their time with vocational activities, such as planning the day or calling colleagues, or with mundane and social activities, such as calling friends. In line with the notion of the commute as "transformation" (Blumen 2000; Mokhtarian and Salomon 2001; Ory et al. 2004), commuters tend to engage in activities related to the destination they are heading to. On their way to work, vocational activities were in the fore, on their way home, the mundane and social dominated. One interviewee who shared a car with a co-worker noticed that even the content of their conversations changes, depending on the destination. This seems to be quite a common pattern. By referring to Nippert-Eng (1996), O'Dell (2009, p. 91) describes this in the context of public transport:

> Take the commute home from work, for example. Compared to the morning commute, this is an emotionally 'lighter' trip. While the morning commute may be characterized by people getting ready for work, reading papers, working on laptops, consuming coffee to get them focused and ready for the hours ahead, the evening commute features buttoned-down collars and loosened ties (which are immediately jettisoned and replaced by leisure clothing upon arrival home). This is a commute in which alcoholic beverages tend to replace

caffeinated ones […], as people talk, joke and prepare themselves mentally for the arrival home.

Thus, no matter what the mode of transport, the destination (work versus home) considerably impacts the way the time is spend (planning, preparing, getting ready vs. being spontaneous, relaxing, being social).

4.3.2 The *Perfect Commute:* The Envisioned Experience

The daily commute is a highly routinized journey. The driving itself (e.g., wayfinding) fades into the background. This creates an empty "pocket of time" for the driver and, thus, an opportunity to engage in personal relevant activities.

For the *Perfect Commute* we assumed this "pocket" to be an important (time) resource for well-being. Thus, not the driving should be made more interesting or challenging to fill the time, but various advanced driver assistant systems should be orchestrated to make the driving even more undemanding. The resulting "pocket" can now be filled with "activities," which do not conflict with the highly procedural task of driving.

Of course, people can be left to do with their free time whatever they enjoy doing, such as listening to music or audiobooks. Thus, the obvious idea for a commuting app may be to provide content and games, which can be consumed while driving. The *ShoutCar* system by Kesson and Nilsson (2002) took this approach by providing an access to online music (way before streaming and the purchase of music online was reality). *Perfect Commute*, however, takes a different approach. It is not about actually filling empty time, but about structuring the commute in a way to make it more meaningful. For this, we chose to learn from those commuters reported about in the literature, who already enjoy commuting. Instead of focusing on the apparent problems of commuting (e.g., boredom), we focused on the benefits of commuting reported by "successful" commuters. Consequently, we emphasize the "interdependence of activities and destination" as well as the theme of "transformation". In other words, we frame the journey to work as marked by the duties ahead, the first meeting to attend, coworkers waiting and a need to be punctual. In contrast, the journey back home is about the evening ahead, friends and loved-ones available to greet, talk to and meet.

4.4 Journey to Work: Predictable, Relaxed, Time for Contemplation and Privacy

Driving to work should be predictable. In fact, the commute already starts at home, before actually being in the car. While commutes are highly routinized, they still depend of a number of external factors. The route may be known, but unforeseen congestions will require an earlier start or taking a different route. Commuters often

Fig. 4.1 Vision of the *Perfect Commute* to work (from Eckoldt et al. 2013)

monitor traffic reports at home to adapt their routine accordingly. Essentially, commuters want to know about what time they are going to arrive at the office. In fact, some commuters use their in-car navigation system just to predict the time of arrival (and get a fully-fledged turn-by-turn navigation). In the case of a traffic problem ahead, drivers need alternatives, often quickly. Suddenly, they require detailed turn-by-turn directions for a route they may have never taken before.

For many commuters, work already begins in the car. They use the time to make work-related calls or even try to answer their email. We find this at odds with the notion of transformation. For well-being, it might be better to spend the time in the car with a more self-directed contemplation of the day ahead instead of instantly jumping into the hassle of the work day. Thus, our *Perfect Commute* attempts to support role transformation by focusing drivers on the work ahead, but at the same time shielding them from intrusions. *Perfect commute* understands the car as a protective bubble, an island of calm to enjoy before the day's grind.

The literature on commuting shows that besides "own-time", the route itself, sights and places, can be a source of enjoyment. Of course, it is not likely for commuters to have a scenic route for their daily commute. In addition, commuters on their way to work may not be especially willing to accept additional commuting time just for picturesque diversions. Thus, getting to know a route, the villages one passes, the idiocrasies of the different places, and how they change throughout the year may be the more appropriate approach to more enjoyment from the environment.

In sum, the guiding themes for the journey to work are predictability, security, time for contemplation, and privacy (*see* Fig. 4.1).

4.5 Journey Home: Spontaneous, Stimulating, Time for Conversations, and Togetherness

Driving home should be different from driving to work. For one, it could be more spontaneous. Of course, there are many scenarios, in which people have to be at home in time as well, such as showing up on the dot for the family dinner.

However, *Perfect Commute* wants people to experience their journey home in a more relaxed, spontaneous, and especially social way. This again rests on the notion of transformation and the idea that the destination sets the stage for how to spend time in the car. However, on the journey home, the destination must not only be a place, but can also be people. Already in the car, commuters may want to get in contact with their spouses and children to feel related and to plan the rest of the day.

Kesson and Nilsson (2002, p. 179) reported an insight, where one "participant said that they would like to know if friends or colleagues were in their car at the moment, and if they were busy or not. If they were not busy, he would then be able to call them without disturbing them." In fact, commuters use the journey home for social calls. This is not always well received by friends and family, since they often do not share idle time. However, some may be commuting as well, being in the perfect situation to talk. Talking to them while on the way home, would be a little like commuting together.

Since the route home is essentially the same as to work, it will not become more or less scenic. However, while on the journey to work time seems of essence, it might be less so on the way home. There could be variations in the way home for discovering interesting new places, do a little shopping on the way, or having a coffee in the newly opened café just 2 min away from the regular route. Similar to the case, when a traffic problem causes the commuter to deviate from the routine route, the commuter may be in need of more detailed turn-by-turn directions to these unknown places. However, while in the former situation the fastest way is still important, in the latter, it may be more about scenic routes, interesting places or even people. *In sum, the guiding themes for the journey home are spontaneity, stimulation, time for others, conversation and togetherness.*

4.5.1 The *Perfect Commute:* Shaping Activity and Experience Through Functionality and Interaction

Since *Perfect Commute* is still in a conceptual stage and not implemented yet, we present it as a smartphone app and a more generic interface to be positioned in the car, for example, in the center stack. The experience ideally resulting from *Perfect Commute* was summarized in a video prototype. The chapter presents screen shots as well as video stills from this design work.

Figure 4.2 shows the main interface of the *Perfect Commute* in its smartphone version.

It is mainly a visualization of the destination and the journey ahead. At the bottom is the actual time and place the commuter happens to be at that moment (left). At the top is the future, that is, the time of arrival and the destination itself (left). In the present example, it is 6:45 in the morning and the commuter Matthias is still at home. He is using the app on a smartphone placed on the kitchen table. He is about to drive to the office, a drive which would take about one hour. If the

Fig. 4.2 *Perfect Commute* at home

commuter would start the journey right now, he would be at the office at about 7:45. This estimated time of arrival is continuously updated. By taking a look at the time displayed at the top, the commuter can decide whether to enjoy the company of his family at the breakfast table for a little longer or to better leave immediately (Fig. 4.3). This increases the predictability of the journey, since traffic problems become instantly visible through updates of the estimated time of arrival.

In between actual place and destination is a visualization of the route, with a focus on the estimated time left needed for the journey (e.g., 60 min) and the villages, cities and personal places on the way (e.g., Café Zweibar, LEVERKUSEN). This structures the journey. In a routine drive, places can have particular meanings, for example, some cities may be prone to traffic congestions or a particular village is where the motorway starts and the journey speeds up. In general, the interface suggests to associate places along the route with meaning. This is for example supported by a feature shown in Fig. 4.4.

Below the actual place is an up-to-the-minute headline displayed related to the place. For example, the commuter is passing the village of BREITSCHEID, where a "Large Fire caused damage running into millions" (Fig. 4.3) or ERKRATH reports "120 deep drillings for a new eco-friendly office building" (Fig. 4.3). Note

Fig. 4.3 Checking the route at breakfast

Fig. 4.4 *En route* in the car

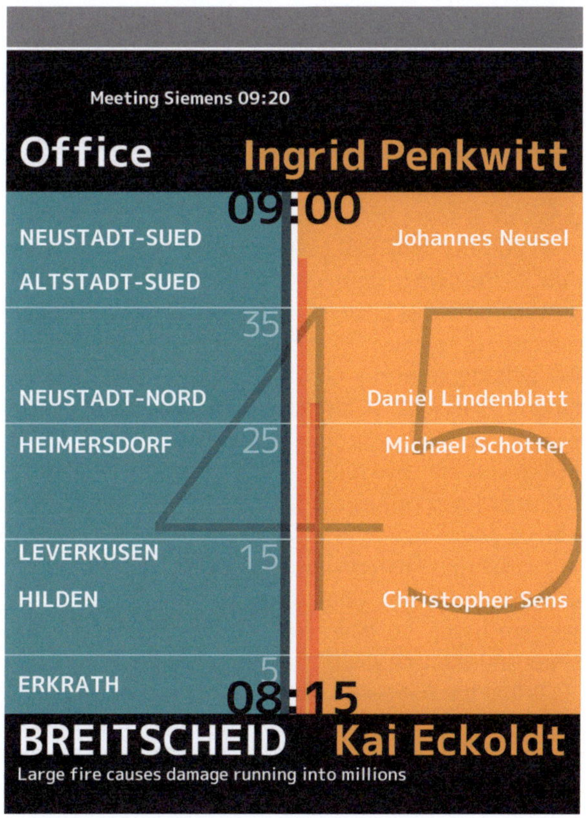

that while these headlines could be linked to the full newspaper report, it is not crucial. This feature is primarily about enriching the bleak places *en route* with layers of story. The literature on commuting shows that sights could be a source of enjoyment. But it seems unlikely to have many breathtaking sights on a standard commute. However, even the most impressive cathedral or valley is more than architecture or a sunset to consume visually. Sights are always associated with stories. Now, while we cannot guarantee visually stunning places on a commute, we are certain that any place, even BREITSCHEID or ERKRATH, has some interesting stories to offer. A further feature to make the route more meaningful is the display of personally relevant places on the route. Café Zweistein for example is one of the favorite cafes of Matthias. Passing it, adds to familiarity. And on the way back, the café may be even worth a stop.

Perfect Commute sets out to support the commuter's transformation by contemplating the activities at the destination. In the case of work, we present the first work-related calendar entry just above the destination (e.g., "Meeting Siemens, 9:20", see Fig. 4.4) as a "reminder" of work. The purpose of this is twofold: First, it seems good practice to prepare the day by contemplating immediate things to do upon arrival. Second, it may help to relax a little. While punctuality seems always important in the context of work, there is a difference between arriving just a little late at one's own desk and being late for an important meeting. The calendar entry helps with sensibly managing the commute. Is there enough time for a quick coffee on the road? The displayed calendar entry answers this question. It is also able to inject some urgency. If the predicted arrival time runs later than the calendar entry, the entry slides onto the route. This will prompt the commuter into action, such as looking for alternative routes to speed up the journey (see Fig. 4.5) or to call the delay in.

While the left, darker side of the interface is dedicated to places, routes, and duties, the right, lighter side is exclusively social. The actual location and the destination is not presented in terms of toponyms but in terms of people available at the respective place. In Fig. 4.2, for example, "Eva" is at home with the commuter, but nobody from the personal address book is in the "office." In Figs. 4.4 or 4.5, "Ingrid Penkwitt" is already in the office. The route itself is presented in terms of people known by the commuter living along the route. Similar to the names of places and the "headlines" function described above, the friends along the route are a further layer of meaning, a way to make the commute more personal. Other than official place names, which are fixed and the same for every driver on a particular route, the route in terms of people is much more depending on the social network of the commuter and, thus, likely to be highly individual. Compare this to the street you are living in. You can think of it in terms of its name, the houses along the street and their numbers, or you can think of it in terms of the neighbors you know.

While the right side frames the commute entirely in social terms, the positioning as side by side with the toponyms creates correspondence. People and places do not remain different entities, but become associated with each other.

The right, "social" side features a number of additional interaction possibilities. Friends and acquaintances available at the current location can be greeted by

Fig. 4.5 Running late and alternative routes

touching the name at the bottom (right side). In Fig. 4.6, the Matthias is on his way home. "Kirsten Neusel" lives or actually is in "NEUSTADT-SÜD".

Touching her name sends a standard text saying that "I was just driving by and was thinking of you. Cheers Matthias." The purpose of this function is not to engage in a fully-fledged conversation with "Kirsten." It is more of a gesture, a nod, a little wave of the hand while passing, a brief moment of feeling related. This is in line with the commute as "own-time" and the *Perfect Commute* as a protective bubble, which allows reaching out, but in an uncomplicated and unintrusive manner.

Messaging is also possible with people available at the destination. A touch on the name sends the standard text: "See you in < remaining estimated travel time > minutes! Cheers < commuter's name >." This provides the recipient with a "sign of life" and some practical information. Again this seeks to balance the need to communicate on a commute with people at the destination (e.g., to announce the time of arrival to waiting coworkers) and the notion of the time in the car as precious "own-time".

A further social function of *Perfect commute* is to allow people to become a dedicated destination. Typically, when it comes to navigation, we tend to think of psychical places, addresses and their GPS coordinates. However, home is as much a

Fig. 4.6 Greeting a friend en route

physical place as it is people, i.e., family member, spouses, or friends. Accordingly, if a commuter tries to get in contact with home (on the journey back from work), she is not actually calling her "house" but her partner. This is already reflected in our design by placing people alongside with places. "Eva" is at home and "Ingrid Penkwitt" is at work. People and particular places are, thus, co-located. However, this must not be necessarily so. This is where the notion of "people as destination" becomes interesting. If the commuter sets a person as a destination, *Perfect Commute* initiates a communication with this person (Fig. 4.7).

The person is displayed on the top right of the screen (see Fig. 4.6). The car invokes a turn-by-turn navigation to the person's current location. However, by default this location is not shared with the commuter. In other words, while the car knows where Eva is, Matthias does not. Matthias can send a message to Eva ("See you in …") by touching her name. However, only if Eva choses to reply by sending a ♥ her current location is briefly shown at the top, left side of the screen, where the destination typically appears. On the one hand, this guarantees the privacy of the person. Eva might not yet be where Matthias expects her to be. On the other hand, it allows for interesting practices. Eva, for example, could surprise Matthias with a cozy dinner in a favorite restaurant. (We guess it's Café Zweibar, though.)

Fig. 4.7 Matthias establishes Eva as a destination

A further social feature is driving together alone. Small orange bars right to the middle line of the center indicate whether friends are in their cars as well (Figs. 4.2 or 4.3). Each bar represents an available friend. The length of the bar roughly indicates the duration of this person's journey with respect to my own journey. A full bar means, for example, that this person will remain available at least as long as the commuter remains in the car. We found this to be important to decide whether to call or not. The bars themselves act as little reminders of other people sharing the same "fate". In the sense of a more ambient sociability this might be enough to feel less isolated, especially early in the morning. Just seeing that some of your friends are on the road at the same time may by comforting already.

Nevertheless, *Perfect Commute* offers to directly start a call with the respective friends. While this goes slightly beyond the notion of maintaining a protective bubble around the commuter's "own-time", it is not just a simple communication functionality. It specifically suggests to get in contact with those friends being in a similar situation to create the experience of a shared commute (Fig. 4.8).

You certainly noticed that while we structured the envisioned experience (pior section) according to the different destinations (work, home), we refrained from simply applying this structure to the actual interface. It seems tempting to just offer two different "pages", each featuring the functionality ideally needed for the respective journey. However, we found this naïve. While it is helpful to create a clear vision of the target experience, in reality, each journey is likely to be a mix of the experiences, we envision. For example, even when focusing on the first meeting at hand when in the office, a quick message to a dear co-worker already there would be great. In the same vein, even the journey back home can involve the need for punctuality and avoiding traffic congestions. Thus, we rather decided to group the

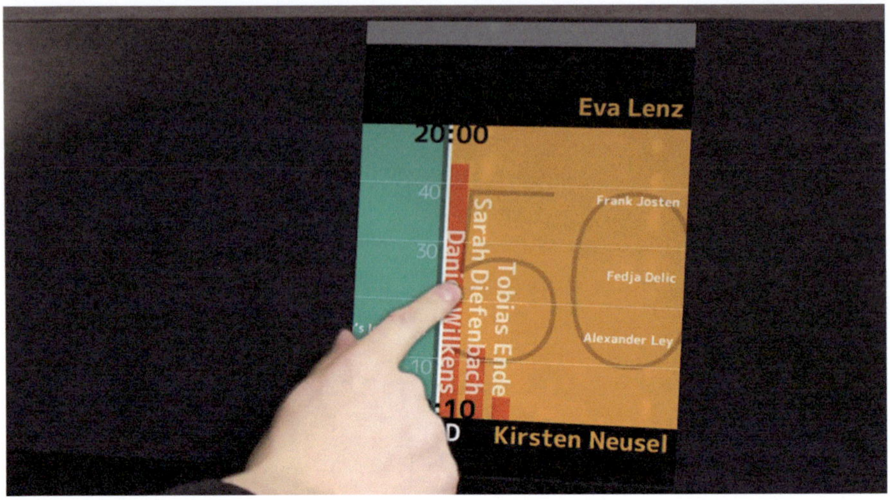

Fig. 4.8 Commuting together over the distance

functions into the predictability/contemplation-related features (left, dark) and the social features (right, light). Thus, while we believe that the commute to work should be ideally led by focusing on the left side of *Perfect Commute*, while the way back home should be led by focusing on the right side, we acknowledge that each journey may be a little more of a mixed experiences. Consequently, each cluster of functionality occupies a similar amount of screen space in the default. However, to support the notion of focusing (on work, on the social), the user can slide the middle line to the right or the left, then using about two-thirds of the screen for the respective functionality (contemplation, social) (see Fig. 4.5). Note, however, the "other" perspective will always remain visible. This reminds of the fact that the other way of thinking about a commute can be sensible and appropriate too. There are two exceptions, though, where *Perfect Commute* requires an explicit change of focus to allow for additional functionality (see Figs. 4.5 and 4.8). In the case of a traffic problem, the left side is forced into focus, presenting alternative routes to be chosen from. Even after having selected one, the focus remains to emphasize the break of routine, the availability of alternatives and the need to regain the predictability of the arrival time. On the social side, a slide to the left invokes the possibility to directly call other commuters. The slight bars indicating the general availability of friends in their cars turn into bigger elements showing the name of the respective friends (Fig. 4.7). In both cases, *Perfect Commute* communicates an opinion about how to actually behave in particular situation. Handling a traffic problem as well as calling a friend need full attention. They are mutually exclusive activities which should not be done simultaneously.

4.6 Discussion and Conclusion

In this chapter, we argued that Automotive Interaction Design (AID) should become more well-being-oriented and experiential. Daily enjoyable and/or meaningful moments are crucial for being happy in and satisfied with life. The fact that people spend considerable time in and with cars creates a responsibility for manufacturers and designers. As Verbeek (2011, Pos. 1859) concludes:

> Even when designers do not explicitly reflect morally on their work, the artifacts they design will inevitably play mediating roles in people's actions and experience, helping to shape […] the quality of people's lives.

At the heart of shaping people's experiences is the material, i.e., functionality and interaction. What is offered in a car and how it is presented and operated, will inevitable influence how people spend their time. AID needs to make sure that the result is more and not less well-being.

We believe this to be a large step for AID, which is rather focused on the low-level operation of technology (button, displays) and concrete tasks than general concepts of how to "be" in the car. While topics, such as driver distraction, attention and response times, remain important, AID could address the more general level as well. The goal should be to create explicit ideas of how to spend time in a car and how to shape this through interactive technology. This may become especially relevant in the face of the disappearance of driving through automation. So far, the vision of autonomous driving seems to rather imply endless boredom (filled with clips from *YouTube*, of course) than a clear understanding of how to spend time sensibly in the car. While academic AID has some interesting concepts and experiences to offer (see above), in practice it does not create the leverage it could.

This is certainly also because of some challenges of engaging in a well-being-oriented AID. In the present chapter, we designed in two steps (see also Hassenzahl et al. 2013). We first tried to understand the meaningful and enjoyable moments, commuting could possibly offer. Based on this we designed the experience, that is, how we want people to ideally act, feel and think while a commute. Only then we materialized this as a system. In many cases in AID, the second step is left out. After an exploration of potential "requirements," designers leap into the concrete design task, drafting wireframes and menu structures. However, designing the experience is a necessary step in between. The experience is not a mere summary of available information but a designed vision of what the system should create. It is as much information as it represents design issues found important and design choices already made. In this sense it acts as a guide to a following materialization. However, materialization very much depends on external factors, such as competing design choices, costs, business opportunities and available technology and skills. The materialization is always a compromise. In this sense, the experience, for example envisioned in form of a story or video, creates a middle ground. It is a reminder of important aspects to consider, a concept, which does not lose itself in a myriad of constraints, always apparent when designing concrete interactions in the car. Of course, materialization matters. A good materialization optimizes the tradeoff between the experience to be created ideally and the external constraints given. But it is still just one of the many possible materializations.

Separating experience from the actual materialization has another advantage. It helps with the handling of critique. It is quite a difference, whether someone disagrees with the envisioned experience or with the current materialization. The former calls for a more general, value-oriented discourse. Do I want to provide commuters with a protective bubble? Do I want them to focus on the destination or rather distract them? Do I share the notion that contemplation helps or do I believe that every minute has to be filled with a game, a puzzle or the latest buzz? These are topics, which touch upon issues of well-being and society in general. Questions of how we want to live. The latter is more about crafting the system itself. Do I find the information architecture to be sensible, the typography to be appropriate? Is the materialization in line with the experience? These are rather technical questions, quite different from the value judgments involved in deciding for the experience.

Obviously, the "experience" envisioned above is not just one, but consists of a number of significant moments, each an experience in itself. For example, "commuting over a distance" as proposed by the *Perfect Commute* is an experience, emphasizing the need for relatedness when being alone in the car. Greeting a friend along the route is about relatedness as well, but framed in a much more noncommittal way. The visualization of the first scheduled meeting at work is about contemplation and security, a source of enjoyment quite different from relatedness. While it makes sense to think about each experience separately—its envisioned purpose, the needs to be fulfilled—all these single experiences need to be integrated into a larger system revolving around broader scenarios. The scenarios determine to a good part, which needs to fulfill and provide hints about the how. The commute, for example, is a particular scenario, which—from our perspective—calls for predictability, contemplation, and a more noncommittal social exchange. However, a trip into the countryside with friends may need adventure and unpredictability instead (Knobel et al. 2013a), while driving together in a motorcade may call for specific ways of relatedness (Knobel et al. 2012). The challenge is to arrange single, well-understood experiences into an overarching "story" and to materialize a system able to tell this story. Instead of talking about "social functions" and providing access to them somewhere in an interface, the social functions are to put into use and to be attuned to particular scenarios, for example, a commute. It is not entirely left to the user to envision uses, but quite the opposite. Functions are arranged in a way to suggest changing and enriching everyday practices. They assume certain uses and not only present themselves as neutral tools.

This leads to another interesting point. Our approach assumes that manufacturers of technology have a certain responsibility for how their products are used. It is puzzling to see, how, for example, addictive use of the smartphone is solely framed as a problem of the respective users. The manufacturer just provided some neutral piece of technology and it is the apparent lack of self-control of people, which creates a problem. Of course, this is not true. Smartphones are deliberately designed to create addictive behavior for the sake of the business models involved. In the same vein, cars are not just neutral. Aggression in traffic is not only a matter of bad driver personalities, but also to a good part made through contemporary car design. Applied to AID, it becomes obvious that we have an enormous responsibility. We are responsible for how fulfilled and happy people spend about 5 h per working week, 220 h in a year, and 8.800 h in a fulfilled working life (a little more than a year). We can turn a full year of people's lifetime into a largely miserable experience or can make it more enjoyable and meaningful. The latter requires assuming responsibility and developing a clear attitude towards how life should be spend in a car. It is not done by leaving all this to the users themselves.

Another challenge is overcoming the in the car-domain quite widespread fascination with technology *per se*. The present example of the *Perfect Commute* certainly poses some technical challenges, but most of it is about understanding technology as a material rather than an end in itself. The navigation system, for example, must not be necessarily thought of as a "device." To us, it is rather a bundle of functionalities, which can be invoked in certain scenarios. In this sense, *Perfect Commute* is a good example, since it makes use of the typical functionality of a navigation system (e.g., estimating time of arrival, finding alternative routes, turn-by-turn navigation), without pretending to be one. There are plenty of ways the same functionality could be redressed to fulfill different needs and, thus, to create different moments of meaning and choice. AID should be more about this, than about designing devices.

Obviously, we did neither implement nor test *Perfect Commute* so far. For the technically-oriented reader, it remains just a bunch of fancy graphics without implementation. For the more human factors-oriented, it lacks evaluation and, thus, proof. However, we believe that there is a place for conceptual designs well-argued for and discussed before we engage in implementation and studies. In many cases, empirical testing is a reflex rather than a thoughtfully applied instrument. Sometimes testing is used to defer design choices to consumers, as an example of misunderstood democratic design. More often it is applied to concepts too hastily drawn-up or already watered-down by myriads of meetings with too many people voicing clearly conflicting views. At best, these tests tease out the already obvious. At worse, they cover a potentially good idea under a bad materialization. Consequently we believe that there is a need for a more analytic-creative approach, where conceptual designs are developed into sufficient detail (through stories, videos, role play), discussed and refined before further processed. This also implies writing more about conceptual designs not yet implemented on the same level of tedious detail, technologists discuss their system architectures and APIs. Through this, the quality of AID will certainly improve.

To conclude: The car is an ubiquitous, fascinating interactive technology. People spend hours and hours of their lifetime in cars. Consequently, AID research and especially practitioners should jump at the chance to improve people's well-being by designing interactions in and through the car in a well-being-oriented, experiential way. This chapter provides an outline of how to approach such an experiential AID and some examples. We hope this to be sufficient to get started.

Acknowledgements This work was made possible by BMW Research, Munich.

References

Blumen, O. (2000). Dissonance in women's commuting? The experience of exurban employed mothers in Israel. *Urban Studies*, *37*(4), 731–748.

Casner, S. M., et al. (2016). The challenges of partially automated driving. *Communications of the ACM, 59*(5), 70–77.

Desmet, P. M. A., & Hassenzahl, M. (2012). Towards happiness: Possibility-driven design. In: M. Zacarias & de J.V. Oliveira (Eds.) *Human-Computer Interaction: The agency perspective* (pp. 3–27). Springer.

Diefenbach, S., et al. (2017). Designing for well-being: A case study of keeping small secrets. *The Journal of Positive Psychology*, *12*(2), 151–158.

Diefenbach, S., et al. (2014). The "Hedonic" in Human-Computer Interaction: History, contributions, and future research directions. In: *Proceedings of the 2014 Conference on Designing Interactive Systems—DIS 14* (pp. 305–314). ACM Press.

Eckoldt, K., et al. (2013). Alternatives: Exploring the car's design space from an experience-oriented perspective. In: *Proceedings of the 6th International Conference on Designing Pleasurable Products and Interfaces—DPPI'13* (pp. 156–164). ACM Press.

Eckoldt, K., et al. (2012). An experiential perspective on advanced driver assistance systems. *IT—Information Technology*, *54*(4), 165–171.

Eckoldt, K., et al. (2015). Soziale Assistenzsysteme—respektvoll handeln im Straßenverkehr. In *Mensch und Computer 2015 Tagungsband* (pp. 2013–2212). Oldenbourg.

Eckoldt, K., et al. (2016). The Gentleman. A prosocial assistance system to promote considerate driving. In: *Proceedings on the 10th Conference on Design and Emotion.*

Eckoldt, K., & Schulz, B. N. N. (2009). Das Auto als Musikinstrument: Gemeinsames Trommeln als positives Erlebnis. *i-com. Zeitschrift für interaktive und kooperative Medien*, *8*(1), 83–85.

Esbjörnsson, M., et al. (2004). Traffic encounters and Hocman: Associating motorcycle ethnography with design. *Personal and Ubiquitous Computing*, *8*, 92–99.

Forlizzi, J., et al. (2010). Where should I turn: Moving from individual to collaborative navigation strategies to inform the interaction design of future navigation systems. In: *Proceedings of the International Conference on Human Factors in Computing Systems—CHI'10* (pp. 1261–1270). ACM Press.

Gatersleben, B., & Uzzell, D. (2007). Affective appraisals of the daily commute: comparing perceptions of drivers, cyclists, walkers, and users of public transport. *Environment and Behavior, 39*(3), 416–431.

Hassenzahl, M., et al. (2013). Designing moments of meaning and pleasure. Experience design and happiness. *The International Journal of Design*, *7*(3), 21–31.

Hassenzahl, M., et al. (2015). The delicacy of handshakes: Reflections on the aesthetics of interaction. In: *Proceedings of the 9th International Conference on Design and Semantics of Form and Movement—DeSForM 15* (pp. 206–214).

Hassenzahl, M., & Klapperich, H. (2014). Convenient, clean, and efficient? The experiential costs of everyday automation. In: *Proceedings of the 8th Nordic Conference on Human-Computer Interaction—(NordiCHI 14)* (pp. 21–30). ACM Press.

Howell, R. T., et al. (2011). Momentary happiness: The role of psychological need satisfaction. *Journal of Happiness Studies, 12*, 1–15.

Inbar, O. & Tractinsky, N. (2011) Make a trip an experience: Sharing in-car information with passengers. In: *Proceedings of the International Conference on Human Factors in Computing Systems—CHI'11. Extended abstracts* (pp. 1243–1248). ACM Press.

Jain, J., & Lyons, G. (2008). The gift of travel time. *Journal of Transport Geography, 16*(2), 81–89.

Kahneman, D., et al. (2004). A survey method for characterizing daily life experience: The day reconstruction method. *Science, 306*(5702), 1776–1780.

Kasser, T., & Sheldon, K. M. (2009). Time affluence as a path toward personal happiness and ethical business practice: Empirical evidence from four studies. *Journal of Business Ethics, 84* (2), 243–255.

Kesson, K. P., & Nilsson, A. (2002). Designing leisure applications for the mundane car-commute. *Personal and Ubiquitous Computing, 6*(3), 176–187.

Knobel, M., et al. (2013a). A trip into the countryside: An Experience Design for explorative car cruises. In: *Proceedings of the International Conference on Human Factors in Computing Systems—CHI '13. Extended abstracts* (pp. 565-570). ACM Press.

Knobel, M., et al. (2013b). Become a member of the last gentlemen: designing for prosocial driving. In: *Proceedings of the 6th International Conference on Designing Pleasurable Products and Interfaces—DPPI'13* (pp. 60–66). ACM Press.

Knobel, M., et al. (2012). Clique Trip: Feeling related in different cars. In: *Proceedings of the 2012 Conference on Designing Interactive Systems—DIS 12* (pp. 29–37). ACM Press.

Kroll, C., & Pokutta, S. (2013). Just a perfect day? Developing a happiness optimised day schedule. *Journal of Economic Psychology, 34*, 210–217.

Latour, B. (1992). Where are the missing masses? The sociology of a few mundane artifacts. In W. E. Bijker & J. Law (Eds.), *Shaping Technology/Building Society: Studies in Sociotechnical Change* (pp. 225–258). MIT Press.

Lyons, G., & Chatterjee, K. (2008). A human perspective on the daily commute: Costs, Benefits and Trade-offs. *Transport Reviews, 28*(2), 181–198.

Lyubomirsky, S., et al. (2005). Pursuing Happiness: The architecture of sustainable change. *Review of General Psychology, 9*(2), 111–131.

Mokhtarian, P. L., & Salomon, I. (2001). How derived is the demand for travel? Some conceptual and measurement considerations. *Transportation Research Part A: Policy and Practice, 35*(8), 695–719.

Nippert-Eng, C. E. (1996). *Home and work: Negotiating boundaries through everyday life [Kindle Edition]*. University of Chicago Press.

Novaco, R. W., & Gonzales, O. I. (2009). Commuting and well-beinf. In Y. Amichai-Hamburger (Ed.), *Technology and psychological well-being* (pp. 174–205). Cambridge University Press.

O'Dell, T. (2009). My soul for a seat: Commuting and the routines of mobility. In: E. Shove et al. (Eds.) *Time, Consumption and Everyday life. Practice, Materiality and Culture* (pp. 85–98). Berg.

Ory, D. T., et al. (2004). When is commuting desirable to the individual? *Growth Change, 35*(3), 334–359.

Perterer, N., et al. (2013). Come drive with me: An ethnographic study of driver-passenger pairs to inform future in-car assistance. In: *Proceedings of the ACM conference on Computer Supported Cooperative Work —CSCW'13* (pp. 1539–1548). ACM Press.

Reckwitz, A. (2002). Toward a theory of social practices: A development in culturalist theorizing. *European Journal of Social Theory, 5*(2), 243–263.

Reis, H. T., et al. (2000). Daily Well-Being: The role of autonomy, competence, and relatedness. *Personality and Social Psychology Bulletin, 26*(4), 419–435.

Ryan, R. M., & Deci, E. L. (2000). Self-determination theory and the facilitation of intrinsic motivation, social development, and well-being. *American Psychologist, 55*(1), 68–78.

Sheldon, K. M., et al. (2001). What is satisfying about satisfying events? Testing 10 candidate psychological needs. *Journal of Personality and Social Psychology, 80*(2), 325–339.

Statistisches Bundesamt: Berufspendler [Commuter]. https://www.destatis.de/DE/ZahlenFakten/GesamtwirtschaftUmwelt/Arbeitsmarkt/Erwerbstaetigkeit/TabellenArbeitskraefteerhebung/Berufspendler.html.

Trösterer, S., et al. (2014). Using a parking assist system over time: Insights on acceptance and experiences. In: *Proceedings of the International Conference on Automotive User Interfaces and Interactive Vehicular Applications—Automotive UI 2014* (pp. 1–8).

Verbeek, P. P. (2009). Ambient intelligence and persuasive technology: The blurring boundaries between human and technology. *Nanoethics, 3*(3), 231–242.

Verbeek, P. P. (2011). *Moralizing technology: understanding and designing the morality of things*. University of Chicago Press.

Chapter 5
Design to Support Energy Management for Electric Car Drivers

Anders Lundström and Cristian Bogdan

Abstract Electric cars (EVs) are a promising alternative to combustion engine cars to lower emissions and fossil fuel dependencies. On the downside, in comparison to internal combustion engine cars (ICE), the user experience (UX) of EVs is seriously compromised due to shorter and more varied driving range depending on driving style other context of use. A further complication is that recovering from unexpectedly low battery levels is tedious due to long charging times. This causes range stress among drivers and research has highlighted a need to improve the information and tools available in order for drivers to better understand range-influencing factors and estimations, leading to increased reliability, and trust in the information. This currently leads to poor UX that may shadow all the benefits and other important environmental and experiential qualities of electric cars. In this chapter, we will provide an introduction to the subject and go through some of our studies and key lessons that have emerged from our research. In particular, we have come to the realisation that we need to energy-empower electric car drivers in order for them to be able to conceptualise how energy is intertwined with their actions and behaviour while driving. This is important, as current tools fail to provide such empowerment, causing unnecessary surprises and worries among the drivers who call the standard tool available in the electric car for the 'guess-o-meter'. Through our designs and discussion we demonstrate how some aspects might be addressed to energy empower electric car drivers.

This chapter is based on Lundströms Ph.D. thesis (Lundström 2016) and a number of research papers referenced in the text.

A. Lundström (✉) · C. Bogdan
KTH Royal Institute of Technology, Stockholm, Sweden
e-mail: andelund@kth.se

C. Bogdan
e-mail: cristi@kth.se

G. Meixner and C. Müller (eds.), *Automotive User Interfaces*,
Human–Computer Interaction Series, DOI 10.1007/978-3-319-49448-7_5

5.1 Introduction

5.1.1 Limitations and Challenges of Electric Cars

Electric cars are a promising alternative to combustion engine cars in the pursuit to lower emissions and fossil fuel dependencies. On the downside, the driving range in today's battery electric cars is limited in comparison to conventional cars, mainly due to the available battery technologies in combination with high battery costs. In practice, this means that driving a battery electric car is somewhat like driving a conventional car with an almost empty fuel tank, with the addition that it takes several hours to recharge. The driving range also varies substantially in comparison to conventional cars depending on speed, driving style, use of the car's climate control unit, and topography, to mention some prominent factors.

Overall, limited driving range is one of the main obstacles for market acceptance of electric cars (Franke et al. 2012; Franke 2012), as people worry about a phenomenon referred to as range anxiety (Eberle and von Helmolt 2010), as they fear running out of battery. However, it is important to point out that energy management and charging are central to electric driving even under less anxious and dramatic circumstances. The driver needs to be battery-aware and to plan for charging, as this is something that might be needed almost daily for some people, and more commonly a few times per week. Therefore, the milder term *range stress* will also be used in this chapter to underline that it does not always have to be such a dramatic and difficult experience.

There is of course always hope for improvements of the underlying technology by making lighter, cheaper and more sustainable batteries, while also having higher storage capacity capable of faster charging. While improvements are made in this area every year, e.g. by lithium-ion batteries that could be charged faster (Pikul et al. 2013), or by optimising the electrical storage system (Chang et al. 2013), or by using biodegradable carbon-based super capacitors made of graphene (El-Kady and Kaner 2013), no sufficiently cheap and manageable solutions have yet been presented. In addition, as lithium—the key component of the most advanced battery technology commercially viable—is a scarce metal, it is unclear whether large-sized lithium ion battery solutions will scale to a broader market (Vikström et al. 2013), indicating that small-sized batteries might be a prerequisite for electric cars in a foreseeable the future.

5.1.2 The Human Side: Driver Trust in the New Technology

In spite of the limitations of electric vehicle technology, studies have shown that the standard battery electric cars have more than enough driving range for the average driver in Europe, USA and Australia to cover daily transportation needs (Franke et al. 2012; Cedersund and Lewin 2005). Furthermore research has highlighted that range anxiety is a psychological barrier that could be addressed with better interface design (Franke et al. 2012), more suited to the needs when driving electric cars (Strömberg and Andersson 2011). Provided that planning of when and how to drive and charge can be done efficiently. This seems particularly relevant for new electric car practitioners as they have been observed to experience difficulties understanding the battery state of the vehicle (Nilsson 2011). It has also been shown that experienced drivers eventually learn to cope with the short driving range in everyday life and perceive the limitations as less problematic (Bühler et al. 2014).

In a recent study of 40 beginners using electric cars for 6 months (Neumann and Krems 2015), the driving range and battery information in the vehicle was concluded to be only moderately reliable and helpful, which indicates a need for improved—and additional—information. They also highlight that their respondents had difficulties understanding electrical units and energy consumption. Other studies have also shown that it is challenging for drivers to deal with range resources in electric cars (Franke et al. 2012; Carroll and Walsh 2010; Franke and Krems 2013).

The main focus when it comes to addressing these issues is generally to improve the automatic range prediction and provide better route guidance by adding more data to the calculations and tips for alternative routes (Demestichas et al. 2012; Neaimeh et al. 2013). In such approaches, trust in the provided information is the key to reducing range anxiety (Hoff and Bashir 2015; Lee et al. 2004; Xu et al. 2014) and improve user experience (Lee et al. 2004). This means that any type of range display or range estimation provided to the driver needs to be reliable and trustworthy (Strömberg and Andersson 2011; Neumann and Krems 2015; Birrell et al. 2014). However, such approaches always require the system to know the full travel plan of the driver, otherwise the system will need to rely on qualified guesses.

Variation of the displayed range was shown to cause confusion and surprises among drivers leading them to call the range estimation the "guess-o-meter" (Lundström 2014). This has also been highlighted by Franke and Krems who argue that subjective range competence, which they define as the "skills to control range-influencing factors as well as predicting remaining range under different conditions" (Franke et al. 2015), is one of the key factors for drivers to manage driving range. This along with route familiarity and reliability of range displays are what they highlight as important areas to focus on to reduce range stress. A crucial need for being able to assess trustworthiness of technology, is to be able to perceive and evaluate important elements of the system itself (Xu et al. 2014).

5.1.3 The Human Side: Driver Skill

In a recent study analysing real electric driving by Birrell et al. it was concluded that range estimations were approximately overestimated by 50% and that driving style had the largest impact on driving range (Birrell et al. 2014). They also conclude "what intelligent systems will not be able to control will be the driver behind the wheel, as the human is still the biggest factor for increasing the available range of an EV". Assuming that there is a human behind the wheel, for autonomous cars the plan and driving behaviour would be in the hands of the system. However, it is important to highlight that then the system might behave in an undesirable way to the passengers causing other types of uncertainties and stress. For instance, if the system decides to drive slow to make the journey possible—by extending the driving range—the passengers might become frustrated if not informed why this is happening or why they would not reach their destination in a desirable time.

Interestingly, the relevance of dealing with aspects like driving speed and climate control have also been picked up by experienced drivers. As an example, drivers of the electric car Nissan Leaf have developed a spreadsheet[1] for estimating remaining driving range in addition to the simplistic estimation provided in the car. This indicates a need for information about how driving behaviour and other aspects (e.g. driving speed, driving style, topography, ice, snow, passengers, luggage, temperature) affects driving range. It also demonstrates a potential for drivers to more actively take control over their remaining driving range. This is especially interesting when driving on low battery or when attempting to reach a destination on the verge of the electric car's capabilities.

Therefore, it is relevant to investigate how drivers could better handle the limited driving range and how to design the interface of the electric car so that it appropriately represents this new and—to many—unfamiliar technology (Strömberg and Andersson 2011). However, while research has repeatedly pointed this out, little has been done to systematically explore what alternative designs might look like and what information could be valuable to drivers.

5.1.4 Design Challenges

In this chapter, we will report on research that addresses the introduced topics through design, by taking on two main challenges. The *first design challenge* is to figure out what is needed for drivers to be able to manage and be in control over the battery while driving. This also requires us to look into the causes of these difficulties and why drivers struggle with the current driving range and battery displays? For instance, what aspects in the displays are missing and hidden? The *second design challenge* is to figure out useful and understandable design concepts that help drivers to overcome and handle these issues.

These are difficult design challenges as there are many interrelated and complex aspects behind energy consumption in electric cars, which lead to a difficult tension between simple, black-boxed and less informative displays—where all aspects are hidden for the driver—and more complex and informative displays that might be due to the complexity too difficult to both design and understand for the driver. A further complication is that studies have shown that drivers have difficulties understanding

[1]Nissan Leaf Range Chart http://www.mynissanleaf.com/viewtopic.php?p=101293.

measures of energy (Neumann and Krems 2015) such as kilowatt-hours. This poses the question of how to design useful displays without using technical terms.

5.2 Summary of Studies

We have addressed this design challenges in a series of five studies, presented in the following sections. While the method used in each study will be described in detail, the overarching approach is the investigation of support for a new use of technology (driving electric cars) through a series of *design exemplars* that help us, by their design process and by testing them in the field, to understand more about this new design space, and to derive new knowledge about electric cars and how we can support their drivers.

5.2.1 Study A: Enough Power to Move: Dimensions for Representing Energy Availability

Our first study in the area of electric cars (Lundström et al. 2012) focused on driving range visualisations with the aim of exploring how more nuanced such visualisation could aid people in understanding electric cars. In our design process, we had noticed that most driving range estimation tools simply provided the driver with a single calculated number for remaining driving range. However, at the time Nissan Leaf had introduced a map-based visualisation showing driving range as a circle on a map. In our analysis, we concluded that the Nissan Leaf driving range visualisation did not account for geographical factors that made the driving range vary to a large extent, such as topography and speed limits, which in turn would be particularly important in hilly areas. Therefore, we decided to explore the topic by building a tool with a more refined algorithm accounting for these dimensions. The result became a tool to role-play electric car driving with our visualisation in a web browser (Fig. 5.1). Today, a similar visualisation is included in BMW electric car interfaces that accounts for topography.[2]

5.2.2 Method

We invited five experienced conventional fuel car drivers with no electric car experience for an opening study. In the study the participants got to identify and

[2]BMW ConnectedDrive—http://www.bmw.com/_common/shared/newvehicles/i/i3/2013/showroom/connectivity/bmw-i-navigation-02-en.jpg.

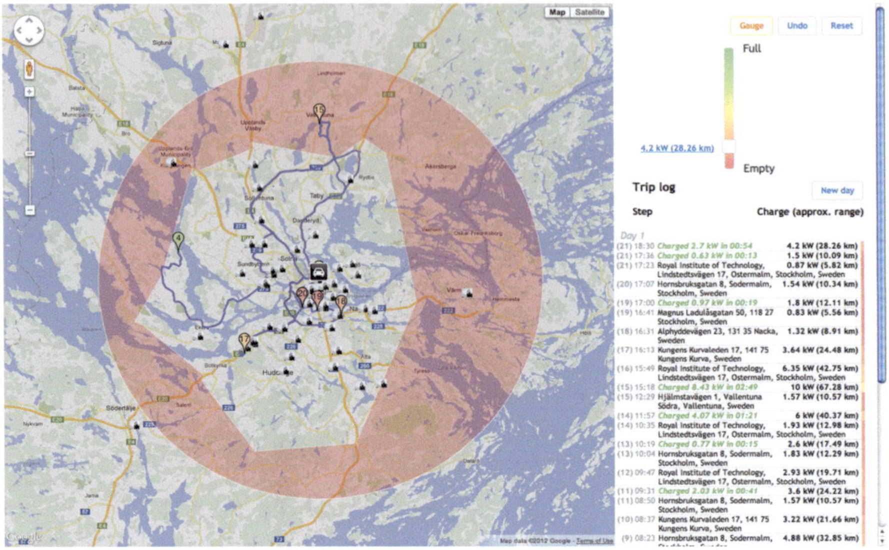

Fig. 5.1 Map-based interactive visualisation used for exploring energy management in real-world scenarios of driving electric cars

role-play four different scenarios using our design (Fig. 5.2 for context of the study). The scenarios were derived from their own driving practice during the study under the guidance of scenario themes. More precisely, they had to role-play driving to their workplace under different circumstances, and using the car for typical car-dependent weekend activities, as well as using the car for a longer, and for them typical, holiday trip. After the role-play we followed up with additional questions. The sessions lasted between 30–90 min.

The purpose of the study was to use our tool to extend our understanding of current driving practices among ordinary fuel car drivers, explore mobility aspects, identify future challenges in electric car usage, get feedback on our design concept, and explore new ideas and challenges. There were many reasons for us choosing this approach. The most prominent was that it was possible to investigate many typical everyday scenarios that would otherwise take months to explore in real life, which also was practically impossible since we did not have access to any electric cars at the time. Another reason for choosing this is that it allowed us to explore the application area much earlier than we could have done if building an in-car prototype. Following this study, that raised interesting questions on energy availability, we engaged with a design process in which we explored how energy availability and infrastructure could be expressed through map visualisations. The results became maps that provided a visual picture of how electric car driving range is connected to infrastructure and mobility.

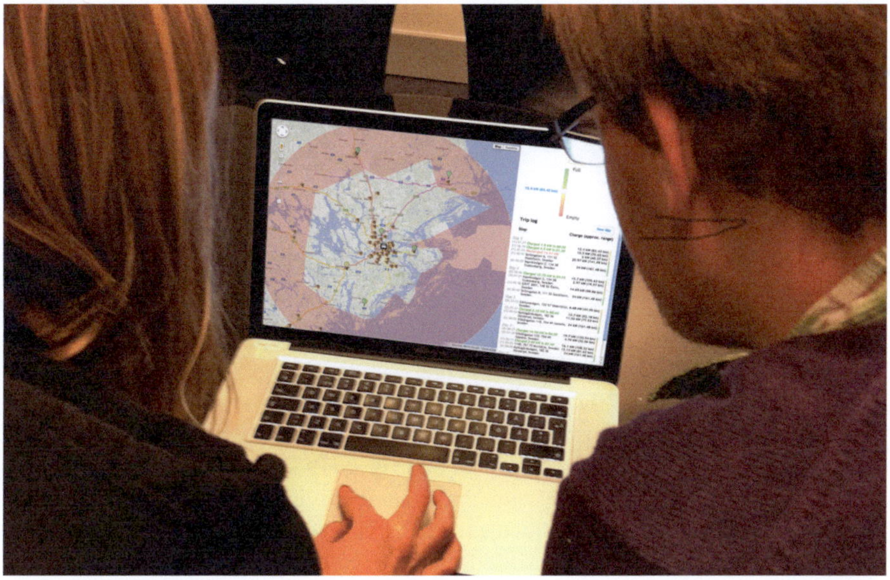

Fig. 5.2 Illustration of the tool in Fig. 5.1 in the context of the study

5.2.3 Results: A Map-Based Interface Exposing Driving Range Relative to Topography and Speed Limits

Our respondents confirmed the values of using more nuanced map-based range visualisation accounting for topography and speed limits. Their input also suggested that we should focus more generally on energy availability visualisations in mobile settings. This argumentation we constructed by showing how energy maps could be designed to reveal different aspects of electric car mobility related to *Public and Private Resources*, *Current and Future States* and *Energy Forecast.* These suggested that maps might be crucial in understanding electric cars from a larger perspective in terms of how they would practically function in society using private (e.g. home, family and friends) and public charging stations.

For instance, the suggested visualisation directly reveals the outer bounds of mobility, unreachable areas, density of charging stations. If connected to smarter infrastructure, social media, weather and geographical data, it could also reveal temporal aspects such as the influence of broken charging stations, how aspects such as social life and the weather influence mobility. In addition, these maps become unique depending on the properties of the specific electric car, as a car with shorter range has a smaller energy mobility map than a car with a longer driving range. In general, we came to see how portraying driving range in this manner and exposing the complex interplay of factors consuming the battery, was an underexplored potential solution that could help our participants answer many of the questions that arose in the role-play study.

5.2.4 Study B: COPE1—Incorporating Coping Strategies into the Electric Vehicle Information System

In Lundström and Bogdan (2012), continuing on from our first study with inexperienced drivers (Study A), we decided to talk to experienced electric car drivers about our work and explore how they manage driving range in their everyday lives. The aim was to look for difficulties and challenges to further engage with and to get a better understanding of the context and what might emerge in real electric car driving. This was judged as important as we hypothesised, and had indications from research that experienced electric car drivers, in contrast to inexperienced drivers, might develop ways of dealing with the limited driving range. We also found it important to observe potential tacit strategies and details that they were using. Neither of these aspects was possible to capture in our first lab study.

5.2.5 Method

This was a full day field study where the first author individually met two electric car drivers with 2 and 5 years experience of owning and driving electric cars. Both could be considered early adopters of electric cars and they were involved in various electric car interest groups and in the development towards a more electrified transportation fleet. Both were board members of the Swedish national electric car interest group[3] and one had created a website that displayed available charging stations in Scandinavia.[4] During each meeting we both went driving and discussed their electric car practices. As things unfolded with one of the participants, and as an impromptu attempt to get into a real driving situation, the first author asked whether he could get a ride to the airport, which turned out to be enough of a challenge and was a good thing to discuss and learn from, in particular about managing situations where the driving range is just about sufficient to reach the destination.

[3]http://www.elbilsverige.se/.

[4]http://www.uppladdning.nu.

5.2.6 Results: Continuously Comparing with a Calculated Overall Mean Energy Efficiency Value

In the airport-driving situation, tacit knowledge surfaced as the informant used and demonstrated a useful coping strategy for managing energy consumption and evaluating more demanding situations. The strategy he used was to calculate a maximum energy consumption that he then could compare with an average energy consumption value for the trip as provided by the car dashboard. Based on this comparison he was able to continuously evaluate how he was doing and adjust the driving speed to keep it below the computed value. The strategy was powerful, but also complicated, as it required calculations (albeit simple) and extensive knowledge about the vehicle capacity. It also needed to be learned and remembered. However, our conclusion was that it was simple and useful enough to make it into a serious interactive tool without the need to know every detail, thus making it suitable for non-expert users. We then conducted a design experiment that resulted in an example (Fig. 5.3) of how such a strategy could be utilised in design to support everyday driving.

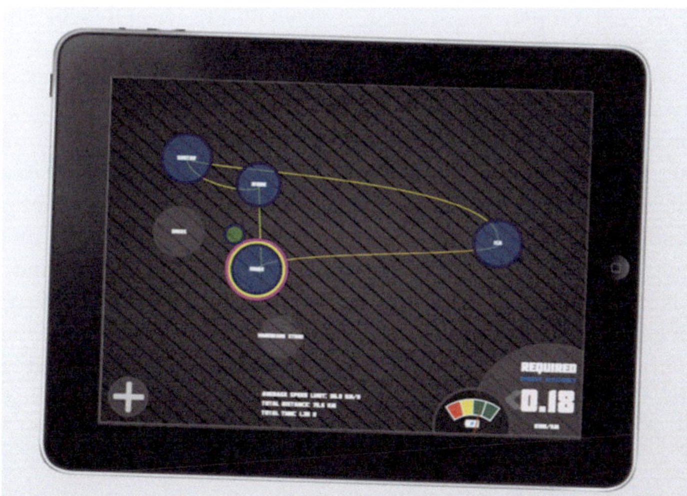

Fig. 5.3 Interactive visualisation on an iPad (9.7 inch screen), based on chosen destinations defined by the user

5.2.7 Study C: Visualising Expected Energy Consumption in Contrast to Real Usage While Driving

This study was also inspired from the experience of expert EV drivers but in a different way. We have learned that driving style plays an important role and that there often is a substantial discrepancy between "typical"—or estimated—energy consumption in comparison to real energy consumption while driving. We therefore investigated in Lundström and Bogdan (2014) the potential value and design of an application that provides an estimation of what might be considered "normal" energy consumption and then compares it to the energy used in the current trip, while actually driving. Having an interface providing such functionality, would potentially enable driver's to compare, plan, learn and adjust while driving in response to how they perform. In a more extreme scenario it could potentially aid the driver to maximize driving range based on the available battery level in order to reach their destination when being on the verge of reaching.

5.2.8 Method

To engage with this topic, we decided to evaluate the feasibility of creating and using such a tool by trying to build a functional prototype connected to a real electric car. As we iterated with our design we conducted formative self-evaluation but we did not come to the stage of testing the design with other drivers. The project team explored how to visualise and sonify expected energy usage in comparison to actual energy usage. The prototypes were tested in short iterations by experiencing and using it while driving and analysing its potential issues and usefulness.

The calculated expected energy consumption for a route was based on properties of the vehicle, e.g. air and rolling resistance, and the topography and estimated mean speeds along the route provided by Google Maps. This data is then displayed in an application running on a smartphone (Fig. 5.4) alongside the "real" energy consumption while driving, calculated from the current and battery voltage values read from the car CAN bus. When using more power than estimated, the curve area in between the real and the estimated gets colored red to indicate "over consumption", and when driving more efficiently than the estimation, the "diff" area turns green and goes below the estimation. Below the actual visualisation there are smaller Sparklines that represent topography (red), estimated speed over the route (yellow), and additionally the actually driven speed while driving. A limitation of the prototype was that the estimated energy consumption is computed for a total of 100 steps of the total route, so if the route is longer the steps also get longer. Each step is then computed taking into account mean speed, properties of the vehicle (weight, frontal area, air and rolling resistance) and elevation and the result is estimated energy consumption for that part of the route.

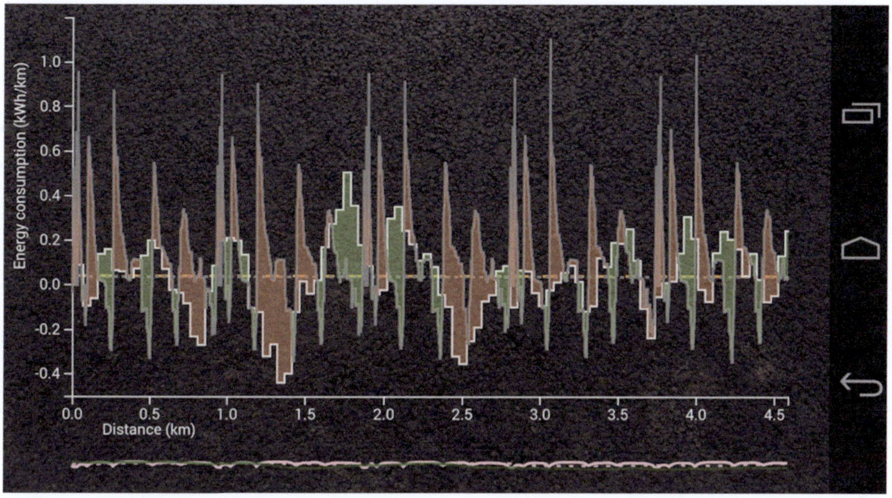

Fig. 5.4 The main screen of the app showing the visualisation while driving, *red* indicates using more energy than estimated on that part of the route, *green* less. Below are sparklines for topography and estimated and actual speed along the route. Note that the driving data in this image is test data for testing the visualisation and not representative for real driving

5.2.9 *Outcome*

From our experiments and test drives, we concluded that this might be a feasible and useful tool, but the visualisation needs to be more accurate (more steps) and provide a zoomed in "mode" for longer trips as the 100 steps quickly provided a too crude and locally misguided picture of the performance. An additional conclusion was that while the momentary discrepancy was interesting and educational, it might sometimes be better to work with a comparison that compared the accumulated performance for the trip so far, in relation to the estimate. This was emphasised as particularly important by the sonification, which sometimes became too intrusive when informing about overconsumption and thereby made us discuss the value of knowing the momentary performance versus knowing the overall accumulated performance. The accumulated performance would be very similar to the strategy used by an experienced electric car user in Study B. However, sonification of momentary performance might still be useful in critical situations, suggesting that the feature and the sensitivity of the sonification could be controlled by the driver and be enabled when needed for eyes-free support. Designing sounds that are pleasant to listen to, even if they convey bad driving, is another challenge that is left for future work.

5.2.10 Study D: Differentiated Driving Range—Exploring a Solution to the Problems with the "Guess-O-Meter" in Electric Cars

Starting from research and general observations that electric car drivers struggle with managing their battery using existing instrumentation, in the next step of our research endeavour we wanted to address these challenges more concretely. The purpose was to explore and suggest alternatives to the prevailing and simplistic "distance left to empty meter"-visualisations, as we at this point had a hunch (based on results from Study B and C) that those failed to account for important driving range effects. In fact, they were commonly referred to as the "Guess-o-meter" among drivers.

5.2.11 Method

The study (Lundström 2014) began with an investigation on why people have come to call the distance-left-to-empty-meter the "Guess-O-Meter", where the state of the art of electric car interfaces was analysed. Interviews were conducted with drivers, and we analysed the discourse by electric car drivers in online user forums. The analysis concluded that the distance-left-to-empty-meter was poor in regard to providing an account of aspects such as driving style, driving speed and use of climate control. We therefore decided to engage in a design experiment aiming to visualise and make these factors visible to the driver, presuming that such information would prevent surprises. In the design process, we worked technically with developing a prototype and calculating energy relations.

5.2.12 Results: Visualising Dynamic Range Predictions

In addition to the identification of a concrete issue with the typical approach of improved driving range interfaces by improving the existing prediction, analysed and articulated potential underlying causes turning the distance-left-to-empty-meter into a "Guess-O-Meter". Based on this articulation, the main results became an alternative novel visualisation concept (Fig. 5.5) that revealed the most important aspects affecting driving range and how they interact with each other. This presentation not only explains how behaviour is directly intertwined with battery use, which is crucial for learning and managing the mechanisms of electric cars, but also brings to the fore central compromises and trade-offs for the driver to balance and judge in different driving situations. In Fig. 5.5, the visualisation shows how driving range varies with the speed for each bar. The grey bars show the driving

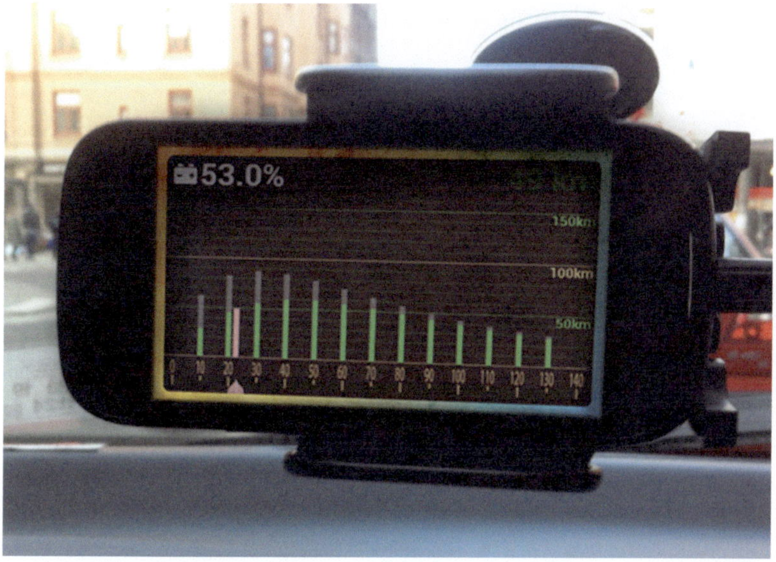

Fig. 5.5 Differentiated Driving Range app showing the correlations between driving speed (current speed just above 20 km/h), driving range and climate control. *Grey bars* show maximum attainable driving range for different speeds with the climate control turned off. The *green bars* overlaying the *grey bars* show estimated driving range with the current climate control setting

range if the climate control is turned off and the green bar the driving range with the current climate control setting.

The visualisation makes it possible to "read" the compromises involved related to these specific aspects and take decisions in relation to needs. For instance, drawn from Fig. 5.5, if you need to drive 70 kilometres then you need to turn off the climate control and stay between 20–60 km/h. This a tough challenge that assumes a steady speed and might require some extra safety margins to account for variations, while these potential variations are not visible in the visualisation. This result also challenges the current electric car dashboard designs and suggests an alternative agenda of research and development. An interesting additional realisation made through this design exercise is how climate control usage interacts with the optimal speed for reaching as far as possible, as more climate control usage pushes the optimal speed for energy efficiency (eco-speed) upwards. This might be crucial information, as driving too slow may reduce driving range. However, in practice, this concept might need some improvement, in order to make speed averages more visible in the design, for easier assessment.

5.2.13 Study E: Getting to Know Electric Cars Through an App

Another lesson learned from experienced EV drivers is that in order for people to transition from combustion engines to electric cars, one important step is to let presumptive drivers see how an electric car performs. This was a realisation that emerged from the fact that experienced electric car drivers learn to manage these vehicles without much problem by learning how they work over time. Hence, they also worry less in their everyday life (although they still have to plan and manage their battery and driving range to a much larger extent than when driving an ordinary car). This also means that practice and exploration is a powerful tool in overcoming scepticism and unfamiliarity with the technology. As there are few who have access to electric cars and letting many of them test EVs is also unfeasible on a larger scale, instead, the proposal was that an alternative approach could be to create a smartphone app that simulates an electric car (Fig. 5.6). Building such an app was also fairly easy for us, as we already had our own libraries for energy estimation of the battery level for any electric car provided that we had the properties of the vehicle (coefficient of rolling resistance, coefficient of air resistance, frontal area,

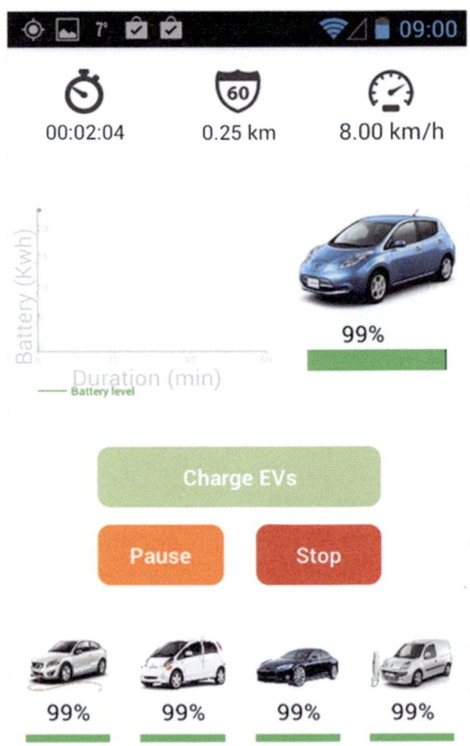

Fig. 5.6 The Virtual Electric Vehicle app. The user can see battery status, pause/resume driving, stop the journey, and "recharge"

mass and battery capacity). In this study (Lundström and Hellström 2015), we aimed to study how such an app could aid the understanding and assessment of electric cars when using their combustion engine car.

5.2.14 Method

For the study, we recruited eight participants who were instructed to use the app that we designed and built (Fig. 5.7 for use in context) over one week when driving their ordinary combustion engine vehicle. During the week, they took notes about the battery level and destination at the beginning and end of each trip. They were also instructed to document questions and thoughts that emerged during the week. We also scheduled a 30-min follow-up semi-structured interview in which we asked about their understanding and attitude towards electric cars, how an electric car would work in their life, what issues and uncertainties still remained, and how they understood the battery and driving range. All interviews were recorded and transcribed. The interviews were then thematically analysed for general themes based on the lessons learnt and understandings of electric cars from a battery and driving range perspective. The focus was not primarily to extract new directions for electric cars, but to research how the app could broaden the understanding of electric cars when used alone.

Fig. 5.7 The Virtual Electric car used in context

5.2.15 *Results: Learning About Electric Cars' Driving Range Is also Learning About One's Habits*

Our results demonstrate how a simulation app like this could be an effective tool for people to assess electric cars in a broad sense, as it created both a multifaceted engagement with electric cars, and triggered a learning process on technology, infrastructure and charging. This demonstrated how a seemingly small piece of information—battery level—could trigger a whole range of other considerations, such as where to charge and how long driving range is in practice. This is important as the purpose with the app is primarily for people to explore electric cars by themselves. In the study, we also highlight that assessing one's own personal driving needs and habits might be equally important as learning about electric cars and infrastructure, as this is essentially what determines if the car is good enough and what compromises might be needed in comparison to combustion engine cars. Interestingly, today the emphasis is more often on explaining the technology. Another finding was that it quickly became easy for our participants to utilise the battery level percentage as a tool for assessing the feasibility of trips, as it was easy to learn and subtract percentages for different routes. While some electric cars do provide the driver with a battery percentage, many still tend to use bars or kWh and we suggest that a strategic shift from a kWh and bar perspective to a percentage-perspective might be a beneficial move in order to aid learning processes. We also identified several uncertainties causing worries among presumptive electric car drivers that should be clarified to prevent misjudgement.

5.3 Discussion: Energy Management in Electric Cars

Throughout our research we have addressed battery management anxieties for electric cars through design. Through our design-driven process, the concrete designs gradually led to a shift in understanding of the underlying issues: we went from a focus on making accurate predictions and visualisations into focusing on energy management and driver empowerment. In this way, our result shifts the attention to the role of the driver in driving range management. Instead of simply striving for an accurate and trustable distance-left-to-empty meter, the work emphasises the need to empower the driver to make crucial decisions concerning plans and driving behaviour—as we believe that they are the only ones who can make the correct compromise in the specific situation. This requires that the driver is in an informed position vis-à-vis the energy mechanisms of the vehicle and is provided with the right tools to plan ahead, as well as follow and reassess plans while driving.

The work presented in this chapter investigates several different aspects of the energy mechanisms of electric driving. In Study A we looked into the impact of topography, weather and speed limits and how those factors could be surfaced in the interaction with the driver through a map-based visualisation. Then, we realised that it would still fail to account for one of the largest factors of driving range variations, which is driving behaviour. This was further strengthened after we encountered a strategy for continuous assessment used by an experienced electric car driver (Study B), which we looked at how to convert into a useful design tool to support everyday driving between common places. Driving behaviour is also addressed in Study C by emphasising in user interface the difference between the current and the 'nominal' driving style.

Following up on that work we reframed our design challenge to focus on what we came to understand as the main challenge for improved battery management, which was to surface how behaviour affected driving as an active asset for battery management. As a reminder, the driving range is reduced up to 50 per cent if the driver changes speed from 60 to 120 km/h. Similarly the driving range is reduced up to 50% if the climate control is on. This is why we designed a visualisation aiming to reveal the inner mechanisms in the form of the concrete correlations between driving range, driving speed and climate control (Study D). The designs presented in this chapter have all demonstrated practical usefulness for assessing the feasibility of different trips. This shows how they could aid the energy management process by highlighting different aspects affecting driving range needed to support planning and assessment. They have also demonstrated the learning required to become an informed electric car driver.

In addition to focusing on tools for driving, we gradually understood how experienced drivers learned to live with and manage electric cars over time as their electric car competence grew. Therefore, we attempted to jump-start people's driving range competence before even driving a real electric car through an app that simulates the battery of an electric car while driving a regular combustion engine vehicle (Study E). In the study, we report how our participants were able to estimate early on the feasibility of different trips and assess whether electric cars would work for them in their everyday life. Tools such as these might be important pedagogical assets for widespread assessment and learning of electric cars among presumptive drivers. Based on our results we can conclude that the battery percentage provided by our design was easy to use for estimating and learning about the driving range. This illustrates the active role that the designed artefact could play.

Although many of our design concepts reveal energy mechanisms and properties of electric cars, there are missing aspects, making them inadequate for maintaining energy plans while driving (as in Study A and D). To capture overconsumption of energy as early as possible in relation to the plan, and to project the implications for the route ahead, we have devised the design described in Study C, inspired also by the strategy used by the experienced driver that we reported on in Study B. His strategy was to create a figure for comparison that he could use for active adjustments and reassessments while driving.

In general, designing for energy management in electric cars could be divided into two main strands of work. First, it is important to design for learning in order to support the driver to conceptualise the vehicle in a good way and gain a so called subjective range competence (Franke et al. 2015). Second, it is important to design concrete tools to support electric driving in critical situations when it is not enough with a general understanding of the vehicle, but where details and tangible compromises to driving behaviour or comfort are necessary. This entails both supporting planning and supporting drivers in assessing and re-assessing their plans.

This in turn relates to a problematic aspect of the studies: as range anxiety happens only occasionally in real-life settings, we had to find a different way to study such phenomena. This is why we used staged, imaginative and role-played scenarios, as well as discussion of past events with the expert drivers. Using the different designs as provocations was another way we approached the range anxiety situation. The studies can be criticised for not studying "real" range anxiety, and thus lacking in terms of what can be learnt from drivers' behaviours. However, the aims behind the studies were not to make claims about typical or atypical behaviours, but to develop a richer understanding to spur relevant design. Had this chapter been in psychology or behaviour science, we would had chosen a different approach, and also arrived at other outcomes when it comes to making claims about driver behaviour. Instead our work concerns *design* knowledge.

One relevant question is whether autonomous cars—self-driving cars—could circumvent these problems as they know the destination and may have an automatic mechanism to adjust the driving style to reach the desired destination selected by the passengers. Here, we would like to highlight that surfacing inner energy compromises is relevant information and considerations even for autonomous cars as they are equally affected by weather, topography, traffic, accidents, jams, driving speed, use of climate control, and so on. This means that the behaviour of the autonomous car is highly depending on external factors not controllable by the system and may need to re-plan and make serious compromises to attain the desired goals by the passengers. For instance, if the passenger turns on the climate control after half the trip the system might need to re-plan completely, e.g. by substantially lowering the speed or by taking alternative routes. Why this happens and what the passengers could do about it (same considerations as for regular cars) then becomes a key concern that needs to be surfaced. The same applies to most other aspects mentioned above that may affect the plan of the autonomous car. The extent and the ways in which the car users are involved in such re-planning situations, based on their understanding of the current autonomous car status and features of the road ahead are an ample field for future research. This is probably true for many other interactions between users and advanced autonomous systems, and we are likely to encounter more such cases in the future.

The explorations conducted have resulted in new design knowledge related to (1) ways of learning key inner mechanisms of electric cars by surfacing how driving speed, speed limits, climate control, weather, and topography affect driving range (mainly Study A, C, D), (2) ways of learning about the driving range of electric cars and gain range competence by learning to use these inner mechanism when planning (Study A, B, D) and by learning about typical needs for everyday routes (Study B, E), (3) demonstrated tools that support planning and assessing the feasibility of trips (all studies) and (4) demonstrated strategies and suggested tools for supporting continuous assessment and dynamic relationship to initial plans while on the move (Study B and C). As all these contributions are directly related to energy management, we have suggested a shift away from understanding range anxiety and stress as a prediction problem. Instead, we have shown that it is crucial to perceive it as energy management issues that needs better design to empower drivers to understand and take control over their driving range.

5.4 Conclusion

We have illustrated the design space for managing energy in electric cars by a series of design studies, involving implemented artifacts that we have reflected upon with users or just within our team. The studies enabled us to extract knowledge related to electric vehicle interaction design. We suggested that the central focus should be on empowering the user, through learning about the new technology in general, and about the specific vehicle in particular, extending the user ability to judge a driving situation from the range perspective. While planning support is important, we also exemplified how the car interface can support the driver during the trip. We believe that many of our lessons also apply to designing user interfaces for autonomous cars.

Acknowledgements This project has been funded by StandUp for Energy and the Swedish Energy Agency (37054-1). We would also like to thank our colleagues Hanna Hasselquist, Filip Kis, Lennart Fahlén, Maria Klingegård, and Ingvar Ohlson, as well as, our students John Turesson, Daniel Molin, Fredrik Hellström, Max Roth, Luan Xueyang for all their contributions to our work.

References

Birrell, S. A., Mcgordon, A., & Jennings, P. A. (2014). Defining the accuracy of real-world range estimations of an electric vehicle. In: *17th International IEEE Conference on Intelligent Transportation Systems (ITSC)* (pp. 2590–2595).

Bühler, F., Franke, T., Cocron, P., & Schleinitz, K. (2014). Driving an EV with no opportunity to charge at home—is this acceptable? In: *Proceedings of the Human Factors and Ergonomics Society Europe* (pp. 369–379).

Carroll, S., & Walsh, C. (2010). The smart move trial: Description and initial results. *Centre of Excellence for Low Carbon and Fuel Cell Technologies Report.*

Cedersund, H.-Å., & Lewin, C. (2005). Män och kvinnor i trafiken: en litteraturstudie. VTI Statens väg- och transportforskningsinstitut.

Chang, W., Lukasiewycz, M., Steinhorst, S., & Chakraborty, S. (2013). Dimensioning and configuration of EES systems for electric vehicles with boundary-conditioned adaptive scalarization. In: *Proceedings of the Ninth IEEE/ACM/IFIP International Conference on Hardware/Software Codesign and System Synthesis* (pp. 1–10). IEEE Press.

Demestichas, K., Adamopoulou, E., & Masikos, M., et al. (2012). Advanced driver assistance system supporting routing and navigation for fully electric vehicles. In: *Advanced Microsystems for Automotive Applications* (pp. 197–206). Springer.

Eberle, D. U., & von Helmolt, D. R. (2010). Sustainable transportation based on electric vehicle concepts: a brief overview. *Energy & Environmental Science, 3*, 689–699.

El-Kady, M. F., & Kaner, R. B. (2013). Scalable fabrication of high-power graphene micro-supercapacitors for flexible and on-chip energy storage. *Nature Communications, 4*, 1475. doi:10.1038/ncomms2446.

Franke, T. (2012). Adapting to the range of an electric vehicle—the relation of experience to subjectively available mobility resources. In: *Proceedings of the European Conference on Human Centred Design for Intelligent Transport Systems, Valencia Spain* (pp. 95–103).

Franke, T., & Krems, J. F. (2013). What drives range preferences in electric vehicle users? *Transport Policy, 30*, 56–62. doi:10.1016/j.tranpol.2013.07.005.

Franke, T., Neumann, I., & Bühler, F. (2012). Experiencing range in an electric vehicle: Understanding psychological barriers. *Applied Psychology, 61*, 368–391.

Franke, T., Rauh, N., & Krems, J. F. (2015). Individual differences in BEV drivers' range stress during first encounter of a critical range situation. *Applied Ergonomics*, 1–8. doi:10.1016/j.apergo.2015.09.010.

Franke, T., Trantow, M., & Günther, M., et al. (2015). Advancing electric vehicle range displays for enhanced user experience—the relevance of trust and adaptability. In: *Proceedings of the 7th International Conference on Automotive User Interfaces and Interactive Vehicular Applications (AutomotiveUI'15)* (pp. 249–256). ACM.

Hoff, K. A., & Bashir, M. (2015). Trust in Automation: Integrating Empirical Evidence on Factors That Influence Trust. *Human Factors: The Journal of the Human Factors and Ergonomics Society, 57*, 407–434. doi:10.1177/0018720814547570.

Lee, J. D., See, K. A., & City, I. (2004). Trust in Automation: Designing for Appropriate Reliance. *Human Factors: The Journal of the Human Factors and Ergonomics Society, 46*, 50–80.

Lundström, A. (2014). Differentiated driving range: exploring a solution to the problems with the guess-o-meter in electric cars. In: *Proceedings of the 6th International Conference on Automotive User Interfaces and Interactive Vehicular Applications (AutomotiveUI'14)* (pp. 1–8).

Lundström, A. (2016). Designing Energy-Sensitive Interactions: Conceptualising Energy from the Perspective of Electric Cars.

Lundström, A., & Bogdan, C. (2012). COPE1—taking control over EV range. In: *Adjunct Proceedings of the 6th International Conference on Automotive User Interfaces and Interactive Vehicular Applications (AutomotiveUI'12)* pp. 17–19.

Lundström, A., & Bogdan, C. (2014). Having a lead foot? exploring how to visualize energy consumption and driving in electric cars. In: *Proceedings of the 6th International Conference on Automotive User Interfaces and Interactive Vehicular Applications* (pp. 1–4).

Lundström, A., & Hellström, F. (2015). Getting to know electric cars through an app. In: *Proceedings of the 7th International Conference on Automotive User Interfaces and Interactive Vehicular Applications (AutomotiveUI'15)* (pp. 289–296).

Lundström, A., Bogdan, C., & Kis, F., et al. (2012). Enough power to move: Dimensions for representing energy availability. In: *Proceedings of the 14th International Conference on Human-Computer Interaction with Mobile Devices and Services* (pp. 201–210). ACM.

Neaimeh, M., Hill, G. A., Hübner, Y., & Blythe, P. T. (2013). Routing systems to extend the driving range of electric vehicles. *Intelligent Transport Systems IET, 7*, 327–336. doi:10.1049/iet-its.2013.0122.

Neumann, I., & Krems, J. F. (2015). Battery electric vehicles—implications for the driver interface. *Ergonomics, 139*, 1–13. doi:10.1080/00140139.2015.1078914.

Nilsson, M. (2011). ELECTRIC VEHICLES: An interview study investigating the phenomenon of range anxiety. Report from Elvire FP7-project. http://www.elvire.eu/IMG/pdf/An_interview_studyinvestigating_the_phenomenon_of_range_anxiety_ELVIRE-2.pdf.

Pikul, J. H., Gang Zhang, H., Cho, J., et al. (2013). High-power lithium ion microbatteries from interdigitated three-dimensional bicontinuous nanoporous electrodes. *Nature Communications, 4*, 1732. doi:10.1038/ncomms2747.

Strömberg, H., & Andersson, P. (2011). Driver interfaces for electric vehicles. In: *Proceedings of the 3rd International Conference on Automotive User Interfaces and Interactive Vehicular Applications (AutomotiveUI'11)* (pp. 177–184). ACM.

Vikström, H., Davidsson, S., & Höök, M. (2013). Lithium availability and future production outlooks. *Applied Energy, 110*, 252–266. doi:10.1016/j.apenergy.2013.04.005.

Xu, J., Le, K., Deitermann, A., & Montague, E. (2014). How different types of users develop trust in technology: A qualitative analysis of the antecedents of active and passive user trust in a shared technology. *Applied Ergonomics, 45*, 1495–1503. doi:10.1016/j.apergo.2014.04.012.

Chapter 6
Cultural User Experience in the Car—Toward a Standardized Systematic Intercultural Agile Automotive UI/UX Design Process

Rüdiger Heimgärtner, Alkesh Solanki and Helmut Windl

Abstract After explaining the motivation and presenting related experiences, an extended Usage-Centered Design approach that integrates standardized process activities from User-Centered Design approach (defined in ISO 9241-210) and uses cultural models is suggested and simultaneously it is also adapted to ASPICE Standard so that the approach is suitable for the design of intercultural user interfaces/experiences in the automotive context. This agile oriented approach makes it possible to track and trace both the culture specific requirements and the design decisions for internationalized HCI in order to produce adequate cultural interaction experiences for users of automotive user interfaces in the car.

6.1 Motivation for a Structured and Quality Oriented Holistic Process for Intercultural HCI Design in the Automotive Context

Within overall product design phase of the product development life cycle, human–computer interaction (HCI) design plays a very crucial role. User interface (UI) and user experience (UX) design are essential parts of HCI design. HCI design also

R. Heimgärtner (✉)
Intercultural User Interface Consulting (IUIC), Lindenstraße 9, Undorf, 93152 Nittendorf, Germany
e-mail: ruediger.heimgaertner@iuic.de

A. Solanki (✉)
Continental AG, Siemensstraße 10, 93047 Regensburg, Germany
e-mail: alkesh.solanki@continental-corporation.com

H. Windl (✉)
WebForUse, Karlsbader Straße 18, 63329 Egelsbach, Germany
e-mail: helmut.windl@webforuse.de

G. Meixner and C. Müller (eds.), *Automotive User Interfaces*, Human–Computer Interaction Series, DOI 10.1007/978-3-319-49448-7_6

incorporates usability engineering. However, "usability" is often misunderstood as just "ease-of-use" or "user friendliness". Rather, it should be viewed as software quality with respect to the context of use as a fundamental element in usability studies (cf. Bevan 1999 and Maguire 2001). However, there are cases where usability professionals and software engineers do not share the same culture and the same perceptive (cf. Seffah et al. 2005). Here it becomes mandatory to improve the collaboration between HCI (usability) engineering and software engineering. The whole product development process *must* be an integrated process with HCI (usability) orientation with collaborating life cycle processes of HCI (usability) engineering and software engineering within it. However, today in general there are just few complete methodologies helping user interface (UI) designers to design user interfaces. Even worse, amongst these there is no systematic and standardized approach that embeds intercultural UI design for the automotive context. Hence, in this chapter a first step toward a holistic approach in order to integrate several types of process models, methods and principles into one individual approach and process is presented in order to cover this gap. The single aspects are just touched as deeply as necessary to grasp the overall idea to be integrated into the holistic approach.

6.1.1 Need for an Integrated Intercultural UI/UX Development Process

In a global industry like Automotive, the UI design and development is distributed over various locations inclusive of locations in emerging economies. Designing a universal UI, which could give the same level of user experience and satisfaction to different user groups, has always been a real challenge. In user interface design specification, it has become essential to take consideration of intercultural issues and human factors into account (cf. Abdelnour-Nocera et al. 2011). Additionally, it does almost become necessary to consider the intercultural issues and human aspects during each phase of the HCI design process (cf. Heimgärtner 2012) as well as of the software (SW) development process which has strong implications on UX. Especially when the user interface design is specified in one region and the product is sold in different parts of the world with users having different cultural and technical background (cf. Honold 2000a). The implementation, nowadays, is mostly done completely or at least for major parts in best cost locations (in emerging markets) by people with completely different cultural background (cf. Solanki and Heimgärtner 2013).

6.1.2 Empathy as Key Factor for Successful Intercultural HCI Design

Successful intercultural communication depends on the personal ability to mutually understand the web of belief of the others using empathic capabilities as shown by empirical examples elsewhere (cf. Heimgärtner et al. 2011). Only by taking over the perspectives of the users, the HCI designer can grasp their needs better and, this can lead to good user interfaces with higher usability and invoking excellent user experiences. Hence, empathy is also a key factor for successful design of intercultural human computer interaction (HCI). Empathy is an essential prerequisite for successful intercultural communication, which promotes intercultural HCI design and intercultural usability engineering and, as a consequence, designing of a good user experience. It facilitates greater project success, better understanding, sustainability, time saving as well as removal of prejudice. Hence, it should be ensured and promoted that usability engineers have, know and apply empathy.

6.1.3 Necessity of Personal Freedom to Increase HCI Design Quality

Creativity needs freedom. This must be considered by project and senior managers of HCI design organizations (cf. Heimgärtner et al. 2011). The improvement of quality through job enrichment and at the same time saving costs demands personal skills, expertise, motivation and creativity. Creating task models leads to a qualitative high-grade HCI design. Usage-centered design is based upon a user interface derived from a series of models containing interrelated task cases. Nevertheless, creative processes must also have been installed. As creativity needs freedom to increase quality and efficiency in general, this should be widely applied. Job enrichment achieved through expertise, motivation, creativity as well as thorough better planning is an essential step for creating qualitative HCI design. Loyalty, motivation and innovation emerge due to the development of synergy effects (e.g., via the feeling of belonging to an organization which is successful for this reason). In this way, personal freedom is also created by the employees themselves—resulting in creative and innovative HCI design.

6.1.4 Intercultural Project Management for Intercultural HCI Design Process

In earlier times technical consumer electronic products with a user interface were first developed for the designers' home market and then exported to other countries by translating the user interface into other languages. The users realized quickly that the product does not fit their needs and therefore do not want to buy it. An alternative would be to develop country-specific products that correspond to the user requirements of the specific target user group. But that would mean that a company must develop several different product user interface designs to correspond to the cultural requirements of the most important key markets. Furthermore, with this strategy, companies are not successful any more in this fast changing globalized world. Therefore, along internationalization additional procedures are needed to reconcile the partly contradictory requirements from the culturally different user groups of one user interface, which fits all relevant user cultures. Global competition requires that new technical products are launched at the same time in all relevant global markets. Therefore, the Human Computer Interface (HCI) product designers need to know all requirements of all global customer groups before starting the global development process, which takes normally place in dispersed intercultural mixed UI designer teams (cf. Schoper and Heimgärtner 2013).

6.1.5 Improving the Quality of the HCI Design Process and Its Outcome for the Automotive Industry Through Standardization

Bevan (2001) described international standards for HCI design. The user-centered HCI design process is defined in ISO 9241-210 and the emerging ISO 9241-220. ISO 9241-171, a standard for accessibility design, can also be used to consider cultural aspects. In addition, the working team "quality standards" of the German UXPA (user experience professionals association) carved out a usability engineering process that can be extended by the necessary roles, tasks, methods, documents and work products, which are necessary to take cultural aspects into account and to fit them into any cultural contexts (cf. also ISO 9241-220). Furthermore, the International Usability and User Experience Qualification Board (UXQB) has created a scheme for CPUX (Certified Professional for Usability and User Experience). Gould and Marcus (2011) suggested a company culture audit to improve development team's collaboration, communication and cooperation. Heimgärtner (2008) developed the intercultural interaction analysis tool (IIA tool) to determine the cultural differences in HCI at interaction level.

6.1.6 Implications and Way Forward to a Standardized Systematic Intercultural Agile Automotive UI/UX Design Process

Until now, however, there is no integrated systematic and standardized intercultural agile HCI design approach fulfilling ASPICE norm, a mandatory process maturity requirement in automotive industry.

Therefore, it is not only necessary to use the empathic design approach (cf. Crossley 2003 and Heimgärtner et al. 2011) to identify the latent needs of end users; but also necessary to consider the distributed participatory (cooperative) design approach involving all the stakeholders (e.g. employees, partners, customers, citizens, etc.) inclusive end users (cf. Beck 2001 and Simonsen 2013) to ensure the product meets user needs and is usable. Prioritizing measures to create personal freedom can also be achieved by adequate development processes integrating the relevant aspects such as freedom, creativity, empathy, lessons learned, etc. Furthermore, to reduce problems and to exploit the synergy effects is necessary to establish an integrated product development process combining usability, software and quality engineering. This can long lastingly reached by integrating all these aspects into a highly quality oriented intercultural and standardized process like the systematic usage-centered design process, the standardized user-centered design process as well as the software development process extended for the intercultural context. For the automotive context, additionally, these standards, processes and tools must be adapted and integrated into ASPICE conform processes acknowledged by automotive industry.

In Sect. 6.2, the influences of cultural aspects that affect standardized processes are considered and analyzed. Section 6.3 presents the systematic usage-centered design approach in a nutshell, followed by Sect. 6.4, introducing the standard of ASPICE briefly. In Sect. 6.5, the available approaches are combined in order to achieve a first draft toward to an agile standardized systematic intercultural automotive UI/UX design process fulfilling the requirements obliged by automotive industry.

6.2 HCI and HCI Design Process Affected by Culture

Several researchers working in the area of taking cultural contexts into account in HCI design have already profited from the results of empirical studies to build well elaborated and comprehensible work products that end in complex but valuable models, theories and tools for further and broader fruitful future research. Culture influences the user's interaction, the usability of the interactive system as well as the approaches, processes and methods for user interface design itself. Therefore, intercultural design approaches such as intercultural usability engineering and intercultural user interface design emerged. After explaining these aspects in the following subsections, finally, the cultural influences to the standardized human-centered design process from ISO 9241-210 will be elucidated.

6.2.1 Influence of Culture on User's Interaction with the UI

Culture as a set of facts, rules, values and norms (structural conditions) representing an orientation system (cf. Thomas et al. 2010)) established by collective

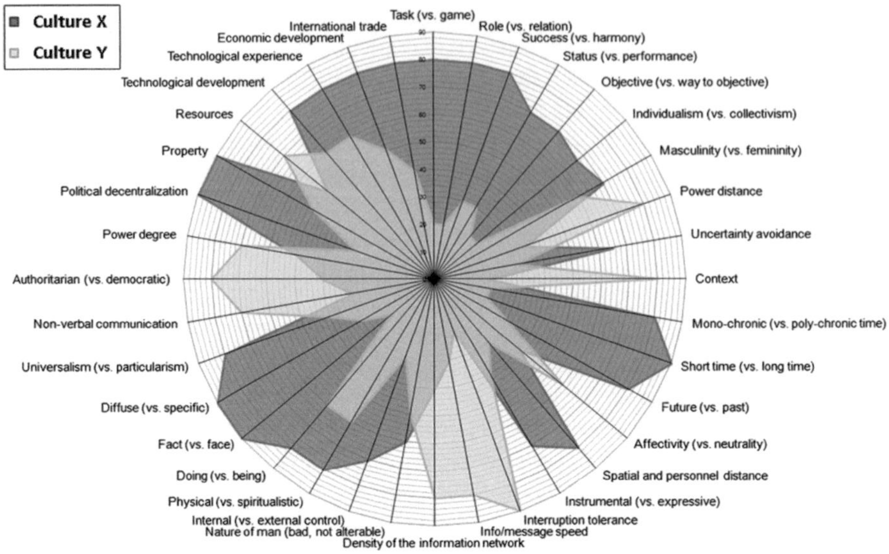

Fig. 6.1 Plenty of cultural models that can be used to analyze the influence of culture on the behavior of users with interactive systems and in living the usability engineering process (indexes are only valid for the cultural models of Hofstede et al. 2010)

programming of the mind (cf. Hofstede et al. 2010) within a group of individuals can influence HCI in different ways.

Figure 6.1 shows cultural models describing cultures based on the compilation from literature by Marcus and Baumgartner (2004) that can be used to analyze the influence of culture on the behavior of users with interactive systems and in living the usability engineering process.

One type of cultural models are cultural dimensions, which serve to describe the behavior and values of members of certain cultures like uncertainty avoidance, individualism or collectivism or power distance (Hofstede et al. 2010). For HCI, those cultural dimensions are most interesting that are directly connected to communication, information, interaction and dialog design, i.e., the cultural dimensions concerning the culturally different concepts of space, time and communication (cf. Sects. 1, 5 and 10 in Hall 2006). Space and time are physical variables influencing the communicative behavior of human beings, which form the social processes of a group of humans and their culture: by learning certain kinds of behavior, the human being matures according to his cultural environment. The influence of cultural imprinting of the user on his behavior in interactions with other communication partners is immense. This is also valid for HCI because communication in HCI is determined by the interaction between user and system (cf. Heimgärtner 2012). Hence, cultural differences in interpersonal communication can and must be transferred to the interaction with technical devices (Röse 2005). Cultural dependencies in user system interactions (HCI) particularly concern interaction and dialog design (cf. Röse 2005). Culture influences the interaction of the user with a computer system or a machine because of the movement of the user in a cultural surrounding (cf. Röse 2005). Therefore, culture has direct influence on the interaction of the user with the system.

The cultural influences on the design process are represented by HCI dimensions, UI characteristics, intercultural variables, and cultural dimensions. HCI dimensions describe the style of human machine interaction expressed by information frequency and density and order as well as interaction frequency and speed (Heimgärtner 2012). Cultural dimensions can be related to HCI dimensions to get a link between the cultural imprint of users to their HCI style (Heimgärtner 2012). User interface characteristics capture the most relevant attributes of user interfaces containing metaphors, presentation, navigation, interaction and mental model (Marcus 2001). Intercultural variables cover the localization levels function, interaction and presentation (Röse and Zühlke 2001). Direct intercultural variables concern HCI directly such as color, icons, language, layout as well as interaction speed and frequency. Indirect intercultural variables embrace HCI margins such as service manual or packaging.

6.2.2 Influence of Culture on Usability/UX and Usability Engineering

The usability of a system strongly depends on how the user can cope with the system (cf. DIN 2010). The user articulates his desires and hence his needs regarding the usability of the system. However, in addition to the common misunderstandings between developers and users, which lead to different product design, there are also misunderstandings because of cultural conditions. There is not only a different comprehension of the requirements of the product but also culturally dependent perspectives and views of them (cf. Heimgärtner et al. 2011). Hence, the developer needs much intercultural knowledge to understand the user of another culture. Furthermore, he needs competency regarding intercultural communication to enable the exchange of information with the user and to know exactly which product the user is likely to have (cf. Honold 1999). Vöhringer-Kuhnt (2002) stated, "Individualism/Collectivism is connected to and has an effect on usability" (Vöhringer-Kuhnt 2002: 17). Therefore, the design, implementation and use of interactive systems should not only meet the general usability criteria but also take into account cultural issues which address relevant topics such as schedule, presence, privacy, authority, control, awareness, safety, error, trust, comfort, coordination, conflict, communication and collaboration as well as interaction style, thinking and action models (cf. Heimgärtner 2013; Liang 2003). The preconditions for intercultural usability engineering are knowledge about the cultural differences in HCI and its considerations in product design and product realization (Honold 1999; Röse 2005; Heimgärtner 2012). "Intercultural" usability engineering is a method for designing products with good usability for users from different cultures. "Intercultural" in this context refers to the special methods that are necessary to do usability engineering for different cultures (cf. Honold 1999). The "interculturally overlapping situation" arranged by a technical system is the most interesting (cf. Honold 1999). Honold (1999) made the approach of Thomas et al. (2010) using "overlapping interaction situations" available for HCI design. These situations arise if a product is defined and formed within one culture and this product is then transferred and used in another culture. A change of cultural environment takes place at the transfer of a technology or a product from the developer's country to another country (cf., e.g., Honold 1999; Röse 2005; Hermeking 2001; Clemmensen and Goyal 2005). Therefore, the work of Honold (1999) deals with the question of whether there is a reduction of the fit between user and product if products of one culture are used in another. Furthermore, the same data can have different meanings in different cultures due to the experiences within one's own, since every culture has its own values, symbols and behavior patterns with meanings and interpretations connected to them. These aspects have an effect on the coding or decoding of news during the communication (cf. Röse 2005). Miscommunication has negative effects on the usability of the product. Therefore, at the collection of culture specific user requirements and culture specific assessment of the concepts used, it has to be examined how far approved methods of usability engineering are suitable. The existing cultural models should be taken into account in the process of product design in the context of intercultural usability engineering. First, the product developers must be sensitized to the difficulties of cultural influences on product development and product use. Then cultural factors influencing HCI must be provided to the developers and considered in the product. This requires knowledge in software ergonomics and intercultural UI design as well as the application of usability engineering methods in the intercultural context. In contrast, if the currently implemented functionality of a system of a certain culture is used as a basis for the analysis of UI characteristics, it may lead to erroneous or simply wrong design guidelines because those requirements need not necessarily match the real needs of the user. Therefore, the user's needs must be collected for every user or at least for the desired user groups (e.g., Chinese and German users).

6.2.3 Intercultural User Interface Design

There are similar concepts taking cultural aspects in HCI design into account. Intercultural user interface design means the process of HCI design in the cultural context (cf. Honold 2000b: 42–43). According to (Röse and Zühlke 2001), intercultural user interface design describes the user and culture oriented design of interactive systems and products taking the cultural context of the user into account with respect to the respective tasks and product usage (Röse 2002: 87). This approach has grown in academic literature from 1990 to 2000 and emerged from the processes of globalization, internationalization and localization of products.

Localization (L10N) means the adaptation of the system to certain cultural circumstances for a certain local market, for example the adaptation of the look and feel of the user interface or the internal data structures to the cultural needs of the user (cf. VDMA 2009). Internationalization (I18N) of a product means that the product will be prepared for its usage in the desired (in the best case for all) countries (cf. International 2003). The internationalization of a software product delivers a basic structure on which a later cultural customization (localization) can be carried out. Globalization (G11N) encompasses all activities with regard to the marketing of a (software) product outside a national market (including I18N and L10N software). The objective is to run successful marketing in one or several regional markets by taking into account the technical, economic and legal conditions there (Schmitz and Wahle 2000). Marcus requested additionally that cross-cultural HCI design should account for dimensions of cultures relating them to user interface characteristics (cf. Marcus 2001). Shen et al. (2006) introduced the culture-centered HCI design process based on research on cross-cultural interface design (Marcus 2006; Röse and Zühlke 2001 and others) and thereby applying iterative analysis to take the target users and their cultural needs into account. Therefore, the topic of intercultural HCI analysis is particularly interesting from the information sciences point of view since this can yield new knowledge, new requirements and goals for the design of information processing systems involving software engineering, software ergonomics and usability engineering.

6.2.4 Cultural Influences on the Standardized User-Centered Design Process

Some ideas have been presented regarding the question how international standards can be valid internationally in Heimgärtner (2014), i.e., worldwide and independent of the different cultures in the world, and how this question can be tackled. Thereby, the impact of culture on the main steps in the usability engineering process is exemplified by analyzing the standard user-centered design process from ISO 9241-210 concerning the requirements of intercultural management and particularly of intercultural project management.

Figure 6.2 shows an overview of the process for designing utilizable interactive systems according to the European Standard EN ISO 9241-210: 2010. This process contains several main steps, which have been analyzed concerning their applicability in intercultural contexts. The weaknesses in every process step if used in intercultural contexts have been identified in order to define and determine recommendations for improvement. Every process step is shortly described, followed by an idea of a possible cultural impact on the step and sometimes also by implications from it on the process, the work products and the product where appropriate or worked out until now respectively (cf. Heimgärtner 2014).

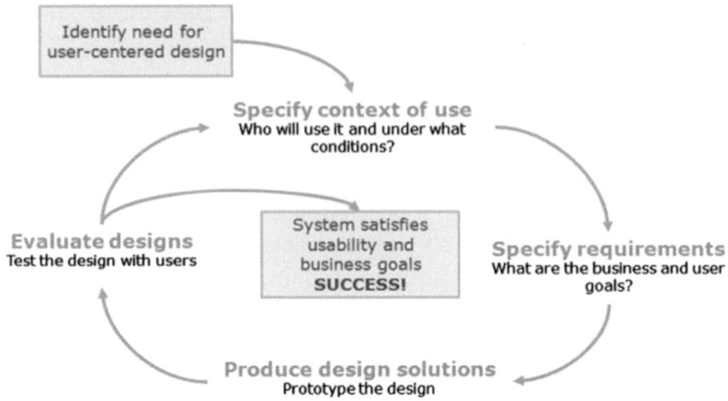

Fig. 6.2 User-centered design process according to ISO 9241-210

The results indicate that the existing HCI development process defined in ISO 9241-210 could be extended by roles and tasks in order to be successful in all cultural contexts worldwide and to fulfill the promised rightness and usability to be valid at least internationally (cf. also Schoper and Heimgärtner 2013). The output of the user-centered design process is influenced by the process itself when applied within an intercultural context. If the process of usability engineering is culturally influenced and different for different cultures, the output of the usability engineering process, i.e., the user interface of the product, is also culturally influenced and different. Furthermore, the results are possibly not as expected by the desired target culture. Therefore, the user interface design methods should be systematically complemented with cultural aspects to ensure that new systems can be designed right from the beginning for one or more cultures while designers better accommodate the diverse global user requirements and respond faster to change using agile methods and parts from usage-centered design (cf. also Windl and Heimgärtner 2013). In any case, it is reasonable that experts in international standardization committees related to HCI have intercultural experience and knowledge in intercultural user interface design and intercultural usability engineering.

6.3 Usage-Centered Design (U-CD) in a Nutshell

Usage-Centered Design (Constantine and Lockwood 1999) is a systematic process using abstract models to design user interfaces for software systems fully and directly supporting all the tasks users need to accomplish. The user interfaces derive directly and systematically from a series of interconnected core models. Center of the process is the robust, fine-grained task model comprising from user perspective the system's functionality expressed in use cases in essential form. Usage-Centered Design has a clear focus on user performance. Systems designed using this approach enable users to accomplish their tasks more accurate and reliable in less time (cf. Constantine 2004). First developed in the early 1990s, it is a proven, industrial strength approach that has been used to design everything from industrial automation systems (cf. Windl 2002) and consumer electronics to banking and automotive infotainment applications (cf. Constantine and Windl 2009). Because it is a streamlined process driven by simple models, it scales readily and has been used on projects ranging from a few person months to a "5 designers, 30 developers, 23 month project" that produced the sophisticated integrated development and award winning environment "STEP 7 Lite" from Siemens AG (Constantine and Lockwood 1999; Ferr et al. 2001).

6.3.1 Process Overview

The U-CD process (Fig. 6.3) can be split in analysis phase and design phase. Role model and task model are the results of a user and task analysis. The content models together with the implementation model are the results of the design phase.

6.3.2 Core Models

UC-D is built around three simple core models that represent the relations between users and system (role model), the work to be accomplished by the users (task model), and the contents and structure of the user interface (content model). The content model derives directly from the task model which is derived from the role model. All models consist each of two parts representing each its description and an additional model representing the relationship between the descriptions (role map, task case map, and navigation map).

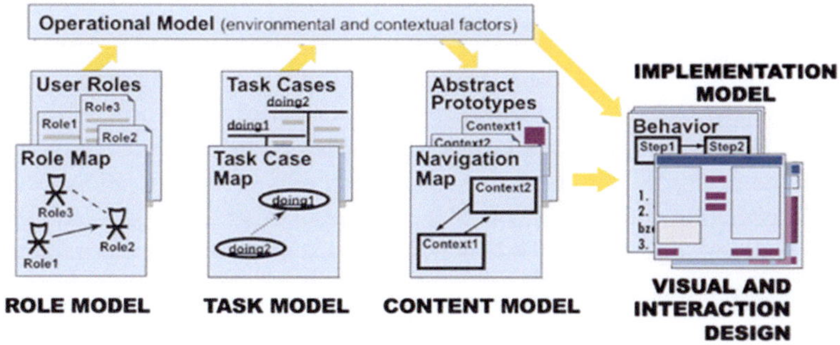

Fig. 6.3 Main constituents of the usage-centred design process

6.3.2.1 Role Model

The role model consists of the user roles and the user role map. User roles are abstract collections of needs, interests, expectations, behaviors, and responsibilities characterizing a relationship between a class or kind of users and a system (Constantine and Lockwood 1999). The user role map is a representation of the interrelationships and dependencies between the user roles.

6.3.2.2 Task Model

The task model combines task cases and the task case map. Task cases are structured narratives, expressed in the language of the domain and of users, comprising a simplified, generalized, abstract, technology-free, and implementation-independent description of one task or interaction that is complete, meaningful, and well-defined from the point of the users in some role or roles in relation to a system and that embodies the purpose or intentions underlying in the interaction (Constantine and Lockwood 1999). The task case map is a representation of the interrelationships and dependencies between the task cases.

6.3.2.3 Content Model

The content model embraces canonical abstract prototypes with the navigation map. Canonical abstract prototypes are abstract representations of user interface contexts modeling the interactive functions, information and basic layout structure needed in the realized user interface utilizing technology-free, and implementation-independent canonical abstract components (Constantine 2003). The navigation map represents the overall architecture of the user interface by modeling the possible transitions between the interaction contexts (Constantine and Lockwood 1999).

6.3.2.4 Additional Models

Two more important models complete the U-CD process holding aspects affecting the design phase (operational model) and the results of the whole effort (implementation model). The operational model comprises the aspects needed to adapt the user interface design to the conditions and constraints of the operational contexts. The implementation model poses a blueprint and construction plan for the final system describing all aspects of interaction and visual design for the implementers realizing the system.

6.4 ASPICE in a Nutshell

Software Process Improvement and Capability Determination (SPICE) or ISO/IEC 15504 is a norm which provides criteria to be fulfilled to achieve high quality software development and software products. Since 2006, it is an international Standard ISO with several parts to perform reviews (assessments) of business processes, initially with a focus on software development (cf. VDA 2013). Key elements of this standard cover the improvement of processes of their own organization (Process Improvement) and the determination of process capability (Capability Determination). Furthermore related activities for the transformation of input products in raw materials are defined. Similar to competing standards such as CMMI, SPICE defines methods to evaluate complete process models and organizations.

6.4.1 Process Model

Automotive SPICE (ASPICE), maintained by VDA and the International Assessor Certification Scheme (iNTACS), describes the part of SPICE which is relevant for automotive industry and acknowledged by automotive OEMs. ASPICE is based on SPICE. The process reference model is ajar to ISO/IEC 12207 AMD 1 and 2 and extended to peculiarities of the automotive industry. The assessment model is ajar to ISO/IEC 15504 and also extended to special features of the automotive industry. Since 2007, the "Manufacturer Initiative Software" (HIS) of the carmaker accept only ASPICE assessments that at least check the HIS scope (cf. Bella et al. 2015). Hence, the ASPICE concept must be integrated in the software development part of the standardized systematic intercultural UI/UX design process in order to be accepted by HIS. Figure 6.4 shows the process areas of ASPICE (adapted from Bella et al. 2015 and HIS scope marked by red dots).

Every process group is structured in sub parts. Every process group consists of one or more processes, which in turn provides several base or generic practices, i.e., activities and requirements that must be done and fulfilled in order to satisfy the ASPICE norm. The more exact these criteria are met the better the rating of the process capability and maturity. For example, a system architecture specifies the elements of the system and a software architecture specifies the elements of the software (cf. Fig. 6.5 from Bella et al. 2015).

Software elements are hierarchically decomposed into smaller elements down to the software components which are at the lowest level of the software architecture. Software components are described in the detailed design. A software component consists of one or more software units. Items on the right side of the V-model are the implemented counterparts of elements and components on the left side. This can be a 1:1 or m:n relationship, e.g. an item may represent more than one implemented element. Evaluation of alternative solutions is required for system and software architectures. The evaluation has to be done according to defined criteria. Such

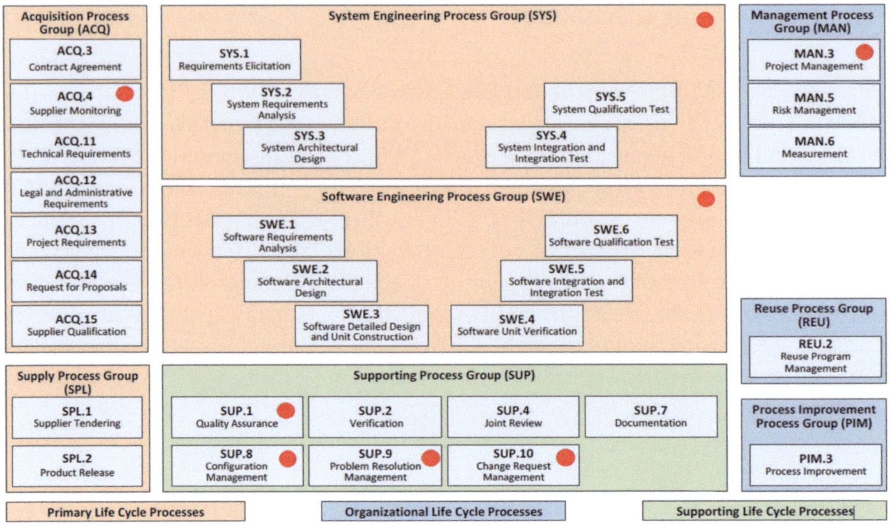

Fig. 6.4 Relationship of ASPICE processes and HIS scope (marked by *red dots*)

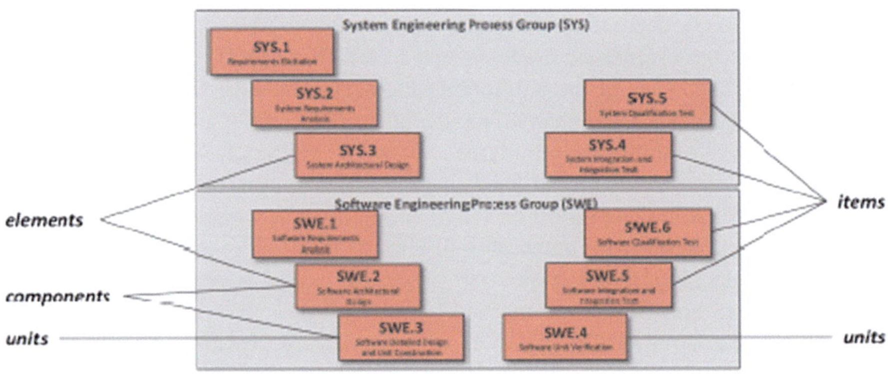

Fig. 6.5 Relationship between elements, components, items and units

evaluation criteria may include quality characteristics like modularity, reliability, security, and usability, or results of make-or-buy or reuse analysis. The evaluation result including a rationale for the architecture/design selection has to be recorded. Compliance with an architectural design (SWE.S.BP3) means that the specified integration tests are capable of proving that interfaces and relevant interactions like e.g. dynamic behavior between

- the software units,
- the software items and
- the system items

fulfill the specification given by the architectural design.

Table 6.1 Process attributes and achieved process capability level

Level	Process	Process attributes (PA)	Description
0	Incomplete		The process is not implemented, or fails to achieve its process purpose
1	Performed	*1.1 Process performance*	The implemented process achieves its process purpose
2	Managed	*2.1 Performance management* *2.2 Work product management*	The previously described performed process is now implemented in a managed fashion (planned, monitored and adjusted) and its work products are appropriately established, controlled and maintained
3	Established	*3.1 Process definition* *3.2 Process deployment*	The previously described managed process is now implemented using a defined process that is capable of achieving its process outcomes
4	Predictable	*4.1 Quantitative analysis* *4.2 Quantitative control*	The previously described established process now operates predictively within defined limits to achieve its process outcomes. Quantitative management needs are identified, measurement data are collected and analyzed to identify assignable causes of variation. Corrective action is taken to address assignable causes of variation
5	Innovating	*5.1 Process innovation* *5.2 Process innovation implementation*	The previously described predictable process is now continually improved to respond to organizational change

6.4.2 Process Rating

The capability of the process areas is evaluated according to the process attributes (Table 6.1, cf. VDA 2013).

The achievement of the processes are rated according to four main categories: fully, largely, partly or not achieved (Table 6.2).

In order to refine the rating, the achievement of the processes can be rated according to the following four extended categories: largely+, largely−, partly+ and partly− (Table 6.3).

6.4.3 Process Outcome

According to the process areas and the base practices representing the tasks to be done within the processes groups, process outcome must be documented in order to perform and fulfill the ASPICE process model (Table 6.4).

Necessary process outcome consists of relevant documents to prove the process is running properly. At least the most significant and useful documents covering the HIS scope of ASPICE must be prepared and regularly updated by the project team beginning with the project plan. In the project plan all relevant documents are

Table 6.2 Basic ASPICE rating categories

Rating	Meaning	Achievement	Rating criteria
F	Fully achieved	>85% to ≤100%	There is evidence of a complete and systematic approach to, and full achievement of, the defined process attribute in the assessed process. No significant weaknesses related to this process attribute exist in the assessed process
L	Largely achieved	>50% to ≤85%	There is evidence of a systematic approach to, and significant achievement of, the defined process attribute in the assessed process. Some weaknesses related to this process attribute may exist in the assessed process
P	Partially achieved	>15% to ≤50%	There is some evidence of an approach to, and some achievement of, the defined process attribute in the assessed process. Some aspects of achievement of the process attribute may be unpredictable
N	Not achieved	0 to ≤15%	There is little or no evidence of achievement of the defined process attribute in the assessed process

Table 6.3 Additional ASPICE rating categories

Rating	Meaning	Achievement	Rating criteria
L+	Largely achieved	>67.5% to ≤85%	There is evidence of a systematic approach to, and significant achievement of, the defined process attribute in the assessed process. Some weaknesses related to this process attribute may exist in the assessed process
L−	Largely achieved	>50% to ≤67.5%	There is evidence of a systematic approach to, and significant achievement of, the defined process attribute in the assessed process. Many weaknesses related to this process attribute may exist in the assessed process
P+	Partially achieved	>32.5 to ≤50%	There is some evidence of an approach to, and some achievement of, the defined process attribute in the assessed process. Some aspects of achievement of the process attribute may be unpredictable
P−	Partially achieved	>15% to ≤32.5%	There is some evidence of an approach to, and some achievement of, the defined process attribute in the assessed process. Many aspects of achievement of the process attribute may be unpredictable

referenced to cover all process areas within software and product development. Documentation regarding software, contracts, architecture, design, measures, plans and strategies (process description, quality management handbook, etc.), specifications, policies, records (review, delivery, results, meetings, quality, etc.), reports (status, etc.), criteria (DOD, release, etc.) and templates must be integrated into the overall configuration management process.

Table 6.4 Performing the ASPICE process model by achieving base practices and output work products according to the ASPICE assessment and reference model (VDA 2013)

Process reference model	Process ID	The individual processes are described in terms of process name, process purpose, and process outcomes to define the Automotive SPICE process reference model. Additionally a process identifier is provided
	Process name	
	Process purpose	
	Process outcomes	
Process performance indicators	Base practices	A set of base practices for the process providing a definition of the tasks and activities needed to accomplish the process purpose and fulfill the process outcomes
	Output work products	A number of output work products associated with each process *Note: Refer to Annex B for the characteristics associated with each work product*

6.4.4 Benefit of Fulfilling the Standardized ASPICE Process Model

Assessments and improvement are an everyday business (cf. van Loon 2007). Each project team member knows the current process and its optimization potential. With a seamless software development process, mistakes can be prevented. Product quality rises because the process output is a high-quality software. Self-assessments are used to continuous process improvement. Each project team member knows what is important in the assessment. As a consequence, time can be saved sustainably and productivity increases.

6.5 Integrating ASPICE, U-CD and UCD into a Systematic Standardized Intercultural Agile Automotive UI/UX Design Process to Optimize User Experience in Car User Interfaces

6.5.1 General

For an intercultural HCI project in automotive context, it was requested to analyze the current status of the commercial product's development process in a very advanced stage of development (actually just before the commercial launch). There were lots of product usability issues, software stability problems and ever increasing customer reported errors (some of them not fixed for one and a half years). Major shortcomings of overall product development process from product usability point of view were result of lack of understanding of usability and erroneous planning. The findings will be briefly pointed out in the following.

6.5.1.1 Missing Common Understanding of "Usability"

The standard definition of usability is given in Sect. 6.3.1 of ISO 9241-11:1998(E) as: "The extent to which a product can be used by specified users to achieve specified goals with effectiveness, efficiency and satisfaction in a specified context of use" (DIN 1999). Unfortunately when the product development process is not usability aware, all the stakeholders involved in product development have different or wrong understanding of product usability.

6.5.1.2 Usability Engineering Activities not Planned Properly

Due to budgeting issues, involvement of usability engineering (UE) specialists could not be planned (Indian vs. German argumentation style, cf. Carmel (1999), Hoffmann et al. (2004). The minimalistic UE process followed was decoupled from SW engineering process. The user interface specification was prepared by the function/system specialists, developed for the very first time within the SW development process and finalized in the advanced phases of SW development. The user interface specification was developed without proper consideration of statutory requirements of various markets as well as of expectations of users from various markets and with varying personality, culture and technical background. The user interface was not designed with usability (in its true sense) in mind. The user interface was developed/implemented *without* human (user) orientation as well as *without* activity (task/usage) orientation. Usability tests, if at all, were performed with main focus being on functionality and only in Germany. The end user was not involved at all or if involved only during certain phases and not throughout the development process as well as not in the initial phases (neither in HCI analysis phase nor in HCI design phase).

6.5.1.3 Software Engineering Process Issues

Apart from above mentioned usability engineering issues there were also other specific issues of SW development process. In the distributed SW development, the SW development process of best cost locations was not at all integrated with the development process of high cost (often strategy driving) locations. Specifics are detailed in the following section.

6.5.2 Lessons Learned

Due to the aspects presented above, the inconsistencies of product design from usability point of view as well as inconsistencies of the chaotic product development process had to be eliminated. Therefore, the following project and process-specific urgent activities had to be carried out in retrospect. This led to additionally more efforts and costs which could have been save if the recommendations and lessons learned presented in the following would have been known and considered in advance. This resulted in delayed product launch and reduced return on investment. Only through the merciless mission of the task force constituted of highly esteemed but very expensive experts, the image of the company could finally be saved.

6.5.2.1 Project Specific Lessons Learned

A structural refactoring of HCI design style guide followed by the SW architecture restructuring was taken up. HCI design style guide was updated for interaction concept, feedback strategy and overall consistency. Application restructuring was done with goals of improving code quality (e.g., commenting, coding guidelines, and naming conventions), architectural quality (e.g., modularity, maintainability, extensibility, and testability) and performance quality (application task schedulability, application task execution time, system/HCI response time, input/key handling algorithm). The application documentation, e.g., architecture and design documents, were also prepared. Preliminary usability tests were performed during design update phase with surrogate users and after implementation with end users. For system architectural issues/optimizations it was suggested to do system performance measurements and analysis. Then, depending on the impact analysis it was decided which modifications/improvements must be implemented. At this later stage of development it was possible to implement only the low impact architectural improvements. Additionally, some graphics resource optimization (e.g., graphics layer, icons, fonts, texts, languages, etc.) and graphics performance optimization (e.g., graphics rendering, screen transitions, animations, etc.) were also done. Unfortunately, the amount of improvements carried out had to be restricted to minimum in order not to overescalate budget and timeframe.

6.5.2.2 Process-Specific Lessons Learned

It was observed that there is an urgent need of:

(a) integrated product development process with HCI (usability) orientation (awareness/maturity) facilitating an effortless collaboration between usability engineering (UE) and software engineering (SE),

(b) all stakeholders having to acquire awareness and knowledge about end users and their usage context,
(c) SW engineers having to understand and master HCI design methods and tools,
(d) deploying of better SW architecture/module design methods and tools for seamless process integration,
(e) usability (HCI) engineering professionals having to start thinking and working like engineers and additionally being able to understand how the technology under use and its limitations affect the product's usability,
(f) management not seeing usability engineering activities as unessential and therefore supporting the usability testing,
(g) setting up correct product usability goals using proper (or to be developed) usability testing methods and tools.

Additionally following usability life cycle processes of product development should also be strengthened (cf. ISO 9241-210/220). Proper project infrastructure is mandatory for enabling better information flow between multilocation teams. It also assists in achieving higher levels of integration for a multilocation development process (cf. Binder 2007). Project management (multilocation/multi-project): as recommended by Curtis and Hefley (1994), an organization must have acquired a high level of capability in key process area of international project management (cf. Hoffmann et al. 2004). Localization management: it is very important to manage region specific issues during the product development. Example, a specific market's legal requirement of having to display a disclaimer was added just before launch. Personalization management: as a mandatory requirement of internationalization it is almost compulsory for each product to have personalization feature (e.g., user specific settings). These issues have also to be managed properly during product development. Example, units' selection options were extended just before launch. Test management was one of the most underestimated activities in the product development planning. Usability tests must be planned by project management throughout the whole product development process.

6.5.2.3 Product Development Process Review and Process Improvement

It was a good decision from management to appoint a person who had good experience of SW platform as well as of HCI development, who also knew the SW development process, methods and tools very well and has the same cultural background as the development team at the offshore location. As an external consultant to the project that is in advance stage of development life cycle, one

- has to gather a quick overview of current status of the project;
- must also understand in very short time the SW system—system platform as well as the application;

- has also to analyze the overall development process and especially the HCI development process (incl. methods and tools) as well as the SW development process.

To support the project's development team it was necessary to provide onsite coaching and consulting in areas such as project management, problem resolution management, SW testing, SW construction, SW integration, SW release, etc., as well as to provide coaching and consulting of the SW development team for usability engineering techniques.

Intercultural HCI design projects are normally executed by multicultural teams (cf. Schoper and Heimgärtner 2013). Typical products are, for instance, automotive driver information and infotainment systems such as instrument clusters, secondary displays, head-up displays, multimedia systems and rear-seat entertainment systems (cf. Heimgärtner 2007b). All phases in the development process are affected (analysis, design, implementation and evaluation phases).

6.5.2.4 Using Agile Methods in Intercultural HCI Design Projects

Since the agile methods require a culture and mindset change, initially there can be great resistance from the team/stakeholders. To overcome the barriers for increased flexibility and adaptation to change is yet another hurdle. If the expectations of transparency and accountability are being raised there is a great amount of insecurity and the increased sense of being continuously monitored (cf. Blankl et al. 2013). Moreover, team members are possibly not used to continuous and high volume communication. For instance, on the one hand, the colleagues coming from a certain type of culture demanded a great amount system engineer's time until they grasped it, but that only concerned the part of, the task they could do. At this point then they worked through it like a robot (without thinking about the overall task). Moreover, they worked on the current tasks rather narrow mindedly and therefore missed the overall view of the task and the system. In fact, short before delivery of the system to customer, integration tests revealed a show stopper requiring necessary system design changes: certain aspects of the system were completely overlooked because of the inability to think of the system at the abstract level and they were therefore unable to inform the team in time, possibly because of high power distance (cf. Hofstede and Hofstede 2010, cf. Fig. 6.6).

In the analyzed project mentioned in Sect. 5.1, for example,, Romanians refused to adopt agile methods in general. This is indebted to their culture: hierarchical thinking; waiting for and following commands; only after consulting the boss were employees willing to hear others. It is not possible to talk directly to employees to steer them: you must go via the boss. Employees do only what they are instructed to do and lack the courage to try out new things; they hold on to well established and introduced methods and processes ("not invented here" behavior). This means risk aversion because of high uncertainty avoidance and individualistic selfishness (cf. Hofstede and Hofstede 2010). Furthermore, some tasks are even unexplainable to

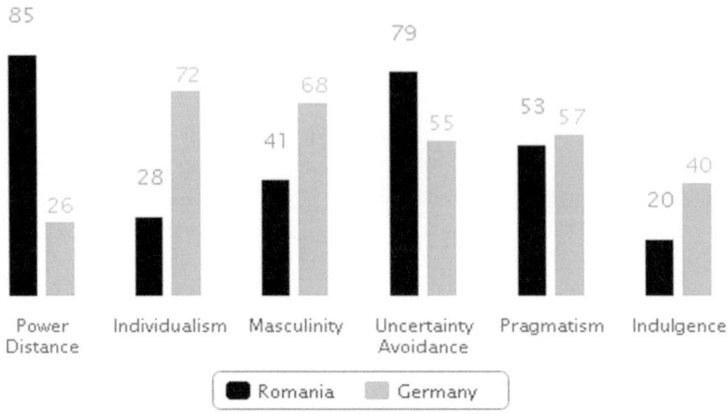

Fig. 6.6 Hofstede's indices for Romania and Germany

these colleague because of different world views. Nevertheless, after several attempts, they accept the task without knowing why and executed the task correctly because of their hierarchical thinking (cf. Hofstede and Hofstede 2010). After on the job training and tips from especially from Scrum Master, Product Owner and Quality Assurance Engineer, the situation changed significantly.

On the other hand, this was not the case with colleagues exposing individualistic behavior because they only do things if they know why in order to be sure of the benefit of their efforts (cf. Hofstede and Hofstede 2010).

These are just some examples of process hurdles that can arise in multicultural teams of intercultural HCI design projects without having to generalize it.

6.5.2.5 Benefit of Using Agile Methods in Intercultural HCI Design Projects

The analyzed examples from intercultural HCI design projects with regard to applying agile methods to expedite the HCI design process while reducing resources (cf. Solanki and Heimgärtner 2013) led to the following conclusions:

- The products and the processes have to be considered with regard to culture.
- The reasons of paradigm shift from Waterfall/V-model to the application of agile methods must put forward.
- At same time, the benefits of using agile methods (SCRUM) in the HCI design process are to be identified.

Using agile methods in intercultural HCI design projects works from our point of view and will definitely be pursued in our future intercultural HCI design projects with the following benefit:

- The team is slowly getting adapted to the new culture and mindset necessary for using agile methods within an intercultural HCI design project.
- Helping each other is taking the first place. By and by it can be felt that everyone is becoming more and more proactive. The team's speed is improving.
- It appears that the team has started enjoying a certain freedom for creativity, experimentation, exploration, and learning new things with and from each other.
- The team's creativity increases. New ideas are getting generated almost every day. Every day something new gets tried out.
- There is no fear of failure. Failures are seen as learning opportunities.
- Team members are starting to take up the collective responsibilities and slowly self-organization is creeping in.

6.5.3 Toward an Agile Intercultural HCI Design Process

Today only few complete methodologies exist that are able to support UI designers while developing user interfaces in an agile, but systematic, structured and guided way in order to create better systems faster. Among these methods there is no systematic approach so far which embeds intercultural UI design. In this section, relevant aspects regarding the phases of the standardized HCI design process in ISO 9241-210 are discussed, followed by the argumentation for a shift in software development from V-model to agile processes, which finally leads to a proposal of an agile intercultural HCI design process.

6.5.3.1 Phases of the HCI Design Process According to ISO 9214-210

The steps of the standardized User-Centered Design (UCD) process of ISO 9241-210 (cf. DIN 2010) have been analyzed concerning its ability to integrate intercultural management aspects (cf. Heimgärtner 2014; Schoper and Heimgärtner 2013) as well as agile methods (cf. Heimgärtner and Solanki 2014). Thereby, the weaknesses in every process were examined and then recommendations for implementing an intercultural UCD process have been defined. In the following, we describe our findings and experiences for the main phases of the HCI design process according to ISO 9241-210.

Analysis

Use agile methods in HCI projects from the very beginning where the uncertainty is very high regarding contractual requirements and thereby commit to a mid- to long-term contract. In agile risk and contract management, bravery and self confidence comes from manageable refactoring cycles. This ensures robust system architecture by allowing change requests by end users and stakeholders. Due to unclear contractual requirements from the customer it is also difficult for the supplier to calculate a realistic and competitive offers. Therefore the supplier is compelled to add a risk premium which makes the offer costly for the customer. Using agile methods generally reduces the risks for both because of iterative and incremental risk management and contracting. Further from the process point of view, as mentioned above, the requirements development has become more effortless with customer's involvement and early feedback.

Produce Design Solution

Since from the beginning the focus has to be on doing right thing, the first tasks/user stories that are taken up are of SW/HCI system architecture (System User Stories) (cf. Walsh et al. 2011). In this process step the culturally diverse UI design team has the task to generate new innovative design solutions. Implementation (extreme) reviews are also being practically implemented with the focus on improving the design for future flexibility and also for testability/maintainability (cf. Highsmith and Cockburn 2001). Research shows that the more diverse the team members are concerning their age, sex, cultural background and education, the larger the chance for innovative new ideas (cf. Kochan et al. 2003 and Sethi et al. 2001). Hence, the following topics are to be considered in intercultural UID teams: Intercultural product design team processes are much more complex to manage than monocultural teams. Problems during the planning of time and budget, managing the project, escalation of problems, conflict management, risk management or a different understanding of quality in the design phase will occur daily (cf. Hoffmann et al. 2004 and Binder 2007). Because of the different cultural imprint and underlying assumptions of the team members, processes such as team development will take longer than in monocultural teams. Communication is a challenge in diverse teams: misunderstandings caused by talking the same language which is, however, not the mother tongue for most of the team members will happen frequently and can lead to anger and frustration. All these potential problems require an interculturally experienced professional team leader to cope with and manage these potential obstacles in a professional way.

Evaluate Design

Tests have to be planned for each iteration at the unit test level as well as at the system test level (cf. Rätzmann 2004). There is still much scope for improving and automating tests. The focus changed from extensive documentation to working software, from processes to people and interpersonal skills and from following a plan to being able to respond to changes. Customer satisfaction increases through continuous delivery. The whole development is accelerated through continuous customer involvement and early feedback. This also helps in finalizing the requirements faster and earlier. Since the development gets broken down into smaller incremental iterations, there are no inhibitions in documenting and analyzing the requirements under discussions. Requirements are immediately implemented and delivered to the customer for final acceptance. The better and more precise the product targets are defined at the beginning of the development project, the easier it is to compare them with the current state of design. These target performance comparisons are to be carried out continuously to make sure that the resulting design solutions correspond to the objectives defined. In intercultural context, it is important to evaluate the design status from the perspectives of the different cultural user groups defined to ensure that the design fits to the different and sometimes even contradictory requirements of all stakeholders.

System Satisfies Usability and Business Goals

The continuous control mechanism of (i) testing the current design by all relevant cultural lead target group testers that know the environmental factors and local user requirements and conditions in detail and (ii) reworking the design on the basis of this feedback is the basis to fulfill the diverse requirements and objectives of all cultural target groups with the new product. In the case of severe conflicts of interests which cannot be solved in one product solution, but where the fulfillment of the requirements is of highest priority for the market success of the product (e.g., incompatible legal prescriptions for the display size of navigations systems) and where it is not possible to reconcile these requirements, another option must be found. In the worst case the requirements cannot be fulfilled by one user interface design and the management must specify how to proceed. Introducing agile methods in the UI design process can help to speed up this process of entangling the manifold and fast changing interests from stakeholders from different cultures. Agile usability (cf. Gundelsweiler et al. 2004) is a reasonable approach to optimize the process of HCI design.

6.5.3.2 From Waterfall/V-Model to Agile Methods

In sequential product development, the software development process often follows the HCI design process (like in waterfall model, cf. Pressman 1992). Due to very high market dynamics and market demands for shorter and shorter time-to-market cycles, the product development has been forced to adapt more and more concurrent product development life cycle processes. And, in practice the development process contains multiple loops of incremental and iterative development phases. In such a concurrent and iterative product development both, the HCI design process and the software (SW) engineering process, do run more or less as two parallel life cycle processes and there are feedbacks between the phases of both processes (cf. Ressin et al. 2011a). The analysis sprints, the design sprints and the evaluation sprints of the HCI concept development process are running in parallel to the implementation sprints of the Hi-fi prototype development process with certain predefined synchronization stages between these two processes (cf. Fig. 6.7).

In the whole product development there are many stakeholders involved (cf. Hoffmann et al. 2004). All these stakeholders like human factors specialists (ergonomists), interaction designers, graphics/sound/haptic designers, software developers and RandD engineers, who are all supposed to be HCI specialists, shall work closely with the end-user and at the same time also with the customer (product engineering and product marketing) (cf. Heimgärtner et al. 2011). The whole

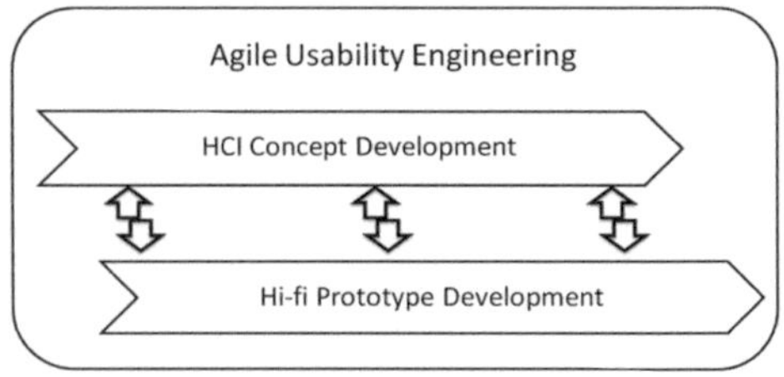

Fig. 6.7 Integrated agile product development/ usability engineering process

product development and especially user interface development is highly people intensive. People involved do have different levels of motivation, creativity and HCI design (usability) orientation (cf. Heimgärtner et al. 2011). This is even truer of people involved in the distributed development process spanned over industrialized and emerging markets. The distributed development process makes it mandatory that all the stakeholders must have the same level of HCI design (usability) awareness and the whole product development process must have achieved higher stage (i.e., at least greater than 3) of usability maturity (Duxu and Marcus 2011).

With competition in mind in the search for flexibility and overall effort reduction in standard (static and rigid) HCI design projects, there have been endeavors in the past exhausting all customization possibilities up to the customer's written approval (cf. Balzert 2008). But the standard V-Model process based on a waterfall model does not fit well when (1) the product requirements are not clear in the beginning (which is even more certain in the case of predevelopment projects), and (2) the requirements get developed during the development phases and therefore the requirements change more often than not (which is always the case for Human Computer Interface Design/Development even up to the last phase of development) (cf. Ressin et al. 2011b). The customer project in question has both the characteristics of a HCI concept development project as well as those of a high-fidelity prototype (demonstrator) predevelopment project.

While trying to avoid such problems arising from strict hierarchies and inflexible process models in our intercultural innovation/platform projects, we had positive experience of with practically using and deploying some of the agile principles and methods (cf. Cockburn and Highsmith 2001) focused on getting (iteratively and incrementally) 'right things' done (to increase product quality) rather than getting things done right (to increase process quality). Having lost a precious 20% of resource capacity and time in the "analysis paralysis" of the product requirements, which were neither complete nor final, it was decided to introduce agile development methods in order not to fail to deliver the values matching the customer's high expectations and to avoid losing the customer for ever. Customers do not trust you as supplier anymore. The supplier loses face (cf. Victor 1998). Different interpretations of the agile manifesto are represented by various methodologies such as SCRUM, XP, Lean Driven, Kanban, and others (cf. Highsmith and Cockburn 2001). All of them have risk management, test and customer validation and acceptance as a major value. Furthermore, they all have artifacts—customer requirements, tests, working software as well as engineering processes that are repeatable and similar companywide: for instance rules the methodology and the customization done in the company (cf. Rätzmann 2004).

6.5.3.3 Proposal of an Agile Intercultural HCI Design Process

On the basis so far, an agile intercultural HCI design process was developed (cf. Heimgärtner and Solanki 2014; Schoper and Heimgärtner 2013; Solanki and Heimgärtner 2013). The resulting UI design approach empowered by cultural aspects makes sure that new systems are designed right from the beginning for the cultural diverse user markets in a time and cost efficient and effective way. As most user-centered design projects are IT or software projects and agile methods are already used in international development projects (cf. Ressin et al. 2011a), the authors suggest applying the methods of agile project management in intercultural user interface design projects. Agile Project Management (APM) is a project management method, which was developed in the software industry (cf. Highsmith and Cockburn 2001). It is an iterative method in engineering and information technology development projects to create innovative new products by using fast feedback loops. By quickly producing rough prototypes and giving them to the customer it is the aim of the method to receive their fast direct feedback and to continuously optimize the product on the basis of the customer/user feedback (cf. Gundelsweiler et al. 2004). The most dominant effect of this project development approach is that the team can be more effective in responding to change. In addition they can dramatically reduce the time between making a decision and seeing the consequences of that decision (cf. Cockburn and Highsmith 2001). These benefits fit exactly to the demands of innovative intercultural user interface design (IUID) projects that require fast feedback loops. In addition, international product development teams are spread all over the world. Agile methods are a way to best use the alleged disadvantage of the time difference. By dividing the design work, e.g., into the programming of the software (which could be carried out, e.g., in Eastern Europe) and into the testing of the software (which could be done, e.g., in Mexico or California by using the time difference of 7–9 h), the software programmer could receive a feedback to his daily programming work over night and optimize it on the next day on the basis of the evaluation. This procedure can lead to a reduction of the development times by 50%. Moreover, the designers would receive feedback from software testers that derive from different cultures than the programmers and therefore know the specific requirement of the customers in their individual markets. Finally, with such a globally dispersed development process it can be ensured that the key relevant customer/user requirements are continuously controlled.

Figure 6.8 shows the concept of an agile intercultural HCI design process by using the methodology of Agile Project Management (APM) combining the best practices of the current HCI design process with the elements of intercultural management. Living this process, the UI designers better accommodate the diverse global user requirements and respond faster to change. In addition, the approach reduces the development time dramatically by fast feedback loops. For instance,

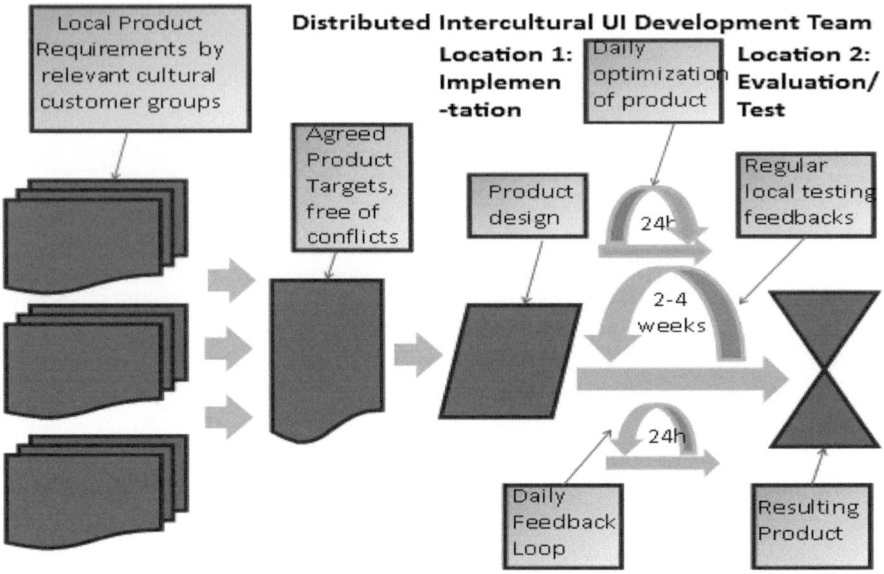

Fig. 6.8 Agile intercultural HCI design process (from Schoper and Heimgärtner 2013)

supplier processes regarding product release and product acceptance are executed by the product owner. Engineering processes such as requirements analysis, software design and software construction as well as software/system integration and testing are at the core of the agile methodology done by the cross-functional core team and managed by the product owner at the system level. Resource management, training and project infrastructure is an intrinsic demand on agile methods managed by the product owner, the scrum master and the stakeholders. Process improvement is continuously in focus in agile development prescribing the continuous monitoring of the learning curve and efficiency. Product evaluation and change management (including problem resolution management) is enhanced by reviewing and reprioritizing requirements and errors within the sprint planning.

6.5.4 Integration of Cultural Aspects in U-CD

We suggest Intercultural Usage-Centered Design (IU-CD) as a method for designing cross-cultural user interfaces, because this approach helps designers of cross-cultural user interfaces to get started, it provides structured guidance throughout the design process, it helps to retain insights in reusable models and thus to create better localized and internationalized systems.

6.5.4.1 Common Aspects for Intercultural UI Design

When looking at the cultural aspects we find a diversity of different aspects. Intercultural variables (Röse and Zühlke 2001), user interface characteristics (Marcus 2001), and HCI dimensions (Heimgärtner 2012) are all sets of culture specific rules and guidelines with more or less overlap that affect the visual and interaction design of a system. Since they are static (i.e., independent of the purpose of the system) and apply almost for all systems they are not specific for a certain system.

6.5.4.2 System Specific Cultural Aspects

This type of cultural aspects is genuine to the system to be designed. The aspects are usually a result of the user and task analysis and can affect user roles as well as task cases. For instance, one user role has different salient backgrounds in different countries and therefore requires different task structures in different countries. Another example would be that one and the same task has a radically different flow in different countries. An example for the latter one would be address input into a GPS navigation system in US and in China. In China the usual strategy to find a location is not to use the postal address but either a street intersection or using a point of interest nearby the desired location. Which method is used is even dependent on the part of China where the navigation system is used. Besides the different kinds of cultural aspects it is important to consider localization as well as internationalization within an intercultural UI design process.

6.5.4.3 Cultural Model

Hoft (1996) developed a cultural model (iceberg metaphor) relevant for international user interface design. Marcus (2000) related user interface characteristics to Hofstede's cultural dimensions (as described before). Shi and Clemmensen (2007) generated a relationship model in cultural usability testing. Kappos and Rivard (2008) postulated a three-perspective model of culture, information systems and their development and use. Other authors based their approach on cultural models to study the relationship between culture and HCI. Shah et al. (2012) studied the influences of culture in global software engineering by thinking in terms of cultural models. Based on the results of Heimgärtner (2012), thoughts on a preliminary model of culturally influenced human computer interaction to cover cultural contexts in HCI design have been given encompassing the relationships between cultural and HCI dimensions (cf. Heimgärtner 2013). According to the results of an empirical study done by Heimgärtner (2007a), some of the correlations between the cultural dimensions and the HCI dimensions as well as their values were determined (cf. Heimgärtner 2010) leading to the concept of HCI style scores, which can be computed for the designated cultural group from the Hofstede's indices.

Table 6.5 HCI styles around the world

HCI style	Cultural characterization using Hofstede's indices	Score
Asian	PDI high, IDV low, MAS middle, UAI low, LTO high	90
Indian	PDI high, IDV middle, MAS middle, UAI middle, LTO middle	70
African	PDI high, IDV low, MAS middle, UAI middle, LTO low	60
Scandinavian	PDI low, IDV high, MAS low, UAI middle, LTO low	40
Slavic	PDI high, IDV middle, MAS middle, UAI high, LTO low	30
Angle-Saxon	PDI low, IDV high, MAS middle, UAI low, LTO low	20
German	PDI low, IDV middle, MAS high, UAI middle, LTO low	10

The HCI style score expresses the average degree of information density and frequency as well as interaction frequency and speed the members in the designated cultural group expect according to this model (cf. Heimgärtner 2013). The lower the normalized HCI style score (ranging from 0 to 100) the lower the expected amount of information and the lower the interaction frequency. The resulting HCI style scores permit the establishment of clusters of countries that have similar HCI scores. According to these cultural clusters identified in the HCI style score continuum it can be expected that these country clusters exhibit a similar HCI style because of their similar cultural characterization defined by PDI, IDV, MAS, UAI and LTO (cf. Table 6.5).

These taxonomic results partially resembles the findings of Galtung (1981) on "Saxonic," "Teutonic," "Gallic," and "Nipponic" styles. However, to generalize the postulated correlations many more studies with other cultural groups are required. To achieve this both the values of the cultural dimensions (using VSM) and the values of the HCI dimensions (such as pieces of presented information per minute, cf. Heimgärtner 2012) must be determined for every desired culture. This can be done for indigenous groups as well by exploiting the same use cases and test settings in the arbitrary cultural groups of interest. A test tool developed by Heimgärtner (2008) can be used to support this. However, until there are no other values for the cultural dimensions than Hofstede's at the national level, those must be used to test the model. In addition, to further confirm findings, factor analysis can be applied to statistically calculate the corresponding loadings to the HCI style by clustering Hofstede's indices according to their HCI style score. The findings should refine the currently assumed rules that describe the relationship between cultural imprint and HCI style of a group (with at least 20 members). The explanatory value of this descriptive model still must be worked out.

The ideas presented represent a reasonable step toward a model of culturally influenced HCI from which the areas of intercultural usability engineering and intercultural HCI design can profit as the model is developed and validated as well as enriched with goal-oriented recommendations for intercultural HCI design. The HCI style of different cultural groups can be compared using the HCI style score, which is computed from the values of cultural dimensions. The hypotheses the model is based on have for the most part been confirmed by the continuum of the HCI style scores. Using these values, the HCI designer can prognosticate the

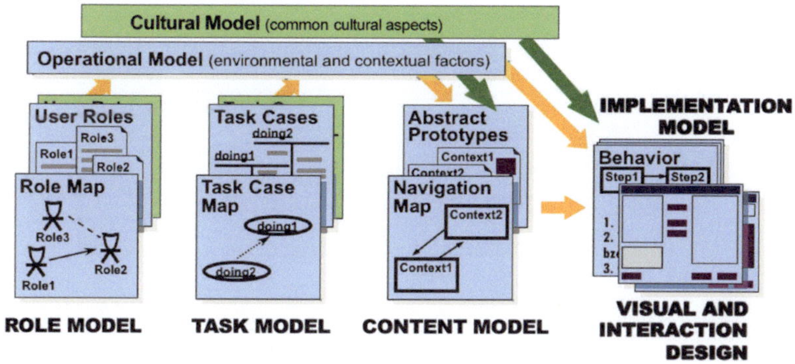

Fig. 6.9 Extended usage-centred design process by a cultural model

localization effort and expenditures. As the connections between HCI and culture will become clearer and more comprehensive, HCI design will also profit from an indigenous perspective. Even if some evidence and rules could be obtained for the core of the model of culturally influenced HCI, the final analysis of the intercultural HCI design process and its related cultural differences as well as recommendations for intercultural HCI design from an indigenous perspective are still outstanding. Nevertheless, the intended model constructed on basic physical dimensions has the power to build the foundation of an overall model for intercultural HCI design, which incorporates many more aspects than merely the relationship between cultural dimensions and HCI dimensions such as HCI design rules, recommendations, processes, and methods, which are applicable for cultural contexts in HCI in general.

To integrate the cultural aspects and the internationalization requirements in U-CD it is necessary to extend and adapt the existing process at different places. The common aspects for intercultural UI design will be included in a cultural model similar to the existing "operational model" (Fig. 6.9).

Since the content in the cultural model affects the visual and interaction design, this new model will affect the transitions from the task model to the content model and also from the content model to the implementation model. The cultural model captures the common rules for cultures the system will be designed for. For each culture one cultural model is used. Table 6.6 shows the qualitative content of the cultural model, which is mainly based on user interface characteristics (cf. Marcus 2007) and HCI dimensions (cf. Heimgärtner 2012) describing a user interface at abstract level filled with the special values for the desired cultures.

6.5.4.4 System Specific Cultural Aspects

The system specific cultural aspects affect the role model and the task model. For the role model and task model in localized systems it is not necessary to have

Table 6.6 Content of the cultural model

Culture
Appearance (Color, Layout, Font, Density…)
Navigation (Hierarchy, Menu Tree, Structure…)
Metaphors (Sound, Icons, Symbols…)
Mental models (Context, Cognitive Style, Metaphysics…)
Interaction (Speed, Frequency…)

explicit cultural extensions. However in internationalized systems to be deployed in different cultures it is necessary to implement different cultures on the level of the role model and task model. There can be culture dependent versions of the same user role and also of the task cases.

Role Model

The user role model is composed of user roles plus the user role map. To cover and include culture specific differences it is necessary to introduce culture specific user roles which will become part of a common user role map. An example could be an inventory control system for canteens and commercial kitchens which is deployed, e.g., in Europe and in Israel. In Israel there exists an additional user role to support the tasks of the Mashgiach who supervises the adherence of the food and ingredients to the Kashrut, the Jewish dietary laws. This means that for a system designed to be deployed in different cultures there can be additional user roles for one or more cultures.

User Roles

The extension to the existing user role model is the introduction of a culture identifier. The notation for the culture identifier is a rhombus with the 2 digit ISO Country Code. To cover also cultural aspects that are not part of the ISO Country Code such as religious denominations, indigenous groups or any other cultural target group (e.g. car driver, elderly people, indigenous groups, etc.), it is possible to introduce project specific culture identifiers (cf. Fig. 6.10).

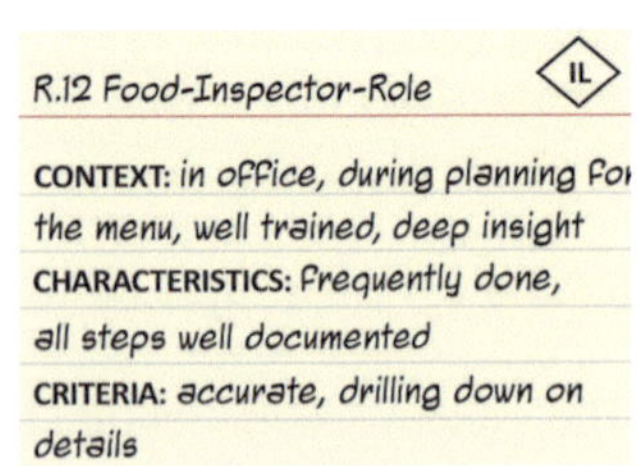

Fig. 6.10 Culture identifiers used as notation for culturally affected user roles

Fig. 6.11 Notation for a culture identifier expressing validity for all cultures except China

The culture identifier is added according to the following rules:

- In its simplest form user roles are described by the context within they are played, the characteristic of performance, and the criteria for support. A culture identifier in the upper right corner complements this description.
- If applicable it is possible that one role is shared between two or more cultures. In this case, the role gets multiple culture identifiers.
- User roles that are common to all cultures remain without culture identifier.

It is useful to introduce an inverse culture identifier to be able to exclude certain culture groups from one user role. An example would be a "Facebook Poster" role that does not exist in China due to legal restrictions. The inverse/not culture identifier is shown by adding a circle to the left corner of the rhombus (Fig. 6.11).

User Role Map

The user role map in general remains unchanged. There is still one user role map for one system; otherwise at this point the design process would split in the design of several different cultures specific systems. Nevertheless the culture specific roles keep the culture identifier and thus still can be identified.

Task Model

Task Cases

Task cases in their basic form are defined by a structured narrative in user's and domain language as two-column abstract dialog representing the user intention and the system responsibility (cf. Fig. 6.12, cf. also Constantine and Lockwood 1999 for details).

In this basic form culture dependency is already included in the abstract dialog, which itself expresses the abstract interaction specifically for a specific culture group. The only change is again the addition of a cultural identifier to be able to track, structure, and organize the task cases according to the following rules (cf. Fig. 6.13):

- Task cases are derived from the user roles, therefore tasks derived from a culture specific role are also culture specific and inherit the same culture identifier as the user role from which they are derived.
- From one user role may different culture specific task cases be derived. These task cases get their culture identifier when they are created.
- It is also possible that a task case can be shared between user roles from different cultures and being marked with multiple culture identifiers.
- Task cases common to all cultures remain without culture identifier.

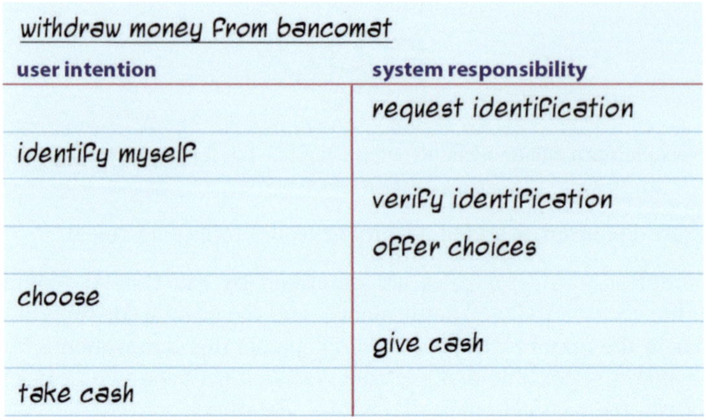

Fig. 6.12 Basic form of task cases

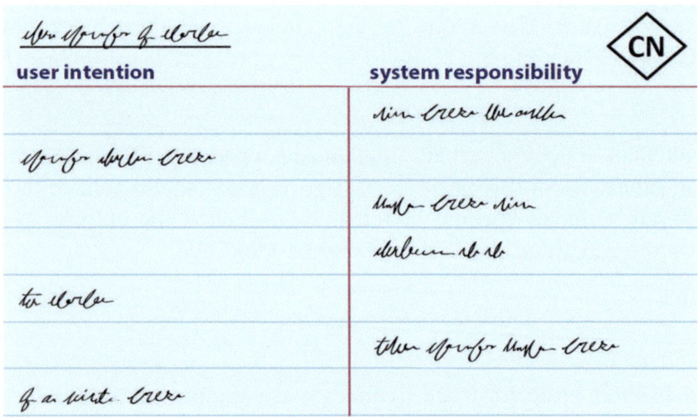

Fig. 6.13 Basic form of task cases indexed by a "cultural identifier"

- Task cases, which are common but exclude one or more culture groups, get the inverse culture identifier for the affected culture(s).

Task Case Map

The task case map depicts the relationship between task cases in a system to guide content organization in the user interface. Thus all task cases including the culture dependent ones are shown in the task case map together, although some of them may be mutually exclusive due to a cultural setting. In the task case map the task cases are shown by their name plus none, one or more culture identifiers (cf. Fig. 6.14).

Fig. 6.14 Task cases with different "cultural identifiers"

Content Model

For consistency, completeness, and traceability it is possible but not necessary to use the cultural identifier also in the abstract prototypes and the navigation map. In both depictions the cultural identifier is used to mark up the contexts for specific cultures.

6.5.5 *Integrating ASPICE*

Several changes from ASPICE 2.5 to ASPICE 3.0 norm should be considered in an integrated process that also have implications for the kind of the designated integrated process. For example, the changes in MAN.3 (project management) include that the base practice "establishing and implementing the project plan" has been deleted as it caused confusion in the past. Instead all aspects of planning have to be identified, monitored and adjusted (estimates, activities, schedules, plans, interfaces and commitments). Across all artifacts consistency has to be established (specific BP)—no traceability, just consistency (cf. Table 6.7).

Scope of work used to contain check of feasibility. In ASPICE 3.0 a specific BP for feasibility has been introduced (BP3). The project plan 08-12 is still an output

Table 6.7 Outcomes and BPs for project management in ASPICE 3.0 in contrast to ASPICE 2.5 (red) (from Bella et al. 2015)

Outcomes V3.0	
As a result of successful implementation of this process	
1	the scope of the work for the project is defined;
2	the feasibility of achieving the goals of the project with available resources and constraints is evaluated;
3	the activities and resources necessary to complete the work are sized and estimated;
4	interfaces within the project, and with other projects and organizational units, are identified and monitored;
5	plans for the execution of the project are developed, implemented and maintained;
6	progress of the project is monitored and reported; and
7	corrective action is taken when project goals are not achieved, and recurrence of problems identified in the project is prevented.

Base Practices V3.0	
1	Define the scope of work.
2	Define project life cycle
3	Evaluate feasibility of the project
4	Define, monitor and adjust project activities
5	Determine, monitor und adjust project estimates and resources
6	Ensure required skills, knowledge, and experience
7	Identify, monitor and adjust project interfaces and agreed commitments
8	Define, monitor and adjust project schedule
9	Ensure consistency
10	Review and report progress of the project

Table 6.8 Outcomes and BPs for quality assurance in ASPICE 3.0 in contrast to ASPICE 2.5 (red) (from Bella et al. 2015)

Outcomes V3.0	
As a result of successful implementation of this process	
1	a strategy for performing quality assurance is developed, implemented, and maintained;
2	quality assurance is performed independently and objectively without conflicts of interest;
3	non-conformances of work products, processes, and process activities with relevant requirements are identified, recorded, communicated to the relevant parties, tracked, resolved, and further prevented;
4	conformance of work products, processes and activities with relevant requirements is verified, documented, and communicated to the relevant parties;
5	authority to escalate non-conformances to appropriate levels of management is established; and
6	management ensures that escalated non-conformances are resolved

Base Practices V3.0	
1	Develop project quality assurance strategy
2	Assure quality of work products
3	Assure quality of process activities
4	Summarize and communicate quality assurance activities and results
5	Ensure resolution of non-conformances
6	Implement an escalation mechanism

work product of MAN.3. The references to risk management were removed, but it is mentioned in the BP's.

The changes in SUP.1 (quality assurance) overall simplify the process with similar content such as that on top of independence in QA objectivity is required and that it is clarified that escalation has to lead to management attention and actions (cf. Table 6.8). The changes from ASPICE 2.5 to ASPICE 3.0 also affected the process outcome: traceability and consistency have been formerly addressed by one single base practice on the right side of the V-model and have now been split into two base practices. Traceability refers to the existence of references or links between work products as well as supports coverage analysis, impact analysis, requirements implementation status tracking etc. Consistency means that all traceability references/links are available (i.e., nothing is missing) and that all traceability references/links are correct (i.e., no links to wrong work products). Consistency has to be proven by technical review of the traceability. Furthermore, new traceability requirements have been added between test cases and test results as well as between change requests and work products affected by these change requests (SUP.10). Finally, both terms "Strategy" and "Plan" are commonly used across the following processes of the ASPICE 3.0 process assessment model:

- SYS.4 System Integration and Integration Test
- SYS.5 System Qualification Test
- SWE.4 Software Unit Verification
- SWE.5 Software Integration and Integration Test
- SWE.6 Software Qualification Test
- SUP.1 Quality Assurance
- SUP.8 Configuration Management
- SUP.9 Problem Resolution Management
- SUP.10 Change Request Management.

This indicates that iNTACS enforces clear and distinct terminology, which is one of the most important aspects of a high quality process. This is illustrated by the relationship between "Strategy" and "Plan." At capability level 1, each of these processes requires the development of a process-specific strategy. The strategy always corresponds to a process-specific "Plan." For each process-specific "Plan" there are process-specific work product characteristics defined (e.g., "08-52 Test Plan," "08-04 Configuration Management Plan"). Furthermore, scheduling like, e.g., old SUP.10 BP10 has been moved to level 2. At capability level 2 or higher, each process-specific "Plan" (WP 08-nn) inherits the work product characteristics represented by the Generic Plan (WP 08-00). This means that for a process-specific "Plan" both the process-specific characteristics (WP 08-nn) and the generic characteristics (WP 08-00) apply. In addition, BPs for proceeding have been deleted.

The reduction of the number of the base practices of SUP.1 from 10 (in ASPICE 2.5) to 6 (in ASPICE 3.0) also reflects that the board members of iNTACS take into account lean approaches as well as agile principles such as the agile manifest (cf. http://agilemanifesto.org/, last access 9|14|2016). Former research by the authors as well as the outcome from several Gate4Spice events by iNTACS support the fact that agile processes cover ASPICE requirements at least up to level 3. Thereby, all relevant work products and base as well as generic practices necessary to reach the desired level must be integrated in the overall mentioned process. Process related documentation and configuration management must be done only to the extent that it satisfies the fulfillment of a desired maturity/capability level. If we map ASPICE (cf. van Loon 2007) to agile methods (such as SCRUM) in intercultural HCI design projects in the automotive context, some of the generic practices of the Process Attributes 3.1 and 3.2 can be mostly ("largely") or fully ("fully") achieved. However, to fulfill all process attributes up to ASPICE level 3 in all process areas, extensive additional documentation as well as intelligent extension of existing SCRUM templates would be necessary (e.g., to achieve bidirectional traceability at level 1). At least the ASPICE criteria (HIS scope) for the systems engineering process group (SYS.2–SYS.5), the software engineering process group (SWE.1–SWE.6) as well as the support processes SUP.1 and SUP.8–SUP.10 must be taken into consideration at creating an accepted development process for the automotive context. Project management (MAN.3) as a process area must be considered in any case as well as supplier management (ACQ.4) in case of subcontractors.

The retrospective reviews of the mentioned projects above also revealed, however, that the version and change management system used was not directly compatible (due to its complexity and inflexibility) with the SCRUM methodology being deployed. Therefore, more flexible configuration management system must be used in agile development (e.g., GIT). Nevertheless, that the SCRUM methodology does work in a distributed development setup has been proved once again in this project. Hence, if the agile development processes are supported by some auxiliary processes, it is possible to achieve up to capability level 3 for most of the process areas according to the ASPICE HIS scope, which is very important for intercultural HCI design projects in the automotive context. The principles of the agile manifesto: individuals and interactions over processes and tools; working software over comprehensive documentation; customer collaboration over contract negotiation are applied at the same time and thereby staying compatible with the process capability model of ASPICE. From this point of view, it is reasonable to determine to what extent the implemented agile development process covers ASPICE process capability levels and that it is deployable in projects with high demand for certified quality.

6.6 Conclusion

On the basis of the existing HCI development process defined in ISO 9241-210 and based on experiences and best practices from intercultural projects in the automotive context, an extended Usage-Centered Design approach integrating standardized User-Centered Design process steps as well as cultural models has been suggested. Thereby, the resulting user interface design method complemented by cultural aspects ensures that new systems can be designed right from the beginning for one or more cultures while designers better accommodate the diverse global user requirements and respond faster to change. The extensions are designed to cover the influences of cultural aspects as well as the quality requirements stated by the ASPICE 3.0 norm for the automotive context. In addition, the suggested agile approach combined with just the most relevant documents to fulfill ASPICE requirements reduces the development time dramatically by fast feedback. Thereby, this approach makes it possible to systematically track and trace the culture specific requirements and design decisions for internationalized user interfaces in order to produce adequate intercultural interaction experiences for users of automotive user interfaces in cars. Even if this suggested overall integrated approach is preliminary and not elaborated enough and must still be exposed to reality for empirical proof in order to be commonly, successfully and long lastingly deployed in everyday projects, it represents a first and reasonable sketch toward a standardized systematic intercultural agile and APSICE oriented automotive UI/UX design process. The next step is to apply the proposed integrated process in practical intercultural HCI design projects to gather data for its empirical validation and, finally, to prove that the suggested lean process is effective, efficient and satisfying, i.e., usable successfully with fun.

References

Abdelnour-Nocera, J., Kurosu, M., Clemmensen, T., Bidwell, N., Vatrapu, R., Winschiers-Theophilus, H., et al. (2011). Re-framing HCI through local and indigenous perspectives. In *Paper Presented at the Lecture Notes in Computer Science, Berlin, New York*.

Balzert, H. (2008). Lehrbuch der Software-Technik. [2], Software management. Heidelberg [u.a.]: Spektrum, Akad. Verl.

Beck, E. E. (2001). On participatory design in scandinavian computing research.

Bella, F., Hoermann, K., Vanamali, B., Herrmann, S., & Müller, M. (2015). Automotive SPICE 3.0 what is new and what has changed? In Kugler Maag CIE GmbH.

Bevan, N. (1999). Quality in use: meeting user needs for quality. *Journal of System and Software* (Serco usability services). *49*(1), 89–96. doi:10.1016/S0164-1212(99)00070-9.

Bevan, N. (2001). International standards for HCI and usability. *International Journal of Human-Computer Studies International Journal of Human-Computer Studies, 55*(4), 533–552.

Binder, J. (2007). *Global project management: Communication*. Collaboration and Management Across Borders: Gower.

Blankl, M., Biersack, P., & Heimgärtner, R. (2013). Lessons learned from projects in Japan and Korea relevant for intercultural HCI development. In A. Marcus (Ed.), *Design, User Experience, and Usability. Health, Learning, Playing, Cultural, and Cross-Cultural User Experience* (Vol. 8013, pp. 20–27). Heidelberg: Springer.

Carmel, E. (1999). *Global software teams: collaborating across borders and time zones*. Prentice Hall PTR.

Clemmensen, T., & Goyal, S. (2005). *Cross cultural usability testing: The relationship between evaluator and test user*. Copenhagen Business School, C.B.S. Institut for Informatik I. N.F. Department of Informatics I.N.F. http://openarchive.cbs.dk/cbsweb/handle/10398/6474.

Cockburn, A., & Highsmith, J. (2001). Agile software development: The people factor. *Computer, 34*(11), 131–133. doi:10.1109/2.963450.

Constantine, L. (2003). Canonical abstract prototypes for abstract visual and interaction design. interactive systems. *Design, Specification, and Verification, 2844*, 1–15. doi:10.1007/978-3-540-39929-2_1.

Constantine, L. (2004). Beyond user-centered design and user experience: Designing for user performance. *Cutter IT Journal, 17*(2), 16–25.

Constantine, L. L, & Lockwood, L. A. D. (1999). *Software for use: A practical guide to the models and methods of usage-centered design*. Reading, Mass.: Addison Wesley [u.a.].

Constantine, L., & Windl, H. (2009). Safety, speed, and style: interaction design of an in-vehicle user interface. In *Paper Presented at the CHI '09 Extended Abstracts on Human Factors in Computing Systems, Boston, MA, USA*.

Crossley, L. (2003). Building emotions in design. *The Design Journal, 6*(3), 35–45. doi:10.2752/146069203789355264.

Curtis, B., & Hefley, B. (1994). A wimp no more: the maturing of user interface engineering. *Interactions, 1*(1), 22–34. doi:10.1145/174800.174803.

DIN. (1999). *DIN EN ISO 9241,11 Ergonomische Anforderungen für Bürotätigkeiten mit Bildschirmgeräten Teil 11, Anforderungen an die Gebrauchstauglichkeit: Leitsätze = Guidance on usability = Lignes directrices concernant lutilisabilité* (Vol. DIN, Teil 11.). Berlin: Beuth.

DIN. (2010). *DIN EN ISO 9241-210 Ergonomische Anforderungen der Mensch-System-Interaktion Teil 210: Prozess zur Gestaltung gebrauchstauglicher Systeme*. Berlin: BeuthVerlag.

Duxu, & Marcus, A. (2011). Design, user experience, and usability theory, methods, tools and practice. In *First International Conference, DUXU 2011, held as part of HCI International 2011, Orlando, FL, USA, July 9–14, 2011, Proceedings. Part II*. doi:10.1007/978-3-642-21708-1.

Ferr, X., Juristo, N., Windl, H., & Constantine, L. (2001). Usability basics for software developers. *IEEE Software, 18*(1), 22–29. doi:10.1109/52.903160.

Galtung, J. (1981). Structure, culture, and intellectual style: An essay comparing saxonic, teutonic, gallic and nipponic approaches. *Social Science Information, 20*(6), 817.

Gould, E., & Marcus, A. (2011). Company culture audit to improve development team's collaboration, communication, and cooperation.

Gundelsweiler, F., Memmel, T., & Reiterer, H. (2004). Agile usability engineering. In R. Keil-Slawik, Selke, H. & Szwillus, G. (Eds.), *Mensch & computer 2004*. Allgegenwärtige Interaktion (pp. 33–42). München: Oldenbourg Verlag.

Hall, E. T. (2006). The silent language.

Heimgärtner, R. (2007a). Cultural differences in human computer interaction: results from two online surveys. In A. Oßwald (Ed.), *Open innovation* (Vol. 46, pp. 145–158). Konstanz: UVK.
Heimgärtner, R. (2007b). *Towards cultural adaptability in driver information and assistance systems* (Vol. 4560, pp. 372–381).
Heimgärtner, R. (2008). A tool for getting cultural differences in HCI. In K. Asai (Ed.), *Human computer interaction: New developments* (pp. 343–368). Rijeka: InTech.
Heimgärtner, R. (2010). Cultural differences in human-computer interaction—Towards culturally adaptive human-machine interaction. Ph.D. Dissertation, Universitätsbibliothek der Universität Regensburg.
Heimgärtner, R. (2012). Cultural differences in human-computer interaction (Paperback B: Einband—flex. (Paperback) (Vol. 1). Oldenbourg Verlag.
Heimgärtner, R. (2013). Reflections on a model of culturally influenced human computer interaction to cover cultural contexts in HCI design. *International Journal of Human-Computer Interaction*.
Heimgärtner, R. (2014). ISO 9241–210 and culture?—The impact of culture on the standard usability engineering process. In A. Marcus (Ed.), *Design, User Experience, and Usability. User Experience Design Practice: Third International Conference, DUXU 2014, Held as Part of HCI International 2014, Heraklion, Crete, Greece, June 22–27, 2014, Proceedings, Part IV* (pp. 39–48). Cham: Springer.
Heimgärtner, R., & Solanki, A. (2014). Using agile methods in intercultural HCI design projects. In A. Marcus (Ed.), *Design, User Experience, and Usability. Theories, Methods, and Tools for Designing the User Experience: Third International Conference, DUXU 2014, Held as Part of HCI International 2014, Heraklion, Crete, Greece, June 22–27, 2014, Proceedings, Part I* (pp. 123-129). Cham: Springer.
Heimgärtner, R., Tiede, L.-W., & Windl, H. (2011). Empathy as key factor for successful intercultural HCI design. In *Paper presented at the Proceedings of 14th International Conference on Human-Computer Interaction, Orlando*.
Heimgärtner, R., Windl, H., & Solanki, A. (2011). The necessity of personal freedom to increase HCI design quality. In *Paper Presented at the 9th International Conference on Human-Computer Interaction, Orlando, Florida, USA*.
Hermeking, M. (2001). *Kulturen und Technik*. München: Waxmann.
Highsmith, J., & Cockburn, A. (2001). Agile software development: The business of innovation. *Computer, 34*(9), 120–122. doi:10.1109/2.947100.
Hoffmann, H.-E., Schoper, Y.-G., & Fitzsimons, C.-J. (2004). Internationales Projektmanagement: Interkulturelle Zusammenarbeit in der Praxis (Orig.-Ausg. ed. Vol. 50883: Beck-Wirtschaftsberater; 50883). München: Dt. Taschenbuch-Verl.
Hofstede, G. H., Hofstede, G. J., & Minkov, M. (2010). *Cultures and organizations: Software of the mind* (3rd ed.). Maidenhead: McGraw-Hill.
Hofstede, G., & Hofstede, G. J. (2010). *Cultures and organizations: Software of the mind*. New York, NY [u.a.]: McGraw-Hill.
Hoft, N. L. (1996). Developing a cultural model. In E. M. Del Galdo & J. Nielsen (Eds.), *International users interface* (pp. 41–73). Wiley.
Honold, P. (1999). Cross-cultural usability engineering: development and state of the art. In *Paper Presented at the Proceedings of HCI International (The 8th International Conference on Human-Computer Interaction) on Human-Computer Interaction: Ergonomics and User Interfaces-Volume I* (Vol. I). http://dl.acm.org/citation.cfm?id=647943.742510&coll=DL&dl=GUIDE&CFID=145069547&CFTOKEN=47400189.
Honold, P. (2000a). Culture and context: An empirical study for the development of a framework for the elicitation of cultural influence in product usage. *International Journal of Human-Computer Interaction, 12*(3/4), 327–345.
Honold, P. (2000b). Interkulturelles usability engineering: Eine Untersuchung zu kulturellen Einflüssen auf die Gestaltung und Nutzung technischer Produkte (Als Ms. gedr. ed. Vol. 647). Düsseldorf: VDI Verl.

International, Dr. (2003). *Developing international software* (2nd ed.). Redmond, Wash.: Microsoft Press.
Kappos, A., & Rivard, S. (2008). A three-perspective model of culture, information systems, and their development and use. *MIS Quarterly, 32*(3), 601–634.
Kochan, T., Bezrukova, K., Ely, R., Jackson, S., Joshi, A., Jehn, K., et al. (2003). The effects of diversity on business performance: Report of the diversity research network. *Human Resource Management, 42*(1), 3–21. doi:10.1002/hrm.10061.
Liang, S. F. M. (2003). Cross-cultural issues in interactive systems. In *Paper Presented at the Proceedings of the International Ergonomics. Ergonomics in the Digital Age.*
Maguire, M. (2001). Context of use within usability activities. *International Journal of Human-Computing Studies, 55*(4), 453–483. doi:10.1006/ijhc.2001.0486.
Marcus, A. (2001). Cross-cultural user-interface design. In M. J. Smith & Salvendy, G. (Eds.), *Proceedings, Human-Computer Interface International (HCII) Conference, USA, LA* (Vol. 2, pp. 502–505). New Orleans: Lawrence Erlbaum Associates.
Marcus, A. (2006). Cross-cultural user-experience design diagrammatic representation and inference (pp. 16–24).
Marcus, A. (2007). Global/intercultural user-interface design. In J. Jacko & A. Spears (Eds.), *The human-computer interaction handbook.* Mahwah: Lawrence Erlbaum Associates.
Marcus, A., & Baumgartner, V.-J. (2004). Mapping user-interface design components vs. culture dimensions in corporate websites. *Visible Language Journal MIT Press, 38*(1), 1–65.
Marcus, A., & Gould, E. W. (2000). Cultural dimensions and global web user-interface design: What? So What? Now What?. http://www.amanda.com.
Pressman, R. S. (1992). *Software engineering: A practitioner's approach.* New York, NY, USA: McGraw-Hill.
Rätzmann, M. (2004). *Software-testing and internationalisierung (2* (Aufl ed.). Bonn: Galileo Press.
Ressin, M., Abdelnour-Nocera, J., & Smith, A. (2011a). Defects and agility: Localization issues in agile development projects. In *Paper Presented at the International Conference on Agile Processes in Software Engineering, Extreme Programming, X.P.*
Ressin, M., Abdelnour-Nocera, J., & Smith, A. (2011b). Lost in agility? Approaching software localization in agile software development. In: *Paper Presented at the International Conference on Agile Processes in Software, Engineering, Extreme Programming, X.P.*
Röse, K. (2002). *Methodik zur gestaltung interkultureller mensch-maschine-systeme in der produktionstechnik* (Vol. 5). Kaiserslautern: Univ.
Röse, K. (2005). The development of culture-oriented human machine systems: specification, analysis and integration of relevant intercultural variables. In M. Kaplan (Ed.), *Cultural ergonomics, advances in human performance and cognitive engineering research* (4 ed.). Netherlands: Elsevier.
Röse, K., & Zühlke, D. (2001). Culture-oriented design: Developers' knowledge gaps in this area. In: *Paper Presented at the 8th IFAC/IFIPS/IFORS/IEA Symposium on Analysis, Design, and Evaluation of Human-Machine Systems, September 18–20, 2001, Kassel, Germany.*
Schmitz, K.-D., & Wahle, K. (2000). *Softwarelokalisierung*. Tübingen: Stauffenburg-Verl.
Schoper, Y., & Heimgärtner, R. (2013). Lessons from intercultural project management for the intercultural HCI design process. In A. Marcus (Ed.), *Design, user experience, and usability. Health, learning, playing, cultural, and cross-cultural user experience* (Vol. 8013, pp. 95–104). Berlin: Springer.
Seffah, A., Gulliksen, J., & Desmarais, M. C. (2005). Human-centered software engineering—integrating usability in the software development lifecycle (Vol. 8). Berlin: Springer.
Sethi, R., Smith, D. C., & Park, C. W. (2001). Cross-functional product development teams, creativity, and the innovativeness of new consumer products. *Journal of Marketing Research, 38*(1), 73–85. doi:10.1509/jmkr.38.1.73.18833.
Shah, H., Nersessian, N. J., Harrold, M. J., & Newstetter, W. (2012). Studying the influence of culture in global software engineering: Thinking in terms of cultural models. In: *Paper*

Presented at the Proceedings of the 4th International Conference on Intercultural Collaboration, Bengaluru, India.

Shen, S.-T., Woolley, M., & Prior, S. (2006). Towards culture-centred design. *Interacting with Computers, 18*(4), 820–852. doi:10.1016/j.intcom.2005.11.014.

Shi, Q., & Clemmensen, T. (2007). Relationship model in cultural usability testing. In N. Aykin (Ed.), *Usability and internationalization. HCI and culture*. (Vol. 4559, pp. 422–431): Berlin: Springer.

Simonsen, J. (2013). *Routledge international handbook of participatory design*. London: Routledge.

Solanki, A., & Heimgärtner, R. (2013). Lessons learned during a HCI design process in intercultural context. In P. L. Patrick Rau (Ed.), *Cross-cultural design. Methods, practice, and case studies* (Vol. 8023, pp. 253–260). Berlin: Springer.

Thomas, A., Kinast, E.-U., & Schroll-Machl, S. (2010). *Handbook of intercultural communication and cooperation. Basics and areas of application*. Göttingen: Vandenhoeck & Ruprecht.

van Loon, H. (2007). Process assessment and ISO/IEC 15504: A reference book. Springer.

VDA. (2013). Automotive SPICE process assessment/reference model: VDA QMC working group 13/ automotive SIG.

VDMA. (2009). Software-Internationalisierung Leitfaden. Frankfurt a.M.: VDMA Fachverband Software.

Victor, D. A. (1998). *International business communication* (7th ed.). New York, NY: Harper Collins.

Vöhringer-Kuhnt, T. (2002). The influence of culture on usability. M.A. master thesis, Freie Universität Berlin.

Walsh, T., Nurkka, P., Koponen, T., Varsaluoma, J., Kujala, S., & Belt, S. (2011). Collecting cross-cultural user data with internationalized storyboard survey. In *Paper Presented at the Proceedings of the 23rd Australian Computer-Human Interaction Conference, Canberra, Australia*.

Windl, H. (2002). Designing a Winner. In *Paper Presented at the ForUSE 2002 1st International Conference on Usage-Centered Design.*

Windl, H., & Heimgärtner, R. (2013). Intercultural design for use—Extending usage-centered design by cultural aspects. In *Paper Presented at the HCII 2013, Las Vegas.*

Part III
Interaction Techniques and Technologies

Chapter 7
The Neglected Passenger—How Collaboration in the Car Fosters Driving Experience and Safety

Alexander Meschtscherjakov, Nicole Perterer, Sandra Trösterer, Alina Krischkowsky and Manfred Tscheligi

Abstract When designing interfaces for a vehicle, the focus often lies on the driver. Since the driver always has a primary task (i.e., maneuvering the vehicle safely), interfaces for secondary tasks (e.g., entertainment systems) are designed to reduce distraction threats to a minimum. However, it is not always only the driver who is interacting with the vehicle; passengers also interact with the car. They may support the driver in the primary task (e.g., by providing navigation information) or take over secondary tasks (e.g., operating the climate control) in order to unburden the driver. Thus, we see a need for interfaces that foster the communication and collaboration between the driver and passengers but also among passengers themselves. Currently, such interfaces are usually neglected in automotive user interface research. Over the last years, we have conducted several studies focusing on communication and collaboration between drivers and passengers inside cars. Following an experience-centered approach, we started with ethnographically informed studies to gain a deeper knowledge on how drivers and passengers interact with each other inside a vehicle. Based on these insights we conceptualized and designed several prototypes that enabled collaboration between drivers and passengers. These prototypes were then studied in different studies both in a simulator

A. Meschtscherjakov (✉) · N. Perterer · S. Trösterer · A. Krischkowsky · M. Tscheligi
Christian Doppler Laboratory "Contextual Interfaces", Center for Human-Computer Interaction, University of Salzburg, Salzburg, Austria
e-mail: alexander.meschtscherjakov@sbg.ac.at

N. Perterer
e-mail: nicole.perterer@sbg.ac.at

S. Trösterer
e-mail: sandra.trosterer@sbg.ac.at

A. Krischkowsky
e-mail: alina.krischkowsky@sbg.ac.at

M. Tscheligi
e-mail: manfred.tscheligi@sbg.ac.at

G. Meixner and C. Müller (eds.), *Automotive User Interfaces*,
Human–Computer Interaction Series, DOI 10.1007/978-3-319-49448-7_7

setup, as well as, in real-traffic situations. In this chapter, we describe five of these research activities in more detail and present implications for designing interaction approaches that foster collaboration in the vehicle.

7.1 Introduction

Most automotive user interfaces are targeted at the driver. They support drivers in their primary task of maneuvering a vehicle from one place to another by providing important information relevant to operate the vehicle. Advanced Driver Assistant Systems (ADAS) are interfaces that support the driver in this primary task. Some ADAS enable or actively support the driver in the stabilization of the vehicle (i.e., steering, accelerating, and breaking). They include electronic stability programs (ESP) or steering assist systems. Other systems help the driver to conduct maneuvers such as overtaking or parking. Such ADAS include adaptive cruise control (ACC) or lane departure warning systems (LDW), while others support the driver in higher order navigation tasks, such as deciding which route to take or the orientation along points-of-interests (POI).

When targeting secondary tasks such as operating in vehicle information or entertainment systems (IVIS), the interaction design must not distract drivers from their primary task. This is comprehensible since safety is the most important factor in automotive interaction design. With the advent of in-car applications and the need of people to be constantly connected and entertained, a plethora of in-car information, entertainment, and communication systems have entered modern cars. They range from weather apps and access to audio streaming services to hands-free kits and Facebook in the vehicle.

Since there is always a driver, ADAS and IVIS interfaces always need to be operable by the driver. Thus, interface elements such as buttons, knobs, touch screens, or gesture interfaces are cluttered around the driver's seat. This is logical, when we imagine a driver in a vehicle and with no passengers to help operating these systems. On the other hand, this focus on the driver neglects that often a driver also has passengers in the vehicle. These passengers—in terms of interaction design—are usually neglected. Only a few studies exist, which incorporate passengers in the actual driving process. This is astonishing since it is often the case that a passenger takes over many of the secondary tasks, such as changing a radio station, entering a navigation destination, or simply by helping the driver to monitor the scenery outside the vehicle. In these situations, communication and collaboration strategies are essential.

Setting aside the fact that there is a lack of interfaces especially designed for passengers, interaction systems for the driver are also often technology driven and do not consider knowledge we have from successful interactions between the driver and the passengers. We believe that it is fruitful for the design of driver interfaces to learn from effective collaboration strategies in a vehicle and apply this knowledge to the design of innovative driver interfaces. These new ADAS and IVIS may be regarded as co-driver for the driver rather than functions. Especially with the emergence of autonomous vehicles, such co-drivers may become even more relevant since driving will become a collaborative activity between a driver and the autonomous vehicle.

In this book chapter, we report on a body of studies we conducted over the last five years with the focus on communication and collaboration between drivers and passengers. We describe this design space within the car from a theoretical perspective including the spatial arrangement inside the vehicle. This includes several ethnographically informed studies, as well as user studies in the driving simulator. Based on these findings, we designed and prototyped different systems that foster collaboration between passengers and driver. We present these prototypes, along with user evaluations both in driving simulator experiments and studies on the road.

Our overall research vision is to thoroughly examine from different perspectives how communication and collaboration inside a vehicle is experienced by drivers and passengers. We target at understanding why, how, and when passengers act as co-drivers and what we can learn from collaboration strategies for the design of empathic driver user interfaces. We aspire to inform the design of passenger user interfaces that help them to support the driver and how in-car communication and collaboration can be enhanced. We aim at identifying challenges and possibilities how collaboration in future autonomous vehicles will be designed.

The rest of this chapter is structured as follows. In the next section, we provide a brief overview of related work when it comes to in-car communication and collaboration. Then, we present an overview of the different studies included in this chapter. Thereafter, we describe five research activities we undertook in more detail. We describe an *ethnographic study* of driver-passenger pairs in real traffic. In the section *co-navigator,* we present a navigation device for front-seat passengers. The *shared-gaze* section describes a system, which visualizes the gaze of the front-seat passenger for the driver. In the *active corners* section, we present a concept that maps the corners of a tablet to four seats inside the car and enables the sharing of information between driver and passengers. In the section *space and place in cars,* we report on a car mock-up study in which we used the active corners prototype to study the interplay of physical in-car properties (i.e., space) on emerging cooperative practices, by highlighting how the pure spatial properties of a car cabin can shape and be place for social encounters. Within the discussion section we combine findings and lessons learned and provide an outlook on collaboration in autonomous vehicles.

7.2 Related Work

Collaboration in the car has been researched in human–computer interaction (HCI) in various ways, though mainly through ethnography (e.g., Brown and Laurier 2012; Juhlin 2013; Leshed et al. 2008; Forlizzi et al. 2010). In particular, Oskar Juhlin's group designed a range of systems based on the findings from ethnographic studies. For instance, the Sound Pryer system by Östergren and Juhlin (2004) enabled a truly joint music listening between road users. Leshed et al. (2008) conducted an ethnographically informed study with the focus on the usage of GPS, showing evidence for practices of disengagement, as well as new opportunities for engagement with the environment.

Contemporary research focuses on how navigation devices and mobile technologies are used in practice through collaboration. Forlizzi et al. (2010) conducted a qualitative design study of navigation among three types of teams, i.e., parents and their teenage children, couples, and unacquainted individuals. Their results show that collaborative strategies often rely on shared knowledge and experiences between speakers and listeners, and were influenced by social roles and task role of the driver and co-driver. This means that teams less experienced in navigating together or less familiarity with the route were more explicit in giving instructions (e.g., turns and lane positions). Couples with more experience in collaborative navigation or familiarity with the route were less explicit when providing information. Based on an analysis of video-recordings of various episodes of car travels, Laurier et al. (2008) have documented what happens during a car journey. Their key argument is that if we want to design better GPS systems, we need to move beyond the notion of a docile driver, who follows GPS commands blindly, to a better understanding of how drivers, passengers, and GPS systems work together. Cycil et al. (2013) examined routine family car journeys and examined how passengers assistance during a mobile telephone call influences the competing demands of handling the vehicle while being engaged in co-located and remote conversations. Furthermore, Gärtner et al. (2014) documented that drivers and passengers often experience problems in effectively communicating with one another.

These studies show that understanding in-vehicle collaboration is crucial for the design of many in-car user interfaces. They show that driving is often a social activity and highlight different aspects of collaboration strategies. Inspired by these research activities we have conducted a series of studies that investigates in-car collaboration from an experience-centered perspective. We have built several prototypes and studied them both in the lab and in the field.

7.3 Research Activities

In the following section, we present five research activities with the focus on in-car communication and collaboration.

With the *car-sharing ethnography,* we aspired to identify why, how, and when passengers support drivers in both primary and secondary tasks. Our approach was to conduct a qualitative in situ study using a car-sharing platform to recruit participants. We wanted to be able to observe communication and collaboration in a natural setting and applied an ethnographically informed approach by participating in the shared ride. This enabled us to experience the interplay between passengers and the driver in different situations.

With the *co-navigator* approach, we provided the front-seat passenger with a tool to support the driver especially in demand situations, such as navigation in confusing situations or to warn about potholes. We prototyped an enhanced navigation application especially designed to be used by the front-seat passenger. The tablet app visualized navigation information and various demand situations in different ways. By means of a real-traffic study, we could show the feasibility and usefulness of our approach and drive insights on the design of such a system.

With the *shared gaze* approach, we enabled the front-seat passenger to point at objects in the environment by visualizing his/her gaze for the driver and thus supporting the driver in a direct way. In two simulator studies, we studied the usefulness and distraction risks of our approach. In the first study, the gaze was visualized as an overlay (i.e., as a dot) in the driving simulation projection. On real roads, this could be achieved through head-up displays (HUDs) covering the full windscreen. In the second study, we further used an LED stripe mounted at the bottom of the car-mock-up windshield to indicate the direction of the co-driver's gaze and compared it to dot visualization.

Active corners is an interaction design that enables the exchange of items between the driver and front-seat passenger but also among the other passengers by mapping the four corners of a tablet to the seats inside a car. This approach also enables the driver to remotely control rear-seat entertainment screens. For the active corners approach, we prototyped a simple card game and studied the prototype in four studies.

The active corners card game prototype was also utilized in an exploratory study with a car mock-up to research how the spatial properties of a car cabin (e.g., seat positions) shape the emergence of cooperative practices among all passengers in the car. This study is reported in the section *space and place in cars*.

Table 7.1 provides an overview of each research activity including its goal, approach, and applied methods. We describe each research activity in more detail along with study settings and results in the following sections.

7.3.1 Car-Sharing Ethnography

The first study was a two-month ethnographic study of nine driver-passenger pairs recruited from two popular European online car-sharing portals. The goal was to develop a deeper understanding of human assistance, and collaboration in general while driving. Of special interest was how in-car technology would influence the interaction between drivers and passengers. Additionally, we aimed at gathering information on how non-car technology (e.g., the mobile phone) is used for collaboration while traveling. The findings from our approach build the basis for future research on collaborative design by means of driver assistance regarding safety issues and experiences. A more detailed description of the study and findings can be found in Perterer et al. (2013).

Table 7.1 Overview of five research activities presented in this chapter including the name of the activity, its overall goals, our approaches to reach the goal, as well as applied methods and studies

Research activity	Goal	Approach	Applied methods
Car-sharing ethnography	Identify why, how, and when passengers support drivers in primary or secondary tasks	Qualitative in situ study using a car sharing platform for participant recruitment	Ethnography in nine vehicles with 18 participants
Co-navigator	Enable front-seat passenger to support driver in demand situations	Enhanced navigation application for front-seat passenger visualizing demand situations	Real-traffic study with ten driver-front-seat passenger pairs utilizing an experience prototype
Shared gaze	Enable front-seat passengers to show the driver points-of-interest in the environment	Visualizing the gaze of the front-seat passenger using eye-tracking and a dot or LED indicator	Two simulator studies with two different visualizations using an experience prototype
Active corners	Enable exchange of items among driver and all passenger and remotely control in-car content	Application that maps the four corners of a tablet to the seats in a car	Four studies with different user groups in standing and driving vehicle using an experience prototype
Space and place in cars	Identify how different seat positions affect social practices when collaborating in a vehicle	Using the active corners prototype to trigger collaboration in a car mock-up for observation and evaluate perceived experience	Exploratory lab-study with a car mock-up involving 56 participants

Car-Sharing Study

The study took place in the winter, lasting from January to March. Nine driver-passenger pairs were recruited either from the online car-sharing community *Compano* (http://www.campano.at) or *Mitfahrgelegenheit* (http://www.mitfahrgelegenheit.at). On the two online platforms, car owners can register to take people with them as passengers in their car and non-car owners can register to take a ride with someone going in the same direction. The varied nature of the car-sharing community enabled us to observe collaboration on various topics and between people with different relationships. To investigate human assistance and the relevant influences, a researcher joined the participants by sitting in the backseat. Thereby, the researcher observed and conversed with the driver-passenger pair. Paper and pencil were used to take notes of the situations. Our aim was to live the experiences and to take part in the trip as a researcher while minimizing the social distance from the participants. Figure 7.1 visualizes a typical situation.

Participants were between 20-and 32-years old (27.9 on average) with seven male and two female drivers. On two of the trips, there was also an additional backseat passenger (not including the researcher). One of the drivers was testing the car-sharing platform for the first time; the others were regular members. All were familiar with taking along other people in their car. Three of the nine driver-passenger

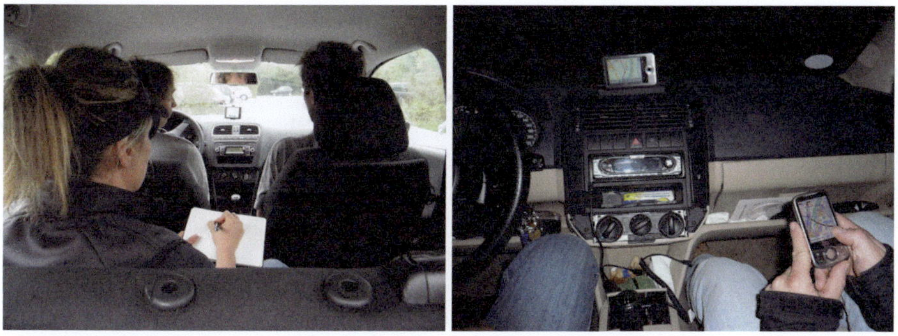

Fig. 7.1 Researcher sitting at the rear seat and taking notes while driver and co-driver are talking to each other (*left*). Co-driver using multiple devices to support the driver in a navigational task during the night (*right*)

pairs were partners. Two of them were either friends or work colleagues. The remaining pairs (four) met each other through the car-sharing community. The drivers used various make of cars such as Audi A3, BMW 1, Peugeot 207, Volkswagen Polo, or Mercedes Vito. Four out of nine drivers used a mobile navigation device, but additionally switched on the radio to hear either music or traffic news.

In total we collected data from nine journeys, lasting in sum over 21 h 30 min. Journey times ranged from 1 h 30 min to 3 h 30 min; the average trip duration took 2 h 23 min. The observed episodes collected in this study were analyzed using an interaction analysis approach (Heath and Luff 2000).

We found that different communication strategies, such as *no*, *delayed*, *frequent and spry* communication were used by the front-seat passenger to support the driver. These various forms require a *common ground* of the context and coordinated actions (i.e., stopping the discussion triggered by a contextual cue). *No communication* was used when front-seat passengers overtook any larger sized vehicle. The *delay* form could be identified if there was a truck and other road users in front of, or behind the car. This identified communication behavior seemed to be dependent on the distance from the moving larger sized vehicle: the closer the car came to the vehicle, the more delayed and shorter the answers from the front-seat passenger were. A *frequent and spry communication* was mainly shown when there were specific circumstances, such as the danger of black ice or snow-covered roads.

As our data shows the process of navigating from A to B was a highly collaborative activity. Especially at night, navigation systems mounted on the windshield are not always sufficient for navigation in unfamiliar areas. In our observations, we could see how the driver used the portable navigation unit (mounted on the center stack) for detailed information, whereas the front seat passenger simultaneously aided him using his smartphone with an overview picture. Customized navigation devices can potentially support this collaboration by providing different information for the driver and the passenger.

7.3.2 *Co-navigator*

Following our ethnographic results from the first study, we developed a tablet-based navigation app, especially designed for the front-seat passenger, in order support the driver in navigation and driving-related tasks. The design of the app was motivated by Wilfinger et al. (2013) who suggested that front-seat passengers could use display solutions that include a much higher information load while still beneficial for collaboration (i.e., for a collaborative navigation task). Therefore, we used sketches as design prototyping technique. During two iterative design circles with two designers and two engineers, we decided to implement the "Co-Navigator" application. It consists of turn-by-turn instructions, a map overview of the route and the visualization of demand situations in form of a bar plot and images. Compared to conventional navigation systems, the prototype provides more details about the route as well as foresighted information that could be communicated (via the front-seat passenger) to the driver. This is achieved by providing diverse landmarks that refer to specific locations and various demanding situations in advance (e.g., a construction site, a pothole, or a narrow road).

Our aim was to investigate how the prototype is used and how it supports the shared navigation interactions. A detailed description of our prototype is described in Perterer et al. (2015). Figure 7.2 shows the co-navigator application as well as a typical demand situation in which we imagine the co-driver to support the driver.

Co-navigator Study

Contrary to experimental studies of GPS navigation in the laboratory (e.g., Medenica et al. 2011), we decided to go into the field. We observed and interviewed ten collaborative in-car navigation device users (i.e., driver-passenger pairs) while using the co-navigator prototype.

We selected a specific route through Salzburg, which was characterized by a number of different maneuvers (e.g., turning left or right, passing through cross walks or road construction). Participants were asked to drive a predefined route

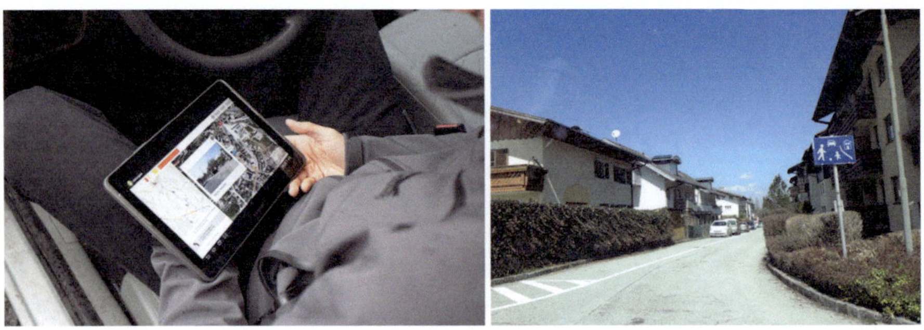

Fig. 7.2 Co-navigator application inspected by the driver (*left*). Demand situation since the car enters a narrow road and a sign indicates a "living street" (*right*)

through Salzburg and its surroundings with the help of the Co-Navigator. During the study the driver-passenger pairs should collaborate in order to successfully reach the destinations. We mounted a customary navigation system into the car, which was mounted on the windshield and helps the driver-passenger pair to find their destination. When the destination was reached, the group took a break and the researcher conducted a brief semi-structured interview.

All participants had experience in using a navigation system while driving as well as were familiar with using Google Maps. Participants were aged between 18 and 39 years, and consisted of six male and four female drivers (M = 26; SD = 4.6), respectively six female and four male front-seat passengers (M = 25.5; SD = 5.6). We chose couples since, according to Forlizzi et al. (2010), they are experienced in navigating together and show more efficient strategies for exchanging information during the navigation task than unacquainted individuals. One couple that drove the predefined route regularly was excluded from our data.

Overall, we collected 6 h 29 min video material. Journey times ranged from 25 to 45 min. The data analysis consists of a combination of individual and group analysis sessions based on the video-based interaction analysis method of Jordan and Henderson (1995).

Our data revealed that the map overview that shows the current location and the driving direction at any time with a red arrow in order to locate the current position while driving supported the co-drivers (5 out of 9) in getting an overview or planning ahead. Especially, at the start of the journeys, co-drivers (n = 6) used the hybrid view (satellite image with overlapped map image) to get a quick overview whether they are familiar with the route or not. In those specific case, the co-drivers moved the window until the first important turn came into view. Furthermore, we could observer a difference between urban and rural environments. In urban traffic, almost half of the co-drivers swiped forward until the next relevant turn in order to get information about the upcoming demands. On rural roads, they always swiped further, at least until the second relevant turn.

Almost half of the co-drivers stated in the post-trip interviews that the bar plot showing the upcoming hazard warnings enabled the co-drivers to estimate the amount of assistance needed and was helpful the driver of the upcoming demand situations in time. For instance co-driver 4 stated, "*It helps me to estimate whether the driver needs help and how much I should be of support.*"

Our results also show that in three cases, the contradictory information of the two navigation devices (e.g., an unexpected indication of a lane change on the driver's display) led to more communication between the couples. Concerning the request of the driver "*Do you have the same information*? *This must be a wrong message.*" The co-driver was incorporated into the navigation tasks in order to justify the information. In this particular case, the 'ad hoc' collaborative practice was supported by a shared display interaction, i.e., by passing the tablet to the driver so both had a shared view. This kind of shared view was also used during longer waiting periods at traffic lights or on rural or requested by the driver in order to gain more information about the upcoming demands.

Co-navigator Summary

The study with the co-navigator app showed the strengths of a navigation device especially designed for passengers. It helped the co-driver to gain an overview of the route and, when necessary zoom into a detailed view. Above that, the visualization of demand situations made navigation and collaboration more relaxed since for both (driver and co-driver) it was obvious when to concentrate on navigation and when to be more relaxed (e.g., on straight roads). It also showed, that the need for the possibility for the co-driver to highlight objects in the environment in a more direct way. Thus, we developed the shared gaze approach, presented in the next section.

7.3.3 *Shared Gaze*

Shared gaze in the car is an approach where the gaze of the front-seat passenger is captured and visualized for the driver. The approach aims at overcoming two main barriers which hinder the face-to-face communication of the driver and the front-seat passenger in the car: (1) the sitting position, which is side-by-side, and (2) the need of the driver to pay attention to the driving task, i.e., to keep his/her eyes on the road. Due to these restrictions, the driver has only limited chances to perceive communication from the front-seat passenger that goes beyond verbal communication (like, e.g., gestures or gaze).

Hence, with the shared gaze approach, we provide a new way of sharing non-verbal information between the front-seat passenger and driver. The driver can immediately see where the front-seat passenger is looking at, without the need to turn the eyes from the forward scenery. The front-seat passenger is provided with a means, which enables the communication of spatial information like, e.g., navigational information or information about upcoming hazards, faster and more efficient to the driver. For example, a navigational instruction by the front-seat passenger like "*you need to turn left at the house over there*" accompanied by a visualization of the front-seat passenger's gaze while looking at the meant house, could make it much easier for the driver to detect this spatial reference.

Indeed, findings from the remote collaboration domain indicate that shared gaze can foster collaboration. Here, it was found that mutual visual search tasks could be completed faster if the gaze of remotely located subjects was visualized to each other. Also, spatial referencing could be done more efficient due to the quick communication of spatial information with the help of gaze (e.g., Brennan et al. 2008; Neider et al. 2010).

So far, our shared gaze approach for the automotive domain has passed through different stages. Inspired by the findings from a probing study, a first design sketch of the approach was outlined in the work by Gärtner et al. (2014) In an exploratory study (Maurer et al. 2014), we validated the technical setup and identified further possible application cases. A detailed investigation of the shared gaze approach was then done in two consecutive driving simulator studies:

Shared Gaze Study 1

The first study was concerned with the general usefulness of the approach and its impact on driving performance, perceived distraction, and workload (Trösterer et al. 2015a). In total, 17 driver/co-driver pairs, who were not familiar with each other, participated in the study. Main task of the pairs was to navigate together along a predefined route. Thereby, the front-seat passenger had the navigational information displayed on a tablet and needed to guide the driver. For the study, we developed and used the LCTNav (Trösterer et al. 2015b) an abstract navigational task based on the principles of the Lane Change Task (LCT) (Mattes 2003).

We compared four different conditions in the study, with each pair performing all conditions in permutated order. In the *solitary* condition, a tablet showing the navigational information was mounted at the middle console and the driver needed to navigate on his/her own. In the *verbal* condition, the front-seat passenger explained verbally to the driver where to go. In the *gaze* condition, verbal instructions by the front-seat passenger were accompanied by a permanent visualization of the front-seat passenger's gaze as a yellow dot in the simulation shown on-screen. In the *gaze activation* condition, front-seat passengers could decide on their own, whether they wanted to show their gaze or not during the task by pressing a button. The gaze of the co-driver was captured with the SmartEye eye-tracking system and the gained data was visualized in real time in the simulation.

We found that driving performance in terms of lane deviation was comparable in all conditions, although errors (i.e., wrong lane changes, missed lane changes, or wrong lane changes that were corrected) happened more often in the collaborative conditions compared to the solitary condition. However, among the collaborative conditions, fewest errors happened in the gaze condition. Compared to the solitary condition, we also found that the collaborative conditions were perceived as less demanding by the drivers in almost all cases. Furthermore, the verbal condition had disadvantages compared to the shared gaze conditions as auditory demand and stress were rated higher. Perceived distraction was also rated lower in the verbal and gaze condition compared to the solitary condition. The results of the study can be seen twofold: On one hand, the results support the general assumption that the collaboration between driver and front-seat passenger can indeed be beneficial. With regard to the shared gaze approach in particular, we could find that the approach enabled front-seat passengers to communicate more easily what is meant to the driver. Showing the gaze to the driver led to less stress, less perceived distraction, and also less driving errors during the navigational task. However, we could also find, that the gaze activation condition was a bit problematic as front-seat passengers quite differed in how and how long they showed their gaze in this condition. Hence, in that condition the degree, to which drivers could benefit from the approach was varying.

Shared Gaze Study 2

Driven by the question how the shared gaze approach could be brought in a real car in a technically feasible way, the second study focused on the comparison of two different visualization alternatives of the front-seat passenger's gaze for the driver, i.e., a dot visualization (which would require a full windshield HUD in a real car) vs. a light emitting diodes (LED) visualization (Trösterer et al. 2015c). In the simulator study, the dot visualization matched the visualization used in the first study. For the LED visualization, we mounted an LED stripe at the bottom of the windshield of our car-mock-up and gaze was visualized with five glowing blue LEDs, using a shape that is perceived to be sinusoidal (see Fig. 7.3).

Twelve driver/co-driver pairs participated in the study. Apart from a static task to determine the perceived accuracy of each kind of visualization, the driver/co-driver pairs also needed to perform a navigational task together, using the LCTNav as in the first study. Front-seat passengers could show their gaze to the driver if they wanted to (i.e., *gaze activation*). In contrast to the first study, they were instructed how they could use it best to support the driver (as the first study had shown that this was not always intuitively clear for the front-seat passengers). The navigational task was then performed once with the dot visualization and once again with the LED visualization of the front-seat passenger's gaze (order permutated). In line with the first study, eye movements of the front-seat passenger were captured with the SmartEye eye-tracking system. Additionally, the eye movements of the driver were captured with a head-mounted eye-tracking system (Dikablis) in order to get deeper insights with regard to driver's distraction.

We found that, in general, the LED visualization was perceived as less accurate than the dot visualization by the driver. Additionally, co-drivers also thought that drivers were less satisfied with their advice if LED visualization was used and that they could communicate the relevant information faster with dot visualization.

Fig. 7.3 LED visualization of the co-driver's gaze for the driver (driver's perspective)

Eye-tracking data of the driver did show that the driver perceived spatial reference points sooner if dot visualization was used. Co-drivers also felt they had less control, when the LED visualization was used (as this visualization was mapped to the driver's viewpoint). However, our data also revealed, that LED visualization was primarily perceived peripherally by the driver. In contrast to the dot visualization, this bears the advantage that the driver's gaze can be guided in a less distractive way. LED visualization also allowed the drivers to easily recognize the intended direction, which helped them to reduce the search space during the navigational task. Taking into account these findings, we are currently working on ways to enhance LED visualization accuracy and improving this visualization approach.

Shared Gaze Summary

In sum, we could find in these two studies that the shared gaze approach is indeed a helpful means for the front-seat passenger to communicate spatial information in a fast and efficient way to the driver, thereby leading to less stress and distraction. In particular, the LED visualization of the front-seat passenger's gaze seems a less distracting indicator. However, LED visualization lacked accuracy and further work is needed to increase the feasibility of this kind of visualization. While the drivers clearly benefitted from the shared gaze approach, the two studies also revealed that the approach needs some improvement with regard to the front-seat passenger's needs. We found in the first study that it was not intuitively clear for the front-seat passengers how to best support the driver with the help of their gaze. In the second study, we took this into consideration by briefing the front-seat passengers accordingly and indeed gained better results. However, the second study also revealed that a visualization of the front-seat passenger's gaze solely mapped to the driver's perspective led to a feeling of less control. Hence, future work also needs to focus on this issue by, e.g., increasing visualization accuracy and subsequently front-seat passenger's trust in the visualization.

Apart from this direct interaction between a co-driver and a driver we identified the need for communication and collaboration tools inside the vehicle. In the next section we present such an approach.

7.3.4 Active Corners

In this section, we present an approach that enables in-car collaboration and communication for all passengers inside a vehicle, as reported in Meschtscherjakov et al. (2016). By mapping the four corners of a tablet to fours seats inside a vehicle (driver, front-seat passenger, rear seat passenger left and right), we allow passengers inside a car to share information with each other. By using the rectangle form factor of a tablet and mapping its corners to the seat positions in the corners of a vehicle, we aspired to design an intuitive, fast, and easy to use communication platform.

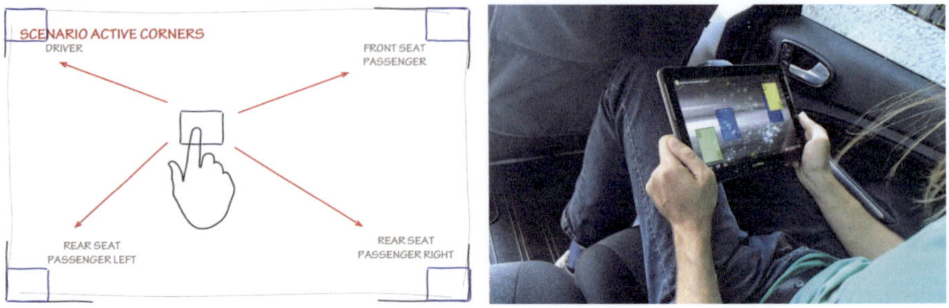

Fig. 7.4 Active corners interaction. An object may be sent to different seats by dragging it into the according corner (*left*). Study participant interacting with the card game prototype in a moving vehicle (*right*)

Sharing information is done by the drag-and-drop of items into the corresponding corner of the vehicle. Thus, each person can share information with every other person inside the vehicle. Items that can be dragged-and-dropped can be the visual presentation of information such as a navigation destination (see Fig. 7.4). Items can be any files that one wants to share such as pictures, drawings or sound files. Items can also represent commands that are sent to another tablet and, thus, active corners can serve as a remote control for the rear seat. For example, a driver could use the active corners approach to send a command to start or stop a movie on the rear seat.

In order to be able to evaluate the active corners approach we built a multiplayer tablet card game. The card game prompts players to share cards and, thus, fosters collaboration among players. Sending cards is done by dragging them into one of the four corners of the tablet. When the card reaches a corner, an expanding red quarter cycle is visualized in that corner, which turns into a green quarter cycle indicating that the card may be sent on release. At the same time, a pulsating red quarter circle in the corresponding corner (i.e., the corner from which the card was sent) indicates the arrival of the card. By tapping on that quarter cycle the receiver can accept the card, which is then shown on the player's canvas.

Regarding gameplay, we decided to implement a simple card game that is played simultaneously. Players had to reach a shared goal within a limited time frame (5 mins). In the beginning, each player had nine random cards on a digital canvas with four colors and unique numbers. Then they had to reach a shared goal by exchanging cards; e.g., "*All green cards to the front-seat-passenger!*". When the mutual goal was reached another shared task was visualized. The card game itself is one possibility for the application of the general interaction concept. To win the game players have to act together and thus communicate and collaborate with each other.

One obvious challenge of the active corners approach is that the driver might not be able to use the system for reasons of distraction. We imagine that the driver may use the system in situations such as traffic jams or may have free resources in future autonomous vehicles. Regarding the good reachability of a tablet for the driver, we imagine a touch screen integrated into the steering wheel as a potential solution as shown in (e.g., Pfeiffer et al. 2010; Osswald et al. 2011).

In order to evaluate the usability and user experience of the approach with different user groups (children and adults), we conducted four consecutive user studies: children in a parked vehicle; adults in a parked vehicle; three passengers under real-traffic conditions; four participants in a driving vehicle on a test track. Below, we briefly describe each study and their results. More details on the concept and the studies can be found in Meschtscherjakov et al. (2016).

Active Corners Study 1

In the first user study, we wanted to evaluate active corners with children sitting in a stationary vehicle. The card game was played with four players including one player sitting on the driver's seat. Study aims were functionality, ease-of-use, and how intuitively the interaction concept was perceived. The study took place at a University exhibition, thus we took a convenient sample of visiting children. Children were introduced to the interaction concept and after playing the card game for 5 mins they were handed over an adapted version of the extended Short Feedback Questionnaire (Moser et al. 2012). It included tag clouds of opposed attributes for describing game control questions that had to be answered on a 5-point Likert scale (1 = "difficult" to 5 = "easy"). Additionally, children were observed from outside the vehicle by two researchers. In the study, 138 children (70f, 67 m, 1n.a.) aged between 6 and 18 years (M = 12.57, SD = 2.24) participated in the study.

Both the questionnaire and observations showed that children immediately understood the interaction concept. Children stated that sending a card was easy (M = 4.40, SD = 0.88) and the coordination with other players was perceived as quite simply (M = 4.12, SD = 0.92). Game control was labeled with the adjectives great (57.3%), simple (52.2%), fun (45.7%), intuitive (40.6%), and exciting (21.7%). A very limited number of children stated that the game control was childish (6.5%), difficult (2.2%), boring (1.5%), or impractical (1.5%). We found no considerable differences found in ratings for the four different seat positions, but logging data showed that considerable more cards were exchanged between the driver's seat and the left rear seat (1,628 exchanges vs. M = 1,349.33, SD = 166.03). Based on this finding we conducted another study on the effect of the seat position on the activity in the game. We will elaborate on this in section space and place in cars.

In summary, the study showed that the active corners approach was easy to use and most children immediately understood the mental model of the approach. Difficulties were reported on the latency during drag and drop gestures and that actively receiving a card by tabbing at the red quarter circle was missed due to overlapping cards. As a next step, we wanted to evaluate if the concept also works for less than four players and for what kind of applications adults can imagine to utilize the active corners concept in their own car.

Active Corners Study 2

For the second study, we changed the card game so it could be also played with two or three players releasing the driver to participate in the game. This study took place at an exhibition in a parked vehicle. Again we used a convenience sample of the visitors of the exhibition as participants. After playing the card game participants were asked to fill in a questionnaire consisting of 26 items regarding the game, the active corners concept and potential interaction scenarios. Items were taken from the CTAM (Osswald et al. 2012) and the SUS questionnaire (Brook 1996) and had to be rated on 5-point Likert scale (1 = "Strongly disagree" to 5 = "Strongly agree"). From the 37 participants (11f, 26 m), aged between 9 and 69 years (M = 29.97, SD = 15.41), seven sat at the driver's seat, 13 at the front-seat passenger's seat, 11 at the left rear seat left, and 6 at the right rear seat.

Again, observations and participant ratings showed that-sending and receiving was straightforward. The interaction modality was perceived as easy to learn (M = 4.76, SD = 0.44) and fast (M = 4.51, SD = 0.56) and the mapping of corners to seat positions was intuitive (M = 4.28, SD = 0.91). Both sending (M = 4.62, SD = 0.55) and receiving a card (M = 4.46, SD = 0.73) was rated on the usability scale and participants were satisfied with the interaction modality (M = 4.49, SD = 0.69) and found it useful (M = 4.08, SD = 0.98). Participants rather disagreed if the active corners concept would be appropriate for the driver (M = 3.18, SD = 1.22) or if it would be distractive for the driver (M = 2.97, SD = 1.26).

Application scenarios that were ranked high were "The driver may manipulate volume in the rear seats." (M = 4.35, SD = 0.79), "The driver may start a game at the rear seat" (M = 4.11, SD = 1.01), and "The front seat passenger may send a navigation target to the driver." (M = 4.14, SD = 1.13). Application scenarios that were ranked low were "The driver starts/stops a movie at the rear seat" (M = 3.77, SD = 1.31) and "A child on the rear seat sends a drawing to the driver" (M = 3.35, SD = 1.42).

The second study showed that the driver was not needed for the game and that the active corners approach could be imagined for different application scenarios inside a vehicle. What we did not find out was, if the interaction is also suitable in a moving car with centrifugal forces, vibrations, and difficult light conditions. Thus, we conducted a third study with participants in a moving vehicle.

Active Corners Study 3

This study took place with one of our research colleagues as the driver and three participants in a car that drove in a real-traffic condition. We were especially interested if the moving of the vehicle had a negative influence on the usability of the approach and if looking at the tablet may induce motion sickness. We used the questionnaire from the second study as a measurement instrument and added questions about the influence of the movement of the vehicle on usability and motion sickness. Answers were given on a 5-point Likert scale (1 = "Strongly disagree" to 5 = "Strongly agree").

Six individuals (2f, 4 m) aged between 21 and 32 (M = 27.67, SD = 3.72) and recruited from our facility took part in the study. We made two trips for approx. 20 min with three participants sitting in the front, left rear, and right rear-seats for each trip. Also, moving vehicle learnability (M = 4.67, SD = 0.52), speed (M = 4.83, SD = 0.41), intuition (M = 4.50, SD = 0.55), and easy-of-use (M = 4.50, SD = 0.55) of the interaction were rated high. As in the first study, sending a card (M = 4.33, SD = 0.52) was perceived as being easier than receiving a card (M = 3.67, SD = 0.52).

Vibrations (M = 2.50, SD = 1.05), turns (M = 2.17, SD = 0.41), breaking, or accelerating (M = 2.33, SD = 0.82) had no negative influence on the interaction with the tablet. Participants reported that sun glare was sometimes negatively affecting the readability. Regarding motion sickness, interestingly for one group of participants this was an issue; the second group had no issues with motion sickness.

Thus, also the third study showed the intuitiveness of the active corners approach along with its good usability during driving. Neither vibrations nor forces caused by driving had a negative influence on the interaction. For some people, motion sickness could be a problem.

Active Corners Study 4

For the fourth study, we asked four participants to drive on a circle-shaped test track. One participant was the driver, the others were again passengers playing the active corners game. Our main research goal for this study was to figure out how distractive active corners could be for the driver.

Besides using the questionnaire from study 3, we video recorded the rides and conducted a group interview after completing the ride with all four participants in each group. Four groups with, a total of 16 participants (10 m, 6f) aged between 19 and 69 years (M = 46.20, SD = 16.70), took part in the study. From the four drivers, three indicated that the active corners game played by passengers was not disturbing during driving; for one driver, it was very disturbing. Three drivers could not imagine sending objects during driving, one neither agreed nor disagreed. Three drivers agreed that active corners are useful for the passengers; one neither agreed nor disagreed.

Passengers' feedback (N = 12) was similar to the other three studies. Active corners was perceived as easy to learn (M = 3.92, SD = 0.90), easy to use (M = 4.08, SD = 0.99), fast (M = 3.92, SD = 0.79), and intuitive (M = 4.08, SD = 0.99). As in study 3, vibrations (M = 1.50, SD = 0.79), bends (M = 1.42, SD = 0.90), acceleration, and deceleration (M = 1.58, SD = 0.99) did not negatively influence interaction. This time, almost no motion sickness issues were reported (M = 1.17, SD = 0.57).

In summary, the active corners were not disturbing for the driver when being used by passengers, but not usable by the driver in its current implementation. We propose that active corners might be suitable for drivers when they have free resources (e.g., in a traffic jam or in an autonomous vehicle). Interaction might be incorporated in a steering wheel, e.g., in form of a touch screen as proposed by Pfeiffer et al. (2010) or Osswald et al. (2011).

Active Corners Summary

Our studies have shown that the active corners approach and the mapping of tablet corners to seats in the vehicle is intuitive and easy to learn. The usability to send objects is high, for accepting objects interaction design may be still improved. The approach is also suitable for a driving car. Vibrations and other forces caused by the movement of the vehicle are no problem. Motion sickness might be an issue for some individuals. We have shown that the approach is applicable for passengers, the integration of the driver is seen rather skeptical in its current form, although it is not seen as being disturbing for the driver. Participants agreed on the usefulness of the approach especially when passengers support the driver (e.g., send a navigation destination) and when front-seat passengers or drivers have the possibility remotely control rear seat entertainment systems (e.g., change volume).

In order to learn more about the different social roles inside a vehicle induced by the seat position we have conducted another study using the active corners game as a facilitator for collaboration. This study is presented in the next section.

7.3.5 Space and Place in Cars

In this study, as reported in Krischkowsky (2016), we used the active corners interaction concept as collaborative trigger for drivers and passengers to explore the impact of spatial properties of the car cabin on social encounters, i.e., using the active corners game as baseline collaboration task. This research is based on the idea that interaction environments, such as cars or specifically car cabins, are characterized by their spatial manifestations, which potentially direct, guide, and provide an opportunity to become a place for social engagements.

Already in 1996 Harrison and Dourish argue that when researching interaction environments and design 'spaces', it is a central necessity to distinguish between 'space' and 'place', by arguing that space is the opportunity while place is the understood reality of given space. All of us live in a three-dimensional world whereby the geometrical structure around us directs and guides our interactions with one another (Harrison and Dourish 1996). With the research presented here, we take a critical step in understanding how the car as an interaction environment becomes a place for social encounters, by investigating the relations between people in spaces in a systematic way. By incorporating the rear-seat space in our investigations as well, we enhance knowledge around the automotive design space by going beyond research on driver and co-driver interactions.

Exploratory Lab-Study About People in Spaces

In order to explore the interplay of space and place in cars (i.e., the impact of spatial properties of a car cabin on collaboration), we conducted an exploratory lab study with 56 participants in groups of four people (i.e., 16 groups with 4 participants each) in a hardware mock-up (i.e., as a pure physical representation) of a real car.

Those groups were gender-balanced and aged between 20 and 31 years (M = 24.8 years). All participants had a driving license and were recruited via different student mailing lists.

The main research question we addressed with this study is as follows: *How do the spatial properties of a car cabin shape emerging collaboration?*

The study was conducted in our laboratory without involving any driving task. The rationale for this setup is threefold. First, we wanted to create a setting that allows to systematically *explore the basic spatial manifestations* of a car cabin. Second, we wanted to explore how people *ascribe structural features* to a given space that inform collaboration, regardless to a given role or activity. Finally, we wanted to explore in what way the given *spatial properties effect emerging collaboration*. Thereby, the car mock-up served its purpose in embodying a real car with its characteristic spatial manifestations (e.g., seats, steering wheel) without any interference or biases from outside the car (e.g., bad weather or pedestrians walking by), as it would be the case when being conducted in the field. Within this car-mock -up, groups of four people had to play the active corners card game collaboratively to fulfill the given tasks. For more details regarding the active corners interaction concept and exemplary game tasks, please see the *active corners* section above. These tasks were executed on basis of two conditions in a fixed consecutive order: C_1 participants could *freely decide* where to sit in the car and collaboration was triggered by the active corners game for 5 min, and in C_2 participants were *reseated by the researchers* in the car mock-up (i.e., driver to rear-seat left and vice versa and front-seat passenger to rear-seat right and the other way around) and collaboration was triggered by the active corners game for 5 min. Those two conditions have been defined in order to investigate how the fixed seating arrangement in a car cabin shapes emerging collaborative practices. Figure 7.5 visualizes the setup.

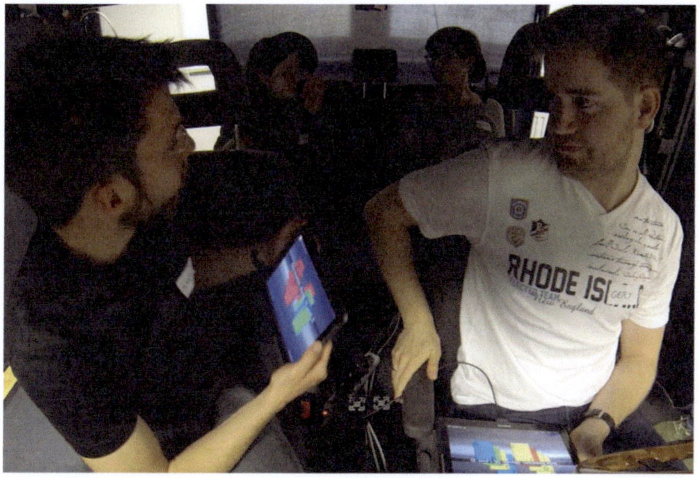

Fig. 7.5 Participants collaborating via the active corners game in the car mock-up in our lab

Concerning the study *procedure and assessment*, we first introduced the participants to the game and they were given a two-minute practice time to test the game, get used to it, and ask questions if needed. Directly afterwards, the group of participants moved to the car mock-up and participants could freely decide where to sit down (C_1). Then they started with the gaming sessions and were given the task to collaboratively fulfill as many gaming tasks as possible. This procedure was then repeated within C_2, where the researchers reseated the participants. Both conditions were recorded via two small cameras that were attached at different places in the car cabin. All in-game events (e.g., card exchange) were logged into a database. For the *data analysis*, we conducted a quantitative video analysis (on basis of Argyle and Cook 1976) on basis of the captured recordings, followed by a collaborative qualitative video analysis (on basis of Jordan and Henderson 1995) that were complemented with data from sociograms (on basis of Moreno 1956) which have been filled out by the participants after each condition.

Within the quantitative video analysis we extracted the key-communication modes that were used by participants to collaborate with one another. Thereby, we considered speech (i.e., verbalizations and their directions), gestures (i.e., gesticulations and their direction), gaze (i.e., in general direction to another's face), as well as posture (i.e., turning one's body towards each other) as the main communication modes in relation to the participants seating position. On basis of this quantitative video analysis, we found that verbal communication is the participants' primary mean to communicate (n = 529), followed by gaze (n = 494). Contrary, gestures (n = 107) and postures (n = 84) were much less frequently used by the participants to communicate. Concerning gaze, we came to understand that especially the participants from the back-seats were 'gazing' at the front-seat positions to gather more information from the front-seat passengers (i.e., observing the front-seat passengers actions and interactions on the tablet). Opposing, the participants sitting at the front positions hardly gazed to the rear positions, as the front seats impeded them in looking behind to the rear-seat passengers. Such exemplary findings emphasize that the front-seats act as communication hindrances between all passengers in a car. A detailed description of the used communication modes with respect to their spatial occurrence and direction can be found in Krischkowsky (2016).

Concerning coordinative strategies that emerge throughout the game play, we found (on basis of the qualitative video analysis) that quite often one participant can be identified as instance that triggers coordination within the group (i.e., a participant started to coordinate subsequent actions within the group). Furthermore, some groups do agree upon a certain strategy they collectively follow to fulfill a certain task. Regarding the normative arrangement of the collaborative practices, we could identify six relevant instantiations: (1) participants giving instructions to *direct* others actions, (2) giving *status updates* to other players, (3) giving *compliments* to motivate the group, (4) articulating *errors* to inform other players about false actions, (5) *confirming* other players about correct actions, and (6) ask *questions* to the group if something was not clear. On basis of the above-described normative arrangement of the groups, we were able to extract three social roles from the data, i.e., 'organizer/leader roles', executor roles', and 'transformer roles'. The organizer/leader roles embody participants that take over a leadership role that is needed to successfully play the game in the entire group. These organizer/leader roles were mostly defined within the group (i.e., based on the group decisions) before the actual game task started. The executor role shows that there are always people needed that execute particular tasks that were mostly assigned to them by the leader/organizer role. The transformer role shows that people are needed in such cooperative processes that take over various functions that are necessary in given situations.

Regarding the impact of spatial properties on emerging coordinative and collaborative strategies, we mainly identified the car seats as being the critical hindrances directing the group interactions. Thereby, mainly the driver position could have been identified as being the '*heart of the matter*'. We identified manifold instantiations, wherein either from or to the driver position group coordination was directed to, i.e., participants at the driver position (1) intensively collaborating with the participants at the front-seat passenger position, (2) collaborating diagonal backwards to the rear-right position. Communication and collaboration from the driver position to the rear-left seating position are appearing rarely since participants at the driver position hardly can turn around.

Space and Place in Cars Summary

On basis of these exemplary findings as well as much more insights we gathered throughout this exploratory study (for details see Krischkowsky 2016), we came to understand that the spatial properties of the car cabin have a strong effect on emerging collaborative practices. The very characteristic spatial make-up of a car cabin shape collaborative practices, thereby creating an inhabited place for people to interact in. Thereby, the driver position can be considered as the 'heart of the matter', playing a fundamental role in cooperative practices, even in a non-driving situation. With our research we accounted for looking not just at the space as being a separate entity of inquiry distinct from collaboration, but considered collaboration as being directly embedded in the spatial surroundings where enacted. With this perspective we emphasize and follow the argument that "*[…] behavior can be framed as much by the presence of other individuals as by the location itself […]*" (Harrison and Dourish 1996, p. 69). Throughout the qualitative and quantitative video analysis as well as the analysis of the sociograms, we found that participants ascribe to the driver position specific expectations of action and roles, even in a non-driving situation, i.e., meaning that people in the driver position are prescribed as being the organizers, coordinators, and leaders of the game.

With the knowledge we gathered throughout this research, we argue that our findings can provide an initial understanding of how systems and technologies may be embedded in the car space itself and/or how we can boost collaborative practices in the car by considering it is inherent structural features in the way we design for collaboration and communication in the automotive domain.

7.4 Discussion

The presented research activities have tackled different aspects of communication and collaboration inside a vehicle. Following our research vision we have examined in-car collaboration from various perspectives. We have investigated how passengers support drivers and designed and evaluated d several prototypes that fosters in-car collaboration. We have shown that a passenger is not necessarily only a passive or even disturbing element during a car ride, but may support the driver in various ways. Based on our research activities, we identified the following five aspects how HCI research can incorporate passengers more into the digital eco-system of a vehicle:

1. Enhanced driver interfaces that are inspired by human passengers
2. Digital co-driver that serves as a substitute for a human co-driver
3. Interfaces especially designed for the co-driver
4. Communication tools for all passengers
5. In-car collaboration in autonomous vehicles

Enhanced Driver Interfaces that are Inspired by Human Passengers

ADAS and IVIS are targeted at the driver and designed from a technological perspective. Whenever, a new system is technically matured, it is incorporated into cars as a new function. These systems are often first integrated in high-end models as a unique selling point. Then, they are released in medium-class vehicles and available for a broader market. It seems that usability and user experience aspects in the interaction with these systems follow as a second step. We argue that drivers could be supported better if such systems are designed in the style of driver-passenger interaction, i.e., rather human than technology driven. The careful observation how passengers and drivers interact with each other in day-to-day real-traffic situations can both, inspire the design of new interfaces for the driver and enrich the usability and user experience of existing approaches.

In the *co-navigator* study, we have observed that not only the design of a navigation device for the front-seat passenger is valuable, but also that a navigation application for a driver could be designed with a human co-driver in mind. For example, one of the main findings from the *car-sharing ethnography* was that passengers usually had a good intuition when and where the driver needed support, but also when there is an inappropriate situation to communicate with the driver. This intuition included both, the observation of environmental parameters as well as the observation of the driver status. For example, contextual parameters (e.g., night ride in an unfamiliar environment) in combination with driver behavior (e.g., the uneasiness of a driver) were interpreted as a prompt for support by the co-driver. Nevertheless, the co-driver did not try to communicate with the driver in situations in which either the situation itself was dangerous (e.g., when entering a highway) or the driver was stressed (e.g., while overtaking another vehicle). This kind of support was often very nuanced. In a hazardous situation, it may have been the case that a co-driver intervened even though neither the context nor the driver allowed for an intervention. For example, a co-driver might shout "caution" if the driver would have elsewise caused an accident.

Thus, novel interaction designs for the driver need to be, not only context aware but also driver-status based, to provide the best help. They need to know when to support the driver and when to minimize communication in order to avoid an unnecessary distraction of the driver. In case of an emergency, they even may consciously intervene and thus distract the driver.

Digital Co-driver that Serves as a Substitute for a Human Co-driver

If we go one step further, we might not only enhance single functions, such as navigation systems by studying a human passenger—it is also imaginable to develop a digital co-driver that serves as a substitute for a human co-driver. A digital co-driver would be a system that behaves like a supportive passenger. By gaining information about the driver's state or the driving context through different sensors (e.g., GPS and physiological data of the driver) it could act and react. In the *car-sharing ethnography,* we experienced that there is a need for a common ground between driver and co-driver, i.e., meaning that a driver and a digital co-driver need to build a relationship. The driver must trust the digital co-driver. It must be able to support the driver in different situations and needs to be context and driver aware. We can imagine that a digital co-driver that serves as an interface to the entire vehicle may be able to build that common ground more easily than a set of different ADAS. This digital co-driver might be a self-learning system that adapts its support strategies based on the successes and failures in collaborating with the driver.

Approaches such as *shared gaze* might be not only a way to provide information on the perception of a human passenger, but also be incorporated in the co-driver design as well. For example, we can imagine that a virtual co-driver might visualize its "gaze" for the driver (i.e., visualization of sensor data) and, thus, help to build a common ground for the driver and the virtual co-driver.

Interfaces Especially Designed for the Co-driver

With the *co-navigator*, but also the *active corners* approach, we have built and studied systems that are not mainly targeted at the driver, but also at the passenger to make him/her become a better co-driver. These systems do not necessarily have to take limitations into account, which are relevant when designing for the driver (e.g., not to be distracting). Such systems might take advantage of the free resources a passenger has to support the driver. We have shown that especially assistance in demand situations or help to gain a broader overview on the current trip is helpful for the driver, especially when he/she is concentrating on turn-by-turn instructions from a typical navigation device.

With the help of the *active corners approach*, we aspire to help passengers to support the driver by operating functions of the vehicle that might only be available for the driver. We are aware of the fact that some functions should not be accessible for the co-driver (e.g., acceleration or steering), but for many other functions, help from the co-driver might be beneficial. On a strategic level (in terms of operating the vehicle), one obvious example is the entering of a navigation destination, which can be a distracting activity since typically the address consists of many letters and number. On an operational level, the co-driver might also have some more information, e.g., on upcoming hazards, such as an icy road, or other contextual environmental information. For example, we can imagine that a co-driver has information about school opening times near a school in order to inform the driver to be especially cautious or get a pre-visualization of how to navigate through a demanding situation. This could be done in the form of a short video, so that the co-driver already knows in advance how to drive and can then inform the driver in a better way. Contrary, the video visualization would not be appropriate for a driver due to distraction issues.

Communication Tools for all Passengers

Another clear result from our studies was that there is a need for communication tools between drivers and passengers, as well as among passengers themselves. Of course, passengers and drivers can talk verbally to each other and can turn to one another to see who is doing what, but as our *space and place* study showed, there seems to be an invisible wall between the front and rear area of the car, i.e., meaning that mainly the front seats act as hindrances for efficient communication and collaboration among all passengers. It is especially difficult for drivers and front-seat passengers to interact with the rear seats. However, in many cases drivers or front-seat passengers need to interact with people sitting in the back, such as e.g., when entertaining children on the rear seat position (e.g., starting or stopping a video on the rear seat entertainment system). When children are older, they would be able to reach a touch screen. However, a toddler is fasten into his/her seatbelt and is not able to reach a touchscreen. In such situations, systems like that exploit the *active corners* approach are a possible solution.

A more subtle solution could be to connect the front and rear seats by including an intercom into the vehicle. Another approach could be to include a video stream, similar to Skype, so that the front-seat passenger or the driver could look at the person who is sitting behind him/her without turning his/her head.

In-Car Collaboration in Autonomous Vehicles

When it comes to in-car collaboration with autonomous vehicles, a whole new range of interfaces can be imagined. In general, we have to specify which level of automation we presume. With an automation level 2 and below (with regard to the SAE taxonomy 2014), the driver needs to monitor the driving environment and needs be able to intervene at any time. When it comes to level 3 and higher, the role of the driver becomes a different one. At level 3 (conditional automation), the driver does not need to observe the environment, but can do other things such as reading a newspaper or surfing the Internet. At this level, it becomes ambiguous who the driver is and who the passenger is. Is it the person sitting on the driver's seat or the person who is first responding to a hand over request? Thus, communication and collaboration becomes an important issue. At level 4 (high automation) or 5 (full automation), the driver might not even be able to intervene in a driving task. Here, we have to ask ourselves who is allowed to command the vehicle? Who enters a destination? How should the autonomous vehicle behave if it gets conflicting instructions?

On the other hand, release of the driving task offers new possibilities to interact inside a vehicle apart from controlling the vehicle. *Space and place in cars* could be perceived differently, when it is possible that the driver and front-seat passenger may turn-around their seats and face passengers on the rear-seats.

Summary

We have investigated in-car collaboration and communication by conducting ethnographic studies and deploying experience prototypes in our lab and in real-traffic situations. Based on our findings of why, how, and when passengers act as co-drivers we have presented several interface approaches for the driver, that is informed by our investigations and are thus more empathic and less technology driven. Moving one step further we imagine future digital co-drivers that acts more like human co-drivers enhancing driving experience and safety. Above that, we have shown that designing interfaces for co-drivers is fertile. We also identified the need for in-car collaboration tools that exploits the spatial properties of a car. Finally, we have discussed how and provided an outlook on which questions have to be addressed when designing for collaboration in future autonomous vehicles.

7.5 Conclusion

In this chapter, we have shown that communication and collaboration inside a vehicle is an underestimated research topic. Based on five different research activities we have conducted over the last years, we identified topics that hold potentials for HCI research and practice in the future. We reported on the setup and results of a *car-sharing ethnography* in which we observed driver-passenger pairs in their collaborative activities in real traffic. Based on these insights we conceptualized and prototyped a navigation device (labeled *co-navigator*) especially designed to be used by a front-seat passenger in order to support the driver in demanding situations. We then reported about the *shared-gaze* approach, which visualizes the gaze of a front-seat passenger for the driver. *Active corners* is another concept that allows the driver and passengers to share information inside a vehicle by drag-and-drop gestures on a tablet. Finally, we reported on an exploratory car mock-up study in which we researched how the spatial properties of the car cabin have an impact on how people collaborate and communicate within a car, shaping their perception of *space and place* inside the vehicle.

We believe that we are only able to provide a high user experience inside the vehicle if we perceive the vehicle holistically, in the sense of incorporating the needs and potentials of all passengers. Future work should further investigate how passengers support the driver in the fast changing world of driving. We believe that with the integration of evermore assistant systems and the increasing autonomy of novel vehicles in-car collaboration will change significantly. In order to design more human like assistant systems further investigations needs to be done. Another aspect will be the design of interfaces especially for passengers. Driving is and will be a highly collaborative activity.

Acknowledgements The financial support by the Austrian Federal Ministry of Science, Research and Economy and the National Foundation for Research, Technology, and Development and AUDIO MOBIL Elektronik GmbH is gratefully acknowledged (Christian Doppler Laboratory for "Contextual Interfaces").

References

Argyle, M., & Cook, M. (1976). *Gaze and mutual gaze*. Cambridge University Press.

Brennan, S. E., Chen, X., Dickinson, C. A., Neider, M. B., & Zelinsky, G. J. (2008). Coordinating cognition: The costs and benefits of shared gaze during collaborative search. *Cognition, 106*(3), 1465–1477.

Brooke, J. (1996). SUS-A quick and dirty usability scale. *Usability Evaluation in Industry, 189* (194), 4–7.

Brown, B., & Laurier, E. (2012). The normal natural troubles of driving with GPS. In *Proceedings of the SIGCHI conference on human factors in computing systems* (pp. 1621–1630). ACM.

Cycil, C., Perry, M., Laurier, E., & Taylor, A. (2013). Eyes free in-car assistance: Parent and child passenger collaboration during phone calls. In *Proceedings of the 15th International Conference on Human-Computer Interaction with Mobile Devices and Services* (pp. 332–341). ACM.

Forlizzi, J., Barley, W. C., & Seder, T. (2010). Where should I turn: Moving from individual to collaborative navigation strategies to inform the interaction design of future navigation systems. In *Proceedings of the SIGCHI CONFERENce on Human Factors in Computing Systems* (pp. 1261–1270). ACM.

Gärtner, M., Meschtscherjakov, A., Maurer, B., Wilfinger, D., & Tscheligi, M. (2014). Dad, stop crashing my car!: Making use of probing to inspire the design of future in-car interfaces. In *Proceedings of the 6th International Conference on Automotive User Interfaces and Interactive Vehicular Applications* (pp. 1–8). ACM.

Harrison, S., & Dourish, P. (1996). Re-place-ing space: the roles of place and space in collaborative systems. In *Proceedings of the 1996 ACM Conference on Computer Supported Cooperative Work* (pp. 67–76). ACM.

Heath, C., & Luff, P. (2000). *Technology in action*. Cambridge University Press.

Jordan, B., & Henderson, A. (1995). Interaction analysis: Foundations and practice. *The journal of the Learning Sciences, 4*(1), 39–103.

Juhlin, O. (2013). *Social media on the road*. London Limited: Springer.

Krischkowsky, A., Trösterer, S., Bruckenberger, U., Mauerer, B., Neureiter, K., Perterer, N., et al. (2016). The impact of spatial properties on collaboration: An exploratory study in the automotive domain. In *Proceedings of GROUP'16*. ACM.

Laurier, E., Lorimer, H., Brown, B., Jones, O., Juhlin, O., Noble, A., et al. (2008). Driving and 'passengering': Notes on the ordinary organization of car travel. *Mobilities, 3*(1), 1–23.

Leshed, G., Velden, T., Rieger, O., Kot, B., & Sengers, P. (2008, April). In-car gps navigation: Engagement with and disengagement from the environment. In *Proceedings of the SIGCHI Conference on Human Factors in Computing Systems* (pp. 1675–1684). ACM.

Mattes, S. (2003). The lane-change-task as a tool for driver distraction evaluation. *Quality of work and products in enterprises of the future* (pp. 57–60).

Maurer, B., Trösterer, S., Gärtner, M., Wuchse, M., Baumgartner, A., Meschtscherjakov, A., et al. (2014). Shared gaze in the car: towards a better driver-passenger collaboration. In *Adjunct*

Proceedings of the 6th International Conference on Automotive User Interfaces and Interactive Vehicular Applications (pp. 1–6). ACM.

Medenica, Z., Kun, A. L., Paek, T., & Palinko, O. (2011). Augmented reality vs. street views: A driving simulator study comparing two emerging navigation aids. In *Proceedings of the 13th International Conference on Human Computer Interaction with Mobile Devices and Services* (pp. 265–274). ACM.

Meschtscherjakov, A., Krischkowsky, A., Neureiter, K., Mirnig, A., Baumgartner, A., Fuchsberger, V., et al. (2016). Active corners: Collaborative in-car interaction design. In *Proceedings of the 2016 ACM Conference on Designing Interactive Systems* (pp. 1136–1147). ACM.

Moreno, J. L. (Ed.). (1956). *Sociometry and the science of man*. Beacon House.

Moser, C., Fuchsberger, V., & Tscheligi, M. (2012). Rapid assessment of game experiences in public settings. In *Proceedings of the 4th International Conference on Fun and Games* (pp. 73–82). ACM.

Neider, M. B., Chen, X., Dickinson, C. A., Brennan, S. E., & Zelinsky, G. J. (2010). Coordinating spatial referencing using shared gaze. *Psychonomic Bulletin and Review, 17*(5), 718–724.

Osswald, S., Meschtscherjakov, A., Wilfinger, D., & Tscheligi, M. (2011, November). Interacting with the steering wheel: Potential reductions in driver distraction. In *International Joint Conference on Ambient Intelligence* (pp. 11–20). Berlin: Springer.

Osswald, S., Wurhofer, D., Trösterer, S., Beck, E., & Tscheligi, M. (2012, October). Predicting information technology usage in the car: Towards a car technology acceptance model. In *Proceedings of the 4th International Conference on Automotive User Interfaces and Interactive Vehicular Applications* (pp. 51–58). ACM.

Östergren, M., & Juhlin, O. (2004). Sound Pryer: Truly mobile joint music listening. In *International Workshop on Mobile Music Technology*.

Perterer, N., Sundström, P., Meschtscherjakov, A., Wilfinger, D., & Tscheligi, M. (2013). Come drive with me: an ethnographic study of driver-passenger pairs to inform future in-car assistance. In *Proceedings of the 2013 conference on Computer supported cooperative work* (pp. 1539–1548). ACM.

Perterer, N., Meschtscherjakov, A., & Tscheligi, M. (2015, September). Co-Navigator: An advanced navigation system for front-seat passengers. In *Proceedings of the 7th International Conference on Automotive User Interfaces and Interactive Vehicular Applications* (pp. 187–194). ACM.

Pfeiffer, M., Kern, D., Schöning, J., Döring, T., Krüger, A., & Schmidt, A. (2010, April). A multi-touch enabled steering wheel: Exploring the design space. In *CHI'10 Extended Abstracts on Human Factors in Computing Systems* (pp. 3355–3360). ACM.

SAE On-Road Automated Vehicle Standards Committee. (2014). Taxonomy and definitions for terms related to on-road motor vehicle automated driving systems.

Trösterer, S., Gärtner, M., Wuchse, M., Maurer, B., Baumgartner, A., Meschtscherjakov, A., & Tscheligi, M. (2015a). Four eyes see more than two: Shared gaze in the car. In *Human-Computer Interaction* (pp. 331–348). Springer.

Trösterer, S., Wuchse, M., Baumgartner, A., Maurer, B., Gärtner, M., Meschtscherjakov, A., et al. (2015b). LCTNav: A method for investigating collaborative navigation. In *Adjunct Proceedings of the 7th International Conference on Automotive User Interfaces and Interactive Vehicular Applications* (pp. 124–129). ACM.

Trösterer, S., Wuchse, M., Döttlinger, C., Meschtscherjakov, A., & Tscheligi, M. (2015c). Light my way: Visualizing shared gaze in the car. In *Proceedings of the 7th International Conference on Automotive User Interfaces and Interactive Vehicular Applications* (pp. 196–203). ACM.

Wilfinger, D., Meschtscherjakov, A., Perterer, N., Osswald, S., & Tscheligi, M. (2013). Passagier computer interaktion. In *Mensch and computer* (pp. 313–316).

Chapter 8
The Influence of Non-driving-Related Activities on the Driver's Resources and Performance

Renate Häuslschmid, Bastian Pfleging and Andreas Butz

Abstract Today, drivers perform many non-driving-related activities while maneuvering the car. To ensure driving safety, the designers of automotive UIs have to respect the driver's available cognitive, perceptual and motor resources to prevent overload and in turn accidents. In this chapter, we look at the different types of driver resources, how they are loaded and limited by the primary driving task, and how this affects the resources available for non-driving-related activities. We discuss aspects such as attention, driver distraction, (cognitive) workload, and other factors such as the driver's physical and mental state to understand the limitation of the driver's resources and how non-driving-related activities affect the primary task performance. To enable the safe execution of non-driving-related activities, we need to design the cockpit and its UI in such a way that it requires a minimal amount of resources. We will provide an outlook towards selected novel technologies such as large head-up displays and also discuss expected effects of the transition to automated driving.

8.1 The Problem with Non-Driving-Related Activities

According to the National Highway Traffic Safety Administration (Ranney et al. 2000), distraction and inattention due to non-driving-related activities were some of the major causes (25%) of traffic accidents in the U.S.A. already in 2000. Mobile phones and route guidance systems are the most widely used devices in non-driving-related activities and have therefore triggered most of the research on distraction and interaction in this context, dating back almost 50 years (Brown et al. 1969).

R. Häuslschmid (✉) · B. Pfleging · A. Butz
Ludwig-Maximilians-Universität München, Lehrstuhl für Mensch-Maschine-Interaktion, Amalienstr. 17, 80333 Munich, Germany
e-mail: renate.haeuslschmid@ifi.lmu.de
URL: http://www.mmi.ifi.lmu.de

B. Pfleging
e-mail: bastian.pfleging@ifi.lmu.de

A. Butz
e-mail: andreas.butz@ifi.lmu.de

G. Meixner and C. Müller (eds.), *Automotive User Interfaces*,
Human–Computer Interaction Series, DOI 10.1007/978-3-319-49448-7_8

The results of this research have led to certain legal or factual usage restrictions: Mobile phones, for example, in many countries are legally restricted to handsfree operation and manufacturers of route guidance systems often limit the most distracting task—entering the destination—to parking or at least standing cars.

8.1.1 No Comprehensive Theory

In general, driving does not always require the full attention of the driver. Most of the time, drivers do not spend their full attention on the driving task, without a noticeable deterioration of driving performance (Lee 2004). However, when uncertainty in the road situation increases, drivers normally shift more of their attention towards driving (Senders and Kristofferson 1966). Although considerable research has been done in this respect (e.g., McKnight and Adams 1970; Treat 1980; Ranney 2008), there is not even a commonly accepted definition, which activities actually are critical for driving. Instead of a general unified theory about driving and distraction, there are different approaches to understanding and modeling it. It is not fully understood how drivers think, feel, act, and react. Each existing model only represents a certain viewpoint, but none of them considers driving in a holistic sense (Knapper et al. 2012). Even the definition of the terms *inattention* and *distraction* and their interrelation has been approached differently and so far there is no commonly accepted definition and shared understanding of these terms in the automotive context.

8.1.2 Will Technology Solve it?

Automated driving and novel displays such as head-up or windshield displays are starting to tremendously change the way in which people interact with the car. Autonomous cars will bring about more infotainment and communication functions to counteract boredom and help drivers to make good use of their time and available resources. As long as the driver is not required to take over and the car drives fully autonomously, the engagement in other—non-driving-related activities—may not even be considered a distraction (from the driving task) anymore. However, if such a driver suddenly has to resume control of the car, a strong focus on other tasks (now a distraction!) will substantially increase the risk, since distraction is generally associated with a breakdown of task timing and drivers will simply not be able to pick up the driving task in an adequate manner quick enough. It is therefore even more important with automated driving to fully understand the cognitive and perceptual mechanisms underlying driver distraction. Compared to other displays, head-up displays guide the driver's gaze direction towards the road scene. This can, for example, increase the safety of mobile phone use if the user interface for making calls is displayed in the HUD (Nowakowski et al. 2002). However, this also brings along an increased risk for other, more subtle, negative effects such as change blindness and inattentional blindness. These subtle but powerful perceptual effects can make drivers miss things they actually look at (Wickens and Horrey 2008). These effects are even more dangerous since drivers are mostly not aware of them.

8.1.3 Will Human Insight Solve it?

Finally, humans habitually act against the law or better knowledge. They systematically overestimate their capabilities. This will always limit the effectiveness of any legal and technical safety measures: Although drivers are generally aware of the risks of texting while driving, many still do it. A British survey (Direct Line (Motor Insurance) 2002) found that drivers correctly consider sending a text message on a mobile phone a very distracting activity. Nevertheless, the Canada Safety council reported (Canada Safety Council/Conseil canadien de la sécurité 2016) that 30% of their participating drivers admitted to texting while driving in the past. In addition, drivers are generally convinced that they perform well, or at least better than average: Half of the drivers think they belong to the safest 20% and almost nine out of ten drivers believe to drive safer than the average driver (Svenson 1981). Almost a third (32%) of the drivers between 18 and 24 claim they can glance away from the road for 3–10 s and 27% of the drivers over 25 share this opinion (Tison et al. 2011). Seo and Torabi (2004) showed that almost all college students who own a cell phone use it at least occasionally while driving. Similar results have been reported by Olson et al. (2005). Pfleging et al. reported frequent phone use of their participants while driving: 27% of the drivers read text messages while driving and about 15% even wrote text messages in the car (Pfleging et al. 2013). The crash statistics, however, speak a different language: Inattention as a cause at least contributes to almost 80% of all crashes and 65% of near-crashes (Neale et al. 2005). Approximately 70% of all distraction-related crashes are single-vehicle collisions or rear-end collisions. More recent surveys (Young et al. 2003) actually find even higher numbers than the 2000 NHTSA estimate (Ranney et al. 2000) which fully attributed 25% of all crashes to inattention and distraction, and it is still expected that distraction as a cause is underestimated in most crash studies (Ranney 2008).

8.1.4 Structure of this Chapter

The goal of this chapter is to structure and analyze the influence of non-driving-related activities on the driver's performance. We will first have a general look at the driver's tasks and activities and then proceed with an analysis of the cognitive and perceptive mechanisms behind distraction and their effects. Building on these two foundations, we will then examine the particular types of and causes for distraction by non-driving-related activities, and see how they can be measured and counteracted or at least mitigated. After this analysis of the current state of the art, we will conclude the chapter with an outlook onto some emerging technologies and the effects, risks, and potentials that can be expected from them.

8.2 The Driver's Tasks and Activities

In order to start from a shared understanding, we will introduce driving-related but also unrelated tasks and activities commonly performed by drivers, and then reflect on the resources and internal processes involved. Driving itself is a complex task consisting of multiple interrelated activities (Lee et al. 2008b) involving the driver's vision, cognition, and motor activity. There is no comprehensive list of activities that are critical for or related to driving. Different lists are presented, in, e.g., McKnight and Adams (1970), Treat (1980), Ranney (2008).

In addition to these tasks, drivers engage in non-driving-related activities such as conversations with co-passengers or daydreaming, but also interact with built-in and brought-in devices such as the central information display or a smartphone. These activities compete with the driving task for attention, or—more precisely speaking—for cognitive, perceptual, or motoric resources. How drivers manage and time these tasks and decide on the allocation of their resources is still a matter of research. The driver's motivation, behavior and tasks are not yet fully understood but several models and theories have been developed to reach a better understanding of each activity as well as the resources and processes involved. Each of these approaches focuses on other aspects of driving, but so far, there is no model that comprises all of them. Below we will introduce some of the most popular approaches briefly.

8.2.1 The Driving Task

Michon (1985) described driving as a problem-solving task that has to be accomplished on three levels: the strategic, tactical, and operational level. The strategic level represents the trip planning tasks, such as route choice and travel patterns, which involve a time horizon of minutes to weeks. The tactical level represents the maneuvering of the car, such as lane choice and obstacle avoidance, and requires actions within seconds to minutes. The operational level comprises all tasks of car control, such as the lateral and longitudinal control and gear shifting, and proceeds within milliseconds to seconds.

Another way to understand driving is to look at the types of control from a rather goal-directed perspective and discriminate beween feedforward, feedback and adaptive control (Lee et al. 2008a). In feedforward control, the driver's anticipation of the future state leads to a certain behavior. Experienced drivers benefit from more precise and accurate expectations and expected events normally lead to lower reaction times (Green 2000). Feedback control means adjusting the current state towards the

Table 8.1 Lee et al. (2008a) lists example challenges for each type of control and each time horizon. These can quickly lead to a breakdown of control or in the long run cause behavior changes

Control Type	Operational: Control Attention to Tasks (milliseconds to seconds)	Tactical: Control Task Timing (seconds to minutes)	Strategic: Control Exposure to Tasks (minutes to days)
Feedback-reactive control based on past outcomes	Time constant of driver response is slower than that of driving demands	Feedback is too delayed or noisy to guide behavior	Poor choices might not affect performance
Feedforward-proactive control based on anticipated situation	Task demands are unpredictable or unknown	Task timing is unpredictable or unknown	Potential demands are unpredictable or unknown
Adaptive-metacontrol based on adjusting expectations, goal state, and task characteristics	Tasks that lack a graded effort/accuracy trade-off	Biological and social imperatives not calibrated to task importance	Poor calibration regarding interaction between driving and IVIS goals

goal state. This relies on timely and precise information regarding the difference and a fast reaction to it. Adaptive control means the behavior to reduce the differences between the current state and the goal state by redefining the goal. Lee et al. (2008a) related these control types to the task levels and describe potential problems for each combination (see Table 8.1). They further state that control suffers from interactions across the time horizons, i.e., a breakdown on one level affects the other levels.

Fuller's task-capability model (Fuller 2000) describes how the interdependency of task demand and the driver's capabilities influences the safe control of the car (see Fig. 8.1). *Capability* refers to the driver's ability to perform at his or her level of competence and depends on the driver's condition (mental and physical characteristics), experience and training (as the upper limit of competence) as well as his or her current state (e.g., stressed, tired). The *task demand* depends on the environment, the other road users, driver communication, the vehicle and especially its speed, road position and trajectory. When *demand* exceeds *capability*, the driver will experience a loss of control and potentially end up in a road accident. Both, demand and capability, can be influenced on the strategic, tactical, and operational level.

The multiple resource theory by Wickens (2002, 2008) is a general theory about multitasking that has been applied to driving. It proposes four dimensions of information processing: stages (perception, cognition, response), codes (verbal, spatial), modalities (auditory, visual) and visual processing channels (focal, ambient). The processing of one piece of information requires at least one level of each dimension; the concurrent processing of several stimuli or tasks is hindered when both use the same levels. The bottleneck of information processing is the response selection on the stage level (Pashler 1994; Strayer and Johnston 2001), which can cause per-

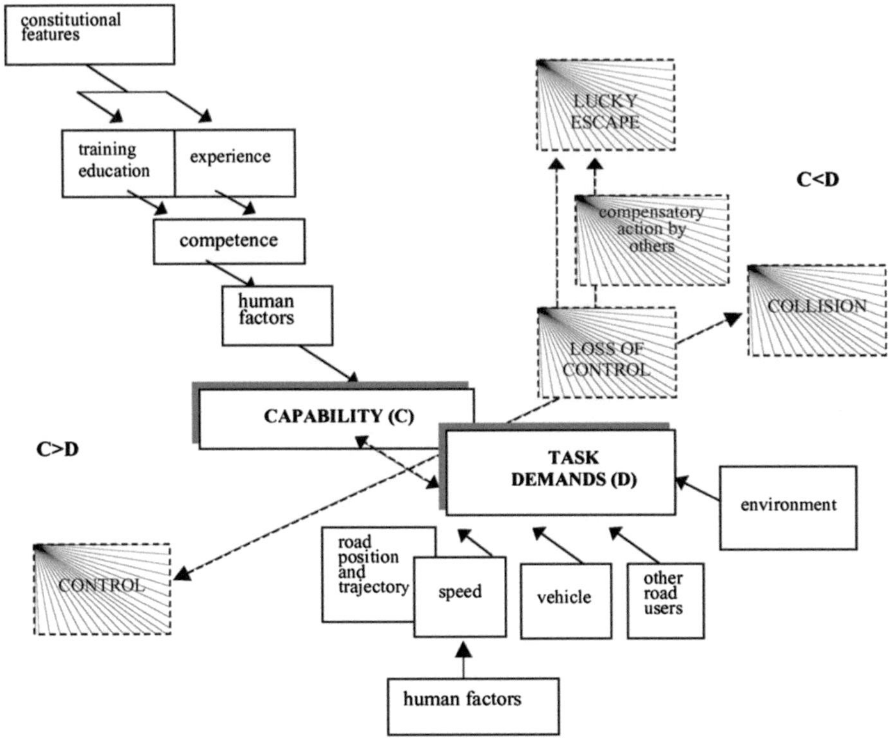

Fig. 8.1 The task-capability interface model predicts the driver's performance from the interplay of his momentary capabilities and the sum of the demands of all ongoing tasks. The driver will potentially fail and cause an accident when demand exceeds capability. He will succeed and reach the destination safely when capability exceeds demand (Fuller 2000). Figure from Knapper et al. (2012). *Source* Fuller (2005)

formance deterioration even when different modes and codes are used (Gladstones et al. 1989). In general, this theory focuses rather on the mechanisms involved in multitasking than the overall resource demand.

These models allow us to understand the interrelations between multiple, concurring tasks and how this concurrence can create problems at different conceptual levels of driving. While they are certainly not a fully comprehensive description of the processes involved, they at least explain the occurrence and allow the prediction of problems in many situations.

8.2.2 Non-driving-Related Activities

In addition to the basic activities for maneuvering the vehicle, drivers often perform additional tasks. These tasks can be classified in different ways. Traditionally, in-vehicle activities have been split into two (Wierwille 1993) or three (Bubb 2003) groups of activities. The *primary driving task* comprises all activities necessary to maneuver the vehicle as explained for instance in ISO 17287 (International Organization for Standardization 2003). When using a bisection of all activities, the *secondary task* refers to all other activities. As an alternative, a trisection of driving activities (Bubb 2003) describes the *secondary driving task* as all tasks and functions that increase driving performance or driving safety. This can, for instance, include tasks such as activating the headlight, enabling cruise control, or adjusting the windshield wipers. In this case, the *tertiary driving task* refers to the remaining activities such as operating comfort, entertainment, or communication features, eating, and drinking; Stutts et al. (2005) identified more non-driving-related activities in a real-world study.

When driving fully automated, these definitions may get confused in discussions since most of the *primary* and *secondary tasks* will disappear. Therefore, we decided to adopt the term *non-driving-related tasks* or *activities* instead (Department of Transportation and National Highway Traffic Safety Administration 2013; Pfleging and Schmidt 2015; Young et al. 2008). This term can be used to describe any kind of activity or task beyond maneuvering the vehicle, no matter whether it is driven manually or in an automated way.

8.2.3 Resources for Cognition and Perception

Driving mainly (up to 90%) relies on the visual channel and can be executed practically without detriments in performance even if all other input modalities are excluded (Cohen and Hirsig 1990). Drivers react to visual input with manual output for controlling and maneuvering the car. Between this input and output, many conscious and unconscious processes and decision-making have to take place—which brings us to the driver's mental capabilities. Driving does not always require the driver's full attention (Lee 2004). While a driver might be overloaded when driving in the center of a mega city, the same person might be bored and unchallenged on an empty, straight highway and actually develop strategies (such as performing non-driving-related activities) to create a certain level of mental load to stay attentive or even awake. Drivers can dynamically adapt the attention devoted to the driving task without a deterioration of performance (Lee 2004). They can move more attention to the driving task when uncertainty in the road situation (Senders and Kristofferson 1966) and/or task difficulty increase. Task difficulty in turn depends on the driver's capability and the task demand (Fuller 2000) as discussed above.

Generally, drivers seem to apply some kind of self- and task-regulation processes to balance task demand and capability (Lee 2004). The multiple resources theory suggests, that the driver's resources are limited in capacity and vary over shorter or longer time frames (e.g., fatigue and aging). When the demand of the sum of all tasks exceeds the driver's capacity, the driver is overloaded and the risk for crashes increases. Drivers increase concentration and attention and thereby decrease the risk and in turn anxiety (Taylor 1964). By investing more effort, drivers might in fact be able to compensate a high workload, but only to a certain extent and for a certain time (Knapper et al. 2012). Over a longer time frame, drivers adopt strategies to keep the subjective risk low ('zero risk model' Näätäänen and Summala 1974; Summala 1988). Risk is subjectively related to task difficulty (Fuller 2011) but according to Fuller (1984, 2005), drivers want to stay within safety margins and therefore try to keep task difficulty below a certain level.

The perceived risk is a dynamic and complex internal representation, which depends not only on the current road situation and the control of the own car but also on the driver himself or herself. Apart from individual differences in the predisposition toward risky behavior, the driver's experience and history also plays a considerable role (Lee et al. 2012). For example, a driver might have engaged in non-driving-related activities such as texting on the smartphone only few times but always without causing a dangerous situation. Since no immediate danger appeared, the driver develops the impression that texting is not dangerous and that he or she is capable of controlling the car and the phone simultaneously—probably without reflecting on the current demand of the road situation. Consciously or not, the driver might be developing a habit and a specific mental model of the associated risk of this behavior due to the poor and often delayed feedback of driving ('conditioning trap') (Strayer et al. 2003; Mccarley et al. 2004; Rasmussen 1997; Lee et al. 2008b; Lee 2004). As the demand of the road situation increases, also the actual risk increases—in contrast to the subjective risk. Although such a behavior often ends in crashes, drivers are found to continue this type of habits (Rajalin and Summala 1997).

The visual channel is highly important for driving but also for the interaction with in-car and brought-in devices. When performed in parallel, the driving task competes with device interaction tasks for the visual resources so that drivers need to apply a task and resource management strategy. These strategies vary from immediate completion of short tasks to the subsequent execution of small partitions of longer tasks (see Wierwille et al. 1993 for a more detailed description of this process). Surprisingly, drivers do normally not exceed a duration of 1.6 s for one glance (varies between 0.62 and 1.66 s) (Wierwille 1993). The overall count of focus switches between the two activities depends on the overall task duration which in turn strongly depends on the interaction task. An easy and intuitively designed interface promotes a short task duration but the completion of a task does not necessarily mean that the driver's cognitive focus will be devoted to the road scene again. Drivers seem to keep thinking about a completed task, which holds them from devoting their entire attention to the road (Lee et al. 2008a). The glance behavior and engagement to accomplish a task and the associated goal can be influenced by a concept called task perseveration. Task perseveration occurs when a driver has spent effort on accomplishing a goal-oriented activity but the goal could not be achieved or when there is an opportunity to continue, e.g., a subsequent task and the driver neglects the broader goals such as safe driving (Fox and Hoffman 2002). Task perseveration considers goal valence, proximal closure and goal emergence as influencing factors (Fox and Hoffman 2002). This means that cognitive inertia and propensity rather lead to continuing than to giving up a goal, that the motivation (or activation) to reach a goal increases toward its achievement, and that the achievement can induce new goals.

When drivers interact with a device or system and switch the center of their visual focus from the road scene toward the display, the road scene is often shifted into the peripheral visual field (due to the position of the display). Peripheral vision is characterized by a lower resolution and less mental processing capacity (Trent 2005) and hence provides slower reaction times (Gish and Staplin 1995). Complex and short-term aspects of driving, such as hazard detection, require central vision and its benefits and break down when performed as a peripheral task. Other tasks, such as lane keeping, can be performed at an acceptable performance with peripheral vision, but increase the overall workload of the interaction task (Wann et al. 2000). However, drivers are probably not aware of their visual limitations: The detail and precision of human vision can induce the wrong impression that the situation is perceived and understood comprehensively (Kalat 2004; Kevin O'Regan and Noë 2001; Findlay and Gilchrist 2003). When the driver's visual and attentional focus are not on the same object or activity (*covert attention*), the processing resources available for each of them are lowered, compared to focusing visually and attentionally on the same object (*overt attention*) (Trent 2005). Phenomena such as change blindness—the 'looked-but-failed-to-see' problem—may occur as a consequence (Trent 2005). Therefore, the driver is probably not aware of the road situation—and potentially neither of the manifest distraction and the lack of awareness (Lee et al. 2008b). Especially in conjunction with a high workload (complex tasks), processing delays of the driving task can occur which in turn create an increased risk.

8.3 Inattention and Distraction

In this section, we will introduce existing concepts and taxonomies for the phenomena of *attention*, *inattention* and *distraction*. The general concept of inattention and distraction describes how the driver splits attention between tasks or distributes tasks over time and how drivers choose to engage in tasks (Lee et al. 2008a). Attention, inattention and distraction are psychological constructs based on theories and observations. Hence, there is no full agreement what these constructs comprise, how they differ and how they relate to each other. Several surveys compared existing definitions and pointed at the differences and similarities, such as the recent survey and classification approach of Regan et al. (2011).

8.3.1 Defining Inattention and Distraction

To describe attention, Regan et al. (2011) referred to the definition of the Macquarie Dictionary in which attention is defined as the concentration of the mind upon an object. As explained earlier, attention also often includes a visual focus on an object (Trent 2005). In the driving domain, drivers are often considered attending, if their eyes are directed toward the road (Liang et al. 2012), although this approach neglects the phenomenon of cognitive distraction. In general, attention does not directly induce conscious awareness (Lamme 2005). Accordingly, inattention means that insufficient or no attention is devoted to a task or an object. In the context of driving, Lee et al. (2008b) suggest that "*inattention represents diminished attention to activities that are critical for safe driving in the absence of a competing activity*." Competing activities include interaction with in-car systems or passengers, eating, thinking, and also non-critical driving activities such as turn signal control. Regan et al. (2011) have a different perspective on this construct and define inattention as a process and its various forms by psychological mechanisms that give rise to the process.

Distraction is one form of inattention (Regan et al. 2011) and stands out by a secondary task that requires focusing on an object, event, or person not related to the driving task (Ranney 2008). Corresponding to the prior definition of inattention, Lee et al. (2008b) state that "*driver distraction is a diversion of attention away from activities critical for safe driving toward a competing activity*." It is further considered as a mismatch between demanded and devoted attention (toward the road) (Lee et al. 2008b) under an overload of resources (Hurts et al. 2011). Hence, it is generally associated with a recognition or processing delay (Pettitt et al. 2005) and a deterioration of driving performance (Regan et al. 2011).

Generally, distraction has been approached from different perspectives, e.g., by focusing on the initiator, the process of distraction, its outcome (Lee et al. 2008b) or the workload and resources. Most research regards distraction as excessive workload (*overload*) and limited attentional resources (Hurts et al. 2011). Workload is defined as the mental resources or information processing capacity devoted to a task (Brookhuis and De Waard 2010). Accordingly, also the NHTSA (Ranney 2008) describes (cognitive) distraction as the (mental) workload associated with another task. Another perspective on distraction considers the dynamics and management of distraction, e.g., in terms of task timing and interruption management. The queuing theory, for example, describes how drivers plan and manage interaction with different tasks (Lee et al. 2008a). It describes the demands in terms of the policy for queuing of tasks, task timing, and how easily the task can be interrupted at the tactical level. This suggests that distraction is caused by failures in timing and prioritization and by switching costs. It further suggests that every—even a short—competing activity will lead to delays in the driving task. However, this is an oversimplification: The driver is not considered as an active individual who determines the timing of tasks. Dynamic adaption of task demand as well as the delay and interruption of tasks is neglected. Distraction and performance are assumed to break down when the task is neither ignorable, predictable nor interruptible (Lee et al. 2008a).

Also the initiator of the distraction influences task timing and interruption. The engagement in a non-driving-related activity can be initiated by the driver (e.g., due to boredom) but also by the task itself being a compelling and salient stimulus (Regan et al. 2011). Some researchers argue that the latter is a controlled mechanism (and a conscious decision) that gives rise to distraction (Lee et al. 2008a). Although this distinction might seem unimportant at first sight, it is useful upon closer inspection: when drivers voluntarily engage in a task, they have more time to adjust their driving behavior and compensate for the increased demand (Regan et al. 2011). Furthermore, drivers can time interaction better: they initiate interaction themselves when the current and predictable task difficulty is manageable. In contrast, when reacting to compelling stimuli (such as a ringing phone), drivers often neglect the upcoming driving demand (e.g., a construction zone ahead) (Nowakowski et al. 2002). Compelling stimuli seem to somehow put pressure on the driver (Knapper et al. 2012; Fisher et al. 2002) that is why Regan et al. (2011) suggest that the psychological mechanisms involved in self-initiated and stimulus-initiated distraction might differ and could lead to different patterns of interference.

8.3.2 Distraction Types and Phenomena

In general, distraction can occur in a single or in several of the following modalities (Knapper et al. 2012; Ranney 2008; Young et al. 2003): visual (not looking on road), auditory (locating the source of a sound), motoric (grasping something), or cognitive (such as letting the mind wander). Visual distraction is further subdivided into blocking of the view, focus shift away from the road scene and loss of visual attentiveness (change blindness) (Young et al. 2003). Regan et al. (2011) propose a more detailed differentiation between inattention and distraction. A critical aspect can not be detected or an activity which is part of of driving is neglected due to:

- restricted attention: physical prevention (e.g., due to biological factors),
- misprioritized attention: focus on the wrong aspect of driving,
- neglected attention: neglect of attending to an object (e.g., due to a rare event),
- cursory attention: cursory or hurried attention leads to failed detection,
- diverted attention: attending a competing activity (commonly referred to as 'distraction'), further divided into non-driving-related and driving-related.

As an extension to these types of inattention, the driver's mind can be absent with or without external reason. Daydreaming and thoughts—driving-related or unrelated (mind wandering)—can keep the driver from attending the forward scene (Regan et al. 2011). Researchers do not agree on whether these count as inattention or distraction. Thoughts can be triggered internally or externally but are not necessarily intentional. Both, thoughts and daydreaming, can induce an involuntary retraction of attention from the driving task and are not easy to ignore. Since this may not be a conscious process, it is not or hardly controllable. Mind wandering increases when the overall task demand (e.g., due to a familiar road or uneventful driving (Lee 2004)) is low and competing activities could further hamper or damp attention (Kane et al. 2007; McKiernan et al. 2006).

Furthermore, the phenomena of *change blindness* and *inattentional blindness* have to be considered. Inattentional blindness refers to the driver's failure in detecting an appearing object when looking elsewhere or even at the region or precise location at which the object appears (Wickens and Horrey 2008). The detection depends on the expectation of the new stimulus and its similarity to the attended object (Ambinder and Simons 2005). Change blindness is very similar to inattentional blindness but refers to the failure in detecting a change of a previously noticed object. Whether or not a change is detected depends on the expectation, its visual transients and its importance for the task or goal (Rensink 2002; Mack and Rock 1998).

8.4 The Influence of Non-Driving-Related Activities

Performing non-driving-related activities can affect driving and potentially distract the driver. People mainly engage in non-driving-related activities because of the associated benefits (Ranney 2008) and do not care so much about distraction. Therefore, UIs for the support of such tasks should keep the influences on the driver minimal. In order to design non-driving-related activities and appropriate automotive UIs for them, it is important to understand how non-driving-related activities influence the driver. This includes understanding the reasons and causes of distraction as well as the associated effects and risks. By measuring certain aspects during the execution of non-driving-related activities, designers can compare the influence of different interfaces and tasks on the driver and the driving situation. Based on many such findings, they can derive a set of best practices for the design of automotive user interfaces and non-driving-related activities.

8.4.1 The Reasons and Causes of Distraction

Knapper et al. (2012) differentiate three major factors which influence the effects of distraction on task performance: Driver characteristics, driving task demand, and secondary task demand (see Fig. 8.2). In addition to that, Young et al. (2008) mention the driver's self-regulation in this context. Generally, distraction and a deterioration of performance are assumed to occur when a competing task makes use of the same (mostly visual) resources as the driving task (Wickens 2002), the driver's resources are overloaded and the processes and task management break down. Once the driver engages in a competing activity, goal-activation and task perservation can further increase distraction due to an increasing fixation on the completion of the competing task and the neglect of the overall driving goal (Fox and Hoffman 2002). When the driving control breaks down on one level, distraction increases further (Lee et al. 2008a).

Most drivers think they drive safer than the average, although many of them engage in non-driving-related activities (Svenson 1981). As mentioned before, performing a secondary task can become a habit, and habits seem to have a greater effect on the engagement than norms and attitudes (Bayer and Campbell 2012). The social role and imperatives can further prompt the driver to attend to other tasks, such as parents who try to calm a child on the back seat (Hancock et al. 2008; Fisher

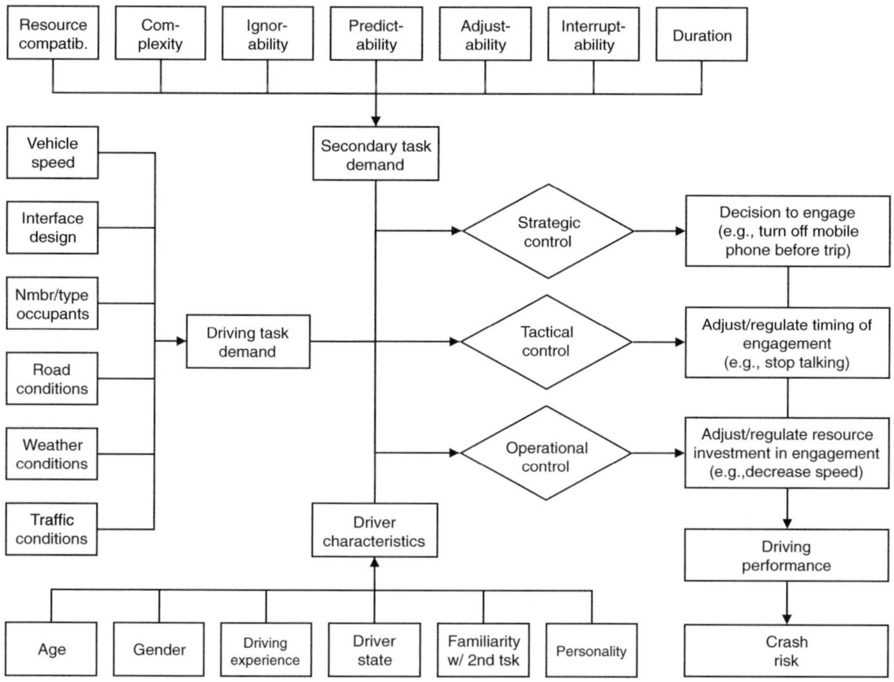

Fig. 8.2 Factors related to the drivers themselves as well as the demands of the driving task and the non-driving-related activities can all influence the effects of distraction on driving performance; adapted from Young et al. (2008)

et al. 2002). Also the driver's experience (e.g., anticipation and expectation, timing and interruption management), but also his or her personality (e.g., the willingness to take risks or to obey to law restrictions, self-esteem), current state (e.g., bored, sleepy, drunk, and aroused) and condition (e.g., age, limited vision) can influence the motivation for an engagement in other activities but also their effects on driving performance (Green 2000; Lee et al. 2008a; Regan et al. 2011).

Finally, also the task difficulty of all ongoing activities as well as their timing affect the extent of distraction (Young et al. 2008; Lee et al. 2008a, b). Non-driving-related stimuli can be compelling and salient (Wickens and Horrey 2008; Trent 2005) since they are often unusual, unpredictable, irritating, sudden, unexpected, or violate expectations (Stutts et al. 2005). Such tasks are difficult to ignore (Regan et al. 2011) and can lead to an unintentional engagement in these tasks—even after the task itself is completed (Lee et al. 2008a). The target of a driver's attention can—to a certain extent—be predicted by means of the SEEV model (Wickens and Horrey 2008) based on the **S**alience of a stimulus, the **E**ffort to switch attention, the **E**xpectancy of what requires attention, and the **V**alue (importance) of a source of information.

8.4.2 The Effects and Risks of Distraction

Looking again at the driving task as a control process at three levels (strategic, tactical, and operational), Lee (2008a) outlines that distraction causes a breakdown at one or multiple levels: On the lowest, the operational level, the driver controls how to invest resources. Here, performing an additional activity could cause a breakdown of visual (eyes off the road), cognitive (mind off of driving), and/or physical demands (e.g., hands off the wheel). The additional activity competes and interferes with lateral and longitudinal control. If the gaze is directed off the road (operational level) longer than expected, this will also influence the tactical level at which the driver makes sure to adapt driving to the current context (e.g., by adjusting speed and gap). On this level, distraction especially leads to the failure of proper task timing. At the same time, also the strategic level is involved since the driver voluntarily decided to perform another activity concurrently to driving the car. Here, the driver might make inappropriate priority calibrations. Wierville (1993) explains the influence of technology use on the driver's resources: (1) Visual interfaces may create a demand and conflict in terms of foveal vision, because the driver uses time-sharing to use vision for different tasks. (2) In contrast, Wierville sees only a reduced influence of additional tasks regarding motor resources (e.g., taking one hand off the steering wheel). (3) Regarding cognitive resources, he argues that cognitive load is expected to draw foveal vision to specific areas which suppresses visual scanning and therefore may reduce the margin of safety. (4) Both, for cognitive load and auditory responses, Wierville sees the risk of perceptional narrowing.

8.4.2.1 The Effects of Specific Non-Driving-Related Activities

A recent large-scale study by Strayer et al. compared the use of ten different *in-vehicle informations systems* (IVIS) and their influence on the drivers' cognitive load (Strayer et al. 2015). Overall, the authors identified a moderate to high level of cognitive load during IVIS use, and they associated workload ratings with system complexity and task completion time. Further, Strayer et al. (2015) highlight the importance to test with older drivers since their workload was found to be significantly higher compared to the younger ones. They also found long-lasting residual costs since it took the drivers about 27 s to return to their baseline level of performance.

Sayer et al. (2005) performed a naturalistic driving study to investigate the circumstances causing the driver to engage in non-driving-related activities. They reported not only the probability of the engagement in the single activities but also the context of use—such as the road situation and daytime—and the effects that these tasks had on driving performance. Very similar to this work, also Stutts et al. (2003) performed a naturalistic driving study but found evidence for more diverse distraction causes, such as the distraction by babies on the backseat. Both Sayer et al. and Stutts et al. considered the context of use—which allows to understand the real-world causes and in turn to counter them in the future—during their analysis of the impact on driver performance. This allowed a direct comparison of the impacts of the diverse tasks. Young et al. (2003) thought one step further and provided a list of driving-related and -unrelated but also more novel tasks together with the measured and expected effects of these activities. The two activities most frequently mentioned and most deeply studied are the use of phones—calling, texting, and reading—and the use of route guidance devices.

Calling while driving is known to distract the driver. The risk of collision is up to four times higher while using a phone (Redelmeier and Tibshirani 1997). As summarized by Young et al. (2003), many publications investigated the manifold influences of mobile phone usage. In their survey, they provide an overview of the different types of distraction that can be caused by mobile phones. This includes visual, auditive, physical, and cognitive distraction, which in turn impairs driving-relevant activities such as visual search patterns, reaction times, and decision-making. Crundall et al. (2005) showed that talking on a mobile phone is different from talking to a passenger in the car: When talking to a passenger, conversation stops in challenging situations since the passenger is aware of the current driving context. This does not happen during phone conversations where the amount of communication can even increase.

Similarly, also *reading and writing text messages* distracts the driver visually, cognitively, and physically as for instance shown in a meta-analysis by Caird et al. (2014). Since an increasing number of accidents was reported in which the phone had been used just before the accident (Redelmeier and Tibshirani 1997), handheld calling (and texting) has been prohibited in a number of countries.[1] In contrast, handsfree calling is allowed in most countries, although studies revealed that the distraction is similarly high (Caird et al. 2005, 2008; Redelmeier and Tibshirani 1997) since the conversation itself is often the distracting part. Also Strayer et al. (2015) studied voice-based (smartphone) user interfaces in the car and recommend to be cautious with introducing and using such technology in the driving context.

The use of *navigation systems* is another frequent activity. The influence of using such systems has been investigated in depth by many researchers. Young et al. (2003) provide a detailed overview of related research and findings (Young et al. 2003). Since most navigation systems use displays to show route instructions or menus for destination entry, the driver's visual attention is required at certain times. Similarly, the auditive channel is occupied when turn-by-turn instructions are given and manual demand occurs while entering the destination; Destination entry is claimed to be one of the most challenging tasks from a distraction point of view. A recent real-road study revealed that the use of navigation devices increases the eyes off the road time (Morris et al. 2015). Interestingly, the detailed analysis of the gaze data showed that the average glance duration when looking at the navigation devices was 0.76 s which is below the officially proposed thresholds.

These examples illustrate that many of the non-driving-related activities distract the driver in one way or another. From the example of calling and texting, however,

[1] http://www.cellular-news.com/car_bans/, last access 2016-07-10.

we already see today that a (legal) ban of highly distracting tasks while driving does not keep the drivers from performing such activities. Even though handheld calling and texting is prohibited in many countries, drivers keep doing it. Therefore, to increase driving safety effectively, we need to investigate how to support these activities in such a way that future systems are less distracting and, thus, increase driving safety. This holds both for integrated in-vehicle information systems and for nomadic devices. The use of the latter is especially challenging—for instance most smartphone and tablet apps are not specifically developed for the driving context, which can negatively impact driver performance even more. One key to success could be supplying and using more context-related data. In the case of calling and texting, such information could for example be used to create an awareness of the driving situation for the remote party (Pfleging et al. 2013). This might positively affect driving safety during the communication.

Also, with increasing driving assistance, an increased engagement with in-vehicle systems could actually help to keep the user alerted. A study conducted by Takayama and Nass (2008) for example revealed that "slightly interactive media" can improve the performance of drowsy drivers. Thus, engaging the driver in non-driving-related tasks could become relevant with increasing driving automation.

8.4.2.2 Car Accident Statistics

In many countries, detailed statistics are collected about road accidents and their consequences. However, often these statistics can only be based on police-reported incidents since the data of other (minor) incidents is just not reported. Most accident statistics provide insights on where (e.g., road type, state), when (day of the week, time, month), how (e.g., while turning left/right, exceeding speed limit, collisions with cars/pedestrians, road exceedances, …), and under which environmental conditions (road situation, illumination, obstacles, parties involved) road accidents occurred. They also document reasons for the analyzed accidents as they were reported by the involved parties and the police. One drawback of most accident statistics is that they rely on self-reported details about facts like pre-accident information which cannot be recorded or measured after the accident. For instance, a crash may be caused by a delayed brake action of a driver following a vehicle in front. While this (final) reason may be present in accident reports, the original reason (e.g., drowsiness, distraction due to texting) often is not recorded or even asked for. Since this information is self-reported, one can assume that in certain cases negative behavior remains unreported as a matter of self-protection of the driver who caused the accident. This hinders the analysis of accident reasons and safe driving research.

In order to overcome the knowledge gap about real accident reasons and to enable a detailed analysis of general driving behavior, a number of so-called naturalistic driving studies (Neale et al. 2005) has been conducted during the last years. For this type of study, a set of cars is equipped with technology to record driving behavior such as speed, acceleration, and other car network information along with cameras to record multiple views (e.g., forward and rear driving scene, drivers cockpit) (Neale et al. 2005). Being installed in privately owned vehicles there is evidence that after a short adaptation phase the drivers disregard the installed technology (Neale et al. 2005) and drive as usually and, thus, provide more detailed insights into driver behavior than it would be possible with explicit test vehicles or with post-crash interviews. Already the first study of this type ("100-Car Naturalistic Driving Study" Neale et al. 2005) revealed that only a small subset of minor incidents was reported to the police. When analyzing the data for driver inattention as a contributing factor for events, this inattention was split up into four categories: secondary task involvement (i.e., involvement in non-driving-related tasks), fatigue, driving-related inattention to the forward roadway (e.g., looking at the rear mirror), and non-specific eye-glance. In total, 78% of the recorded crashes and 65% of the near-crashes were preceded by an event that falls into one of these four categories of inattention. The analysis showed that the distraction due to non-driving-related activities was the largest category of inattention that contributed to crashes and near-crashes. Among distraction by non-driving-related activities, the use of wireless devices (mainly cell phones) was the most frequent activity observed, followed by passenger-related activities (e.g., conversation), and internal distraction. For 93% of all crashes with lead vehicles and minor collisions, inattention was a contributing factor (Dingus et al. 2006).

8.4.3 Operationalizing and Measuring the Influence of Non-Driving-Related Activities

For users at home who operate a desktop computer or mobile phone, task and interface evaluations often focus mainly on usability, task performance (errors, task completion time), and user experience. Typically only a single task is performed throughout the observation. In contrast, non-driving-related activities in a car are generally additional activities that the driver performs concurrently to the original driving task. In order to understand the influence of non-driving-related activities, researchers therefore need to analyze both the non-driving-related activity and the driving task of this dual-task situation. This includes for example measuring the task performance (errors, completion time and response time) and usability of a secondary task as well as its influence on the primary task performance (Green 2012). Driving (task) performance measures comprise for example information about lateral control (e.g., lane deviation, steering wheel activity), longitudinal control (e.g., maintaining speed, braking behavior), and driver reaction (e.g., recognition time for unexpected incidents) (Green 2012; Bach et al. 2009). In this section, we discuss the evaluation of non-driving-related activity with a special focus on those tasks that make use of interactive in-vehicle systems. This includes the choice of the environment where to test user interfaces and tasks as well as the selection of appropriate measurements.

8.4.3.1 Choosing the Right Experimental Environment

Often, an analysis of performance-related measures is not sufficient, since it may only reveal extreme interface issues (Pauzié 2014). Therefore, often advanced methods and measures are necessary (Pauzié 2014). Lee (2004) explained one limitation of current driving and distraction studies: Many experiments make the assumption that drivers constantly devote their entire attention toward the driving task, but it is not appropriate to consider a non-distracted driver as a fully attentive driver. Due to the experimental environment, the participants might spend more attentional resources on driving than the typical undistracted driver in the real world. Lee therefore suggests that a disengaged driver might be more realistic since more than 70% of all drivers report a lack of concentration on driving (McEvoy et al. 2006).

The differences between lab experiments and the real-world regarding attention devoted to driving and additional tasks shows the importance of choosing the right experimental environment. By only conducting lab studies, one might miss important insights. Therefore, different evaluation methods should be chosen, for instance based on the current step of the design cycle or the maturity of the system under test. As outlined by Burnett (2009), the distinguishing factors for such methods to test in-car systems are related to environment (where does the experiment take place?), task manipulation (which tasks does the driver need to perform, e.g., single/multiple task), and the choice of the dependent variable. Figure 8.3 gives an overview how experiments can be conducted in a variety of environments—from simple lab setups to real-world driving situations. As also shown in this figure, Burnett points out that the choice of the environment also influences reproducibility and ecological validity: The closer an experimental condition is to the real world, the more confidence we have that the observed data corresponds with real phenomena (ecological validity). In contrast, the opposite holds for internal validity: In real-road experiments on public roads it is very difficult to control all variables and, thus, difficult to replicate experiments. In practice, initial evaluations of early prototypes are therefore often tested in lab environments and driving simulators. Later on, with progressing system maturity, real-road experiments are conducted to understand real-world influences (Broy et al. 2015; Santos et al. 2005).

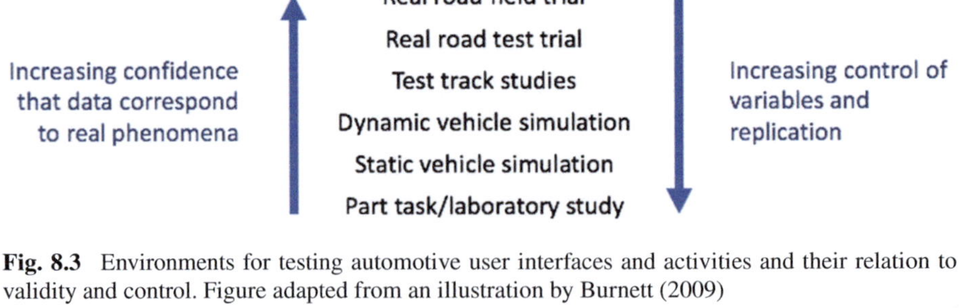

Fig. 8.3 Environments for testing automotive user interfaces and activities and their relation to validity and control. Figure adapted from an illustration by Burnett (2009)

8.4.3.2 Which Mectrics to Choose?

Given the (manifold) definitions of *(in)attention*, *distraction*, and *workload*, there is no commonly accepted direct way to measure either of these aspects. Instead surrogate measures are used as indicators to infer for instance *workload* and *cognitive load*. Metrics for assessing workload can be classified as subjective (e.g., questionnaires), physiological, or performance-based (Gawron 2008; Miller 2001; Wickens and Tsang 2015). Since we are mainly interested in the influence of non-driving-related activity in this chapter, we do not report on metrics related to user experience and subjective usability/user satisfaction.

Performance-based metrics can either relate to the drivers' performance with driving tasks (primary task performance, e.g., standard deviation from lane position as an example for lateral control, standard deviation of gap as an example for longitudinal control, or hazard detection), the use of primary vehicle controls, or the drivers' performance and/or demand of the non-driving-related activity (secondary task performance, including errors, task-completion time, reaction time, and eye gaze such as eyes off the road time) (Burnett 2009). Green (2012) provides an overview of driving-specific usability and performance measures and detailed, unified definitions of many of these metrics are part of the SAE standard J2944 (SAE International 2015).

Besides measuring driving performance or performance on the non-driving-related activity, the detection-response task (DRT) provides an interesting method to assess driver distraction, mental workload and (if using a visual version) visual distraction (Harbluk et al. 2013). This artificially sustained attention task shall be performed by the driver in addition to driving and the non-driving-related activity. The drivers' task is to quickly respond (by pressing a button) to a visual, tactile, or auditive stimulus which the system presents frequently but randomly. Measuring reaction time and hit rate then provides insights into distraction and workload. Another example is the so-called "n-back" task (Kirchner 1958) or the related delayed digit recall task (Mehler et al. 2009). In both tests a certain stimulus is presented to the user and needs to be repeated n steps later.

While secondary task performance might indicate a certain cognitive load associated with a task, Lee (2004) points out that this is only one aspect of distraction. He recommends to also assess how drivers decide to engage and disengage in certain (non-driving-related) activities. In addition to questionnaires for investigating user experience and satisfaction, a variety of *subjective methods* exists to assess the driver's workload. These range from unidimensional metrics such as the Bedford workload rating scale (Roscoe and Ellis 1990) to multidimensional questionnaires such as NASA TLX (Hart and Stavenland 1988) or the driver activity load index (DALI) (Hart and Stavenland 1988). Both questionnaires investigate different dimensions of workload such as auditory, tactile, and visual demand.

Physiological metrics have an advantage over subjective ones since physiological sensors allow for continuous and rather unobtrusive sensing of the drivers' physiological state (de Waard 1996). The goal of these measurements is to deduce the level of activation and global arousal. Various sensors and signals such as heart rate, heart rate variability, brain activity, speech measures, eye activity, respiration, and electrodermal activity have been shown to change with the drivers' workload (Kramer 1990; Miller 2001; de Waard 1996).

8.4.4 Countermeasures and Mitigation Strategies

For unwanted driver behavior, a typical countermeasure is to introduce laws, enforcement, and sanctions—which has been used successfully for example to prevent driving under influence or without wearing a seat belt (Goodwin et al. 2013). However, as presented in the same publication, the effectiveness of these means for distracted and drowsy driving are only of little effectiveness. Lee and Strayer (2004) relate this to social norms governing acceptable risk. These for instance define whether it is socially acceptable to use the mobile phone while driving. NHTSA therefore proposes to convince drivers to pay attention to the driving task, reduce underlying causes, and promote awareness of the associated risks (Goodwin et al. 2013).

Nevertheless—as already stated before—mobile phone usage is limited to specific use cases by law in many countries (e.g., only handsfree calling). Stutts et al. (2005) propose two ideas to reduce the risk of distracted driving: By implementing rumble strips as lane markings, roadways shall be made safer for distracted and drowsy driving. In addition to that, they recommend to improve the feeling of safety for resting areas. From an interface side, Ranney (2008) highlights the importance of interface design guidelines and standards. He also highlights the advantage of assistive technology, which could for instance warn the driver in risky situations. Another approach related to the user interface is presented by Donmez et al. (2008) who recommend to provide feedback to the driver in order to *enhance immediate driver performance* as well as to *initiate positive behavior*.

An ergonomically well-designed vehicle cockpit interface, which minimizes workload, will give the driver more capacity to attend to competing tasks, and hence reduce overall interference between the tasks (Regan et al. 2011). If information gathering can be *chunked* at (i.e., organized into chunks of) about 1 second or less, the driver will do so and will then return the glance to the forward scene. Chunking can be defined as breaking an information gathering task into segments, which together provide equivalent information gathering. On the other hand, if chunking takes longer, the driver will continue to glance at the location for a bit longer. However, in doing so, the driver at least senses time pressure to return to the forward scene (Wierwille 1993).

8.4.4.1 Designing Non-Driving-Related Activities

For the mitigation of driver distraction on the vehicle side, the design of activities and their associated user interfaces play an important role (Ranney 2008). With this regard, a body of guidelines, rules, and standards has been created over the last decades. The goal of all these documents is to facilitate interaction with in-vehicle systems, to ensure certain characteristics of automotive UIs, and to limit driver distraction. For a detailed discussion of the available documents, we refer to the work of Schindhelm et al. (2004) and Green (2008) who provide overviews on standards and guidelines from a European and an American point of view, respectively.

Major guidelines with the goal to reduce distraction while interacting with technology in the car have been created in Europe (European Statement of Principles, Commission of the European Communities 2007), Japan (JAMA Guidelines, Japan Automobile Manufactures Association 2004), and the United States (AAM Guidelines, AAM Driver 2006). These guidelines borrow specific aspects from each other and cover especially in-vehicle communication systems and integrated but also nomadic devices. Special principles addressed in these guidelines relate to overall design, installation, information presentation, interaction, system behavior, and information about the system. Heavily based on these existing guidelines, the NHTSA Visual-Manual Guidelines (Department of Transportation and National Highway Traffic Safety Administration 2013) provide various additions and also integrate newer research results. The guidelines relate only to the use of integrated IVISs for non-driving-related activities when performed as visual-manual activities, i.e., when the driver looks at the interfaces, performs manual input with his/her hand, and finally waits for a (visual) response. The guidelines also distinguish between a set of inappropriate and potentially suitable non-driving-related activities. For the latter, design recommendations are given and for novel activities testing procedures are proposed.

8.5 Outlook

8.5.1 New Resources for Non-Driving-Related Activities?

Vehicle manufacturers and suppliers envision that automated driving functions will be available in production vehicles approximately by the end of this decade.[2] In our context, we understand automated driving as driving at the automation levels 3, 4, or 5 as defined in SAE J3016 (SAE International 2014). At these levels of automation the driver does not need to maneuver or monitor the car in most situations. Thus, the driving task is almost fully removed. Consistent with Carsten et al. (2012), we expect that the removal of the driving task will cause drivers to engage in various types of non-driving-related activities. In fact, we believe that the possibility to do so is key for the success of automated driving. However, this might pose the driver at risk whenever the automation cannot accommodate and thus requires the driver to take over control (Lee 2004). This holds especially for those levels of automation at which

[2]See for instance http://www.driverless-future.com/?page_id=384 for current predictions, last access: 2016-07-01.

only parts of the driving task are automated (assisted and partly automated driving, SAE levels 1 and 2), where a take over by the driver (Gold and Bengler 2014) and manual driving are the fallback when automation fails (SAE level 3) and/or when not all driving situations are supported by the automation (SAE levels 3 and 4).

The way in which the transitions from automation to manual operation are designed and supported can affect and restrict the driver and thus make such transitions very difficult to perform as shown for example in medical environments (Cook et al. 1991). Hence, research on distraction and design guidelines will remain necessary even in the age of automated driving. Applying the Yerkes-Dodson-Law (Yerkes and Dodson 1908) to the driving context, we expect the human operator performance to decrease both when the driver is overloaded or underloaded (Coughlin et al. 2009). With increasing assistance and automation, we expect that underload will happen more frequently. In such situations, an additional non-driving-related activity might actually be helpful to maintain or regain optimal driver performance as already shown in prior research (Takayama and Nass 2008).

A variety of experiments investigated how vehicle automation changes the driver's behavior and perception regarding attention and distraction during automated driving. For instance, Young and Stanton (2002) identified a decreased availability of mental resources when the mental workload is reduced, e.g., by increasing the level of driving automation. With regard to non-driving-related activities during automated driving, it is not yet fully clear, (1) which activities are most desired by the drivers, (2) which of these activities can be allowed during automated driving, and (3) which of them will actually be accepted by the drivers. In mixed levels of automation, it is also important to consider smooth transitions between those levels, ideally without having to interrupt a non-driving-related activity. The aspect of motion sickness could become more important for the driver of the future when automation allows to perform more and more activities such as reading or watching movies (Sivak and Schoettle 2015).

It will be particularly interesting to see how the driver's workspace (cockpit) may be adapted for a better support of such activities. First explorations such as the carinsurance.com survey[3] investigated what drivers would "do with their newly freed time". Texting and talking was the most frequently stated activity (26%), followed by "other" (21%, including enjoying or observing the road), and reading (21%). Less frequently, also sleeping (10%), movies (8%), playing games (7%), and working (7%) were mentioned. Schoettle and Sivak (2014) examined the public opinion regarding self-driving vehicles in the U.S., the U.K., and Australia in a web survey. Besides the expected benefits and concerns, their survey also asked how participants would spent their extra time in a self-driving car. The results of this survey show that most respondents would "watch the road" (41%) while the second most frequent answer reflects the skepticism of the participants of not wanting to ride in a fully automated car (22%). Asked about (non-driving-related) activities, the most frequent responses are reading (8%), texting or talking with friends or family (8%), and sleeping (7%).

[3]http://web.archive.org/web/20150910142026/http://www.carinsurance.com/Articles/autonomous-cars-ready.aspx, last access: 2016-07-10.

Similar findings were presented by Cyganski et al. (2015). Pfleging et al. (2016) investigated this issue not only by retrieving information from a web survey. Instead, they also did in-situ interviews and observations in public transportation. They found that future non-driving-related activities should cover the domains communication, entertainment, productivity, and relaxation.

8.5.2 A Safe Display for Non-Driving-Related Activities?

Head-up displays (HUDs) let us keep our heads up instead of looking down and away from the road. Technically, they are small displays or other image sources such as projectors placed below the windshield and reflected into it. Their image appears as a floating, transparent display in the driver's foveal field of view (Haeuslschmid et al. 2016). Additional mirrors or lenses are placed between the light source and the windshield to increase the distance between the driver's eyes and the virtual image. Common HUDs provide an image distance of approximately 2 m. Due to the increased image distance and the proximity between the HUD location and the driver's line of sight, the driver can switch faster between the road scene and the display. This allows either longer information uptake (at constant glance time) or shorter glance times (at constant information uptake) compared to head-down displays, e.g., in the center stack or the instrument panel (Yung-Ching and Ming-Hui 2004; Gish and Staplin 1995). It also speeds up reaction times to road events and hence improves driving performance (Gish and Staplin 1995). In addition, HUDs have been found to be specifically beneficial for elderly drivers (Flannagan and Harrison 1994), under limited sight conditions and in complex driving situations (Charissis et al. 2009), and to increase eyes on the road time (Gish and Staplin 1995).

The complex construction and the large space requirements limit the display size and consequently the amount of information presented on it. Standard HUD images measure about 7×20 cm at 2 m distance (Haeuslschmid et al. 2016). So far, the presented information is almost exclusively devoted to the tasks critical for or related to driving. A larger HUD—often called a windshield display (WSD)—provides not only more space for information presentation but also facilitates augmented reality (AR) in the car. In AR, information that refers to the surroundings can be positioned close to its referent and integrated naturally into the real world. This is expected to feel naturalistic and promote fast information uptake and understanding (Gabbard et al. 2014; Haeuslschmid et al. 2015; McCann et al. 1993; Haeuslschmid et al. 2016). AR windshield displays increase the time a driver monitors the car ahead and produce reaction times to hazards in the driver's field of view, that are equivalent to common HUDs and better than without any display aid (Haeuslschmid et al. 2015). However, as explained earlier, distraction is a relatively complex phenomenon and lab-based studies—as many of the ones reporting on the benefits of HUDs—do not necessarily reflect the driver's behavior and attention allocation in the real world. In fact, HUDs do have drawbacks and distract the driver from critical driving tasks. Despite the proximity between display and road scene, drivers can not process the digital information and the road scene simultaneously. Just as with every other display or object, they need to focus their attention on either and switch it to perceive both (McCann et al. 1993; Haines 1991). On the contrary, the location of the display can even hamper the perception of the driving scene. Despite its transparency, the display can partially occlude important parts of the road scene (corresponding to *restricted attention*). In addition, the display can capture the driver's cognitive and visual attention and cause a deteriorated perception of the surroundings (referred to as *cognitive capture* and *tunnel vision*, corresponding to *diverted attention*) (Weintraub 1987; Trent 2005). In this context, also change blindness and inattentional blindness occur and let the driver miss critical driving events (Wickens and Horrey 2008).

HUDs and WSDs will further develop and change the way drivers interact with their car and brought-in devices. Once HUDs are not limited to driving-related information anymore, these displays could present information drivers usually access on the central information display or on their smartphones. However, it is hard to predict how this will influence the driver's behavior and use patterns, especially when this type of information access is not restricted by law and hence assumed to be safe. Drivers may not be aware of the cognitive distraction or underestimate it, since their vision is still directed toward the road and the detail and precision of the perception of a scene are generally overestimated (Findlay and Gilchrist 2003; Kalat 2004; Kevin O'Regan and Noë 2001). HUDs may lower the difficulty of information uptake for the competing activity and as a result increase safety—but as mentioned before, this may encourage the driver to engage more often in competing tasks and to turn this into a habit. Moreover, considering Fuller's assumption that drivers strive for constant task difficulty and the theory of task perseveration, the question is whether drivers will initiate additional (or more demanding) tasks when the demand of the HUD and the driving activities allows them to do so. This means that, although the interaction task might be less demanding on the HUD compared to a head-down display, it remains questionable whether a HUD actually increases overall safety or whether human behavior will interfere.

If (large-sized) head-up displays remain limited to driving-related information and make the driving task considerably easier (e.g., in conjunction with novel ADAS systems, critical warnings could guide attention with less effort for the driver), the driver will be less strained and have more resources available for competing tasks. Drivers will in all likelihood then pick up additional competing tasks—leading to similar problems as explained above. In summary, it remains unclear how HUDs and WSDs will change the drivers' behavior and attention allocation and how this will influence road safety. It remains questionable whether the problem of a driver who is not attending the road situation can be solved by introducing new technology and making tasks and information uptake easier.

8.6 Summary

In this chapter, we structured and analyzed the influence of non-driving-related activities on the driver's performance. After a general look at the driver's tasks and activities, we analyzed the cognitive and perceptive mechanisms behind distraction and their effects. On this basis, we then examined the particular types of distraction caused by non-driving-related activities, and discussed how they can be measured, compared and counteracted or at least mitigated. In the last section, we provided an outlook onto two emerging technologies (automated driving and large head-up displays) and the effects, risks, and potentials that can be expected from them.

References

AAM Driver Focus-Telematics Working Group. (2006). Statement of principles, criteria and verification procedures on driver interactions with advanced in- vehicle information and communication systems. Technical report alliance of automobile manufacturers. http://www.autoalliance.org/index.cfm?objectid=D6819130-B985-11E1-9E4C000C296BA163.

Ambinder, M. S., & Simons, D. J. (2005). Attention capture: The interplay of expectations, attention, and awareness. In *Neurobiology of attention* (pp. 69–75). doi:10.1016/b978-012375731-9/50016-1.

Bach, K. M., Jæger, M. G., Skov, M. B., & Thomassen, N. G. (2009). Interacting with in-vehicle systems: Understanding, measuring, and evaluating attention. In *Proceedings of the 23rd British HCI Group Annual Conference on People and Computers: Celebrating People and Technology. BCS-HCI '09, Cambridge, United Kingdom* (pp. 453–462). Swinton, UK, UK: British Computer Society. http://dl.acm.org/citation.cfm?id=1671011.1671070.

Bayer, J. B., & Campbell, S. W. (2012). Texting while driving on automatic: Considering the frequency-independent side of habit. *Computers in Human Behavior*, *28*(6), 2083–2090. doi:10.1016/j.chb.2012.06.012.

Brookhuis, K. A., & Waard De, D. (2010). Monitoring drivers mental workload in driving simulators using physiological measures. *Accident Analysis and Prevention*, *42*(3), 898–903. doi:10.1016/j.aap.2009.06.001.

Brown, I. D., Tickner, A. H., & Simmonds, D. C. (1969). Interference between concurrent tasks of driving and telephoning. *Journal of Applied Psychology*, *53*(5), 419–424. doi:10.1037/h0028103.

Broy, N., Guo, M., Schneegass, S., Pfleging, B., & Alt, F. (2015). Introducing novel technologies in the car: Conducting a realworld study to test 3D dashboards'. In *Proceedings of the 7th International Conference on Automotive User Interfaces and Interactive Vehicular Applications, Nottingham, United Kingdom. Automotive UI '15* (pp. 179–186). New York, NY, USA: ACM. ISBN: 978-1-4503-3736-6. doi:10.1145/2799250.2799280.

Bubb, H. (2003). Fahrerassistenz—primaer ein Beitrag zum Komfort oder fuer die Sicherheit? In *Der Fahrer im 21. Jahrhundert: Anforderungen, Anwendungen, Aspekte für Mensch-Maschine-Systeme; Tagung Braunschweig, 2. und 3. Juni 2003, Braunschweig, Germany. VDI-Berichte; 1768* (pp. 25–44). Düsseldorf, Germany: VDI-Verlag. ISBN: 3-18- 091768-7.

Burnett, G. E. (2009). On-the-move and in your car: An overview of HCI issues for in-car computing. *International Journal of Mobile Human Computer Interaction (IJMHCI)*, *1*(1), 60–78. doi:10.4018/jmhci.2009010104.

Caird, J. K., Johnston, K. A., Willness, C. R., Asbridge, M., & Steel, P. (2014). A meta-analysis of the effects of texting on driving. In *Accident Analysis and Prevention*, *71*, 311–318. ISSN: 0001-4575. doi:10.1016/j.aap.2014.06.005.

Caird, J. K., Scialfa, C. T., Ho, G., & Smiley, A. (2005). A meta-analysis of driving performance and crash risk associated with the use of cellular telephones while driving. In *Proceedings of the 3rd International Driving Symposium on Human Factors in Driver Assessment, Training, and Vehicle Design. DA '05, Rockport, Maine, USA* (pp. 478–485). http://drivingassessment.uiowa.edu/DA2005/PDF/70_Cairdformat.pdf.

Caird, J. K., Willness, C. R., Steel, P., & Scialfa, C. (2008). A meta-analysis of the effects of cell phones on driver performance. In *Accident Analysis and Prevention*, *40*(4), 1282–1293. ISSN: 0001-4575. doi:10.1016/j.aap.2008.01.009.

Canada Safety Council/Conseil canadien de la sécurité. Canada Safety Council/Conseil canadien de la sécurité. http://www.safetycouncil.org/info/traffic/distract.html.

Carsten, O., Lai, F. C. H., Barnard, Y., Jamson, A. H., & Merat, N. (2012). Control task substitution in semiautomated driving: Does it matter what aspects are automated? In *Human Factors: The Journal of the Human Factors and Ergonomics Society*, *54*(5), 747–761. doi:10.1177/0018720812460246.

Charissis, V., Papanastasiou, S., & Vlachos, G. (2009). Interface development for early notification warning system: Full windshield head-up display case study. In J. A. Jacko (Ed.), *Human-Computer Interaction. Interacting in Various Application Domains: 13th International Conference, HCI International 2009, San Diego, CA, USA, July 19–24, 2009, Proceedings, Part IV* (pp. 683–692). Berlin, Heidelberg: Springer. ISBN: 978-3-642-02583-9. doi:10.1007/978-3-642-02583-9_74.

Cohen, A. S., & Hirsig, R. (1990). Zur Bedeutung des fovealen Sehens für die Informationsaufnahme bei hoher Beanspruchung. In *Sicht und Sicherheit im Straßenverkehr* (pp. 47–58). Cologne, Germany: Verlag TÜV Rheinland GmbH. ISBN: 3-88585-867-3.

Commission of the European Communities. (2007). Commission Rcommendation of 22 December 2006 on safe and efficient in-vehicle in formation and communication systems: Update of the European Statement of Principles on human machine interface (2007/78/EC). In *Official Journal of the Europe*, *50*(L32), 200–241. ISSN: 1725-2555. http://eur-lex.europa.eu/legal-content/EN/TXT/?uri=uriserv:OJ.L.2007.032.01.0200.01.ENG.

Cook, R. I., Woods, D. D., Mccolligan, E., & Howie, M. B. (1991). Cognitive consequences of clumsy automation on high workload, high consequence human performance. In *Fourth Annual Workshop on Space Operations Applications and Research. SOAR 90* (pp. 543–543). http://ntrs.nasa.gov/search.jsp?R=19910011398.

Coughlin, J. F., Reimer, B., & Mehler, B. (2009). Driver wellness, safety and the development of an awarecar. Technical report. Cambridge, MA, USA: Massachusetts Institute of Technology. http://web.mit.edu/reimer/www/pdfs/coughlinwellness2009.pdf.

Crundall, D., Bains, M., Chapman, P., & Underwood, G. (2005). Regulating conversation during driving: A problem for mobile telephones? In *Transportation Research Part F: Traffic Psychology and Behaviour*, *8*(3), 197–211. ISSN: 1369-8478. doi:10.1016/j.trf.2005.01.003.

Cyganski, R., Fradedrich, E., & Lenz, B. (2015). Travel-time valuation for automated driving: A use-case-driven study. In *Annual Meeting of the Transportation Research Board*, *15*(4259), 11–15.

de Waard, D. (1996). The measurement of drivers' mental workload. Ph.D. thesis. Haren, The Netherlands: University of Groningen. ISBN: 90- 6807-308-7.

Department of Transportation and National Highway Traffic Safety Administration. (2013). Visual-manual NHTSA driver distraction guidelines for in vehicle electronic device. In *Federal Register*, *78*(81), 24817–24890. https://federalregister.gov/a/2013-09883.

Dingus, T. A., Klauer, S. G., Neale, V. L., Petersen A., Lee, S. E., Sudweeks, J., et al. (2006). The 100-car naturalistic driving study, phase II—Results of the 100-car field experiment. Technical report, DOT HS 810 593. Washington, DC, USA: National Highway Transportation Safety Administration. http://www.nhtsa.gov/DOT/NHTSA/NRD/Multimedia/PDFs/Crashtraction/100CarMain.pdf.

Direct Line (Motor Insurance). (2002). The mobile phone report: A report on the effects of using a hand-held and a hands-free mobile phone on road safety. In *United Kingdom: Direct Line Insurance Croydon.*

Donmez, B., Boyle, L., & Lee, J. D. (2008). Designing feedback to mitigate distraction. In M. A. Regan, J. D. Lee, K. L. Young (Eds.), *Driver distraction theory, effects, and mitigation* (Chap. 29, pp. 519–531). Boca Raton, FL, USA: CRC Press. ISBN: 978-1- 4200-0749-7. doi:10.1201/9781420007497.ch29.

Findlay, J. M., & Gilchrist, I. D. (2003). Active vision: The psychology of looking and seeing. *Oxford University Press, 41*(08). doi:10.5860/choice.41-4646. REFERENCES 27

Fisher, D. L., Laurie, N. E., Glaser, R., Connerney, K., Pollatsek, A., Duffy, S. A., & Brock, J. (2002). Use of a fixed-base driving simulator to evaluate the effects of experience and PC-based risk awareness training on drivers' decisions. *Human Factors: The Journal of the Human Factors and Ergonomics Society*, *44*(2), 287–302. doi:10.1518/0018720024497853.

Flannagan, M., & Harrison, A. K. (1994). The effects of automobile head-up display location for younger and older drivers. Report No. UMTRI-94-22. Ann Arbor.

Fox, S., & Hoffman, M. (2002). Escalation behavior as a specific case of goal-directed activity: A persistence paradigm. *Basic and Applied Social Psychology*, *24*(4), 273–285. doi:10.1207/s15324834basp2404_3.

Fuller, R. (2011). Chapter 2—Driver control theory: From task difficulty homeostasis to risk allostasis. In B. E. Porter (Ed.), *Handbook of traffic psychology* (pp. 13–26). San Diego: Academic Press. ISBN: 978-0-12-381984-0. doi:10.1016/B978-0-12-381984-0.10002-5.

Fuller, R. (1984). A conceptualization of driving behaviour as threat avoidance. *Ergonomics*, *27*(11), 1139–1155. doi:10.1080/00140138408963596.

Fuller, R. (2000). The task-capability interface model of the driving process. *Recherche—Transports—Sécurité*, *66*, 47–57. doi:10.1016/s0761-8980(00)90006-2.

Fuller, R. (2005). Towards a general theory of driver behaviour. *Accident Analysis and Prevention*, *37*(3), 461–472. doi:10.1016/j.aap.2004.11.003.

Gabbard, J. L., Fitch, G. M., & Kim, H. (2014). Behind the glass: Driver challenges and opportunities for ar automotive applications. *Proceedings of the IEEE*, *102*(2), 124–136. ISSN: 0018-9219. doi:10.1109/JPROC.2013.2294642.

Gawron, V. J. (2008). Human workload. In *Human performance, workload, and situational awareness measures handbook* (2nd ed., Chap. 3, pp. 87–230). CRC Press. ISBN: 978-1-4200-6450-6. doi:10.1201/9781420064506.ch3.

Gish, K. W., & Staplin, L. (1995). Human factors aspects of using head up displays in automobiles: A review of the literature. In *Washington, DC: National Highway Traffic Safety Administration (NHTSA)* (report no: DOT HS 808 320).

Gladstones, W. H., Regan, M. A., & Lee, R. B. (1989). Division of attention: The single-channel hypothesis revisited. *The Quarterly Journal of Experimental Psychology Section A*, *41*(1), 1–17. doi:10.1080/14640748908402350.

Gold, C., & Bengler, K. (2014). Taking over control from highly automated vehicles. In N. A. Stanton, S. Landry, G. Di Bucchianico, A. Vallicelli (Eds.), *Advances in Human Aspects of Transportation: Part II. AHFE Conference* (pp. 64–69). ISBN: 978-1-4951- 2098-5.

Goodwin, A., Kirley, B., Sandt, L., Hall, W., Thomas, L., O'Brien, N., & Summerlin, D. (2013). Countermeasures that work: A highway safety countermeasure guide for state highway safety offices. Technical report DOT HS 811 727. Washington, DC, USA: University of North Carolina Highway Safety Research Center. http://www.nhtsa.gov/staticfiles/nti/pdf/811727.pdf.

Green, P. A. (2008). Driver interface safety and usability standards: An overview. In M. A. Regan, K. L. Young, J. D. Lee (Eds.), *Driver distraction theory, effects, and mitigation* (Chap. 25, pp. 445–461). Boca Raton, FL, USA: CRC Press. doi:10.1201/9781420007497.ch25.

Green, P. A. (2012). Motor vehicle-driver interfaces. In J. A. Jacko (Ed.), *Human-computer interaction. fundamentals, evolving technologies, and emerging applications* (3rd ed., Chap. 32, pp. 749–770). Boca Raton, FL, USA: CRC Press. ISBN: 978-1-4398-2943-1. doi:10.1201/b11963-37.

Green, M. (2000). How long does it take to stop? Methodological analysis of driver perception-brake times. *Transportation Human Factors*, *2*(3), 195–216. doi:10.1207/sthf0203_1.

Haeuslschmid, R., Pfleging, B., & Alt, F. (2016). A design space to support the development of windshield applications for the car. In *Proceedings of the 2016 CHI Conference on Human Factors in Computing Systems. CHI '16* (pp. 5076–5091). Santa Clara, California, USA: ACM. ISBN: 978-1-4503-3362-7. doi:10.1145/2858036.2858336.

Haeuslschmid, R., Schnurr, L., Wagner, J., & Butz, A. (2015). Contact-analog warnings on windshield displays promote monitoring the road scene. In *Proceedings of the 7th International Conference on Automotive User Interfaces and Interactive Vehicular Applications. Automotive UI '15* (pp. 64–71). Nottingham, United Kingdom: ACM. ISBN: 978-1-4503-3736-6. doi:10.1145/2799250.2799274.

Haines, R. F. (1991). Presbyopia research: From molecular biology to visual adaptation. In: G. Obrecht, L.W. Stark (Eds.), *A breakdown in simultaneous information processing* (pp. 171–175). Boston, MA: Springer. ISBN: 978-1-4757-2131-7. doi:10.1007/978-1-4757-2131-7_17.

Hancock, P., Mouloua, M., & Senders, J. (2008). On the philosophical foundations of the distracted driver and driving distraction. In M. A. Regan, J. D. Lee, K. L. Young (Eds.), *Driver distraction theory, effects, and mitigation* (Chap. 2, pp. 11–30). Boca Raton, FL, USA: CRC Press. ISBN: 978-1-4200-0749-7. doi:10.1201/9781420007497.pt2.

Harbluk, J. L., Burns, P. C., Hernandez, S., Tam, J., & Glazduri, V. (2013). Detection response tasks: Using remote, headmounted and tactile signals to assess cognitive demand while driving. In *Proceedings of the Seventh International Driving Symposium on Human Factors in Driver Assessment, Training, and Vehicle Design. DA '13, Bolton Landing, NY, USA* (pp. 78–84). http://drivingassessment.uiowa.edu/sites/default/files/DA2013/Papers/013_Harbluk_0.pdf.

Hart, S. G., & Stavenland, L. E. (1988). Development of NASA-TLX (Task load index): Results of empirical and theoretical research. In P. A. Hancock, N. Meshkati (Eds.), *Human mental workload* (Vol. 52, pp. 139–183). North-Holland. doi:10.1016/S0166-4115(08)62386-9.

Hurts, K., Angell, L. S., & Perez, M. A. (2011). The distracted driver: Mechanisms, models, and measurement. *Reviews of Human Factors and Ergonomics*, *7*(1), 3–57. doi:10.1177/1557234x11410387.

International Organization for Standardization. (2003). Road vehicles—Ergonomic aspects of transport information and control systems - Procedure for assessing suitability for use while driving. ISO 17287:2003. Geneve, Switzerland: International Organization for Standardization.

Japan Automobile Manufactures Association. (2004). JAMA: Guideline for invehicle display systems—Version 3.0. Technical report, Japan Automobile Manufactures Association.

Kalat, J. W. (2004). *Biological psychology*. Belmont, CA: Thomson Wadsworth.

Kane, M. J., Brown, L. H., Mcvay, J. C., Silvia, J., Myin-Germeys, I., & Kwapil, T. R. (2007). For Whom the mindwanders, and when: An experience-sampling study of working memory and executive control in daily life. *Psychological Science*, *18*(7), 614–621. doi:10.1111/j.1467-9280.2007.01948.x.

Kirchner, W. K. (1958). Age differences in short-term retention of rapidly changing information. *Journal of Experimental Psychology*, *55*(4), 352–358. doi:10.1037/h0043688. Apr.

Knapper, A., Hagenzieker, M., & Karel, B. (2012). Driver distraction from some theoretical perspectives. Delft, Netherlands: Faculty of Technology, Policy, Management, Department of Transport, and Logistics, Delft University of Technology.

Kramer, A. F. (1990). Physiological metrics of mental workload: A review of recent progress. Technical report, ADA223701. San Diego, CA, USA: Navy Personnel Research and Development Center. http://oai.dtic.mil/oai/oai?verb=getRecord&metadataPrefix=html&identifier=ADA223701.

Lamme, V. A. F. (2005). The difference between visual attention and awareness: A cognitive neuroscience perspective. In *Neurobiology of attention* (pp. 167–174). doi:10.1016/b978-012375731-9/50033-1.

Lee, J. D. (2004). Dynamics of driver distraction: The process of engaging and disengaging. *Annals of Advances in Automotive Medicine*, *58*, 24–32. http://www.ncbi.nlm.nih.gov/pmc/articles/PMC4001670/.

Lee, J. D., Regan, M. A., & Young, K. L. (2008a). What drives distraction? Distraction as a breakdown of multilevel control. In M. A. Regan, J. D. Lee, K. L. Young (Eds.), *Driver distraction theory, effects, and mitigation* (Chap. 4, pp. 41–56). Boca Raton, FL, USA: CRC Press. ISBN: 978-1-4200-0749-7. doi:10.1201/9781420007497.ch4.

Lee, J. D., Young, K. L., & Regan, M. A. (2008b). Defining driver distraction. In M. A. Regan, J. D. Lee, K. L. Young (Eds.), *Driver distraction theory, effects, and mitigation* (Chap. 3, pp. 31–40). Boca Raton, FL, USA: CRC Press. ISBN: 978-1-4200-0749-7. doi:10.1201/9781420007497.ch3.

Lee, J. D., Roberts, S. C., Hoffman, J. D., & Angell, L. S. (2012). Scrolling and driving: How an MP3 player and its aftermarket controller affect driving performance and visual behavior. *Human Factors: The Journal of the Human Factors and Ergonomics Society*, *54*(2), 250–263. doi:10.1177/0018720811429562.

Lee, J. D., & Strayer, D. L. (2004). Preface to the special section on driver distraction. *Human Factors: The Journal of the Human Factors and Ergonomics Society*, *46*(4), 583–586. doi:10.1518/hfes.46.4.583.56811.

Liang, Y., Lee, J. D., & Yekhshatyan, L. (2012). How dangerous is looking away from the road? Algorithms predict crash risk from glance patterns in naturalistic driving. *Human Factors: The Journal of the Human Factors and Ergonomics Society*, *54*(6), 1104–1116. doi:10.1177/0018720812446965.

Mack, A., & Rock, I. (1998). *Inattention blindness*. Cambridge: MIT Press.

McCann, R. S., Foyle, D. C., & Johnston, J. C. (1993). Attentional limitations with head-up displays. In R. S. Jensen (Ed.), *Proceedings of the Seventh International Symposium on Aviation Psychology* (pp. 70–75).

Mccarley, J. S., Vais, M. J., Pringle, H., Kramer, A. F., Irwin, D. E., & Strayer, D. L. (2004). Conversation disrupts change detection in complex traffic scenes. *Human Factors: The Journal of the Human Factors and Ergonomics Society*, *46*(3), 424–436. doi:10.1518/hfes.46.3.424.3784.

McEvoy, S. P., Stevenson, M. R., & Woodward, M. (2006). The impact of driver distraction on road safety: Results from a representative survey in two Australian states. *Injury Prevention*, *12*(4), 242–247. doi:10.1136/ip.2006.012336.

McKiernan, K. A., D'Angelo, B. R., Kaufman, J. N., & Binder, J. R. (2006). Interrupting the "stream of consciousness": An fMRI investigation. *NeuroImage*, *29*(4), 1185–1191. ISSN: 1053-8119. doi:10.1016/j.neuroimage.2005.09.030.

McKnight, A. J., & Adams, B. B. (1970). Driver education task analysis: Volume I: Task descriptions. Alexandria: VA: Human Resources Research Organization, Final Report, Contract No FH 11-7336.

Mehler, B., Reimer, B., Coughlin, J. F., & Dusek, J. (2009). Impact of incremental increases in cognitive workload on physiological arousal and performance in young adult drivers. *Transportation Research Record: Journal of the Transportation Research Board*, *2138*, 6–12. doi:10.3141/2138-02.

Michon, J. A. (1985). A critical view of driver behavior models: What do we know, what should we do? *Human Behavior and Traffic Safety*, 485–524. doi:10.1007/978-1-4613-2173-6_19.

Miller, S. (2001). Literature review. Workload measures. Technical report, N01-006. Iowa City, IA, USA: University of Iowa. http://www.nads-sc.uiowa.edu/publicationStorage/200501251347060.N01-006.pdf.

Morris, A., Reed, S., Welsh, R., Brown, L., & Birrell, S. (2015). Distraction effects of navigation and green-driving systems—Results from field operational tests (FOTs) in the UK. *European Transport Research Review*, *7*(3), 1–9. ISSN: 1866-8887. doi:10.1007/s12544-015-0175-3.

Näätäänen, R., & Summala, H. (1974). A model for the role of motivational factors in drivers' decision-making. *Accident Analysis and Prevention*, *6*(3–4), 243–261. doi:10.1016/0001-4575(74)90003-7.

Neale, V. L., Dingus, T. A., Klauer, S. G., Sudweeks, J., & Goodman, M. (2005). Driven to distraction: An overview of the 100-car naturalistic driving study and findings. Washington, DC. http://www.nhtsa.gov/DOT/NHTSA/NRD/Multimedia/PDFs/Crash.

Nowakowski, C., Green, P. A., & Friedman, D. (2002). An experimental evaluation of using automotive HUDs to reduce driver distraction while answering cell phones. *46*(22), 1819–1823. doi:10.1177/154193120204602209.

Olsen, E. C. B., Lerner, N., Perel, M., & Simons-Morton, B. G. (2005). In-car electronic device use among teen drivers. *Transportation Research Record: Journal of the Transportation Research Board*, 1–21.

O'Regan, J. K., & Noë, A. (2001). A sensorimotor account of vision and visual consciousness. *Behavioral and Brain Sciences Behavioural Brain Sciences*, *24*(05), 939–973. doi:10.1017/s0140525x01000115.

Pashler, H. (1994). Dual-task interference in simple tasks: Data and theory. *Psychological Bulletin*, *116*(2), 220–244. doi:10.1037/0033-2909.116.2.220.

Pauzié, A. (2014). Evaluation of the driver's mental workload: A necessity in a perspective of iIn-vehicle system design for road safety improvement. *Cognition, Technology and Work*, *16*(3), 299–302. ISSN: 1435- 5558. doi:10.1007/s10111-014-0276-0.

Pettitt, M., Burnett, G. E., & Stevens, A. (2005). Defining driver distraction. In *Proceedings of the 12th ITS World Congress, San Francisco, USA*.

Pfleging, B., Rang, M., & Broy, N. (2016). Investigating user needs for non-driving-related activities during automated driving. In *Proceedings of the 15th International Conference on Mobile and Ubiquitous Multimedia (MUM' 16). Rovaniemi, Finland, December 12-15, 2016* (pp. 91–99). New York, NY, USA: ACM. ISBN: 978-1-4503-4860-7. http://dx.doi.org/10.1145/3012709.3012735.

Pfleging, B., & Schmidt, A. (2015). (Non-) Driving-related activities in the car: Defining driver activities for manual and automated driving. In *Workshop on Experiencing Autonomous Vehicles: Crossing the Boundaries between a Drive and a Ride at CHI '15, Seoul, South Korea*.

Pfleging, B., Schneegass, S., & Schmidt, A. (2013). Exploring user expectations for context and road video sharing while calling and drivings. In *Proceedings of the 5th International Conference on Automotive User Interfaces and Interactive Vehicular Applications, Eindhoven, Netherlands. AutomotiveUI '13* (pp. 132–139). New York, NY, USA: ACM. ISBN: 978-1-4503-2478-6. doi:10.1145/2516540.2516547.

Rajalin, S., & Summala, H. (1997). What surviving drivers learn from a fatal road accident. *Accident Analysis and Prevention*, *29*(3), 277–283. doi:10.1016/s0001-4575(96)00081-4.

Ranney, T. A. (2008). Driver distraction: A review of the current state-of- knowledge. Technical report, DOT HS 810 787. Washington, DC: National Highway Traffic Safety Administration, 2008. http://www.nhtsa.gov/DOT/NHTSA/NRD/Multimedia/PDFs/Crash+Avoidance/2008/810787.pdf.

Ranney, T. A., Mazzae, E., Garrott, R., & Goodman, M. J. (2000). NHTSA driver distraction research: Past, present, and future. Technical report, Washington, DC: National Highway Traffic Safety Administration, 2000. http://www-nrd.nhtsa.dot.gov/departments/Human+Factors/driver-distraction/PDF/233.PDF.

Rasmussen, J. (1997). Risk management in a dynamic society: A modelling problem. *Safety Science*, *27*(2–3), 183–213. doi:10.1016/s0925-7535(97)00052-0.

Redelmeier, D. A., & Tibshirani, R. J. (1997). Association between Cellular-telephone calls and motor vehicle collisions. *New England Journal of Medicine*, *336*(7), 453–458. PMID: 9017937. doi:10.1056/NEJM199702133360701.

Regan, M. A., Hallett, C., & Gordon, C. P. (2011). Driver distraction and driver inattention: Definition, relationship and taxonomy. *Accident Analysis and Prevention*, *43*(5), 1771–1781. doi:10.1016/j.aap.2011.04.008.

Rensink, R. A. (2002). Change detection. *Annual Review of Psychology*, *53*(1), 245–277. doi:10.1146/annurev.psych.53.100901.135125.

Roscoe, A. H., & Ellis, G. A. (1990). A subjective rating scale for assessing pilot workload in flight: A decade of dractical use. Technical report, RAE TR 90019. Bedford, UK: Royal Aerospace Establishment. http://www.dtic.mil/dtic/tr/fulltext/u2/a227864.pdf.

SAE International. (2014). Taxonomy and definitions for terms related to on-road motor vehicle automated driving systems. SAE J3016_201401. Warrendale, PA, USA: SAE International. http://standards.sae.org/j3016_201401/.

SAE International. (2015). Operational definitions of driving performance measures and statistics. SAE J2944_201506. Warrendale, PA, USA: SAE International, Jun. 2015. http://standards.sae.org/j2944_201506/.

Santos, J., Merat, N., Mouta, S., Brookhuis, K., & de Waard, D. (2005). The interaction between driving and in-vehicle information systems: Comparison of results from laboratory, simulator and real-world studies. *Transportation Research Part F: Traffic Psychology and Behaviour*, *8*(2). The relationship between distraction and driving performance: Towards a test regime for in-vehicle information systemsIn-vehicle information systems, 135–146. ISSN: 1369-8478. doi:10.1016/j.trf.2005.04.001.

Sayer, J. R., Devonshire, J. M., & Flannagan, C. A. (2005). The effects of secondary tasks on naturalistic driving performance. Report No. UMTRI-2005-29. Ann Arbor, Michigan: The University of Michigan, Transportation Research Institute.

Schindhelm, R., Gelau, C., Keinath, A., Bengler, K., Kussmann, H., Kompfner, P., et al. (2004). Report on the review of the available guidelines and standards. Technical report, IST-1-507674-IP D43.1. Information Society Technologies (IST) Programme—AIDE—Adaptive Integrated Driver-vehicle InterfacE. http://www.aide-eu.org/pdf/sp4_deliv_new/aide_d4_3_1.pdf.

Schoettle, B., & Sivak, M. (2014). A survey of public opinion about autonomous and self-driving vehicles in the U.S., the U.K., and Australia. Technical report, UMTRI-2014-21. Ann Arbor, MI, USA: The University of Michigan Transportation Research Institute.

Senders, J., & Kristofferson, A. (1966). The attentional demand of automobile driving. Highway Research Record (Vol. 195, pp. 15–33).

Seo, D.-C., & Torabi, M. (2004). The impact of in-vehicle cell-phone use on accidents or near-accidents among college students. *Journal of American College Health*, *53*(3), 101–108. doi:10.3200/jach.53.3.101-108.

Sivak, M., & Schoettle, B. (2015). Motion sickness in self-driving vehicles. Technical report UMTRI-2015-12. Ann Arbor, MI, USA: The University of Michigan Transportation Research Institute.

Strayer, D. L., Cooper, J. M., Turrill, J., Coleman, J. R., & Hopman, R. J. (2015). Measuring cognitive distraction in the automobile III: A comparison of ten 2015 in-vehicle information systems. Technical report, Washington, DC, USA: University of Utah. http://newsroom.aaa.com/wp-content/uploads/2015/10/Phase-IIIResearch-Report.pdf.

Strayer, D. L., Cooper, J. M., Turrill, J., Coleman, J. R., & Hopman, R. J. (2015). The smartphone and the driver's cognitive workload: A comparison of Apple, Google, and Microsoft's intelligent personal assistants. Technical report, Washington, DC, USA: University of Utah. http://newsroom.aaa.com/wp-content/uploads/2015/10/Phase-III-Research-Report.pdf.

Strayer, D. L., Drews, F. A., & Johnston, W. A. (2003). Cell phone induced failures of visual attention during simulated driving. *Journal of Experimental Psychology: Applied*, *9*(1), 23–32. doi:10.1037/1076-898x.9.1.23.

Strayer, D. L., & Johnston, W. A. (2001). Driven to distraction: Dual-task studies of simulated driving and conversing on a cellular telephone. *Psychological Science*, *12*(6), 462–466. doi:10.1111/1467-9280.00386.

Stutts, J., Feaganes, J., Rodgman, E., Hamlett, C., Gish, K., Mercandante, M., et al. (2003). The causes and consequences of distraction in everyday driving. In *47th Annual Proceedings/Association for the Advancement of Automotive Medicine* (pp. 235–251).

Stutts, J., Knipling, R. R., Pfefer, R., Neumann, T. R., Slack, K. R., & Hardy, K. K. (2005). Guidance for implementation of the AASHTO strategic highway safety plan: A guide for reducing crashes involving drowsy and distracted drivers. Technical report, NCHRP Report 500. Transportation Research Board. http://onlinepubs.trb.org/onlinepubs/nchrp/nchrp_rpt_500v14.pdf.

Stutts, J., Feaganes, J., Reinfurt, D., Rodgman, E., Hamlett, C., Gish, K., et al. (2005). Driver's exposure to distractions in their natural driving environment. *Accident Analysis and Prevention*, *37*(6), 1093–1101. doi:10.1016/j.aap.2005.06.007.

Summala, H. (1988). Risk control is not risk adjustment: the zero-risk theory of driver behaviour and its implications. *Ergonomics*, *31*(4), 491–506. doi:10.1080/00140138808966694.

Svenson, O. (1981). Are we all less risky and more skillful than our fellow drivers? *Acta Psychologica*, *47*(2), 143–148. doi:10.1016/0001-6918(81)90005-6.

Takayama, L., & Nass, C. (2008). Assessing the effectiveness of interactive media in improving drowsy driver safety. *Human Factors: The Journal of the Human Factors and Ergonomics Society*, *50*(5), 772–781. doi:10.1518/001872008X312341.

Taylor, D. H. (1964). Drivers' galvanic skin response and the risk of accident. *Ergonomics*, *7*(4), 439–451. doi:10.1080/00140136408930761.

Tison, J., Chaudhary, N., & Cosgrove, L. (2011). National phone survey on distracted driving attitudes and behaviors. In *PsycEXTRA*. doi:10.1037/e562822012-001.

Treat, J. R. (1980). A study of precrash factors involved in traffic accidents. *The HSRI Review*, *10*(1).

Trent, V. (2005). Keeping eye and mind on the road. Uppsala University. ISBN: 91-554-6425-4.

Wann, J. P., Swapp, D., & Rushton, S. K. (2000). Heading perception and the allocation of attention. *Vision Research*, *40*(18), 2533–2543. doi:10.1016/s0042-6989(00)00115-2.

Weintraub, D. J. (1987). HUDs, HMDs and common sense: Polishing virtual images. *Human Factors Society Bulletin*, 1–3.

Wickens, C. D., & Horrey ,W. (2008). Models of attention, distraction, and highway hazard avoidance. In M. A. Regan, J. D. Lee, K. L. Young (Eds.), *Driver distraction theory, effects, and mitigation* (pp. 57–69). Boca Raton, FL, USA: CRC Press. ISBN: 978- 1-4200-0749-7. doi:10.1201/9781420007497.ch5.

Wickens, C. D., & Tsang, P. S. (2015). Workload. In D. A. Boehm-Davis, F. T. Durso, J. D. Lee (Eds.), *APA handbook of human systems integration* (Chap. 18, pp. 277–292). APA handbooks in psychology. Washington, DC, USA: American Psychological Association. doi:10.1037/14528-018.

Wickens, Christopher D. (2002). Multiple resources and performance prediction. *Theoretical Issues in Ergonomics Science*, *3*(2), 159–177. doi:10.1080/14639220210123806.

Wickens, C. D. (2008). Multiple resources and mental workload. *Human Factors: The Journal of the Human Factors and Ergonomics Society*, *50*(3), 449–455. doi:10.1518/001872008X288394.

Wierwille, W. W. (1993). An initial model of visual sampling of in-car displays and controls. Amsterdam: National Highway Traffic Safety Administration (NHTSA) (report no: DOT HS 808 320) (pp. 271–282).

Wierwille, W. W. (1993). Demands on driver resources associated with introducing advanced technology into the vehicle. *Transportation Research Part C: Emerging Technologies*, *1*(2), 133–142. ISSN: 0968-090X. doi:10.1016/0968-090X(93)90010-D.

Yerkes, R. M., & Dodson, J. D. (1908). The relation of strength of stimulus to rapidity of habit-formation. *Journal of Comparative Neurology and Psychology*, *18*, 459–482.

Young, K. L., Lee, J. D., & Regan, M. A. (2008). Defining driver distraction. In K. Young, J. D. Lee, M. A. Regan (Eds.), *Driver distraction theory, effects, and mitigation* (Chap. 3, pp. 31–40). Boca Raton, FL, USA: CRC Press ISBN: 978-1-4200-0749-7. doi:10.1201/9781420007497.ch3.

Young, K. L., Regan, M., & Hammer, M. (2003). Driver distraction: A review of the literature. Technical report, 206. Monash University: Accident Research Center.

Young, K. L., Regan, M. A., & Lee, J. D. (2008). Factors moderating the impact of distraction on driving performance and safety. In M. A. Regan, J. D. Lee, K. L. Young (Eds.), *Driver distraction theory, effects, and mitigation* (Chap. 19, pp. 335–351). Boca Raton, FL, USA: CRC Press. ISBN: 978-1-4200-0749-7. doi:10.1201/9781420007497.pt6.

Young, M. S., & Stanton, N. A. (2002). Malleable attentional resources theory: a new explanation for the effects of mental underload on performance. *Human Factors: The Journal of the Human Factors and Ergonomics Society*, *44*(3), 365–375. doi:10.1518/0018720024497709.
Yung-Ching, L., & Ming-Hui, W. (2004). Comparison of head-up display (HUD) vs. head-down display (HDD): Driving performance of commercial vehicle operators in Taiwan. *61*(5), 679–697. doi:10.1016/j.ijhcs.2004.06.002.

Chapter 9
Eye and Head Tracking for Focus of Attention Control in the Cockpit

Mohammad Mehdi Moniri and Michael Feld

Abstract The driver's focus of attention is a key factor to be considered for building novel, intuitive user interaction concepts, and enhancing the current infotainment and safety applications in the vehicle. In this chapter we present several topics related to the development of application and systems that incorporate the user's visual focus of attention. In the presented real-life experiments, 3D representations of both the vehicle's interior and the outside environment are used. A real-time evaluation concerning the object in the driver's visual focus in these environments is also performed. We describe the functionality and the accuracy of the presented systems, which is integrated in a fully functional vehicle in an actual traffic setting. In addition, several analyses concerning accuracy of the off-the-shelf eye trackers regarding peripheral vision or direct interaction with urban objects are presented.

9.1 Introduction

The functionality of current assistance systems regarding the extraction of drivers' focus of attention is very limited. This leads to the lack of useful safety applications like a recognition of the driver's awareness of a potential danger on the road. In other words, current assistance systems cannot build any contextual or logical link between the information acquired from their numerous sources (sensors, maps, etc.) and the focus of attention of the driver. With the commercially available eye trackers and head trackers it is now possible to acquire and integrate the information about visual attention in different environments. This integration provides an opportunity for developing novel attention-based applications. The vehicle environment provides ideal conditions for integrating stationary eye and head trackers. The car passengers, especially the driver, generally do not move their head out of a constrained area within the vehicle and they are looking forward for most of the time.

M.M. Moniri · M. Feld (✉)
German Research Center for Artificial Intelligence, Saarbruecken, Germany
e-mail: michael.feld@dfki.de

M.M. Moniri
e-mail: moniri@dfki.de

G. Meixner and C. Müller (eds.), *Automotive User Interfaces*,
Human–Computer Interaction Series, DOI 10.1007/978-3-319-49448-7_9

On the other hand, these devices can provide valuable information for safety or infotainment applications in Advanced Driver Assistant Systems (ADAS) as they can reveal the user's focus of attention in real time. There are reports that eye monitoring has resulted in a 70% decrease in both driver fatigue and distraction in the mining domain.[1] This information will be even more valuable when it is merged with other available information within the ADAS (e.g., traffic sign recognition, pedestrian recognition, etc.).

A driver's visual focus of attention can be determined by measuring two main indicators: head pose and eye gaze. In order to develop automotive applications these two information sources have to be combined with spatial information like position of the vehicle or the relative distance between the vehicle and other objects in the environment. This information can be collected using a variety of sensors: Eye tracking and head tracking sensors deliver data about the direction of the visual focus, whereas GPS, radar, and camera systems determine the absolute position of the vehicle on the road and its relative distance toward other objects. All of the aforementioned sensors have various accuracies, precisions, resolutions, and sensitivities. These variations cause a fluctuation in the sensors' data output, which in turn affects the functionality of the automotive applications that depend on this data.

In this article we present a series of studies about test and evaluation of different eye tracking/head tracking experiments and applications in vehicles. Section 9.2 gives an overview of relevant research and applications in this area. In Sect. 9.3 we describe EyeBox, an application which is developed for infotainment purposes. Here the car passengers can ask for information about different buildings in the environment by merely looking at the target building and asking the car. In Sect. 9.4 we describe the more advanced EyeVIUS system, the successor to EyeBox. In that section we also cover several applications of EyeVIUS for safety and infotainment purposes. Section 9.5 describes several experiments to determine the accuracy of automotive applications which use off-the-shelf eye trackers. This section also presents further analyses regarding human peripheral vision. Finally, Sect. 9.6 presents a brief conclusion of the presented studies and applications.

9.2 State of the Art

Determining the focus of attention of the user by extracting the eye gaze or the head pose information is an active field of research in HCI (Human–Computer Interaction). Previous studies have shown that the most precise modality for referring to the outside environment from within a moving vehicle is eye gaze (Moniri 2011). For determining the head pose of the user, there are several approaches regarding the combination of available hardware and the algorithm. There are systems which are based on stereo or mono cameras (for example Gernoth et al. 2010; Jiménez

[1] http://www.dssmining.com/what-we-do/results/.

et al. 2012). These systems rely on the available ambient light, thus they are not very appropriate for in-vehicle applications. There are also research studies, which use infrared or depth cameras for this purpose (Fanelli et al. 2011). These systems work in darkness as well. Regarding the eye tracking, in the automotive domain, eye trackers are used mainly in conjunction with simulators for driver distraction studies. There are also research studies which have used eye gaze as an input modality (like for example Kang et al. 2015; Kern et al. 2010; Moniri and Christian Müller 2014).

Referring to the outside environment from within a moving vehicle can be performed using different modalities such as EYE GAZE, HEAD POSE, POINTING GESTURE, CAMERA VIEW and the user's VIEW FIELD (Moniri 2011). Fusing different information sources for extracting the object in the focus of the driver has been topic of a few research studies. Fletcher and Zelinsky (2009) describe an assistant system that decides on the basis of the driver's eye gaze together with the detected road signs and pedestrians whether the driver has seen a specific object or not. Moniri et al. (2012) describe a system which uses a combination of eye gaze and environment model to deliver context-sensitive information to the drivers. In the context of more recent work, Masayuki et al. (2014) describe an information presentation method for head mounted displays which is based on different gaze situations and surrounding environments. Alt et al. (2014) show that it is possible to use eye tracking systems in 3D stereoscopic displays with good accuracy. Lee et al. present an interactive system that uses a user's gaze in an augmented reality application. Toyama et al. present different systems based on eye tracking for assisting reading activity (Toyama et al. 2013), facilitating interaction with virtual elements such as text or buttons by measuring eye convergence on objects at different depths (Toyama et al. 2014), and for attention engagement and cognitive state analysis (Toyama et al. 2015).

The studies mentioned above either use eye gaze or head pose in their implemented prototypes. With eye gaze, it is possible to obtain detailed information about the driver's focus of attention; however, the operational angle of an eye tracker is very limited (mostly up to 35° for each side). Systems that calculate the head pose, on the other hand, have a high operational angle (up to 90° for each angle). However, it is not straightforward to exactly determine the focus point of the user based on the direction of his head. In addition, the environment inside of the vehicle is not considered in these studies, and hence it is not possible to extract information about how long the driver has been looking at the navigation system or other parts of the dashboard.

9.3 EyeBox: Interaction with the Outside Environment from Within a Moving Vehicle

Imagine the following scenario: Two travelers rent a car and start their trip. When they enter the car, they are greeted by the system with a spoken welcome message from the rental service "Welcome to AVIS. What is your destination?" In the EyeBox

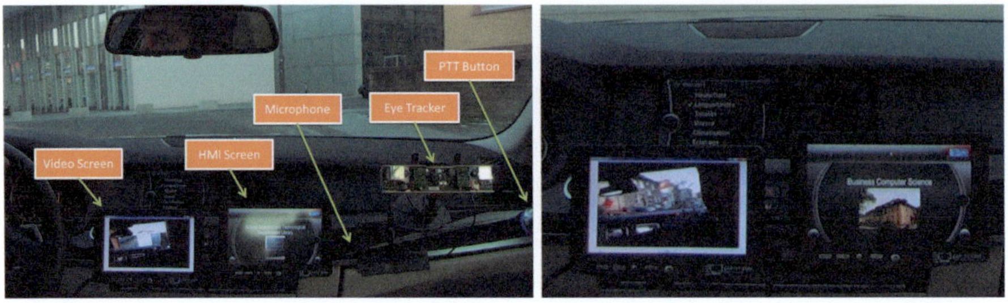

Fig. 9.1 The user selects a building by looking at it and saying "What's this building"? The system responds to the user with information about it. *Left* The overview of the whole setup. *Right* User feedback and live view windows

system, speech input is always initiated by a push-to-talk button. This button can be integrated into the vehicle and can be accessed over a vehicle bus such as CAN. In this prototype, it is an external push button mounted to the dashboard (see Fig. 9.1 (Left)). After activating the button and naming the destination, the system performs the route planning and switches to navigation mode, taking the role of a traditional navigation system, showing the current position, route, and guidance. For safety reasons, considering the prototype state of the system, it has been decided to implement the functions for the front seat passenger only. However, there are no technical limitations preventing the application to the driver as well for a final system in a series car. While the user (i.e., the front seat passenger) is navigating through the city, he can use voice commands to obtain information about the environment. With the eyes on the object of interest, he might for example ask "What is this building?" or "What is this object?" for particularly interesting buildings or landmarks. This works for every big visible object in the surroundings of the vehicle at each moment during the tour. Using an eye tracker and a physical object reference resolution technology that is further described in the following section, the system determines the target object. It can then retrieve information from an internal database or the Internet and present it to the user. In our showcase, the information is accompanied by a close-up picture of the building (see Fig. 9.1 (Right)) shown on the center stack screen.

For purposes of visualization, the setup in the demonstration vehicle features a video camera that is mounted to the passenger's seat. It covers approximately the same area as the eye tracker. It can be used to match the recorded road scene with the eye gazes and overlay the one on top of each other. The result is a video stream where an observer can see the regions where the passenger was looking (indicated by red circles). This stream is shown in our video on the left screen (see also Fig. 9.1). It can serve as a means of validation for building references.

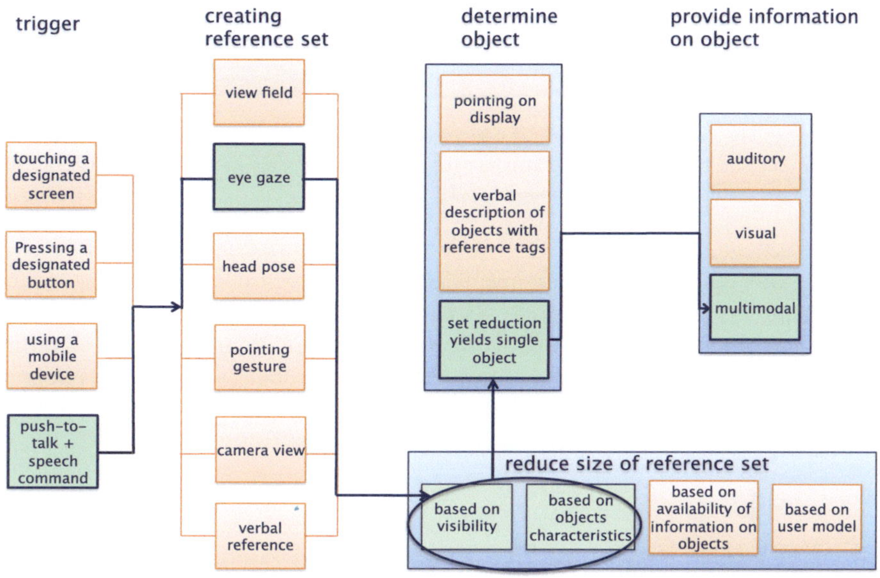

Fig. 9.2 The four stages used for different user interaction possibilities. The *dark line* shows the implemented interaction path in the presented system

9.3.1 Resolving References to Buildings from the Eye Gaze

The interaction of the user with the outside environment can be divided into four stages (see Fig. 9.2): (1) Triggering the system. Here, the user can use a designated button. (2) Creating a reference set for the potential target objects, for which a combination of eye gaze and head pose tracking is used. Note that information contained in the user's utterance could also be useful here ("what is the blue building?"). However, this deictic information is not used in the demonstration system. (3) Reduce the size of the reference set by taking into account the visibility of the buildings. (4) Provide information about the highest ranked hypothesis. Figure 9.2 illustrates this process highlighting the design choices of the system described here and showing alternatives at each stage.

While the user is briefly looking at a specific object, a series of gaze points is collected by the eye tracker. Each of these gaze points is then matched to the position of the vehicle at the gaze time. Then they are compared with a series of pre-calculated scan points (distinct tuples of position and direction). At each of these scan points the system cast a large number of rays in a 2.5D model in 180 directions horizontally and 70 directions vertically using a spatial database. For each ray the collision object is registered. Hence, the panoramic view of the user is converted into a matrix with 181 columns and 71 rows as shown in Fig. 9.3 (Left). Using this matrix the system knows which (part of a) building in the environment is visible to the user at each

Fig. 9.3 *Left* The panoramic view of the user is transformed into a matrix with 181 *columns* and 71 *rows*. *Right* The panoramic view matrix is used to calculate at each moment which part of each building in the environment is visible to the user

moment. This information is then matched with the eye gazes in order to determine which buildings are currently in the focus of the user (see Fig. 9.3 (Right)). Then information about this building is acquired from the environment model. EyeBox presents a showcase that points out the interplay of personalized in-vehicle infotainment system and interaction with visible outside environment. The system offers the possibility of exploring the environment via eye gaze and speech commands.

9.4 EyeVIUS: Fusion of Eye Tracking and Head Tracking Data for Safety and Infotainment Applications

Based on the experience with the EyeBox system, the EyeVIUS (Intelligent Vehicles in Intelligent Urban Spaces) was developed from scratch. EyeVIUS supports all of the functionality of EyeBox, and adds some new concepts to it. In this system the environment reconstruction and data processing are fundamentally different. Instead of using spatial database, in EyeVIUS the Unity 3D Game Engine[2] is used. This solution provides the flexibility and the granularity that is needed to refer to smaller objects in the environment (like for example traffic signs or city furniture). EyeVIUS aims to explore how precisely one can refer to small objects in the outside environment and also in the interior of the vehicle. For this purpose the precise 3D models of the environment and the vehicle (interior) play a major role in this system. EyeVIUS is designed in a way that it can be integrated also with driving simulators. In the following, we will first describe the system architecture together with details about the 3D Environment Reconstruction. Then, some examples of different implemented applications of in-car use of eye tracking and head tracking are provided.

9.4.1 System Architecture

EyeVIUS is divided into three logical parts: Information Source, Processing Engine, and Visualization Module (see Fig. 9.4). Information Source contains all modules for capturing information in real-time or other components, which already contain data

[2] https://unity3d.com/.

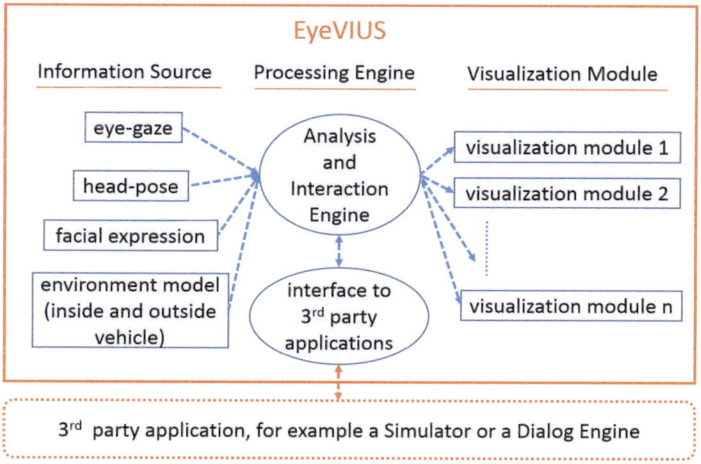

Fig. 9.4 The architecture of EyeVIUS and its communication channel with third-party applications

about the environment, for example 2D or 3D model of the environment. In addition, Information Source includes sensors, which capture different features of the user's head. These features include eye gaze, head pose, and various facial expressions of the user. Processing Engine, another logical part of EyeVIUS, receives this information and computes the relation between the different sources. For example, if the driver looks at the navigation system, the Processing Engine uses the dashboard model and the gaze direction of the user to calculate the object, which is in the driver's view (in this case the navigation system). In addition, it also extracts the facial expression of the driver. This information is then forwarded to the third-party application interface and the Visualization Module. The third-party application interface prepares this information for other modules outside EyeVIUS, for example a dialog system. It is up to these applications to act upon receiving this information. Processing Engine also includes an analysis part, which accumulates data over time and attempts to infer useful information about a specific period of time. This information can be, for example, the number of times or the overall time that the driver has looked at the speedometer.

The Visualization Module illustrates the output of the Processing Engine using various modules. Each of these modules is responsible for showing the information in an appropriate format depending on the type of the data. For instance, if the analysis is based on a 2D map of the city and the eye gaze of the driver, a visualization module can produce a 2D heat map of the city highlighting the places which have been most in the focus of the driver on the map (specific buildings or monuments). In another use case, if the analysis is based on the 3D model of the vehicle's interior and the driver's focus of attention, a 3D model of the driver's floating focus can be produced using a game engine. In the following, technical details about the functionality of EyeVIUS regarding eye tracking and head tracking are revealed.

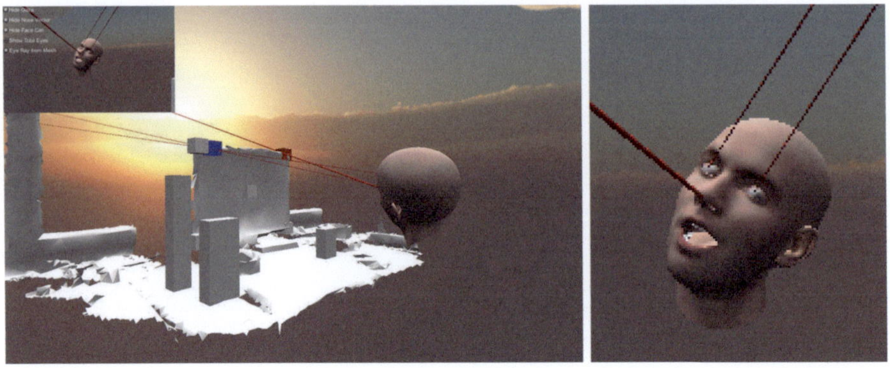

Fig. 9.5 *Left* A virtual character representing the user. The extracted information are eye gaze, head pose, and facial expression. The *colored cubes* on the display show the intersection point of the vectors with the environment. *Right* The virtual character shown from the *front*

Figure 9.5 (Left) shows a virtual character representing the user, who is looking at a display with open mouth in a scanned 3D environment (without texture). The projected line from the nose shows the pointing direction of the head. The projected lines from the eyes show the direction of each eye's gaze respectively. The colored cubes on the screen show the intersection of the vectors with the environment. Figure 9.5 (Right) shows the head from the front view in more detail.

9.4.1.1 Eye Gaze

In order to calculate the gaze direction of the user, the EyeX Controller from Tobii is used.[3] For this purpose, no video analysis is performed; instead, a 3D vector is calculated from the user's eye toward the environment. This solution provides the flexibility to calculate the intersecting object independent of their position in the vehicle or outside the vehicle.

9.4.1.2 Head Pose and Facial Expressions

In order to compute the head pose of the user in conjunction with the facial expression, a depth camera from Asus[4] is used together with the faceshift software.[5] This combination provides the opportunity to position the head of the driver in the car and also to get the values yaw, pitch, and roll relating to the head pose. In addition,

---

[3]http://www.tobii.com/xperience/.

[4]https://www.asus.com/3D-Sensor/Xtion_PRO/.

[5]http://www.faceshift.com/.

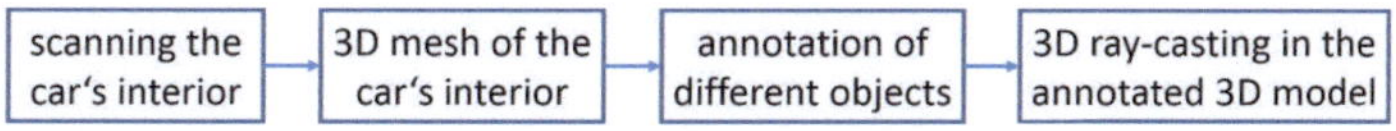

Fig. 9.6 Different phases for the 3D environment reconstruction from scanning to ray casting

various raw data about the facial expression of the driver is provided in real time (see Fig. 9.5 (Left and Right)).

9.4.1.3 Environment Modeling Procedure

As described before, for the in-vehicle environment modeling, a depth camera in conjunction with surface reconstruction software is used. This way it is possible to create a color 3D model of the car's interior, which has the required precision for different interaction and analysis applications. This 3D mesh will then be annotated in a 3D editor. In the annotation phase, every object in the car's interior model will be marked. This annotated model will be used as a basis for further processing. Depending on the head pose and eye gaze of the user, a ray will be cast in this model to reveal the object that is in the focus of attention. Figure 9.6 illustrates the described process. For modeling the outside environment two alternative approaches can be used. In the first approach the environment is scanned using a very precise 3D scanner. Then, based on this scan a mesh model the environment is developed. In the second approach, an online map is imported in the 3D game engine (for more information on this topic you can consult Dinh 2015). In both techniques the same ray casting method is used to reveal the object in the outside environment which is in the focus. In the following, we provide more details on the described scanning procedure for 3D reconstruction of the outdoor environment and also the interior of the vehicle. Figure 9.7 shows an example of the outside environment model.

In order to acquire an exact model of the vehicle's interior and the outside environment, these environments have been scanned with two different techniques. The outside environment is scanned with a professional 3D laser scanner. For this purpose more than three hectares of the university campus is scanned with centimeter accuracy (see Fig. 9.7 (top)). The resulting point cloud was then used as a basis for a 2.5D polygon model (see Fig. 9.7 (bottom)). This 2.5D model is then used together with a GPS map-matching algorithm to position the vehicle in the environment in real time. In addition, this model also contained different buildings in the environment as well as other smaller objects, such as bus stops and traffic signs and even small city garbage cans. As the point cloud model included all of these objects with high detail, it was possible to place them at the right position (relative to other objects) in the 2.5D model. Other available online maps, for example Google Maps or OpenStreetMap, do not contain these details. As EyeVIUS aims to be able to map the driver's focus to any (static) object in the environment, the described approach was selected over the other available solutions.

Fig. 9.7 *Top* A screenshot of the 3D point cloud of the Saarland University campus. *Bottom* The 3D polygon model of the campus based on the point cloud data

In order to get an exact 3D model of the vehicle, its interior is scanned with a commercially available depth sensor (the described Asus or Structure Sensor[6]). The resulting polygon-based model is then colored with respect to different regions in the vehicle (see Fig. 9.8 (bottom)). These colors were then used to distinguish each specific area in the vehicle from another area while performing ray casting depending on the direction of the user's focus. As the hardware of the EyeVIUS was a part of the scanned environment in the vehicle, its position was known. Relative to its coordinates, the information about the position and pose of the driver's head in the

[6]http://structure.io/.

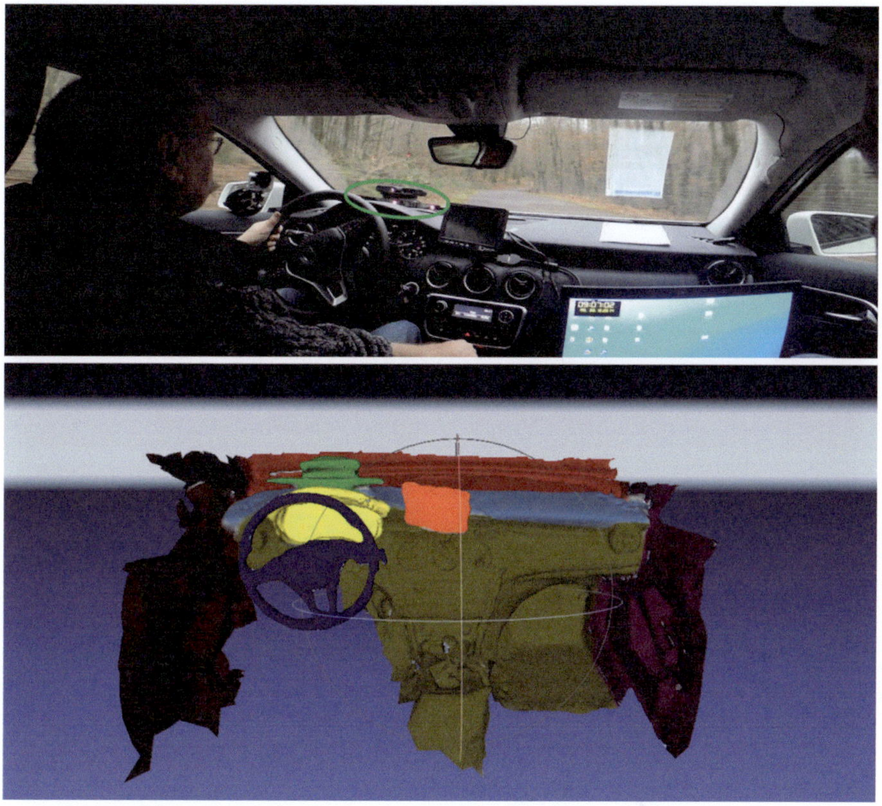

Fig. 9.8 An example car setup and the corresponding annotated 3D model. *Different colors* represent different annotated regions

vehicle was reported by the software. The whole interior model of the vehicle was then placed in the 2.5D model of the outside environment (see Fig. 9.9). The position of the vehicle was updated with a 10 Hz GPS positioning system together with a map-matching algorithm in real time (see Fig. 9.9 (bottom)). The information from EyeVIUS about the driver's attention was then used together with the vehicle's interior model and the 2.5D model of the outside environment to identify which object in which environment (outside or inside the vehicle) was in the focus of the driver in real time. For this purpose three rays were cast in the directions of the driver's left eye, right eye, and head pose (nose). The objects which were hit by the rays first were then logged (see Fig. 9.9). The intersecting objects can be divided into two categories: objects inside the vehicle and objects outside of the vehicle.

Fig. 9.9 An example scene for precision analysis in outdoor environments containing three target objects. The environment model, the vehicle model, and the direction of the focus are all integrated in one environment. The *rays* represent the direction of the eye gaze and head pose

9.4.2 Use Cases

EyeVIUS can be applied to various automotive scenarios. It can be used in applications containing analysis or interaction use cases, both in simulated environments and in the real world. The use cases of this platform can be divided into four categories that are shown in Table 9.1.

9.4.2.1 Analysis in Simulator Setup

In many evaluation studies, the analysis (for example for eye tracking) of driving simulator runs has to be performed manually by reviewing the recorded video and

Table 9.1 Various use case categories for EyeVIUS

Domain	Application	
	Analysis	Interaction
Simulator	Automated focus-of-attention analysis in simulator including off-screen setup	Extending the simulation setup to include intuitive interaction
Real World	Analyzing everyday taffic regarding the driver's focus-of-attention	Adding intuitive interaction to current existing interaction types

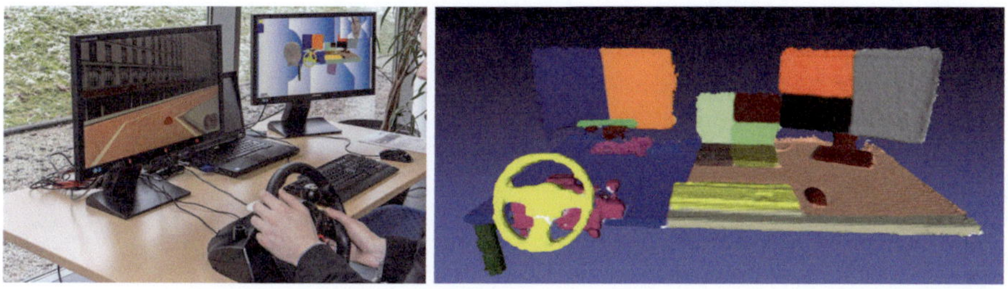

Fig. 9.10 An experiment setup with OpenDS integration and corresponding 3D annotated model. *Different colors* represent different annotated regions

annotating different segments of the video. Using EyeVIUS, it is possible to perform this analysis automatically. The platform provides the ability to scan the whole simulator setup and perform an automatic analysis to discover for example, how long the user has looked at relevant objects in the setup. Figure 9.10 shows an example simulator setup. In this example, EyeVIUS is integrated with OpenDS (Math et al. 2012). OpenDS (Math et al. 2012) is a cross-platform, open-source driving simulation software. EyeVIUS can act as a plug-in for this driving simulator in order to widen its possibilities for user interaction and analysis both inside the virtual world and also in the physical simulator setup.

As the user drives through the virtual environment in OpenDS, the eye tracker component of the system sends the eye gaze data to this driving simulator. OpenDS then uses this information to perform a real-time ray casting in the virtual environment. The intersected object together with the timestamp is logged by OpenDS in the database. In addition, EyeVIUS uses the eye gaze and the head pose data to perform a ray casting in the 3D model of the experiment setup. Here too, the intersected object (defined by color) together with the timestamp is logged in the database. If the user looks at another object besides the main display of OpenDS, this information is logged by the EyeVIUS as well, provided that it is positioned within the opening angle of the eye tracker. In any case the direction of the head together with the intersecting region is logged. We then access both aforementioned logs in the database to calculate the fixations of the user both in the virtual world (in OpenDS) and in the real world (experiment setup). For this purpose the algorithm described in

Duchowski (2007) is used. As data from the same eye tracker is used in the virtual and real environment, it is possible to determine the fixations in both worlds with analyzing the timestamps of the gaze data.

9.4.2.2 Analysis in Real-World Applications

EyeVIUS can be used together with a positioning system and a city model to preform an analysis of the driver's focus of attention. In this use case it is possible to extract useful information for urban planing. For instance, we might want to find an answer for questions like "At intersection 'A,' which buildings attract the attention of the users most? How long do drivers look at the billboard 'B' on the highway? How distracting are the advertisements on the specific part of road 'C'?"

9.4.2.3 Interaction in Simulator Setup

EyeVIUS provides third-party simulators the possibility to add interactive components to their off-screen setup. In other words, with EyeVIUS it is possible to map the focus of attention of the driver to each object in the physical environment and also get feedback in real time. For example, it is possible to add several small screens (e.g., small tablets or other objects) to the environment, and as soon as the user looks at each screen (or object), a message is sent by the EyeVIUS to the third-party application (see Fig. 9.10).

9.4.2.4 Interaction in Real-World Applications

As described in the simulation part, with EyeVIUS it is possible to make different objects in a scene (for example the interior of a car) interactive. By scanning the car's interior via the described technique, it is possible for EyeVIUS to send messages to a third-party application as the driver looks at each predefined object in the scene. This object can be a part of car console, e.g., the navigation screen or the speedometer. The third-party application can then use this information to deploy various use cases in real time. For example, if the third-party application is a dialog system, it can provide the user with useful information as the driver looks at different parts of the car and asks questions about their functionality.

9.4.3 Applications

The following sections present a number of actual automotive applications which have been realized with EyeVIUS.

9.4.3.1 Interaction with Buildings

Similar to EyeBox, it is possible to use EyeVIUS to interact with the buildings in the environment in real time. Instead of a spatial database, the physics engine in the Unity 3D Game Engine is used as a basis for reference resolution. Therefore, there is no need for predefined interaction points (and the related scanning and ray casting).

9.4.3.2 Interaction with Billboards

In addition to interaction with buildings in the environment, it is possible to use EyeVIUS to interact with–digital or analog–billboards in a controlled environment. Figure 9.11 depicts an example setup. This setup resembles a situation in which a car is waiting at a traffic light. The driver sees an advertisement of two different movies on a billboard. Using the EyeVIUS system it is possible for the driver to look at different parts of the billboard and ask the car about more information about the respective movie.

9.4.3.3 Interaction with In-Car Functions

With EyeVIUS, the driver has the opportunity to control in-car functions like the turn lights or the front windows. For this purpose the driver has to look in the appropriate direction of the physical actuator to be manipulated and activate the operation through speech. A command is then sent to the CAN bus of the vehicle, and performed by the car's internal actuators.

Fig. 9.11 Interaction with a digital billboard in a controlled environment. The driver can look at one of the two advertised movies and ask for more information it

9.4.3.4 Analysis of Driver's Focus Direction

Besides the interaction applications, the system also performs an analysis by evaluating the eye gaze and head pose data in the reconstructed environments. One of these environments was user study experiment setups. Figure 9.8 depicts an example setup for such a user study. Here, the role of EyeVIUS is to determine if the driver is looking at the road or at some object inside the vehicle.

9.4.3.5 Analysis of Eye Tracking Accuracy

One of the main purposes of the development of the EyeVIUS system was to analyze the limits of accuracy for mobile eye tracking applications in vehicles. For this purpose several tests have been performed in a lab and also in a vehicle in real traffic. Section 9.5 provides details of these experiments.

9.5 Precision in Eye Tracking Applications

In order to determine the precision of the off-the-shelf eye trackers for automotive applications, we proceed as follows. First, the precision of the same eye tracker is determined indoors for a 2D surface and also for a 3D environment. Then a similar experiment is conducted outdoors in a vehicle in actual traffic. Finally, the results of the eye tracking experiment are analyzed together with the characteristics of the different regions of the human peripheral vision. In the following each of these analyses is presented in a separate section.

9.5.1 Experiment Setups

The experiment setup constitutes of two parts: an indoor and an outdoor part. In both cases the real environment is reconstructed in a 3D virtual setting. This 3D reconstruction is performed in each case with special scanners, so that a precise model of the environment is acquired. The eye tracking data is then evaluated in this environment. This way, we ensure the precision of the approach in the 3D space. In the following each of these setups and subsequent analysis are described in detail.

9.5.1.1 Indoor Eye Tracking Data Analysis

The setup for indoor eye tracking data analysis consisted of three tables placed behind each other. Above each table a number of markers were placed in different horizontal and vertical distances (see Fig. 9.12). This setup was used for the 3D experiment. For

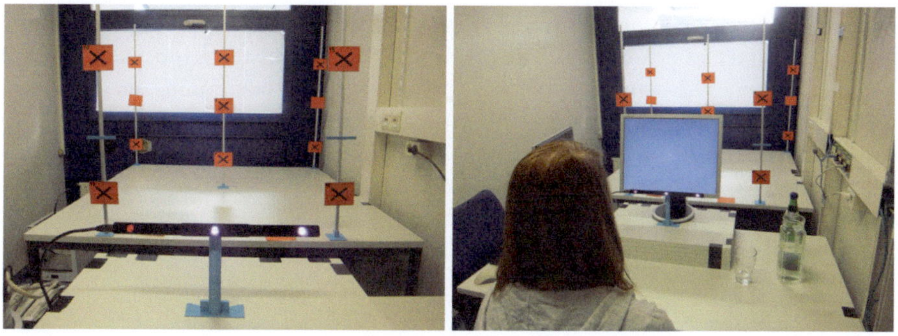

Fig. 9.12 Setup for indoor eye tracking data analysis. *Left* 3D analysis in room. *Right* 2D analysis on display

the 2D experiment, a display was positioned in front of the table. On this display the eye tracker was fixed on a printed pole. The participant was seated in front of the first table in both setups (see Fig. 9.12). The whole environment was scanned (similar to the methods described before) for the precise 3D analysis. The eye tracker used for the experiment was an EyeX controller from Tobii.

There were 22 participants involved with the experiment, 16 without optical aids, 3 using glasses, and 3 using contact lenses. Two measurement sets, one from a participant without optical aids and one from a user of glasses, were discarded due to the eye tracker being unable to detect their eyes at all times, meaning that the final set of measurements consists of recordings from 20 users. For each user there were about 5000 to 7000 individual data points collected. The experiment consisted of different test series combining several calibration levels (including no calibration) with 2D and 3D analyses. When the outliers where excluded the accuracy range for the 3D eye tracking with free head movement and without calibration was between 0.5 and 5°. The median accuracy for this combination was 1.98°. The vertical and horizontal median accuracies where −1.06 and 0.2°, respectively. For comparison, the median accuracy for the 2D case with free head movement and 9-point calibration was 0.48°. As here eye tracking for the automotive applications are interesting for us, we will consider the 3D analysis without a former calibration and with a free head movement, as today this is the practicable combination for easy eye tracker installation in the vehicles (for more information please consult Dinh 2015). In the tests inside the vehicle, the user has also performed a 3D analysis without a former calibration including free head movement.

9.5.1.2 Vehicle Eye Tracking Data Analysis

This analysis is aimed to determine the accuracy of the system when referring to small objects in the outside environment. This analysis is performed in real traffic in a fully functional vehicle while maintaining safety measures. This accuracy analysis is aimed to determine the horizontal and vertical measurement errors of the system when referring to an object outside of the car. The analysis does not include the raw or the map-matched position information of the vehicle. Instead, the exact position of the car is entered manually to avoid propagating any positioning measurement error in the calculations. Regarding the eye gaze, the presented information is the mean over the results of the right eye and the left eye. If measurement data for any eye was not available, only the available information was used.

In this test, the driver looked at five different objects on the campus. The car was standing still at the middle of the street. The five objects were two traffic signs, two advertisement billboards, and one garbage can. Figure 9.7 depicts three of these five objects. While the driver was looking at the center point of the objects, the data from the eye trackers and head trackers was logged. For each of these five objects, 60 measurements were logged. After the experiment, the logged data was checked against the reference data. The reference data was calculated by manually defining a reference ray from the driver's head in the model to the center of the target object. The horizontal and the vertical differences between these vectors were then calculated in degrees.

Figure 9.13 shows the position of each target object towards the driver in vertical and horizontal degrees. It also depicts the measurement error of the EyeVIUS when referring to each of these objects by eye gaze or head pose. As can be seen, there is no data for the eye tracker when the driver has referred to billboard 2. The reason becomes clear when we look at the position of this billboard. It is located horizontally in an angle more than 35° relative to the driver. This position is outside of the opening angle of the eye tracker (35° horizontally on each side). The measurement error of the head pose is also very high (more than 22° horizontally) for this object. For this modality, it seems that one can observe lower measurement errors as the horizontal distance of the target objects towards the driver becomes less. Besides billboard 2, the horizontal and vertical errors for all other objects for the eye gaze modality are always less than 10° and in some cases even as low as 2 to 3°. For this modality, there seems to be no relation between the position of the target object and the amount of the measurement error. Considering measurement error of the eye gaze and head pose in this analysis, it can be concluded that with the obtained accuracy it is possible to refer to big urban objects (like buildings, etc.). However, the resolution is not sufficient to refer to smaller object like traffic signs.

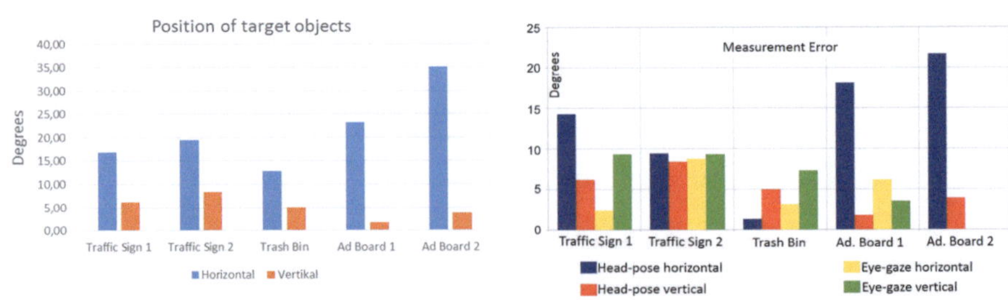

Fig. 9.13 *Left* The position of the target objects relative to the driver in horizontal and vertical degrees. *Right* The measurement errors of eye gaze and head pose for each target object

9.5.2 *Effect of Eye Tracking Imprecision on Reference Resolution for Small Urban Objects*

As it is described in the previous sections, interaction with big urban objects is possible using the off-the-shelf eye trackers in the vehicle cockpit. In the tests on the Saarland University campus, the precision of the developed reference resolution algorithm to determine the target object was more than 90 %. However, the 3D eye tracking inside the vehicle (with the off-the-shelf components) has high imprecision and therefore cannot be used to refer to small objects in the outside environment. In this section we want to measure this imprecision relative to the projection of the object's layout on the different regions of human peripheral vision. This information is valuable for developing infotainment or safety applications which need to be aware of the objects located in each peripheral region of the driver (pedestrians, tragic signs, etc.). If this shift is big, such applications cannot be implemented with the described setup. The amount of this shift also depends on the size of the object as well as the distance of the observer to the object. This explains why the presented eye tracking applications preformed well for the big objects in the environment despite the measurement errors.

For the following tests, we always consider the normal distance[7] to each of the listed objects. The size of the objects is also clear by category. As a base for the analysis the Hatada model of peripheral vision (Hatada et al. 1980) is considered. In the following this model is described and then the results of our analysis are presented.

9.5.2.1 3D Model of the Human Visual Field

The Hatada peripheral view model is very suitable for this purpose due to its detailed descriptions with respect to the characteristics of the defined regions and its exhaustive capture of the HVF following a 2D angular parametrization. The model divides the visual field into the following four regions with corresponding angular boundaries (see Fig. 9.14):

- The *discriminatory* visual field (3° circular).
- The *effective* visual field (3° to 15° horizontally on each side, 8° upwards, and 12° downwards).
- The *induced* visual field (15° to 50° horizontally on each side, 8° to 35° upwards, 12° to 50° downwards).
- The *supplementary* visual field (50° to 100° horizontally on each side, 35° to 50° upwards, 50° to 75° downwards).

As the name states, in the *discriminatory* visual field, an observer has high-precision discriminatory capabilities and perceives detailed information accurately with a visual acuity of over 0.5. Within the *effective* visual field, the visual acuity

[7] The object is visible and not too far or too close.

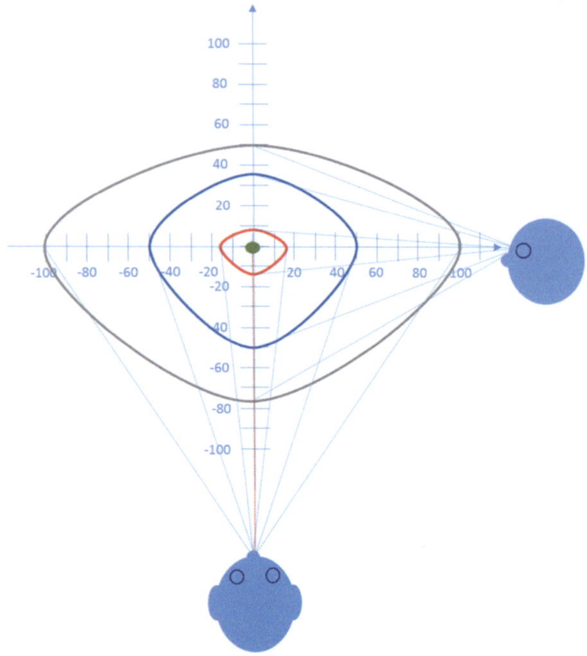

Fig. 9.14 Model of Hatada et al. using horizontal and vertical angles for parametrization

falls to about 0.1, while the discrimination of a simple figure can still be accomplished in a short period of time. This is the range within which an observer looks naturally at an object without head movement and is able to effectively process the information perceived. The *induced* visual field constitutes the range within which an observer has discriminatory capabilities to the extent of being able to recognize the existence of a visual stimulus. Hence, information displayed to the user which falls in this range should feature a reduced level of detail in terms of minimalistic representations. The HVF is complemented in terms of the *supplementary* visual field which states a range with no direct functional role in the perception of visual information. All it provides is a supplementary function in the sense that a shift of the observer's gaze can be aroused in response to abrupt stimuli.

9.5.2.2 Shift of Object Position on Peripheral Vision

As described, when using an off-the-shelf eye tracker, performing a 3D eye tracking without calibration in a real traffic situation includes several degrees of measurement error. In the presented test scenario, the layouts of several objects in the environment are projected on the different regions of the human peripheral vision. For this purpose a baseline is used. Baseline here is the user's exact eye gaze toward the object. The measured sensor data is the measured gaze from the eye tracker. Because of

Table 9.2 Shift of the projection of different objects on the regions of human peripheral vision. This shift is due to the measurement error of the eye tracker

Object	Type of data	Object occupation in different peripheral view regions (in %)			
		Discriminatory	Effective	Induced	Supplementary
Information board 1	Baseline	59.76	29.88	0	0
	Sensor data	0	9.21	80.42	0
Information board 2	Baseline	89.64	0.36	0	0
	Sensor data	0	6.27	83.72	0
Bus plan 1	Baseline	74.04	22.51	2.35	0
	Sensor data	0	1.12	97.60	0
Traffic sign 1	Baseline	99.6	0	0.4	0
	Sensor data	0	0	100	0
Traffic sign 2	Baseline	99.6	0	0	0
	Sensor data	0	0	99.6	0
Traffic sign 3	Baseline	100	0	0	0
	Sensor data	0.4	0	99.59	0
Traffic sign 4	Baseline	100	0	0	0
	Sensor Data	0.4	0	99.59	0
Traffic sign 5	Baseline	85.79	13.8	0	0
	Sensor data	0.09	29.66	69.84	0
Traffic sign 6	Baseline	99.6	0	0.4	0
	Sensor data	0	0.6	99.39	0
Road lamp 1	Baseline	99.99	0	0	0
	Sensor data	0.41	1.62	97.96	0
Road lamp 2	Baseline	100	0	0	0
	Sensor data	0.4	5.6	94	0
Road lamp 3	Baseline	99.6	0	0	0
	Sensor data	0	0.4	98.39	0
Traffic sign 7	Baseline	100	0	0	0
	Sensor data	0.42	0	99.57	0

the measurement error of the eye tracker, the place of the projected object on the peripheral vision differs from the baseline. Table 9.2 shows this shift for different objects.

The test consists of 16 small urban objects (see Fig. 9.15 for the categories). For each of these objects 250 data points have been collected. The presented results are the average of these values. As can be seen in Fig. 9.2, the percentage for the baseline in the Discriminatory field of view is high. This is due the fact that the user has always looked at the middle of the object. This way, we can be sure that the baseline ray has been calculated correctly. Regarding the sensor data, the percentage in the induced

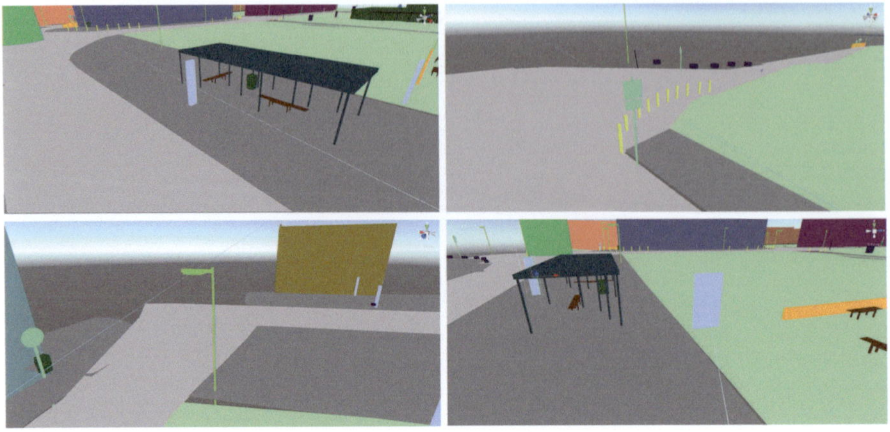

Fig. 9.15 Examples of information board (*up left*), traffic sign (*up right*), road lamp (*button left*), and bus plan (*button right*). The *line* shows the baseline gaze data

field of view is high. This means that because of the measurement errors, the field of view has been shifted in a way that the focus object (object in the discriminatory field of view) has landed in the induced field of view. As this pattern occurs for all the listed objects, it is a strong indicator that with this measurement error no statement can be made regarding the object in focus or the position of the objects in the different locations of the peripheral view field.

9.6 Conclusion

In this chapter we have presented an overview of the usage and applications of the eye and head tracking for focus of attention control in modern vehicles. Besides the review on relevant research, we have presented different attempts for integrating safety and infotainment applications based on focus of attention. Regarding interaction with the outside environment from within a vehicle, the presented systems have reached good results for interaction with big urban objects. Considering the smaller urban objects, due to the high measurement error of the integrated off-the-shelf eye trackers, it is not possible to build a reliable safety or infotainment application which involves these kinds of objects. For this purpose, a custom eye tracker with the option of calibration should be integrated in the vehicle. The utilized eye trackers in the presented studies reach an accuracy of about 0.5° with 9-point calibration on 2D surfaces. In order to be able to implement reliable applications based on eye tracking in a vehicle, the quality of integrated devices should match this accuracy in 3D environment. Without such accuracy, any attempts for building precise models are not very useful. In the presented EyeVIUS system the underlying models for the vehicle and the environment had centimeter accuracy; however, due to the measurement errors in the eye tracker, the potential of these models could not be unleashed. However, it should be mentioned that with these off-the-shelf eye trackers and also online map data, it is possible to build applications which represent a basis for interaction with large urban objects.

Regarding in-car functions, all the implemented features were functional when the user was looking above (upper region) the eye tracker. Therefore, for addressing devices in the lower region of the vehicle's interior, the head pose of the user is used. Using head pose instead of eye gaze reduces the accuracy and also the number of implemented functions. However, it is possible to monitor the focus of the driver and extract information about the direction in which he is looking (in the car or on the road). The combination of head tracker and eye tracker offers a high potential for building applications that do not always need ultimate accuracy at the object level.

References

1. Alt, F., Schneegass, S., Auda, J., Rzayev, R., & Broy, N. (2014). Using eye tracking to support interaction with layered 3D interfaces on stereoscopic displays. In *Proceedings of the 19th international conference on Intelligent User Interfaces* (p. 267272). ACM.
2. Dinh, B. -T. (2015). *Accuracy measurement of low-cost eye trackers in 2D and 3D environments*. Bachelors thesis, Computer Science Institute, Saarland University, Saarbrücken, Germany.
3. Dinh, B. -T. (2015). *Framework for analyzing drivers focus-of-attention based on dynamic 3D Map Data*. Bachelors thesis, Computer Science Institute, Saarland University, Saarbrücken, Germany.
4. Duchowski, A. (2007). Eye tracking methodology: Theory and practice (Vol. 373). Springer Science & Business Media.
5. Fanelli, G., Weise, T., Gall, J., & Gool, L. J. V. (2011). Real time head pose estimation from consumer depth cameras. In R. Mester & M. Felsberg (Eds.), *DAGM-Symposium*. Lecture Notes in Computer Science (Vol. 6835, p. 101110). Springer.
6. Fletcher, L., & Zelinsky, A. (2009). Driver inattention detection based on eye gaze—road event correlation. *International Journal of Robotics Research*, *28*(6), 774801.
7. Gernoth, T., Martínez, K. A., Gooßen, A., & Grigat, R. -R. (2010). Facial pose estimation using active appearance models and a generic face model. In P. Richard & J. Braz (Eds.), *VISAPP* (Vol. 2, p. 499506). INSTICC Press.
8. Hatada, T., Sakata, H., & Kusaka, H. (1980). Psychophysical analysis of the "sensation of reality" induced by a visual wide-field display. *SMPTE Journal*, *89*(8), 560569.
9. Jiménez, P., Bergasa, L. M., Nuevo, J., & Alcantarilla, P. F. (2012). Face pose estimation with automatic 3D model creation in challenging scenarios. *Image and Vision Computing*, *30*(9), 589602.
10. Kang, S., Kim B., Han, S., & Kim, H. (2015). Do you see what I see: Towards a gaze-based surroundings query processing system. In *Proceedings of the 7th International Conference on Automotive User Interfaces and Interactive Vehicular Applications (AutomotiveUI '15) (pp. 93–100.)*. New York, NY, USA: ACM. http://dx.doi.org/10.1145/2799250.2799285
11. Kern, D., Mahr, A., Castronovo, S., Schmidt, A., & Müller, C. A. (2010). Making use of drivers' glances onto the screen for explicit gaze-based interaction. In A. K. Dey, A. Schmidt, S. Boll, & A. L. Kun (Eds.), *AutomotiveUI* (p. 110116). ACM.
12. Math, R., Mahr, A., Moniri, M. M., & Müller, C. (2012). Opends: A new open-source driving simulator for research. *Adjunct Proceedings of the 4th International Conference on Automotive User Interfaces and Interactive Vehicular Appilcations* (p. 78). Portsmouth: NH, USA.

13. Merkel, D. (2015). *System für die Aufmerksamkeitsanalyse des Fahrers basierend auf Blickdaten*. Bachelors thesis, Computer Science Institute, Saarland University, Saarbrücken, Germany.
14. Moniri, M. M. (2011). Multimodal reference resolution for mobile spatial interaction in urban environments (Masters thesis, Computer Science Institute, Saarland University, Saarbrücken, Germany).
15. Moniri, M. M., & Müller, C. (2014). EyeVIUS: intelligent vehicles in intelligent urban spaces. In *Adjunct Proceedings of the 6th International Conference on Automotive User Interfaces and Interactive Vehicular Applications (AutomotiveUI '14)* (pp. 1–6). New York, NY, USA: ACM. http://dx.doi.org/10.1145/2667239.2667265
16. Moniri, M. M., Feld, M., & Müller, C. A. (2012). *Personalized in-vehicle information systems: Building an application infrastructure for smart cars in smart spaces* (p. 379382). In Intelligent Environments: IEEE.
17. Nakao, M., Terada, T., & Tsukamoto, M. (2014). An information presentation method for head mounted display considering surrounding environments. In *Proceedings of the 5th Augmented Human International Conference*, (p. 47). ACM.
18. Toyama, T., Dengel, A., Suzuki, W., & Kise, K. (2013). Wearable reading assist system: Augmented reality document combining document retrieval and eye tracking. In *2013 12th International Conference on Document Analysis and Recognition (ICDAR) (p. 34)*. IEEE.
19. Toyama, T., Sonntag, D., Orlosky, J., & Kiyokawa, K. (2014). A natural interface for multifocal plane head mounted displays using 3D gaze. In *Proceedings of the 2014 International Working Conference on Advanced Visual Interfaces* (p. 2532). ACM.
20. Toyama, T., Sonntag, D., Orlosky, J., & Kiyokawa, K. (2015). Attention engagement and cognitive state analysis for augmented reality text display functions. In *Proceedings of the 20th International Conference on Intelligent User Interfaces* (p. 322332). ACM .

Chapter 10
From Car-Driver-Handovers to Cooperative Interfaces: Visions for Driver–Vehicle Interaction in Automated Driving

Marcel Walch, Kristin Mühl, Johannes Kraus, Tanja Stoll, Martin Baumann and Michael Weber

Abstract As long as automated vehicles are not able to handle driving in every possible situation, drivers will still have to take part in the driving task from time to time. Recent research focused on handing over control entirely when automated systems reach their boundaries. Our overview on research in this domain shows that handovers are feasible, however, they are not a satisfactory solution since human factor issues such as reduced situation awareness arise in automated driving. In consequence, we suggest to implement cooperative interfaces to enable automated driving even with imperfect automation. We recommend to consider four basic requirements for driver–vehicle cooperation: mutual predictability, directability, shared situation representation, and calibrated trust in automation. We present research that can be seen as a step towards cooperative interfaces in regard to these requirements. Nevertheless, these systems are only solutions for parts of future cooperative interfaces and interaction concepts. Future design of interaction concepts in automated driving should integrate the cooperative approach in total in order to achieve safe and comfortable automated mobility.

M. Walch (✉) · M. Weber
Institute of Media Informatics, Ulm University, Ulm, Germany
e-mail: marcel.walch@uni-ulm.de

M. Weber
e-mail: michael.weber@uni-ulm.de

K. Mühl · J. Kraus · T. Stoll · M. Baumann
Institute of Psychology and Education, Department of Human Factors, Ulm University, Ulm, Germany
e-mail: kristin.muehl@uni-ulm.de

J. Kraus
e-mail: johannes.kraus@uni-ulm.de

T. Stoll
e-mail: tanja.stoll@uni-ulm.de

M. Baumann
e-mail: martin.baumann@uni-ulm.de

G. Meixner and C. Müller (eds.), *Automotive User Interfaces*,
Human–Computer Interaction Series, DOI 10.1007/978-3-319-49448-7_10

10.1 Introduction

Automated driving is not just a vision, it is already reality. Currently a large amount of research is being conducted in the domain of automated driving and many self-driving prototypes have already been introduced by both industry and academia (e.g. Aeberhard et al. 2015; Kunz et al. 2015; Trimble et al. 2014; Ziegler et al. 2014). SAE International (2014) defines six levels to categorize automated driving: First, the lower levels ranging from *no automation* (level 0) over *driver assistance* (level 1) to *partial automation* (level 2) require the driver to monitor the driving environment and to be available as a fallback. Second, in the higher levels *conditional automation* (level 3), *high automation* (level 4), and *full automation* (level 5), the entire *dynamic driving task* is performed by the system, at least temporarily (level 3 and 4). In consequence, the driver does not have to monitor the driving environment. However, vehicles that can handle an entire journey (level 5) on their own are far away in the public market (Shladover 2016). Accordingly, current interface technologies and interaction strategies for both highly and fully automated vehicles focus predominantly on de-escalation and handover strategies to overcome system limitations or uncertain states depending on technical or environmental circumstances (e.g. Gold et al. 2016; Payre et al. 2016; Zeeb et al. 2016). This requires the driver to take back control and drive manually. Hence, the effect of automation and fail of automation on human drivers need to be considered in interaction design. The *out-of-the-loop effect* is one of the most attended problems in context of handovers. Removing people from the control loop results in loss of skills and loss of awareness of the state that are related to decreased performance and reduced safety (Endsley and Kiris 1995). Accordingly, researchers focused on different handover strategies in order to get the driver back in the loop appropriately (see Sect. 10.2 for more details).

However, a control transition from highly or fully automated driving to manual driving is difficult to perform safely and reliably and requires a substantial amount of time. Moreover, it might result in a substantial decrease of driver's trust in the automation, especially when interpreted as automation failure. If the cause for the request for control transition could be clarified by a simple answer of the driver, an entire transition might be annoying. In our view highly and fully automated vehicles require a better, more flexible interaction concept to be accepted. The automation should behave as a cooperative agent that supports the driver as much as possible. Situations in which systems encounter their limits are also often situations in which drivers require most support, thus a stereotypical transition to manual control—ergo depriving the human driver of all capabilities of the automation, even those still available—should be avoided. Therefore, an integrative strategy for the development of *cooperative human–machine interfaces* is necessary. These should aim to facilitate communication between the human driver and the automation, as well as allocating tasks between the agents in a beneficial and safe manner. Cooperative interaction enables real-time communication and maneuver planning between the two agents aiming to increase shared situational awareness and bilateral understanding of intentions and actions.

First, we review recent research on control transitions from highly automated vehicles at system boundaries. The review of car-driver handovers shows that this concept is only a solution for a very narrow problem and a more adaptive and advanced approach is necessary for driver–vehicle interaction in highly automated driving. As a result, we discuss driver–vehicle cooperation to overcome human factor issues. Moreover, we highlight the need to implement directability, mutual predictability, shared situation representation, and trust in future human–machine interaction (HMI) concepts. Next, systems that already implement parts of these requirements are presented. Finally, we conclude by highlighting future research directions.

10.2 Current State-of-the-Art in Handover Strategies

Conditional and high automation require the driver to drive manually from time to time when the situational circumstances do not allow automated driving. These automation levels allow that either the automated vehicle or the driver is in the entire control of the vehicle for a certain time. Nevertheless, in level 3 automation the driver has to be available as fallback even on short call. In consequence, recent research focused on control transitions between high automation vehicles and drivers. These transitions occur in cases where an automated vehicle reaches its boundaries and the human driver has to get back in control to ensure maintenance of safety. Transitions are necessary to allocate driving responsibilities between the two separate agents—automated system and driver—since both are possible *control authorities* (Flemisch et al. 2012). Lu and de Winter (2015) categorized these transitions depending on the initiator (*control change authority* (Flemisch et al. 2012)) and the party who is in control after the transition. This results in four *control authority transition* categories: automation-initiated automation control (AIAC), automation-initiated driver control (AIDC), driver-initiated automation control (DIAC), and driver-initiated driver control (DIDC). AIAC and DIDC are considered as *active transitions* since the initiator is getting control. As a result, the agent is prepared to take over. On the other hand, in AIDC and DIAC the party who initiates the transition is not the same agent who should take over control. Thus these transitions are considered as *passive transitions* (Lu and de Winter 2015). In consequence, AIDC transitions are challenging for the driver who is likely lacking situation awareness and being out-of-the-loop. When there are transitions between automation levels there can also be a change in the driver's monitoring status, e.g., from level 2 to level 3, these transitions are called *monitoring transitions* (Lu and de Winter 2015). An interface should support these transitions and provide coherent information in order to maintain mode awareness. Most research so far has focused on AIDC transitions and is discussed in the following.

In certain cases, a conditional or highly automated vehicle reaches a system boundary the automation initiates a shift of the control to the driver. Gold and Bengler (2014) defined such *takeover situations* with three phases: starting in a (1) highly automated driving mode the control is shifted within a (2) transition area to (3) manual driving. When the autonomously driving vehicle senses its system limits, it asks the driver to take over with a *takeover request (TOR)* and the transition area is entered. Lorenz et al. (2014) used the term *transition phase* instead of transition area. The amount of time the driver is provided to take over (appearance of the TOR until reaching the system boundary) is called *time budget* (Gold and Bengler 2014). Moreover, Gold and Bengler (2014) defined the period of time from the appearance of the TOR until the driver starts maneuvering actively as *takeover time*. Besides the term *takeover* there can be found other expressions in literature to describe this kind of control transition: *manual control recovery (MCR)* (e.g., Payre et al. 2016) or *handovers* (e.g., Walch et al. 2015). Whereas the terms takeover and (manual) control recovery are formulated from the perspective of the party who is in charge after the transition; handover is formulated from the opposite point of view (from the perspective of the agent who initiates the transition). The term handover can be more precise by adding the direction of the control transition, e.g., *car-driver handover* (c.f. Walch et al. 2015).

Research on handovers began mainly with the focus on the question: *How long do drivers need to take over control from an autonomously driving vehicle?* Thus, participants in driving simulator studies were asked to take over within various time budgets. Damböck et al. (2012) challenged 32 participants with 4, 6, and 8 s time budgets and analyzed whether these participants could manage situations with different complexities. They found that they were able to handle the easiest situation (taking over due to lacking lane markings) even in the 4 s condition. But, in more challenging situations that required the driver to change the lane, participants showed only in the 8 s condition a comparable performance to a manual driving baseline condition. However, Damböck et al. (2012) stated, that their criteria were very restrictive and that it could be argued that results indicate that participants were able to handle the takeovers within 6 s as well when less restrictive criteria would be applied. The subjective comfort ratings of their participants revealed that they perceived the takeovers with 6 and 8 s time budgets as comfortable. Gold et al. (2013) showed that participants in their driving simulator study reacted faster with shorter time budgets (hands on steering wheel: 1.45 s in 5 s time budget condition, 2.89 s with a 7 s time budget) and applied the brake less frequently. But, no statistical analysis was reported. Petermann-Stock et al. (2013) examined takeover times measured in a driving simulator study as well. 70 of 72 participants took over within 10 s with an average takeover time of 3.2 s (max. 8.8 s) in a uncritical highway scenario at a speed of about 35 km/h where the activated traffic jam assistant needed to hand over. In contrast to the above-described studies that were conducted in autobahn scenarios, Walch et al. (2015) conducted a driving simulator study with a rural road scenario with bad weather conditions. Their participants reacted (hands-on) approximately after 1.8 to 2.8 s.

It has been shown that there are factors that affect the takeover time. Gold et al. (2016) observed that their participants maneuvered about 1 s earlier after a takeover when there was no traffic compared to two conditions with 10 or 20 vehicles per kilometer. In contrast, there was no significant effect of traffic density on the hands-on time. But, they found an impact of traffic density on the driving performance (higher accelerations, shorter time to collisions, and even crashes). These findings are in line with the results of Radlmayr et al. (2014): they observed that a situation where the driver has to resume control on a blocked middle lane (three-lane highway), while traffic on the other lanes is blocking a lane change, is very critical, especially when the participants were engaged in a distracting visual task (Surrogate Reference Task). Overall, it can be stated that higher demands due to higher complexity of the scene or additional tasks impair performance of takeovers.

It is probable that drivers engage in non-driving related tasks since the automated system relieves them from the driving task (Carsten et al. 2012; de Winter et al. 2014). Consequently, researchers focused on the question: *How does distraction affect performance of automated driving?* Gold et al. (2015) investigated the effect of different secondary tasks on the takeover performance of drivers in uncritical situations. They found that participants distracted with a pure cognitive task (n-back task) were able to take over and pass an obstacle (a voice suggested the left or right side to overtake) quicker and with a longer time-to-collision than participants dealing with tasks that included visual or motoric aspects. These findings are in line with Petermann-Stock et al. (2013) who observed longer reaction times when participants were challenged with an auditory, cognitive, motoric, and visual demanding task compared to a task with only auditory and cognitive aspects. In contrast, Radlmayr et al. (2014) found no different effects of the visual and motoric task (Surrogate Reference Task) and the cognitive n-back task on the takeover performance except for the occurrence of more collisions in a complex situation in the group challenged with the Surrogate Reference Task. Zeeb et al. (2015, 2016) found that secondary tasks did not affect the time participants in their studies needed to attain motor readiness (e.g., hands on steering wheel), but they found effects on the takeover quality. In conclusion, drivers have to be assumed as out-of-the-loop and distracted by non-driving related tasks. These tasks can have a negative impact on the takeover performance. Moreover, the ability to take over seems to be user individual, for instance, Körber et al. (2015) found correlations between the takeover time and the multitasking performance.

Automated driving has also been shown to effect post-automation behavior of drivers: Merat et al. (2014) found that their participants performed a lot of steering corrections 10 s after control was handed over. These frequent steering corrections ended after 35–40 s after the transition. Post-automation effects of platooning like a decreased headway have also been shown (Brandenburg and Skottke 2014; Gouy et al. 2014; Skottke et al. 2014).

Moreover, characteristics of drivers were focused, for instance age: *How does age affect automated driving?* Whether the age of drivers has an influence on the ability to take over control was investigated by Körber et al. (2016). Older participants (60–79 years) reacted as fast as younger ones (19–28 years). But their takeover behavior differed: the older participants performed the takeovers more carefully and safely (longer minimum time-to-collision, stronger and more frequently brake application). Körber et al. (2016) did not find a difference between the two age groups when they were challenged with a secondary task nor with an increased traffic density. These findings are in line with other studies that did not find any effect of age on takeover performance (Naujoks et al. 2015; Petermann-Stock et al. 2013). Petermann-Stock et al. (2015) conducted three studies, each with a different modality as a takeover request (speech, seat vibration, and a LED bar). Effects of the age of participants could be observed for the seat vibration where the older participants needed more time to touch or activate the relevant input (brake pedal or steering wheel). The same tendency could be observed for the optical cue, however the difference was not significant. Moreover, Petermann-Stock et al. (2015) state, that a pure optical cue using a LED bar was not suitable since some mainly elderly participants did not perceive the cue.

The survey of recent studies shows that drivers are able to take over within a time budget of less than 10 s. However, it has also been shown that the driver state and situational circumstances have impact on the takeovers due to lacking situation awareness. Moreover, there seem to be differences between users in the takeover ability. Future systems should adapt to the context and to the user to ensure safe transitions of control. Nevertheless, it should be kept in mind that the results are based on simulator studies and thus more research under more realistic conditions has to be conducted.

10.3 From Task Substitution to Cooperative Driver–Vehicle Interaction

The research reviewed in the previous section on the interaction between human driver and automated vehicles focuses on system initiated handovers. And one might have the suspicion that one of the main goals of this research is to determine the maximum amount of time that is necessary to bring the driver back into the loop as one of the major parameters automated vehicles need to comply with. If technological development guarantees that automated vehicles will not need the driver as a fallback within this period of time, most of the issues in the design of the interaction between human driver and automated vehicles would be solved.

Whereas the question of takeover times is certainly of great importance for the interaction design in automated vehicles, the reduction of the interaction design to the design of system initiated handovers and the development and evaluation of design solutions for them seems to be a too restricted perspective. The interaction with automated vehicles can be envisioned to be more dynamic and continuous during a drive. Besides the planning and the realization of handovers there can be situations in which the driver trains the automation, in which maneuvers are planned collaboratively (between driver and the vehicle as well as with other road users) and in which uncertainties are tackled together by the driver and the vehicle. These examples illustrate the need for a more complex, dynamic, and overlapping interaction concept for automated driving. In this concept, implementation of handovers is only one issue that has to be solved in the design of an integrated interaction concept for automated driving. Automated systems and their interaction concepts should adapt to the context and driver states as well as cooperate with the driver.

This is the starting point for different authors who support the case of a concept named *human–machine cooperation* as a solution for the complex coordination of human–machine interaction with highly automated systems taking over complex tasks in highly dynamic situations (e.g., Hoc 2000; Hoc et al. 2009). We build on a concept of automated systems that does not only dichotomize between a fully manual and a fully automated mode, but also works in "intermediate, cooperative modes of interaction, which allow human operators to focus the power of the automation on particular subproblems, or to specify solution methods that account for unique aspects of the situation which the automated agent may be unaware of" (Christoffersen and Woods 2002, p. 8).

For organizing the cooperation between humans and autonomous agents in general two basic complexities have to be considered: *uncertainty* and *risk* (Hoc 2000). Dynamic situations such as in road traffic are ever-changing and cannot always be predicted in full by mathematical algorithms incorporated in automated driving systems. The remaining uncertainty has to be addressed by some kind of adaptability of the automation to rapid changes of the interpretations of sensory input. Road traffic is also risky as wrong decisions and actions can lead to extremely expensive or even fatal accidents. Hoc (2000) concluded that a closed system in which certain functions are clearly allocated between human driver and automation is dangerous and would not be able to deal with the complex dynamics of real traffic.

While *function allocation* is a "fundamentally uncooperative system architecture in which the interface between human and machine has been reduced to a trivial 'you do this, I do that' barter" (Dekker and Woods 2002, p. 243), recently the focus of more and more authors has shifted to investigate means to design *automation as an intelligent team player* who cooperates with drivers and rather provides support for them instead of replacing them (Christoffersen and Woods 2002; Dekker and Woods 2002; Hoc et al. 2009; Klein et al. 2004; Young et al. 2007). Unlike *task substitution*, *driver–vehicle cooperation* aims at designing interaction patterns in which the human and the machine work together and find solutions dynamically that both fit the current situation, the current status, and the respective skill set of the both interaction partners (Flemisch et al. 2014).

Hoc (2001) and Hoc et al. (2009) proposed a functional framework for driver–vehicle cooperation: "Two agents are in a cooperative situation if they meet two minimal conditions. (1) Each one strives towards goals and can interfere with the other on goals, resources, procedures, etc. (2) Each one tries to manage the interference to facilitate the individual activities and/or the common task when it exists" (Hoc 2001, p. 515). Within this approach, all goals are related and agents have the ability to facilitate or disrupt each other. Based on this, we will provide essential assumptions of cooperative systems in driving that need to be met to make human drivers and driving automation team players in the following.

10.3.1 Making Human Drivers and Driving Automation Team Players

Considering various cooperative research approaches (e.g., Christoffersen and Woods 2002; Dekker and Woods 2002; Hoc et al. 2009; Klein et al. 2004) we propose four basic requirements for a cooperation between a human and automated technology that have to be met in order to make them effective team players: *mutual predictability*, *directability*, *shared situation representation* and *calibrated trust in automation*. While the first two requirements are mainly realized through interface design, the latter two are based in a dynamic and complicated psychological process. For this reason, shared situation representation and calibrated trust are discussed in more detail.

Mutual Predictability. Team players need to obtain mutual predictability. This can be achieved by knowledge of what the other party is currently doing and is planning to do in the near future. Agents need to be able to predict the next state of the process as well as future activities of the other party. This is crucial for being able to plan own actions (Christoffersen and Woods 2002; Klein et al. 2004). Furthermore, partners should be supported in collaborative planning (Klein et al. 2004).

Directability. Team players should be able to assess and direct the actions of the other, in order to intervene or adapt strategies to changing situations or priorities. Accordingly, machines need to be flexible and easy to direct based on the human operator's goals, state, and capabilities (Christoffersen and Woods 2002; Dekker and Woods 2002; Klein et al. 2004).

Shared Situation Representation. Successful team players maintain a common ground with the mutual intention to work together at the same long-term task. Knowledge, beliefs, and assumptions are shared by involved partners and enable effective coordination within the team (Klein et al. 2004). The underlying basic concept is called *situation representation* or *situation awareness* which is held individually by every agent. It is based on three levels proposed by Endsley and Kiris (1995): (1) the perception of the status, attributes, and dynamics of relevant elements in the environment, (2) the integration of these elements into a holistic picture to comprehend the meaning of the situation elements, and (3) the projection of their future status (Endsley and Kiris 1995). Level 2 and 3 are presumed to be the precondition for efficient action in dynamic and complex environments (Rauch et al. 2009). This situation representation needs to contain both information about the status, plans, goals, and activities of the partners as well as information about the current task status, the context and the situation (Christoffersen and Woods 2002).

The driver and the automated vehicle each develop a situation representation. However, these representations differ for humans and machines because of completely different sensory systems, different scanning strategies, different processing of sensory data and due to the restricted capabilities of machines to integrate the different pieces of information in a coherent and integrated situation representation. This holds true both for the part representing the current task situation, that is the current traffic situation, and for the part representing the status, goals, plans, and activities of the interaction partner. Therefore, humans and machines possess different situation representations that might be overlapping in some case but can be complementary in other cases if these representations are shared. Transactions of relevant information from these situation representations have to take place in order to increase performance (Stanton 2016) within the human–machine system. In order to hold a shared representation of the situation and to solve occurring problems together, an updating of strategies and the motivation to communicate with each other are required (Klein et al. 2004).

Trust and Calibrated Reliance on the System. Users tend to fail adapting their reliance to the actual capabilities of an automation (Lee et al. 2004). Thus, in order to make team play possible, the automation and the associated interfaces should be designed in a manner that fosters *calibrated trust*. Therefore, both *misuse* and *disuse* of driving automation have to be avoided (Parasuraman and Riley 1997). Misuse is the case when a system is used excessively beyond its capabilities (*overtrust*) while disuse describes a usage behavior in which automation is not used in all cases when it would actually work and provide positive effects (*distrust*) (Lee et al. 2004). Trust is dynamic and affected by factors on the side of the system, the person and the situation. First, on the side of the system variables like level of automation (Muir and Moray 1996) transparency (Mark and Kobsa 2005), performance (e.g., the occurrence and timing of failure (e.g., Parasuraman and Riley 1997)) or even surface characteristics of the system's GUI like voice (Large and Burnett 2014) or the degree of anthropomorphism (Waytz et al. 2014) have been found to play a role in trust generation. Thus, it is not only the system behavior but also the design of the systems interaction devices (e.g., displays and controls) that play an important role in how the system is perceived and which effects proper functioning and malfunctions have. Second, several general and specific personality traits have been repeatedly related to differences in trust in automation (a rough overview is provided by Hoff and Bashir (2015)). In regard to situational specifics influencing the evolution of trust, Hoff and Bashir (2015) named task difficulty, workload, perceived risks and benefits, organizational setting, and the framing of a task as factors that could have an effect on trust. If we want to understand how all these variables play together in establishing a way of interacting with a specific automated driving system, we have to closely investigate which factors of the system, the interface, the situation, and the person who uses the system influence this process at which stage. This understanding has to provide an informed basis for system design that pays respect to factors on all sides of this interactive system. In order to optimize an automated driving system (its behavior and its appearance), the psychological process of *dynamic trust generation* has to be kept in mind and different levels in the process of familiarization with the situation have to be carefully planned by both system designers and those responsible for the communication about the system which channels the expectations and the system perception of the users before they first interact with the system.

Taken together it can be concluded with Walker et al. (2016) that "[t]rust, however, is not simply present or absent. It is a dynamic phenomenon, moving along a continuum, spiraling upwards or downwards based on perceptions of how the vehicle system operates, beliefs about what those perceptions mean, and the positive or negative attitudinal attribution that arises" (Walker et al. 2016, p. 4). Hence, trust affects cooperation within a team of human and automated vehicle in different ways and needs to be considered in the interface design.

Overall, in order to support cooperation in human–machine interaction the design of the interface is required to keep the human in the loop and exchange information based on a collaborative approach. This is the base for the development of directability, mutual predictability as well as shared situation awareness. Communication is necessary to provide and share information and negotiate on common goals. Interfaces should enable the display and transfer of these information.

10.3.2 Evaluating Current Handover Strategies in the Context of Cooperation

Current research on handover strategies (see Sect. 10.2) predominantly focused on transitions from high or full automation to manual driving with an emphasis on takeover times and the effect of distraction on takeover and post-automation behavior. In most cases only one-way communication with vehicle-initiated handover signals was considered. Actually, the driver has no chance to reject the prompt or to negotiate about it. Human driver and automated vehicle do not work together to overcome weaknesses of the partner. Instead of adapting task sharing to driver states as provided by steady monitoring, call of actions are just implemented independent of the degree of difficulty for the driver. Taken together, the published studies this far do not sufficiently consider the relevance of driver monitoring as a core feature of mutual predictability. Additionally, most studies did not consider adaptivity of the automated systems. However, this is relevant to fulfill the requirement of directability. In order to share a common ground, the consideration of the current situational awareness of both the driver and the automation is of main interest. Some studies investigated situational awareness, however, they only focused on the driver and did not pay respect to a shared situation representation. Finally, current research on handovers does not include the investigation of (calibrated) trust in handover situations.

Overall, we can summarize that the currently investigated handover strategies have a restricted perspective that does not pay respect to the dynamic and ever-changing nature of the driving task and thus is only insufficiently in line with the ideal of human–machine cooperation. In order to build cooperative interaction concepts for automated driving, the perspective has to be expanded to integrative concepts that allow for real-time collaboration between the driver and the automated vehicle who act as team players in tackling the dynamic challenges of the road. Exchange of information and ways of communication need to be supported. In this regard, the interaction has to be designed in a way that actions of both the driver and the automation are predictable and directable from the perspective of the other part, shared situational awareness is facilitated and that the driver holds a calibrated level of trust towards the system that prevents both mis- and disuse of the automation.

Taken together, automated driving is more than just handover situations. In order to provide a perspective for research aiming at promoting integrative and cooperative interaction concepts for automated driving, in the following we focus on several interfaces that already implemented aspects of cooperation.

10.4 Design of Cooperative Interfaces

After discussing arising human factor issues and presenting the psychological background of human–computer interaction in this domain and the resulting suggestion for the design of cooperative interfaces, we continue with a more applied perspective. The previous section highlights the need to elaborate cooperative interfaces and interaction concepts. In consequence, we survey practical approaches which can be used as basis and inspiration for the implementation of future cooperative interfaces. First, we sketch the design space of in-vehicle interfaces by taking the example of takeover requests. Second, we present concrete interfaces that can be seen as a step towards cooperative interfaces because they implement parts of the basic requirements (mutual predictability, directability, shared situation representation, and calibrated trust).

10.4.1 The Design Space of In-Vehicle Interfaces Using the Example of Takeover Requests

The design space for takeover requests and other in-vehicle interfaces is manifold, they can be unimodal acoustic (Damböck et al. 2012; Gold et al. 2016; Körber et al. 2015), tactile (Politis et al. 2015), and visual (Petermann-Stock et al. 2015; Politis et al. 2015), as well as multimodal (Naujoks et al. 2015; Payre et al. 2016; Zeeb et al. 2016). Moreover, beside alerting the driver with abstract signals instructions and information can be communicated via iconographic (Gold et al. 2015; Naujoks et al. 2014, 2015) or textual cues (Melcher et al. 2015; Payre et al. 2016; Petermann-Stock et al. 2013), as spoken messages (Gold et al. 2015; Payre et al. 2016; Walch et al. 2015) or as a meaningful real-world sounds (e.g., a car horn Ho and Spence 2005). Nevertheless, interface designers have to keep the potential driver states in mind: for instance, while spoken messages are a fast and straightforward way to communicate on the one hand, on the other drivers may overhear these because they are masked by other auditory signals or they may confuse drivers already engaged in linguistic tasks (Ho and Spence 2005). The location of visual cues can also be varied: in the instrument cluster (Gold et al. 2013; Lorenz et al. 2014; Radlmayr et al. 2014), the center console (Naujoks et al. 2014; Payre et al. 2016; Walch et al. 2015), and other displays mounted at the dashboard (Naujoks et al. 2015; Zeeb et al. 2016); in the head-up or windshield displays (Gold et al. 2015; Walch et al. 2015); on nomadic, mobile, or wearable devices (Melcher et al. 2015; Politis et al. 2015). Light sources in the cockpit can also be used as interface (Petermann-Stock et al. 2015), in particular in the periphery to relieve focal vision (Löcken et al. 2015). Besides presenting information redundantly using different modalities, the information can also be displayed at multiple devices using each the same modality (e.g., displaying a TOR in the center console and in the head-up display (Walch et al. 2015)), for instance to reduce the risk that drivers miss a cue. The driveability of the vehicle can also be used to alert drivers, e.g., a brake jerk (Melcher et al. 2015). Moreover, the urgency of cues can be changed by manipulating parameters like pulse rate, base frequency, color, and count of modalities (Politis et al. 2013, 2014). Furthermore, the direction of which a cue is presented can be used to direct the visual attention of drivers (Ho and Spence 2008). Several of these possibilities were tested predominantly in driving simulator studies.

Politis et al. (2015) investigated all uni- and multimodal combinations of auditory, tactile, and visual cues as takeover requests. Moreover, the urgency of these cues was varied. The auditory cues were spoken messages that began with different words depending on the urgency: *danger* (high), *warning* (medium), and *notice* (low). The voice saying these messages also varied the tone depending on the urgency level. Participants recognized the urgency and took over faster when there were highly urgent warnings. On the basis of two experiments, (Politis et al. 2015) suggested to use informative multimodal takeover requests and to avoid unimodal visual cues. This recommendation is in line with Petermeijer et al. (2016): they emphasize that cues can be missed (e.g., missed hearing of an auditory cue or no contact to a device that presents a tactile cue), thus multimodal cues should be used to minimize this risk.

As long as automated vehicles are equipped with the traditional in-vehicle input interface—steering wheel, pedals, gear shift knob, levers and buttons—these can still be used as inputs for future interaction concepts. In case of handovers, the input to signal the vehicle that the driver takes over can be for instance the touch of the steering wheel (Walch et al. 2015) or an actual maneuver above a certain threshold using steering wheel or pedals (Gold et al. 2013). Control transitions from the driver to the vehicle should also be considered, for instance button presses are a common method to engage automation (Damböck et al. 2013; Körber et al. 2015; Naujoks et al. 2014). In other use cases, for example infotainment, speech input, touch screens, rotary knobs, gesture, or gaze input can be used. Beside these explicit input modalities implicit input modalities from sensors that measure the driver's state (Begum 2013; Braun et al. 2015; Li et al. 2014; Riener et al. 2009; Sahayadhas et al. 2012) gain growing importance due to automation and the resulting human factors implications. In particular, knowing the driver state is a major precondition to enable adaptive interfaces.

Martens and van den Beukel (2013) formulated several design recommendations to enable transitions that are safe and smooth: First, mode confusion should be avoided and drivers should be assumed as engaged in non-driving tasks (out-of-the-loop). Second, the driver should be enabled to come back in the loop quickly and easily by interrupting secondary tasks and by using fast information transferring input types like inputs with force feedback. Moreover, system boundaries should be explained (e.g., by naming reasons for transitions), the level of automation should be selected and adapted by the driver actively, and warnings should be self-explaining and well timed. Koo et al. (2015) recommend that an highly automated system should communicate the reason and the resulting behavior in critical situations. Walch et al. (2015) proposed a handover process, that began with alerting the driver, interrupting other (non-driving related) tasks, and decelerating. Subsequently, the automation should explain the situation followed by a takeover request. Finally, if the driver complied the control can be handed over, otherwise the highly automated system has to de-escalate on its own.

10.4.2 Steps Towards Cooperation: Interfaces Implementing Basic Requirements

A central part of projects focusing on *task sharing* is established by designing the interaction concept. Norman (1990) stated that "[a]ppropriate design should assume the existence of error, it should continually provide feedback, it should continually interact with operators in an effective manner, and it should allow for the worst of situations" (Norman 1990, p. 1). Usually the interaction is based on some interface. Zimmermann and Bengler (2013) summarize the purpose of an interface in the human–machine cooperation (HMC) as "interfaces between humans and machines are designed to (1) display and infer the users' and machines' intentions and to (2) convey a dynamic adaptive system reconfiguration" (Zimmermann and Bengler 2013, p. 1286). Bengler et al. (2012) resume that instead of reducing the communication between human and machine to visual interfaces, a multimodal user interface should be established including mainly the auditory and tactile channel alongside the visual channel. Abbink et al. (2012) shared this point of view. They listed different experimental evidence that haptic shared control (controlling with a haptic interface both system and user can apply forces to) can optimize the short-term performance. It can lead to faster and more accurate vehicle control, lower levels of control effort, and reduced demand for visual attention. They hypothesized that haptic shared control can help to maintain situation awareness on a higher level (Abbink et al. 2012).

As already discussed, drivers of highly automated vehicles are likely out-of-the-loop. Nevertheless, system partners need a shared situation representation that enables cooperation. Telpaz et al. (2015) built a haptic seat that mapped the position of vehicles behind the ego-vehicle on a 3×9 grid of vibrating motors in the backrest. The three vertical columns were used to map the lane on which the approaching vehicle drives. The distance was mapped by using the rows: the lowest vibrating motor started to vibrate when a vehicle was 270 m away, with every 30 m the next motor above was activated and the previous motor stopped vibrating. Telpaz et al. (2015) observed an positive effect on takeover performance. A similar approach with a shape changing backrest rather a vibrating one was presented by Grah et al. (2015). These tactile interfaces enable the vehicle to tell the driver whether and where there are vehicles approaching from behind. Another suitable solution for this issue is an ambient light display which was presented by Löcken et al. (2015). They suggest that ambient light displays can close the gap between simple interfaces like abstract auditory cues and complex demanding interfaces like GUIs. Thus, they implemented and evaluated an ambient light display to support drivers during a lane change by indicating a vehicle approaching on the target line. Besides enabling shared situation representation their interface implemented another key aspect of cooperation, namely adaptation: the display adapted its brightness to the driver state. The brightness was increased when a simple certainty model predicted that the driver is unsure which action to perform. This adaption had had the effect that their participants decided faster accompanied by less probable violations of safety gaps. Their design goal was to achieve that the display does not disturb but catches the attention of drivers when they would benefit of it.

Guiding the attention of drivers towards hazards utilizing augmented reality (AR) through windshield displays has already been investigated in driving simulator studies. For instance, Haeuslschmid et al. (2015) investigated the augmentation of a moving hazard compared to a warning in a head-up display. While they did not find any effect on the reaction time, there was an effect on the gaze behavior: participants shifted their attention less often away from the road to perceive hazards when they were supported by the contact-analog augmentation. This is another example for an interface that tries to establish a shared situation representation by guiding the attention of the driver towards an entity that has been sensed by the system.

Another AR approach was presented by Lorenz et al. (2014). They investigated whether AR can help drivers to regain control in a transition phase. In one condition, a red augmentation of an accident and the lane in front of it, which should not be entered by the vehicle, was displayed. In the other condition, a green augmentation highlighted a trajectory passing the accident which the driver can follow. The participants of their driving simulator study maneuvered after about 3 s regardless whether they were supported with one of the AR systems or not. However, Lorenz et al. (2014) found that the augmentations effected takeover behavior. The green augmentation had a positive impact, because more participants decelerated by braking compared to the group of participants who were not supported with AR information. Moreover, the majority followed the suggested trajectory. In contrast, in the red AR condition the participants only received the information where they should not drive which resulted in more diverse behaviors, for instance four participants (25%) stopped entirely and thereby they posed a serious danger in traffic. This illustrates that inappropriate designed interfaces and interaction concepts can have negative effects—the color red may be associated with stopping for some users. Nevertheless, the green augmentation is a good implementation of directability—the system guides the driver.

Biester (2008) investigated an overtaking scenario on highways where drivers had to overtake a truck under varying automation levels (manual, cooperative, semi-automated, and automated). In the cooperative condition the system (Wizard-of-Oz) and the driver conferred with each other about the traffic situation, the overtaking, and the evaluation of their interaction. An experiment has shown that the cooperation produces a higher situation awareness and trust than the semi-automated mode in which there were shifts between manual and automatic control. The system and the driver planned together the overtaking maneuver: for instance the plan (overtaking) and the current activity (e.g., searching a gap) was communicated by the driver which led to predictability and the system responded by helping the driver finding a suitable gap (shared situation representation).

Another ergonomic framework of cooperative guidance and control for highly automated vehicles is provided by Flemisch et al. (2014), who proposed the concepts *Conduct-by-Wire* and *H-Mode*. With Conduct-by-Wire the driver delegates maneuvers to the automation using a specialized maneuvers interface, the *pieDrive*, which was introduced first by Franz et al. (2012). This interface is based on a combination of a contact-analog head-up display and a touchpad. The H-Mode, based on the H(orse)-Metaphor, compares the relationship between a driver and a highly automated vehicle in relation to the driver and a horse cart. Three levels of automation were defined: Tight Rein, Loose Rein, and Secure Rein. During the Tight Rein, which is the assisted mode, the driver is in charge and automation reduces its influence. The Loose Rein represents the highly automated mode, in which the driver still has to be in the loop. In the third level, Secure Rein, the system is temporarily in full control of driving, while the driver is out of the loop (Flemisch et al. 2014). Conduct-by-Wire and H-Mode are two really promising cooperation frameworks, especially as they are not limited to a special maneuver.

Zimmermann et al. (2014) proposed a cooperative driving concept that involves *in-vehicle cooperation* between driver and vehicle as well as other vehicles and their drivers (*traffic cooperation*) to proceed a cooperative lane change on highways. The interaction is divided in five steps: First, in the *request* phase vehicle A informs its driver that there is an obstacle and that it is searching for a partner to cooperate with. Another vehicle (B) which is a suitable partner is requested and thus asks its driver to accept the request. Second, in the *suggest preparation* phase it is suggested to open a gap. When the driver of B confirms, vehicle B opens a gap in the third phase *prepare* and informs the drivers of both vehicles. When the prepare phase is completed, the lane change action is suggested to the driver of vehicle A (*suggest action* phase). Again, after confirming the next phase is entered: in the *action* phase, the lane change is proceeded accompanied by informing both drivers. The interface was designed multimodal: besides visual (AR) and auditory cues the steering wheel and pedals were used. In the request phase the driver of vehicle B can use the accelerator pedal to decline and the brake pedal to accept the request (opening a gap). In contrast, in vehicle A the brake pedal is used to abort cooperation and an actuation towards the left lane on the steering wheel triggers the lane change. These interactions are designed to match the drivers mental model since they are comparable to actions drivers would do in a manual condition in the same situation. Zimmermann et al. (2014) state that future implementations of their system should adapt to the driver's state (i.e., guiding driver's attention when the driver is overloaded). In addition, cooperative interaction should not only consider lane changes. Nevertheless, this implementation shows a general idea how several partners can negotiate with each other.

The requirement *mutual predictability* is already implemented in some systems. Payre et al. (2016) implemented a handover process for anticipated control transitions, e.g., leaving a highway when driving with an automation that only supports driving on highways. They started to inform the driver 30 s prior the control transition using four visual and spoken messages: “end of FAD [fully automated driving] zone coming soon, end of FAD zone, please resume vehicle control, autopilot mode deactivated when manual control was resumed” (Payre et al. 2016, p. 233). A similar stepwise approach was used by Toffetti et al. (2009): they used a *pre-warning*, a *first warning* and a *final warning*. The stepwise information helps drivers to anticipate that the system will reach a system limitation and so they can plan their own action to solve this issue.

Naujoks et al. (2015) used *soft takeover requests* in situations where a handover is likely to prepare drivers for potential handovers. The icon of their assistant was displayed in yellow accompanied by an auditory signal for soft takeover requests, whereas it was displayed in red together with a red visual warning that consisted of a brake pedal, a steering wheel, and a text that asked to take over accompanied with a more urgent auditory signal as *hard takeover request* when the driver has to take over control. The early guidance of attention towards the driving task enables drivers to get back in the loop and develop situation awareness appropriately. A similar approach to deal with system uncertainties is presented by Gold et al. (2013). They prepare the driver for a potential handover with *monitoring requests* in situations with an unpredictable course, e.g., passersby who might cross the lane next to the road (Gold et al. 2013). Helldin et al. (2013) found a positive effect of a system uncertainty representation (a bar which is full when the system is able to drive autonomously and empties when the system becomes uncertain) on takeover time in a system failure scenario. In contrast, Beller et al. (2013) used only a binary system uncertainty representation (a icon of a face with an unsurely expression and gesture), however they found also positive effects on performance (higher minimum time to collision), situation awareness, trust and acceptance of the uncertainty representation. This kind of system transparency—showing the (un)certainty state—helps drivers to calibrate their trust level and as a result to adapt their own state to the capabilities of the automation.

Davidsson and Alm (2009) see a way for cooperation in varying the degree of information drivers get from their cars as well as the adaptability of this information. Due to new sources such as radars or sensors, the available information is constantly growing. On one hand, the given information may help to improve safety, environmental friendliness or transport efficiency and some drivers might even enjoy to have a lot of information. On the other hand, the presentation of all information from these gadgets or functions simultaneously can lead to a high visual and cognitive distraction (Davidsson and Alm 2009). The central idea of their approach is the adaption of the given information depending on the current driving scenario. They presume for example that while driving out of your garage a speedometer is less helpful than a 360 degree camera. So far this framework has not been tested yet but a problem seen by the authors themselves is that drivers using this system would need special training.

The given overview shows that a lot of work was conducted in the field of car-driver interaction and cooperation. However, the majority of presented systems are only isolated applications for a very specific use case and address only parts of the requirements of cooperative interfaces. This highlights that the work that has already been conducted is only the base for more evolved interaction concepts. Multimodal, adaptive, and cooperative driver–vehicle cooperation has to be investigated to become more integrative interaction concepts and HMIs in automated driving.

10.5 Conclusion

The technological basis for automated driving is almost established. However, there are still a lot of unsolved questions regarding driver–vehicle interaction that arise due to automation. The role of the driver changes and thus the classical in-vehicle interface and interaction paradigms have to be reconsidered. The automation relieves the driver from the driving task, thus new human factor challenges arise such as lacking situation awareness and the resulting out-of-the-loop problem, automation bias, and an unsubstantiated degree of trust in automation. Beside the technological advances that enable automated driving these issues have to be taken into account before highly automated vehicles can be introduced to the market.

Recent research investigated control authority transitions between automated system and human driver at system boundaries. It has been shown that such handovers are feasible within relative short time budgets. However, the suggested concepts are just a solution for a very narrow part of future driver–vehicle interaction. We suggest to have a more human-centered perspective on automation. The human should come into focus in order to enable cooperation between human and machine instead of a simple task substitution through automation. Automation should be implemented as a cooperative team player instead of an agent that receives or gives only directives or commands. This has the potential to diminish automation bias, since drivers do not merely become passengers, but rather team partners of automation with a shared task and goal.

We recommend to consider four basic requirements for driver–vehicle cooperation: mutual predictability, directability, shared situation representation, and calibrated trust in automation. There are already systems that implement parts of these requirements, and thus can be used as basis and inspiration for tomorrow's interfaces and interaction concepts. Future cooperative systems should be multimodal and adaptive. Moreover, researches, engineers, and designers should keep the big picture of human–machine cooperation in automated driving in their eyes rather than focusing only on small parts to design integrative interaction concepts and HMIs.

Acknowledgements The authors thank the Carl-Zeiss Foundation for the partial funding of this work.

References

Abbink, D. A., Mulder, M., & Boer, E. R. (2012). Haptic shared control: Smoothly shifting control authority? *Cognition, Technology and Work*, *14*(1), 19–28. doi:10.1007/s10111-011-0192-5.

Aeberhard, M., Rauch, S., Bahram, M., Tanzmeister, G., Thomas, J., Pilat, Y., et al. (2015). Experience, results and lessons learned from automated driving on germany's highways. *IEEE Intelligent Transportation Systems Magazine*, *7*(1), 42–57. doi:10.1109/MITS.2014.2360306.

Begum, S. (2013). Intelligent driver monitoring systems based on physiological sensor signals: A review. In *16th International IEEE Conference on Intelligent Transportation Systems (ITSC 2013)* (pp. 282–289). doi:10.1109/ITSC.2013.6728246.

Beller, J., Heesen, M., & Vollrath, M. (2013). Improving the driver-automation interaction: An approach using automation uncertainty. *Human Factors: The Journal of the Human Factors and Ergonomics Society*, *55*(6), 1130–1141. doi:10.1177/0018720813482327.

Bengler, K., Zimmermann, M., Bortot, D., Kienle, M., & Damböck, D. (2012). Interaction principles for cooperative human-machine systems. *It—Information Technology*, *54*(4), 157–164. doi:10.1524/itit.2012.0680.

Biester, L. (2008). Cooperative automation in automobiles. Psychologie Schweizerische Zeitschrift Für Psychologie Und Ihre Andwendungen.

Brandenburg, S., & Skottke, E. M. (2014). Switching from manual to automated driving and reverse: Are drivers behaving more risky after highly automated driving? In *17th international IEEE conference on intelligent transportation systems (ITSC)* (pp. 2978–2983). doi:10.1109/ITSC.2014.6958168.

Braun, A., Frank, S., Majewski, M., & Wang, X. (2015). CapSeat: capacitive proximity sensing for automotive activity recognition. In *Proceedings of the 7th international conference on automotive user interfaces and interactive vehicular applications, AutomotiveUI '15* (pp. 225–232). New York, NY, USA: ACM. doi:10.1145/2799250.2799263.

Carsten, O., Lai, F. C. H., Barnard, Y., Jamson, A. H., & Merat, N. (2012). Control task substitution in semiautomated driving: Does it matter what aspects are automated? *Human Factors: The Journal of the Human Factors and Ergonomics Society*, *54*(5), 747–761. doi:10.1177/0018720812460246.

Christoffersen, K., & Woods, D. D. (2002). 1. How to make automated systems team players. *Advances in Human Performance and Cognitive Engineering Research*, *2*, 1–13. doi:10.1016/S1479-3601(02)02003-9.

Damböck, D., Bengler, K., Farid, M., & Tönert, L. (2012). Übernahmezeiten beim hochautomatisierten Fahren. *Tagung Fahrerassistenz. München*, *15*, 16.

Damböck, D., Weißgerber, T., Kienle, M., & Bengler, K. (2013). Requirements for cooperative vehicle guidance. In *16th international IEEE conference on intelligent transportation systems (ITSC 2013)* (pp. 1656–1661). doi:10.1109/ITSC.2013.6728467.

Davidsson, S., & Alm, H. (2009). *Applying the "Team player" approach on car design*. Lecture Notes in Computer Science (including subseries Lecture Notes in Artificial Intelligence and Lecture Notes in Bioinformatics) 5639 LNAI (pp. 349–357). doi:10.1007/978-3-642-02728-4_37.

Dekker, S. W. A., & Woods, D. D. (2002). MABA-MABA or abracadabra? progress on human-automation co-ordination. *Cognition, Technology and Work*, *4*(4), 240–244. doi:10.1007/s101110200022.

de Winter, J. C., Happee, R., Martens, M. H., & Stanton, N. A. (2014). Effects of adaptive cruise control and highly automated driving on workload and situation awareness: A review of the empirical evidence. *Transportation Research Part F: Traffic Psychology and Behaviour*, *27*, 196–217. doi:10.1016/j.trf.2014.06.016.

Endsley, M. R., & Kiris, E. O. (1995). *Situation awareness global assessment technique (SAGAT) TRACON air traffic control version user's guide*. Technical report: Texas Tech University, Lubbock.

Endsley, M. R., & Kiris, E. O. (1995). The out-of-the-loop performance problem and level of control in automation. *Human Factors: The Journal of the Human Factors and Ergonomics Society, 37*(2), 381–394. doi:10.1518/001872095779064555.

Flemisch, F., Heesen, M., Hesse, T., Kelsch, J., Schieben, A., & Beller, J. (2012). Towards a dynamic balance between humans and automation: Authority, ability, responsibility and control in shared and cooperative control situations. *Cognition, Technology and Work, 14*(1), 3–18. doi:10.1007/s10111-011-0191-6.

Flemisch, F. O., Bengler, K., Bubb, H., Winner, H., & Bruder, R. (2014). Towards cooperative guidance and control of highly automated vehicles: H-Mode and Conduct-by-Wire. *Ergonomics, 57*(3), 343–60. doi:10.1080/00140139.2013.869355.

Franz, B., Kauer, M., Bruder, R., & Geyer, S. (2012). PieDrive—a new driver-vehicle interaction concept for maneuver-based driving. In *2012 IEEE intelligent vehicles symposium workshops (IV).*

Gold, C., & Bengler, K. (2014). Taking over control from highly automated vehicles. *Advances in Human Aspects of Transportation: Part II, 8*, 64.

Gold, C., Berisha, I., & Bengler, K. (2015). Utilization of drivetime—performing non-driving related tasks while driving highly automated. *Proceedings of the Human Factors and Ergonomics Society Annual Meeting, 59*(1), 1666–1670. doi:10.1177/1541931215591360.

Gold, C., Damböck, D., Lorenz, L., & Bengler, K. (2013). "Take over!" How long does it take to get the driver back into the loop? *Proceedings of the Human Factors and Ergonomics Society Annual Meeting, 57*(1), 1938–1942. doi:10.1177/1541931213571433.

Gold, C., Korber, M., Lechner, D., & Bengler, K. (2016). Taking over control from highly automated vehicles in complex traffic situations: The role of traffic density. *Human Factors, 58*(4), 642–652. doi:10.1177/0018720816634226.

Gold, C., Lorenz, L., Damböck, D., & Bengler, K. (2013). Partially automated driving as a fall-back level of high automation. 6. Tagung Fahrerassistenzsysteme. Der Weg zum automatischen Fahren **28**(29.11), 2013.

Gouy, M., Wiedemann, K., Stevens, A., Brunett, G., & Reed, N. (2014). Driving next to automated vehicle platoons: How do short time headways influence non-platoon drivers' longitudinal control? *Transportation Research Part F: Traffic Psychology and Behaviour, 27*, 264–273. doi:10.1016/j.trf.2014.03.003.

Grah, T., Epp, F., Wuchse, M., Meschtscherjakov, A., Gabler, F., Steinmetz, A., & Tscheligi, M. (2015). Dorsal haptic display: A shape-changing car seat for sensory augmentation of rear obstacles. In *Proceedings of the 7th international conference on automotive user interfaces and interactive vehicular applications, AutomotiveUI '15* (pp. 305–312). New York, NY, USA: ACM. doi:10.1145/2799250.2799281.

Haeuslschmid, R., Schnurr, L., Wagner, J., & Butz, A. (2015). Contact-analog warnings on windshield displays promote monitoring the road scene. In *Proceedings of the 7th international conference on automotive user interfaces and interactive vehicular applications, AutomotiveUI '15* (pp. 64–71). New York, NY, USA: ACM. doi:10.1145/2799250.2799274.

Helldin, T., Falkman, G., Riveiro, M., & Davidsson, S. (2013). Presenting system uncertainty in automotive uis for supporting trust calibration in autonomous driving. In *Proceedings of the 5th international conference on automotive user interfaces and interactive vehicular applications, AutomotiveUI '13* (pp. 210–217). New York, NY, USA: ACM. doi:10.1145/2516540.2516554.

Ho, C., & Spence, C. (2005). Assessing the effectiveness of various auditory cues in capturing a driver's visual attention. *Journal of Experimental Psychology: Applied, 11*(3), 157–174. doi:10.1037/1076-898X.11.3.157.

Ho, C., & Spence, C. (2008). The multisensory driver: Implications for ergonomic car interface design. Ashgate, Aldershot [u.a.]

Hoc, J. M. (2000). From human-machine interaction to human-machine cooperation. *Ergonomics, 43*(7), 833–843.

Hoc, J. M. (2001). Towards a cognitive approach to human-machine cooperation in dynamic situations. *International Journal of Human-Computer Studies*, *54*(4), 509–540. doi:10.1006/ijhc.2000.0454.

Hoc, J. M., Young, M. S., & Blosseville, J. M. (2009). Cooperation between drivers and automation: Implications for safety. *Theoretical Issues in Ergonomics Science*, *10*(2), 135–160. doi:10.1080/14639220802368856.

Hoff, K. A., & Bashir, M. (2015). Trust in automation: Integrating empirical evidence on factors that influence trust. *Human Factors*, *57*(3), 407–434. doi:10.1177/0018720814547570.

Klein, G., Woods, D., Bradshaw, J., Hoffman, R., & Feltovich, P. (2004). Ten challenges for making automation a "Team Player" in joint human-agent activity. *IEEE Intelligent Systems*, *19*(06), 91–95. doi:10.1109/MIS.2004.74.

Koo, J., Kwac, J., Ju, W., Steinert, M., Leifer, L., & Nass, C. (2015). Why did my car just do that? Explaining semi-autonomous driving actions to improve driver understanding, trust, and performance. *International Journal on Interactive Design and Manufacturing (IJIDeM)*, *9*(4), 269–275. doi:10.1007/s12008-014-0227-2.

Körber, M., Gold, C., Lechner, D., & Bengler, K. (2016). The influence of age on the take-over of vehicle control in highly automated driving. *Transportation Research Part F: Traffic Psychology and Behaviour*, *39*, 19–32. doi:10.1016/j.trf.2016.03.002.

Körber, M., Weißgerber, T., Blaschke, C., Farid, M., & Kalb, L. (2015). Prediction of take-over time in highly automated driving by two psychometric tests. *DYNA*, *82*(193), 195–201. doi:10.15446/dyna.v82n193.53496.

Kunz, F., Nuss, D., Wiest, J., Deusch, H., Reuter, S., Gritschneder, F., Scheel, A., Stübler, M., Bach, M., Hatzelmann, P., Wild, C., & Dietmayer, K. (2015). Autonomous driving at Ulm university: A modular, robust, and sensor-independent fusion approach. In *2015 IEEE intelligent vehicles symposium (IV)* (pp. 666–673). doi:10.1109/IVS.2015.7225761.

Large, D. R., & Burnett, G. E. (2014). Letter from the Editors the effect of different navigation voices on trust and attention while using in-vehicle navigation systems. *Journal of Safety Research*, *49*, 69–75. doi:10.1016/j.jsr.2014.02.009.

Lee, J. D., See, K. A., & City, I. (2004). Trust in automation: Designing for appropriate reliance. *Human Factors*, *46*(1), 50–80. doi:10.1518/hfes.46.1.50_30392.

Li, L., Werber, K., Calvillo, C.F., Dinh, K.D., Guarde, A., & König, A. (2014). Multi-sensor soft-computing system for driver drowsiness detection. In V. Snášel, P. Krömer, M. Köppen, G. Schaefer (Eds.), *Soft Computing in Industrial Applications: Proceedings of the 17th online world conference on soft computing in industrial applications* (pp. 129–140). Springer International Publishing, Cham (2014). doi:10.1007/978-3-319-00930-8_12.

Löcken, A., Heuten, W., & Boll, S. (2015). Supporting lane change decisions with ambient light. In *Proceedings of the 7th international conference on automotive user interfaces and interactive vehicular applications, AutomotiveUI '15* (pp. 204–211). New York, NY, USA: ACM. doi:10.1145/2799250.2799259.

Lorenz, L., Kerschbaum, P., & Schumann, J. (2014). Designing take over scenarios for automated driving: How does augmented reality support the driver to get back into the loop? *Proceedings of the Human Factors and Ergonomics Society Annual Meeting*, *58*(1), 1681–1685. doi:10.1177/1541931214581351.

Lu, Z., & de Winter, J. C. (2015). A review and framework of control authority transitions in automated driving. *Procedia Manufacturing*, *3*, 2510–2517. doi:10.1016/j.promfg.2015.07.513.

Mark, G., & Kobsa, A. (2005). The effects of collaboration and system transparency on CIVE usage: An empirical study and model. *Presence*, *14*(1), 60–80. doi:10.1162/1054746053890279.

Martens, M. H., & van den Beukel, A. P. (2013). The road to automated driving: Dual mode and human factors considerations. In *16th International IEEE Conference on Intelligent Transportation Systems—(ITSC 2013)* (pp. 2262–2267). doi:10.1109/ITSC.2013.6728564.

Melcher, V., Rauh, S., Diederichs, F., Widlroither, H., & Bauer, W. (2015). Take-over requests for automated driving. *Procedia Manufacturing*, *3*, 2867–2873. doi:10.1016/j.promfg.2015.07.788.

Merat, N., Jamson, A. H., Lai, F. C., Daly, M., & Carsten, O. M. (2014). Transition to manual: Driver behaviour when resuming control from a highly automated vehicle. *Transportation Research Part F: Traffic Psychology and Behaviour, 27*, 274–282. doi:10.1016/j.trf.2014.09.005.

Muir, B. M., & Moray, N. (1996). Trust in automation. Part II. Experimental studies of trust and human intervention in a process control simulation. *Ergonomics, 39*(3), 429–460. doi:10.1080/00140139608964474.

Naujoks, F., Mai, C., & Neukum, A. (2014). The effect of urgency of take-over requests during highly automated driving under distraction conditions. In *Proceedings of the 5th international conference on applied human factors and ergonomics AHFE 2014*.

Naujoks, F., Purucker, C., Neukum, A., Wolter, S., & Steiger, R. (2015). Controllability of partially automated driving functions—Does it matter whether drivers are allowed to take their hands off the steering wheel? *Transportation Research Part F: Traffic Psychology and Behaviour, 35*, 185–198. doi:10.1016/j.trf.2015.10.022.

Norman, D. A. (1990). The 'Problem' with automation: Inappropriate feedback and interaction, not 'Over-Automation'. *Philosophical Transactions of the Royal Society of London B: Biological Sciences, 327*(1241), 585–593. doi:10.1098/rstb.1990.0101.

Parasuraman, R., & Riley, V. (1997). Humans and automation: Use, misuse, disuse, abuse. *Human Factors: The Journal of the Human Factors and Ergonomics Society, 39*(2), 230–253. doi:10.1518/001872097778543886.

Payre, W., Cestac, J., & Delhomme, P. (2016). Fully automated driving: Impact of trust and practice on manual control recovery. *Human Factors, 58*(2), 229–241. doi:10.1177/0018720815612319.

Petermann-Stock, I., Hackenberg, L., Muhr, T., Josten, J., & Eckstein, L. (2015). "Bitte übernehmen Sie das Fahren!": Ein multimodaler Vergleich von Übernahmestrategien. AAET.

Petermann-Stock, I., Hackenberg, L., Muhr, T., & Mergl, C. (2013). Wie lange braucht der Fahrer? Eine Analyse zu Übernahmezeiten aus verschiedenen Nebentätigkeiten während einer hochautomatisierten Staufahrt. 6. Tagung Fahrerassistenzsysteme. Der Weg zum automatischen Fahren.

Petermeijer, S. M., de Winter, J. C. F., & Bengler, K. J. (2016). Vibrotactile displays: A survey with a view on highly automated driving. *IEEE Transactions on Intelligent Transportation Systems, 17*(4), 897–907. doi:10.1109/TITS.2015.2494873.

Politis, I., Brewster, S., & Pollick, F. (2013). Evaluating multimodal driver displays of varying urgency. In *Proceedings of the 5th international conference on automotive user interfaces and interactive vehicular applications, AutomotiveUI '13* (pp. 92–99). New York, NY, USA: ACM. doi:10.1145/2516540.2516543.

Politis, I., Brewster, S., & Pollick, F. (2015). Language-based multimodal displays for the handover of control in autonomous cars. In *Proceedings of the 7th international conference on automotive user interfaces and interactive vehicular applications, AutomotiveUI '15* (pp. 3–10). New York, NY, USA: ACM. doi:10.1145/2799250.2799262.

Politis, I., Brewster, S.A., & Pollick, F. (2014). Evaluating multimodal driver displays under varying situational urgency. In *Proceedings of the 32Nd annual ACM conference on human factors in computing systems, CHI '14* (pp. 4067–4076). New York, NY, USA: ACM. doi:10.1145/2556288.2556988.

Radlmayr, J., Gold, C., Lorenz, L., Farid, M., & Bengler, K. (2014). How traffic situations and non-driving related tasks affect the take-over quality in highly automated driving. *Proceedings of the Human Factors and Ergonomics Society Annual Meeting, 58*(1), 2063–2067. doi:10.1177/1541931214581434.

Rauch, N., Gradenegger, B., & Krüger, H. P. (2009). Darf ich oder darf ich nicht? Situationsbewusstsein im Umgang mit Nebenaufgaben während der Fahrt. *Zeitschrift für Arbeitswissenschaft, 63*(1), 3–16.

Riener, A., Ferscha, A., & Aly, M. (2009). Heart on the road: HRV analysis for monitoring a driver's affective state. In *Proceedings of the 1st international conference on automotive user interfaces and interactive vehicular applications, AutomotiveUI '09* (pp. 99–106). New York, NY, USA: ACM. doi:10.1145/1620509.1620529.

SAE International: Automated Driving Levels of Driving Automation are Defined in new SAE International Standard J3016 (2014).

Sahayadhas, A., Sundaraj, K., & Murugappan, M. (2012). Detecting driver drowsiness based on sensors: A review. *Sensors* (Basel, Switzerland) *12*(12), 16,937–16,953. doi:10.3390/s121216937.

Shladover, S. E. (2016). The Truth about "Self-Driving" cars. *Scientific American, 314*(6), 52–57. doi:10.1038/scientificamerican0616-52.

Skottke, E. M., Debus, G., Wang, L., & Huestegge, L. (2014). Carryover effects of highly automated convoy driving on subsequent manual driving performance. *Human Factors: The Journal of the Human Factors and Ergonomics Society*, *56*(7), 1272–1283. doi:10.1177/0018720814524594.

Stanton, N. A. (2016). Distributed situation awareness. *Theoretical Issues in Ergonomics Science*, *17*(1), 1–7. doi:10.1080/1463922X.2015.1106615.

Telpaz, A., Rhindress, B., Zelman, I., & Tsimhoni, O. (2015). Haptic seat for automated driving: preparing the driver to take control effectively. In *Proceedings of the 7th international conference on automotive user interfaces and interactive vehicular applications, AutomotiveUI '15* (pp. 23–30). New York, NY, USA: ACM. doi:10.1145/2799250.2799267.

Toffetti, A., Wilschut, E., Martens, M., Schieben, A., Rambaldini, A., Merat, N., et al. (2009). CityMobil. *Transportation Research Record: Journal of the Transportation Research Board*, *2110*, 1–8. doi:10.3141/2110-01.

Trimble, T.E., Bishop, R., Morgan, J.F., & Blanco, M. (2014). Human factors evaluation of level 2 and level 3 automated driving concepts: Past research, state of automation technology, and emerging system concepts.

Walch, M., Lange, K., Baumann, M., & Weber, M. (2015). Autonomous driving: investigating the feasibility of car-driver handover assistance. In *Proceedings of the 7th international conference on automotive user interfaces and interactive vehicular applications, AutomotiveUI '15* (pp. 11–18). New York, NY, USA: ACM. doi:10.1145/2799250.2799268.

Walker, G. H., Stanton, N. A., & Salmon, P. (2016). Trust in vehicle technology. *International Journal of Vehicle Design*, *70*(2), 157–182. doi:10.1504/IJVD.2016.074419.

Waytz, A., Heafner, J., & Epley, N. (2014). The mind in the machine: Anthropomorphism increases trust in an autonomous vehicle. *Journal of Experimental Social Psychology*, *52*, 113–117. doi:10.1016/j.jesp.2014.01.005.

Young, M., Stanton, N., & Harris, D. (2007). Driving automation: Learning from aviation about design philosophies. *International Journal of Vehicle Design*, *45*(3), 323–338. doi:10.1504/IJVD.2007.014908.

Zeeb, K., Buchner, A., & Schrauf, M. (2015). What determines the take-over time? An integrated model approach of driver take-over after automated driving. *Accident; Analysis and Prevention*, *78*, 212–221. doi:10.1016/j.aap.2015.02.023.

Zeeb, K., Buchner, A., & Schrauf, M. (2016). Is take-over time all that matters? The impact of visual-cognitive load on driver take-over quality after conditionally automated driving. *Accident; Analysis and Prevention*, *92*, 230–239. doi:10.1016/j.aap.2016.04.002.

Ziegler, J., Bender, P., Schreiber, M., Lategahn, H., Strauss, T., Stiller, C., et al. (2014). Making Bertha drive—An autonomous journey on a historic route. *IEEE Intelligent Transportation Systems Magazine*, *6*(2), 8–20. doi:10.1109/MITS.2014.2306552.

Zimmermann, M., Bauer, S., Lütteken, N., Rothkirch, I. M., & Bengler, K. J. (2014). Acting together by mutual control: Evaluation of a multimodal interaction concept for cooperative driving. In *2014 international conference on collaboration technologies and systems (CTS)* (pp. 227–235). doi:10.1109/CTS.2014.6867569.

Zimmermann, M., & Bengler, K. (2013). A multimodal interaction concept for cooperative driving. In *2013 IEEE intelligent vehicles symposium (IV)* (pp. 1285–1290). doi:10.1109/IVS.2013.6629643.

Chapter 11
Driver in the Loop: Best Practices in Automotive Sensing and Feedback Mechanisms

Andreas Riener, Myounghoon Jeon, Ignacio Alvarez and Anna K. Frison

Abstract Given the rapid advancement of technologies in the automotive domain, driver–vehicle interaction has recently become more and more complicated. The amount of research applied to the vehicle cockpit is increasing, with the advent of (highly) automated driving, as the range of interaction that is possible in a driving vehicle expands. However, as opportunities increase, so does the number of challenges that automotive user experience designers and researchers will face. This chapter focuses on the instrumentation of sensing and displaying techniques and technologies to make better user experience while driving. In the driver–vehicle interaction loop, the vehicle can sense driver states, analyze, estimate, and model the data, and then display it through the appropriate channels for intervention purposes. To improve the interaction, a huge number of new/affordable sensing (EEG, fNIRS, IR imaging) and feedback (head-up displays, auditory feedback, tactile arrays, etc.) techniques have been introduced. However, little research has attempted to investigate this area in a systematic way. This chapter provides an overview of recent advances of input and output modalities to be used for timely, appropriate driver–vehicle interaction. After outlining relevant background, we provide information on the best-known practices for input and output modalities

A. Riener (✉)
Faculty of Electrical Engineering and Computer Science & CARISSMA
(Center of Automotive Research on Integrated Safety Systems and Measurement Area),
University of Applied Sciences Ingolstadt, Esplanade 10, 85055 Ingolstadt, Germany
e-mail: andreas.riener@thi.de

M. Jeon
Department of Cognitive and Learning Sciences & Department of Computer Science,
Michigan Technological University, 1400 Townsend Drive, Houghton, MI 49931, USA
e-mail: mjeon@mtu.edu

I. Alvarez
Intel Labs, Intel Corporation, 2111 N.E. 25th Avenue, Hillsboro, OR 97124, USA
e-mail: ignacio.j.alvarez@intel.com

A.K. Frison
Faculty of Electrical Engineering and Computer Science, University of Applied Sciences Ingolstadt, Esplanade 10, 85055 Ingolstadt, Germany
e-mail: Anna-Katharina.Frison@thi.de

G. Meixner and C. Müller (eds.), *Automotive User Interfaces*,
Human–Computer Interaction Series, DOI 10.1007/978-3-319-49448-7_11

based on the exchange results from the workshop on practical experiences for measuring and modeling drivers and driver–vehicle interactions at AutomotiveUI 2015. This chapter can help answer research questions on how to instrument a driving simulator or realistic study to gather data and how to place interaction outputs to enable appropriate driver interactions.

11.1 Introduction

Simulator experiments and naturalistic driving studies aimed to validate new forms of driver–vehicle interaction and automotive user interfaces have recently become a major topic in human–computer interaction (HCI) with thousands of papers published only in the past few years. Underpinned is this evolution by the adoption of new conferences in the field. For example, the "Automotive User Interfaces and Interactive Vehicular Applications" (AutoUI) conference was founded in 2009 and put into practice with 40 papers submitted (22 accepted) and ca. 50 attendees (Fig. 11.1). In the coming years, both the number of submitted papers, the categories of the conference as well as the number of attendees rose continuously, reaching a peak of 88 full papers submitted in 2016 and more than 200 people attended in 2014 (Seattle, US) and in 2015 (Nottingham, UK) respectively.

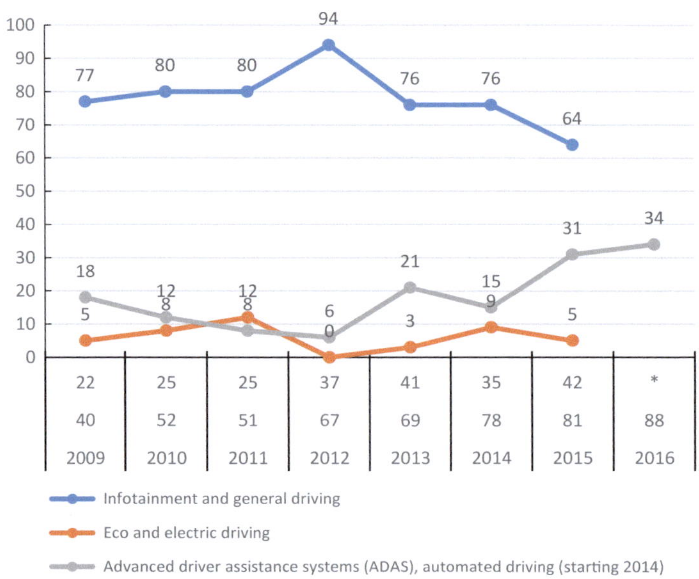

Fig. 11.1 Focus of papers published at AutoUI between 2009 and 2016 (2016: no. of submissions only) shows that interest in advanced driving assistance systems and automated driving is increasing (adapted from Kun et al. (2016) and based on own statistics)

Looking holistically at papers published in the conference proceedings of AutoUI, and other related conferences and journals, the problems which researchers, particularly, those new automotive user interfaces field, are exposed to clearly appears. In particular, for papers addressing topics related to ADAS (Advanced Driver Assistance Systems) and automated driving, it turns out that many decisions regarding the research setting, (e.g., lab/field, low-/high-fidelity simulator, within/between subjects, sample size, biased subjects, learning effect, sensor technology, mobile hardware, synchronization issues, briefing, etc.) need to be defined early in the design phase without the reference of guidelines and best practices to support them in identifying the optimal solution to answer their research question of interest.

To achieve a complete understanding of driver–vehicle interaction loop and modeling of it, we need to identify how we can best monitor and measure the driver states, how we can provide appropriate feedback depending on the states, and again how we can estimate the driver's response to that feedback. This chapter tries to shed light on the matter by sharing our experiences in executing studies to measure drivers' quantitative (driving performance, driver behavior, vehicle conditions) and qualitative (subjective workload, driver–vehicle interaction experience, etc.) data. Furthermore, we review and present available sensor technology and its environments. Readers should be able to properly instrument the interior of the cockpit with different types of sensors, such as camera, infrared, ultrasonic, capacitive proximity-based, physiological, etc. (Lequellec and Lerasle 2000; Boverie et al. 1998) after reading this chapter. Then, we offer the status quo in displaying techniques, specifically, auditory, and tactile feedback.

In addition to this chapter, a good starting point for researchers starting with work in the field are the books of Lazar et al. (2010), Lazar (2007) or Stanton et al. (2013).

Topics covered in this chapter surged from the workshop on "Measuring and Modeling of Driving and Driver-Vehicle Interactions." As seen below, the topics include measuring and sensing (e.g., attention and workload) and displaying (e.g., tactile/haptic and auditory). We summarize the contributions of this workshop as an introduction on state-of-the-art researchers' concerns.

Ignacio Alvarez and Laura Rumbel analyze in "How my car got to know me: reflection on in-vehicle user modelling" the efforts made in order to sense and model human behavior, which is becoming a fundamental research area for automotive UI. Many studies, in particular those on a large-scale tend to use simulation and simulation models to tackle via machine-learning techniques individual human behavior, unpredictability when validating safety issues of self-driving cars. The paper picks up on that issue, presents background on user modeling, and discusses approaches that support modeling of drivers in a car based on a necessary standardized sensor instrumentation.

Nikolas Martelaro presents in his paper "CRUISE: Measuring and Smoothing Driver Behavior Through Haptic Feedback" results from the CRUISE system, a driver behavior modification tool based on the pedal input (acceleration, break), GPS, and CAN data. The system provides real-time haptic feedback via vibration patches attached to the pedals so that drivers can correct immediately inefficient driver behaviors (harsh breaking, steep accelerations, etc.). Results from a field test on the University campus seem to provide evidence of behavior modification as users consistently learned to maintain behaviors below the set thresholds during the road event.

Andreas Löcken describes in the workshop paper "Experiences with User Studies when Investigating Light Displays" his experiences of the research on in-vehicle ambient light displays. The paper outlines the core elements of human-centered design: requirements, design, and experiments. Given that, the paper has a specific design application domain, it is expected to facilitate lively discussions with workshop attendees who are interested in automotive HMI studies and the use of (ambient) light in user studies.

The paper "Report on the In-vehicle Auditory Interactions Workshop: Taxonomy, Challenges, and Approaches" by Myounghoon Jeon, Pavlo Bazilinskyy, Jan Hammerschmidt, Thomas Hermann, Steven Landry, and Katieanna Wolf is an excellent framework on auditory interactions. The paper focuses on four areas of investigation: auditory displays for BEV vehicles and automated driving, fuel efficiency, infotainment, and collision warning. This work presents suggestions and starting points in each of the four areas for further development of user interfaces based on auditory inputs and displays.

Katharina Oeltze and Mandy Dotzauer, the authors of the paper "Towards a best practice for multi-driver simulator studies" discuss best practices that they have implemented in the field of multi-simulator studies. In such a setting, several subjects can drive in the same virtual environment and influence each other (i.e., studying driver–driver interaction). This type of user studies is not very common today, but can be expected to be used more and more in the future.

In the paper "Nudge: Haptic Pre-Cueing to Communicate Automotive Intent," the authors Nikhil Gowda, Srinath Sibi, Sonia Baltodano, Nikolas Martelaro, Rohan Maheshwari, Dave Miller, and Wendy Ju evaluate three haptic cuing prototypes in two different simulated car designs. In summary, subjects were found to favor haptic cues presented via the "Pneumatic floorboard" and the "Pneumatic shoulderpads," while effective also, appeared to induce anger in subjects.

The paper "Multi-Dimensions Motivational Factors in Autonomous Driving" by Nidzamuddin Md Yusof and Juffrizal Karjanto describes an experimental plan for designing takeover scenarios based on the varying motivational factors of drivers. Such motivational factors include being in "hurry," "pressure," and "thrill." The objective is to explore the target best feeling between the occupant demands and the systems performance as well as evaluate the different scenarios derived during the takeover of a vehicle.

Lewis Chuang and Heinrich Bülthoff's paper "Towards a Better Understanding of Gaze Behavior in the Automobile" gives a concise introduction into gaze tracking, attention, and EEG and relates them to show how they can be used together to infer user behavior. This is a topic of high interest for the automotive UI research and the focus of the actual workshop.

Finally, Andreas Riener and Jürgen Noldi discuss in the paper "Cognitive load estimation in the car: Practical experience from lab and on-road tests" findings on estimations of driver mental workload derived from seat pose under simulator and real-drive scenarios. The hypothesis to be tested was that movement dynamics of a driver in the car seat dependent on the effective level of cognitive load. Most interesting for the workshop are the correlations and differences between the controlled and real-life settings, as it is critical to the validity of simulation testing.

11.2 Measurement Metrics

This sections lists all the relevant metrics used in the literature. We then, take a closer look at some of each looking at furthering the reader understanding of the driver–vehicle interaction loop.

Methods for Evaluating Automotive User Interfaces

- HCI in general: Task completion time, Errors, Ease of Use Rating, Physiological measures, Subjective measures
- Measures for automotive HCI: Driving performance, Workload Ratings, Situation Awareness, Object, and Event Detection

Measuring Driver Workload (Overload, Underload)

- Subjective measurements, e.g., NASA Task Load Index (TLX), Driver Activity Load Index (DALI)
- Physiological measurements, including: Heart rate/heart rate variability (HRV), Skin conductance level, Respiration, Task-evoked pupillary response (pupil size)

Measuring Emotions and Satisfaction

- Subjective ratings on specific Interfaces, usually rated in a Likert scale
- Standardized questionnaires, such as System Usability Scale (SUS): subjective perception of system usability; intuitive interaction (INTUI): effortlessness, verbalizability, gut feeling, magical experience, AttrakDiff: hedonic, and pragmatic dimensions of user experience, etc.
- Neurophysiological measurements: fNIRS, EEG, Heart rate/heart rate variability (HRV), Emotiv's Engagement Level.

- Behavioral measurements: facial expression detection, voice recognition, grip strength, body posture, pupil size, etc.

Measuring Driving Performance

- Maxiumum, minimum, mean, variation of: speed, lane deviation, lane departure, torque, steering wheel angle, via CAN bus, in real-driving environments, or from the driving simulator software, head way distance in the car-following task and number of missed events (e.g., signal indications for lane change, street lights, etc.)

11.3 Sensing Techniques and Research Environments

It is always a challenge for researchers to decide on the proper instrumentation of sensors in their driving simulator, the correct way of calibration and the appropriate storage solutions. Most of the times they need to accommodate to the available resources, time period for instrumentation and expertise. Figure 11.2 provides an overview of available sensors in a vehicle cockpit.

In this chapter, we introduce what we call classical sensing which includes performing raw data captures and secondary metrics that can be applied for in-cabin monitoring, context-awareness measurements when linked to dynamic driver interfaces can lead to reactive and adaptive driver user interfaces.

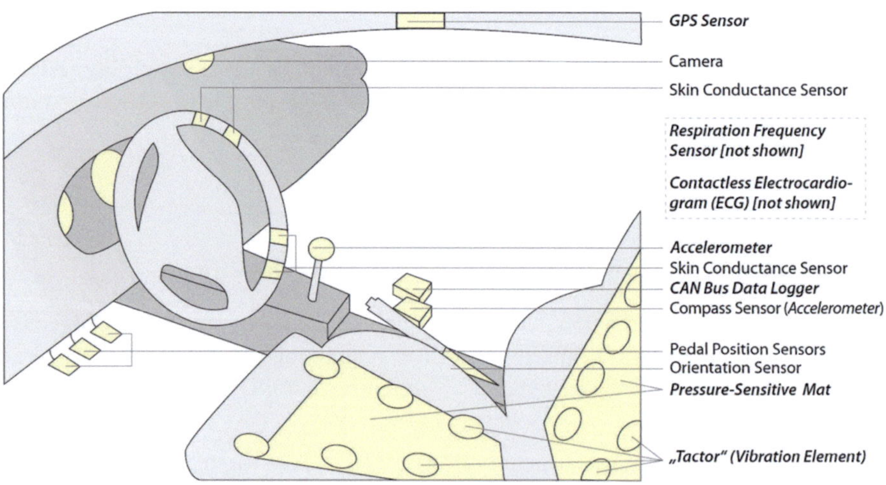

Fig. 11.2 Vital context in vehicles as described by Riener in (2010), Fig. 9.3

11.3.1 Classical Sensing

We understand classical sensing as the instrumentation needed to acquire raw data related to driver performance. In naturalistic and high-fidelity motion-enabled driving simulators this include GPS positioning, acceleration forces, speed, steering wheel deviation angles, pedal and gear shift positioning, and other multitude of vehicle or sensor logs which include exterior cameras, radar and sonar. Other driving simulator setups include capacity proximity sensing, microphone arrays, and an array of interior facing cameras, which range from RGB to depth sensing and infra red (Kinect, RealSense, LeapMotion).

The values obtained through this instrumentation by itself do not provide actionable data points on in-vehicle interactions or driver performance but when observed and correlated in time deltas they can be linked to fundamental human responses, such as reaction time, perception, attention, and engagement. Most driving simulators provide tools that offer driving performance descriptive metrics such as average, standard deviation, min., and max. driving speed, lateral lane deviation or break force. Some of them even include automatic calculation of reaction times, however, they are typically not linked to peripheral in-cabin sensors.

Researches in automotive have commonly instrumented vehicle cabins with cameras for documentation and driver attention monitoring, but only in the recent years RGB, depth and infrared (IR) cameras are starting to make their way commercially as in-vehicle experience differentiators.

Such inward facing sensors have shown promising results toward automotive interfaces in distracted warning systems based on the head/gaze tracking systems. Tawari et al. presented a robust distributed camera framework for head movement analysis in Tawari & Trivedi (2014). Pelaez et al. built a similar system using inexpensive sensors, such as Kinect Pelaez (2014). This setup has been also explored in driving simulators for measuring gaze and physiological data by Gable et al. (2015) and Wuhe et al. applied it in real-driving conditions to measure driver awareness Wuhe et al. (2014). Open data sets are also available for researchers at the Drivers Motion Depth Database, which provides an API to detect merge and lane changes motions (University of Florida 2016).

Besides the driver monitoring use-cases, Xu and Fujimora proposed a generic real-time driver activity recognition system based on the random forest able to detect behaviors, such as phone usage based on depth camera input (Xu and Fujimura 2014). It is becoming evident that camera sensor technology can empower a wide variety of in-vehicle interactions, such as freehand pointing for identification of distant objects (Kern and Schmidt 2009). However, there is still a clear gap between the development of in-vehicle sensor systems for measuring and classifying in-vehicle behaviors and the creation of novel HMI interaction concepts. Prototyping such systems usually requires deep expertise in low-level sensor technology and signal processing as well as HMI software development skills.

Furthermore, raw data coming from in-vehicle cameras, microphones, or wearable sensors need to be synchronized with the driving simulator metrics and normalized in order to find correlations with other driver performance metrics.

11.3.2 Emotion Sensing

There are a number of disciplines in automotive research that have looked at high-level cognitive processes and their effect on driver performance. Human Factors and Computer Science, through the field of affective computing, are looking at emotion measurement. Given that real-driving conditions have been reported to produce fundamental emotional responses that affect negatively driving performance, namely road rage, researchers in automotive user interfaces have looked at measuring and using emotional responses as an input value for the interactions.

For more natural interactions, continuous tracking of the driver's emotions is a requirement, which seems to be hard to implement, but turns out to be rather simple and effective when using the "right" method and tool chain (in particular, if there is no need for real-time classification). Basically, there are a number of techniques out there, but we will focus on three technologies that can be relatively simply (i.e., in terms of equipment needed and without the need to consult medical doctors) but effectively applied to the vehicle context. First, video cameras used to track variation in facial expression, second, IR (infrared; thermal) cameras employed to track temperature variance in the face and relate this to excitement, mental/physical stress, etc., and third, ECG (electrocardiogram) devices used to continuously monitor the heart rate (and derivations) of subjects.

The first approach is based on RGB images recorded with standard Webcams, Smartphones, or sports cameras like GoPro HERO and preprocessed afterwards. Fraunhofer SHORE[1] is a powerful engine for facial recognition and identification of facial characteristics (eye–eye distance, gazing direction, etc.). In addition, SHORE allows to detect age (less than 7 years mean deviation), gender (ca. 95% recognition accuracy) and the four basic facial expressions: angry, happy, sad, and surprised. The Microsoft Emotion API[2] (Project Oxford) takes (a facial expression in) an image as input, and returns the confidence across a set of emotions for each face in the image. The emotion API classifies each frame as either neutral face or one of the seven basic emotions: happiness, surprise, sadness, fear, disgust, contempt, and anger. For (longer) videos, the entire process can be automatized, e.g., using a C# program to automatically extract single frames from the video, call the service routines, and return the classified emotion as a result. In a recent trial, we used video footage from a driving simulator study ($48 \times 10 = 480$ min), extracted

[1]http://www.iis.fraunhofer.de/en/ff/bsy/tech/bildanalyse/shore-gesichtsdetektion.html (November 29, 2016).

[2]https://www.microsoft.com/cognitive-services/en-us/emotion-api (November 29, 2016).

individual frames and used the Emotion API to classify the images. Even when using a standard camcorder with changing lighting conditions in a driving simulator (movement platform) setting, 97.8% of all images could be classified successfully.

The use of thermal imaging technology to measure an individual's physiological response to mental states, such as stress, discomfort, or anger, has great potential for the general recognition of these states. The theory behind is that the surface temperature on our face is mostly dependent on the temperature of the blood circulating in the outer layers of the skin and variations in blood pressure and the fact that blood flow has a direct connection to stress. Stress affects the cardiovascular system in the way that the heart rate increases and the rate of blood flow speeds up, increasing as a consequence also blood pressure (Nees and Walker 2011). Areas of the face exhibiting highest temperature (temples, nose, inner corners of the eyes) offer the most immediate indications of a mental reaction as they relate to areas of high blood flow. It has to be noted, however, that these characteristics do not only depend on the subjects' physiological condition, but also their dermatological health and external factors such as the ambient temperature.

Classifier specifically trained for this task (using, e.g., Gaussian mixture models, GMM (Nees and Walker 2011) or Hidden Markov Models, HMM) can be used to detect a human's thermal reaction to external stimuli, which in turn makes it then possible to link it back to the mental effort required for a specific set of tasks. With this setting, the authors could show that increased levels of workload (e.g., using a PASAT test) result in higher face surface temperature—a first indication that thermal imaging might be a feasible approach for detecting (driver) stress. The main drawback of such a setting is the initial costs for the IR sensor—high-precise thermal cameras with temperature resolution in the range of 0.05 °C, e.g., FLIR SC 655, cost about 25,000 Euro (temperature variation in the face between low and high levels of workload is in the area of only 0.2–0.3°).

Electrocardiogram measures are normally used by medical doctors to detect/assess cardiac diseases, but are also a very useful tool for physiological research in driving studies. A lot of devices are available these days, ranging from cheap to upscale and offering various functionalities (APIs, Matlab toolboxes, etc.). Most devices come with Bluetooth connectivity and are equipped with long-term recording functions on internal memory cards. Today, first devices are offered that allow contact-free measurements directly in the car seat, (e.g., Scholles 2016). Therefore, metal plates are built into the driver's seat and form a capacitor with the skin. These systems operate reliably despite several layers of clothing and movements of the contact surfaces. Typical skin irritations developed during long-time monitoring can therefore be avoided.

Based on our own experiences with devices from HeartBalance ("HeartMan", about 2,000 Euro, discontinued; http://www.heartbalance.org/), GL Neurotechnologies (BioRadio-Wireless Physiology Monitor, about 7,500 Euro; https://glneurotech.com/bioradio/bioradio-wireless-physiological-monitor) and g.tec (g.USBAMP system, about 20,000 Euro; http://www.gtec.at/Products/Hardware-and-Accessories/g.USBamp-Specs-Features), the more expensive the device is, the better it compensates for measurement errors in dynamic environments. One issue is the movement of limbs (arms) while steering a car, another issue is caused by vibrations from engine and roadbed in on-road experiments. Thus, ECG recordings are (highly) affected by limb movements and roadbed vibration, i.e., when driving a winding road, the ECG signal might become deteriorated so as the HRV signal. These problems need to be considered when ordering such a system in order to be later able to use recordings for meaningful analysis. For driving simulator studies (seat box or movement platform), also the cheaper devices work quite well for both ECG and HRV analysis.

Heart rate data alone is rather useless in the context of driving, but heart rate variability (HRV) carries great potential to identify stressful situations, monitor workload of drivers, etc., see, for example, (Riener et al. 2009; Manseer and Riener 2014). It is calculated by analyzing the time series of beat-to-beat intervals from the ECG signal. After a Fourier transform (FFT), and extraction of LF and HF frequency bands, cardia sympathovagal balance can be calculated and the result gives at least a coarse indication of stress or mental load of the driver while performing a task. For the process required to make use of ECG data, see (Riener et al. 2009). There is a lot of discussion about the accuracy of this stress metric, and if it actually represents stress (Billman 2013). Independent from this discussion, it is important to note that HRV normally requires long recordings in order to derive stable results. With "long," a time frame of optimally 24 h is meant, but a minimum of a few minute recording is often enough to derive interpretable measure. Based on the application, time windows of one to five minutes (Riener et al. 2009; Manseer and Riener 2014) are enough to extract at least a stress tendency. However, the FFT method described before has some problems, e.g., apparition of undesired frequency components. Wavelet analysis might be a stronger tool to derive stable HRV results as it provides more specific information about autonomic activity. The wavelet packet transform (WPT) method was found to provide good results of sympathovagal balance, which is the variance (or power) of the ECG signal changes as a function of frequency (Moldovan and German-Sallo 2014).

In a recent work (Wintersberger 2016), we performed a comparative analysis of quantitative data (level of stress from HRV measurements, emotions detected from camera images) and qualitative data (PANAS, AffectGrid, and others; see below) and found out that results are similar for the three forms of measurement. This confirms that these measures are comparable at least for simulated driving.

11.3.3 Identification and Personalization

Another emerging sensory feature is eye gaze control. High-fidelity eye trackers (e.g., Tobii Pro Glasses 2, SMI Eye Tracking Glasses 2) are available at reasonable costs (about 25,000 Euro) and allow for gaze tracking (eyes-off-road time), attention analysis (drowsiness, attention detection), but also driver identification and personalization. The intuitive assumption that gaze is the equivalent of (covert) user attention does not hold true. Gaze tracking is indeed not a reliable indicator of the spatial and temporal allocation of covert attention, however, authors in Chuang and Bülthoff (2015) suggest to combine eye tracking with electroencephalography (EEG) to overcome this issue and get a more stable system to infer driver behavior.

If driver identification (and related topics, such as authentication or personalization) is the focus, different sensor technologies might be used. A special (unusual) use case was demonstrated by Riener and Ferscha (2008) in using pressure sensitive array mats in the driver seat to unobtrusively identify a driver. The authors conclude that this technology is not comparable to methods, such as retina scan or DNA analysis, but is good enough to differentiate between a group of drivers typically share a vehicle (the authors have shown that such a system could differentiate between ca. 30 people). In another field trial, a similar setting was successfully used to detect driver behavior (Riener 2011). The authors demonstrated that sitting position (in particular, the deviation from sitting upright caused by centrifugal forces while cornering) can be used as an indicator for attentive driving. They showed that the driver is implicitly compensating for upcoming lateral acceleration forces (effective in curves) already few hundred milliseconds before actual forces become effective. Detecting misbehavior (i.e., no compensation for an upcoming cornering situation) can be interpreted as inattentiveness of the driver and used to automatically execute safety functions in the car (slow down or stop the car, etc.).

11.3.4 Lab Studies and Field Studies

The problem of the setting described before is, that it only can be tested in naturalistic driving studies, neither in the lab nor in mid-fidelity movement platform simulators (totally different lateral/longitudinal acceleration forces; gravity often used to emulate acceleration—works for cheating the human operator, but not sensor technology). (Riener and Noldi 2015) describes this problem and concludes that it is problematic to directly compare (and draw conclusions from) lab-/field studies due to different environmental natures. For the naturalistic driving setting, the correlation between steering wheel angle and the deflection of the driver in lateral direction (expressed by the center of pressure, COP) is inverse for the majority of subjects (centrifugal force is effective opposite to the direction of steering). On the other hand, for a simulated setting using a seat box and Logitech G27 wheel, the correlation is direct (same direction), as no centrifugal force/lateral acceleration is affecting the driver. Thus, the main reason for different results is the missing centrifugal force in the simulator study, and the fact that drivers automatically (and proactively) compensate centrifugal forces. Figure 11.3 visually highlights these differences.

Another problem in comparing results from lab-based user studies with on-road tests might occur from the underground/road surface. In a lab setting, the test vehicle might not move at all (e.g., seat box) or only moves smoothly and controlled (in case of hexapod or moving platform simulators). In contrast, when running field trials, the

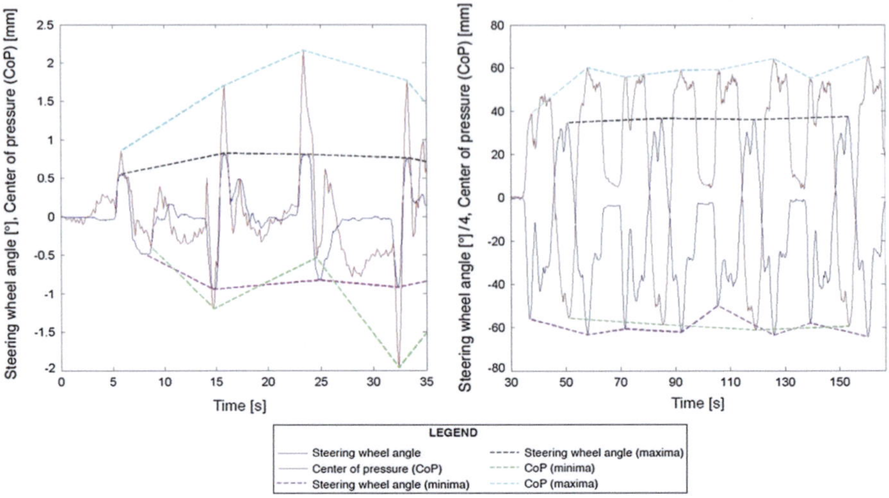

Fig. 11.3 Course of steering wheel angle and center of pressure (COP; driver seat). *Left* Simulator study, *Right* On-road test. (*Source* Riener and Noldi 2015)

condition of the road surface plays a significant role. In case of pressure mat use as pointed out before, vibrations from the underground (e.g., gravel, potholes) significantly impact the recordings. Similar influence might be detected for other sensors highly prone to vibrations/movements, such as, for instance, electrocardiogram (ECG) sensor (see below). The obvious idea for compensation is to record vibrations using highly precise inertial measurement units (IMU's) installed at the car and apply data cleaning/filtering methods on sensors' data based on the IMU data. Unfortunately, in many cases this is not possible (synchronization issues, placement of sensors in the car, accuracy of IMU data, etc.).

Another option (maybe used in addition to IMU recordings) is to carefully select the test track, i.e., account for and avoid irregularities (tarmac and no gravel, no potholes, newly tarmaced street, flat surface, etc.). Simulation software of mid- to high-fidelity driving simulators (e.g., IPG Carmaker) offers in recent versions import functions of routes via the Google API or others. This would allow for a realistic and comparable simulation environment. In addition, IMU road surface recordings (from on-road trial) might be fed into the simulation and create a more realistic driving experience (i.e., inducing road structure, surface condition). This way, the problem of data cleaning gets less critical, as interferences will be similar in lab and field environments.

Sensors brought into the (real) car and used in naturalistic driving studies might further cause serious problems if connected to the car's internal power system. As shown in Fig. 11.4, harmonic waves induced by the dynamo machine led to significant disturbance of pressure image (Riener and Noldi 2015). Similar problems might also occur with other sensors—that is why it is of utmost importance to check data validity in a pilot study (and compare it to lab-recorded data) before executing

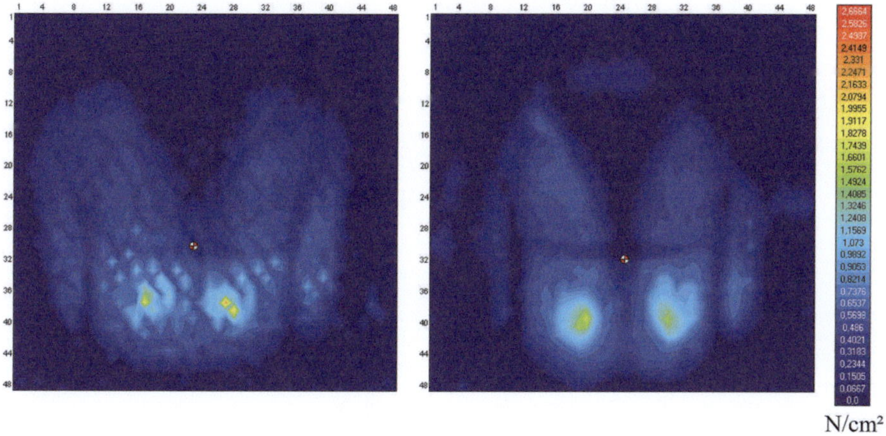

Fig. 11.4 *Left* Correct pressure image (lab setting; field setting after correction with separate power supply). *Right* Pressure image disturbed by electrical interference (caused by the dynamo machine). (*Source* Riener and Noldi 2015)

the real experiment! Depending on the type of car used for the study, there might occur similar problems causing unwanted harmonics to destroy data received from the individual sensors used in an experimental setting. Authors of Riener and Noldi (2015) propose to install an extra power supply (second car battery not connected to internal energy system) as a universal solution. The capacity of typical car batteries is high enough to run sensory systems for many hours. (Extrapolation for notebook computers: 65 W typical consumption car battery typically 60 Ah; results in 60 * 12 = 720 Wh/65 W = 11 h of operation).

With ever increasing driving demands and more and more IVIS/ADAS in the car, estimating and monitoring the momentary cognitive load of the driver will be a necessity in the future, in particular in the coming age of mixed traffic (i.e., manual and automated vehicles sharing our roads simultaneously). Besides objective measures (e.g., EEG, fNIRS), subjective evaluation (NASA-TLX, PANAS, etc.; see Sect. 11.4) of mental demand is an often used tool to quantify an individual's perceived mental demand. When comparing the subjective mental demand (NASA-TLX) and the performance achieved with the secondary task (PASAT) (Fig. 11.5), it can be seen that the average mental demand in the simulator study (blue, bold-faced line) is significantly higher as compared to the on-road experiment (red, bold-faced line). Looking at an objective measure—the performance achieved in the auditory secondary task (PASAT; same configuration for both series)—it can be indicated that the average performance is higher for the on-road part of the study compared to the simulator part, suggesting that lower mental demand results in higher dual-task performance. It is assumed that the reason for it is that the driving simulator study introduced a new, relatively unknown setting/environment for most of the subjects, and this might have finally caused higher mental demand and lower

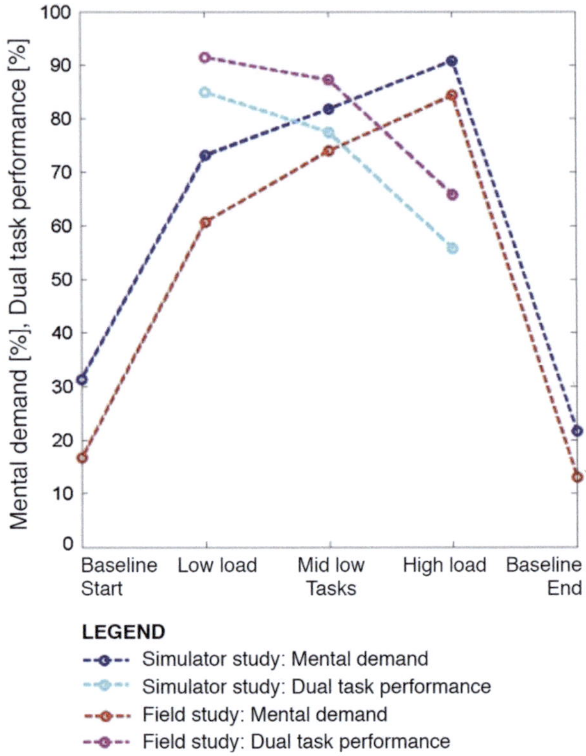

Fig. 11.5 Mental demand (Nasa TLX) versus dual-task performance (PASAT) in comparison of simulator and field studies (*Source* Riener and Noldi 2015)

performance for the secondary task. On the other hand, the on-road study was perceived as familiar and largely automated activity for the drivers (all of the subjects possess a driving license and have several years of driving experience). Due to this fact, the mental demand is lower in the field study, i.e., more resources available for ancillary tasks.

11.4 Behavioral and Qualitative Metrics

Evaluating automotive UIs needs a holistic contemplation of users' experiences including attitudinal user-generated feedback as well as behavioral aspects. A number of different methods and qualitative metrics exist. In the following, a few will be presented by arranging them in a 2D space. We are differentiating between attitudinal (what people say) versus behavioral (what people do) methods and between measuring hedonic versus pragmatic aspects of User Experience. Hedonic quality means the consideration of nontask-orientated attributes, i.e., "Why someone uses a product?" while pragmatic quality refers to the question "How someone uses the product?," and so both usability and utility of products will be respected (Hassenzahl et al. 2008) (Fig. 11.6).

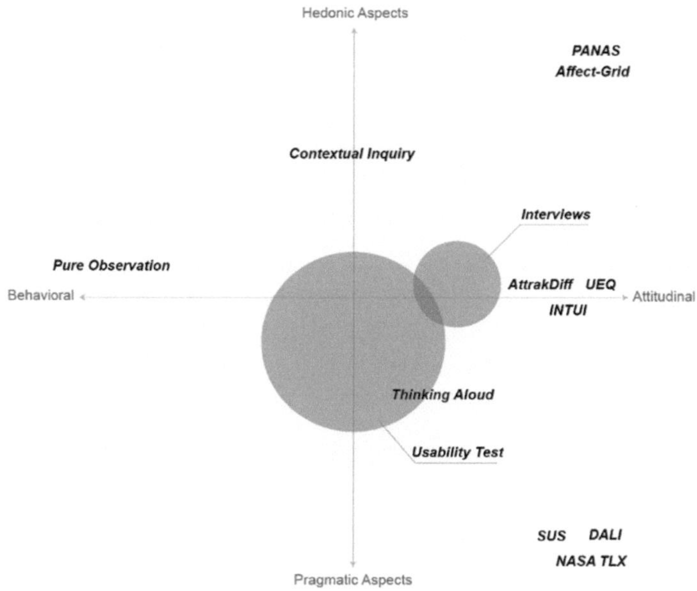

Fig. 11.6 Classification of qualitative methods in two dimensions (attitudinal vs. behavioral and hedonic vs. pragmatic aspects). (*Source* A.-K. Frison)

Attitudinal evaluation methods are usually subjective ratings by the subjects of a study. To assess the hedonic qualities of automotive user interfaces we describe two standardized questionnaires *PANAS* (Positive and negative affect scale) and *Affect-Grid*. The *PANAS* is a self-reporting tool to get insights about the current mood of subjects. Study subjects are asked to rate in a five-point Likert scale their felt intensity concerning ten positive (e.g., enthusiastic, interested, determined) and ten negative adjectives (e.g., scared, afraid, upset). The questionnaire builds on the two-factor model of positive (PA) and negative affect (NA) as distinctive orthogonal dimensions for analyzing affect (Watson et al. 1988). By using *Affect-Grid*, developed by Russel et al. (1989), the emotional state of subjects is collected through a two-dimensional "9 × 9" grid, in which users have to mark a point, which describes their current mood. The x-axis represents pleasure and the y-axis arousal. (e.g., high arousal can be experienced as positive but also as negative).

An attitudinal metric to measure the User Experience of automotive UIs is the standardized questionnaire *AttrakDiff*.[3] Subjects assess the experienced hedonic and pragmatic quality as well as the desirability of a system in the form of seven-point Likert scales of semantic differentials, e.g., ugly—attractive. The value for desirability is dependent from the hedonic and pragmatic quality.

[3]http://attrakdiff.de/index-en.html (November 29, 2016).

The *UEQ*[4] (User Experience Questionnaire) also uses semantic differential to measure the UX of an interactive system by regarding attractiveness, perspicuity, efficiency, dependability, stimulation, and novelty. In relation to a given benchmark dataset the results of the UEQ can be used to rank the rated user experience of the evaluated automotive UI. Another standardized questionnaire to assess the intuitiveness of interactions is called *INTUI*.[5] It is based on the assumption that intuitiveness underlies subcomponents illustrated by 16 questions in form of semantic differentials to get insights about effortlessness, verbalization, gut feeling, and the magical experience of the asked users.

A questionnaire to analyze the workload is the *NASA TLX*.[6] Here mental, physical and temporal demand are focused as well as the own performance, effort, and frustration of the evaluated system. Subjects have to rate these subscales in a 21-point Likert scale. The procedure results in an overall workload score, which is computed by the weighted averages of the subscales.

Another questionnaire to measure users' subjective workload is the method called *DALI* (Driver Activity Load Index) (Pauzié 2008) which is based on the NASA TLX but adapted specially to the driving context. Therefore, people are asked to rate on a six-point Likert scale the level of constraint concerning the global attention, visual, auditory, tactile, and temporal demand. Moreover, they have to rate the level of stress during the task and to assess the level of disturbance of the driving task while doing secondary tasks simultaneously.

For a quick usability evaluation *SUS* (System Usability Score) (Brooke 1996) can be used in an easy way. This method is a survey approach to rank the usability of a system to compare it across a range of contexts. Ten item-scales, using a five-point Likert scale can deliver a global view on the subjective assessment of usability, e.g., "I think that I would need the support of a technical person to be able to use this system" (1 = strongly disagree, 5 = strongly agree). Using SUS for comparing different systems should be avoided because of the context specificity of usability.

Surveys' results are often limited, you get only information from questions, which are asked. To get more detailed answers *interviews* about subjects' subjective assessment can yield new interesting perspectives of the considered UI. Insights may include feedback about hedonic as well as pragmatic qualities. It is important to carefully think about what, whom and how you ask the interview questions (Lazar et al. 2010).

Attitude and behavior are related but what people say and what they actually do are sometimes varying, and so observing in either a *lab* or *field study* subjects' behavior is an alternative approach for evaluating UIs, e.g., if you want to know which errors happen when the subject uses the speller of an infotainment system. Doing automotive UI research in a *lab* is useful if variables need to be controlled,

[4]http://www.ueq-online.org (November 29, 2016).

[5]http://intuitiveinteraction.net (November 29, 2016).

[6]http://humansystems.arc.nasa.gov/groups/tlx (November 29, 2016).

e.g., the traffic scenario should be equal for each subject. Performing *usability tests* in the lab, subjects are observed while interacting with the product. If you want to understand the users' context and the usage of the automotive UI in real world, *field studies* can help you to get insights. The observation can be done without (e.g., *pure observation*) or with conversation (e.g., *contextual inquiry*). Methods usually do not stand-alone. A usability test can also include attitudinal methods, e.g., *thinking aloud* or special surveys and behavioral observation methods. Subjective feedback about hedonic and pragmatic qualities will be collected. Moreover, quantitative metrics, such as error rates or physiological measures are often included in addition (Courage and Baxter 2005).

In conclusion, each single method has its relevance for a certain case. For researchers it is important to know what they want to learn. From there, by choosing the right methods, the appropriate study setting can be derived.

11.5 Displaying Techniques

Display is the essence of Automotive User Interface design. What does it mean? we are instrumenting and building complex sensory systems and control mechanisms to emerge information that is useful to people. In this section, we review several modalities of display techniques with a focus on the dependencies to sensor instrumentation and how displays can achieve interaction goals.

11.6 Reactive Displays

In our understanding, a display is any kind of feedback sensorial mechanism used in the car, be it a visual/graphical display (screen, dashboard icons, indicator lamps, head-up display, AR glasses), auditory display (speech output, beeps or more advanced forms, such as spearcons, earcons, etc.), tactile display (vibrations in seat, steering wheel, safety belt, pedals), or olfactory display (scent of burning oil to indicate problem with the engine, lavender or lemon to calm down or arouse the driver, etc.).

In a reactive display (Joyce and Cianciolo 1967), the entire display content is decomposed into parts and the parts are updated individually. The concept is the same if the display contains several sensory modalities, e.g., a visual display with tactile feedback. If the visual content keeps the same and only tactile output changes, then it is called a reactive display. Given the development of multimodal feedback systems, we will in the future see more and more reactive displays employed for selective feedback/stimuli.

11.7 Visual Displays

Visual displays are without doubt the most restricted display mechanism in automotive user interfaces. Previous research focused on visual displays has established knowledge that driving is a task requiring visual input, and it is recommended to display additional visual information in limited fashion. These design efforts stem from an attempt to minimize distraction while driving (Kern and Schmidt 2009). For example, international standards, such as specifications and compliance procedures for in vehicle visual presentation provide interaction design recommendations (DIN 2003). Furthermore, other guidelines offer recommendations for designers during the design process, which describes how to make entertainment and infotainment systems safe and easy to use for all drivers (Kern and Schmidt 2009). The guidelines include, text size, the placement of displays, with heavy consideration given to safety and usability measures. These guidelines in effect promote designs that require less taxation of visual resources (e.g., interfaces with high contrasts, low spatial frequency, and legible and large fonts) (Stevens 2002).

Even for novel see-through head-up displays (HUDs), that were put forward in the previous years as "the ultimate technology to cope with eyes-off-road time," HUDs yield new problems, such as increased workload or overseeing of obstacles in the field of view due to distraction (Fischer et al. 1980), and even change blindness might occur.

That is why we want in the rest of this chapter to highlight the potential and drawbacks of other forms of displays, suitable for in-car use.

11.8 Auditory Displays

Given that driving is a visually demanding task, research on use of auditory displays has been heavily conducted and a number of applications have been practically applied to actual vehicles (Riener and Anzengruber 2012). In-vehicle auditory displays range from the traditional collision warning sounds for a driving task, voice, and beeps for personal navigation devices, to a recent attempt for fuel efficiency driver interfaces, sounds for in-vehicle infotainment systems, and speech interaction with an intelligent in-vehicle agent.

The traditional warning sounds include forward collision warnings, lateral collision warnings, and lane departure warnings, etc. These warnings are mostly composed of a beep or multiple beeps. In addition, auditory icons (Gaver 1986) (representative sounds of objects or events) have also been introduced for the warning purpose and shown promising results Belz et al. (1999). For example, to represent an approaching motorcycle, a motorcycle engine sound can be used. For a bicycle, a bicycle bell sound can be used.

Navigation Devices have adopted diverse types of voices such as different languages, genders, synthesized text-to-speech (TTS) and real human voice, and even celebrity voices (http://www.garmin.com, http://www.tomtom.com). For efficient implementation, navigation devices often integrate human voice for introduction and conjunctions and TTS for the POI (Point-Of-Interest) database. Jeon et al. (2009) has tried to improve navigation sounds using Spatial Turning Sound and Leading Tones for Turning. With the Spatial Turning Sound, even though drivers miss the message of the voice prompt due to a dialogue with passengers or radios, they could identify the next turn direction from the spatialized location of the audio cue. Leading Tones for Turning generates tones of increasing duration, like "Pip.. Piip.. Piiip.. PiiiiP." Adding contextual sounds before the exact moment is expected to help drivers sense the appropriate timing. One of the most important research topics regarding the use of spatial sounds is "Auditory-Spatial Stroop paradigm" (Barrow and Baldwin 2009). This experiment examines which one between spatial cue and semantic cue is more influential in incongruent cue combination (e.g., speech cue saying, "left" generated from the right speaker). Successive research studies consistently show that it would be more difficult to ignore spatial location information than semantic verbal information when the two pieces of information conflict with each other.

In contrast to use of auditory displays for the primary or secondary tasks, (Jeon et al. 2009) attempted to apply multiple auditory cues for infotainment menu navigation. Auditory menus usually speak out menu items using TTS. Researchers added spearcons (Walker et al. 2013) (speech + earcons: compressed speech), spindex (Walker et al. 2013) (speech index: using a phoneme of the first letter of the item), or both cues before the speech clip. In their experiments, subjects were asked to search target menu items with different combinations of auditory cues while they drove in the simulator. As a result, the spindex cues showed better performance in driving and menu navigation than other cues, were preferred more, and reduced perceived workload.

Fairly recently, researchers have designed sonification (Jeon et al. 2012) (translating data dimension into audition, specifically using nonspeech sounds) systems for eco-friendly driving as a type of fuel efficiency driver interface (Hermann 2008; Nees et al. 2014). Speech-based auditory displays can be used by offering spoken alerts and advice to improve fuel economy. However, speech might interfere with concurrent conversation and create annoyance in the form of a virtual backseat driver. Researchers of Hammerschmidt et al. (2014) have developed an app that can extract all the driving performance data (speed, lane deviation, torque, steering wheel angle, pedal pressure, crash, etc.) from the simulator (NADS miniSim). All these data can be mapped onto sound parameters. Depending on drivers' driving style and performance, different soundscapes can be generated, encouraging them to drive in an eco-friendly way.

Not just Apple Siri, but many vendors try to implement intelligent in-vehicle agents as a co-driver. Here, the driving concept has been rapidly changing from an independent task into a collaborative task. To design a more natural communication with an agent, a couple of research groups have conducted research on emotional interactions. For example, Nass et al. (2005) showed that when the in-vehicle agent's voice emotion matched the driver's emotional state (e.g., energetic to happy and subdued to upset), drivers had fewer accidents and attended more to the road (actual and perceived), and even spoke more with the car. Harris and Nass (2011) showed that drivers in a *reappraisal-down* speech condition (e.g., "heavy traffic results from limited routes, not the behavior of other drivers") had better driving behavior and reported less negative emotions than subjects in a reappraisal-up speech condition (e.g., "the behavior of overly aggressive and inconsiderate drivers leads to traffic congestion") or a silent condition. For angry drivers, Jeon et al. (2015) used emotion-regulation voice prompts and situation awareness voice prompts. Both conditions improved driving performance, but overall situation awareness voice prompts improved performance better and were preferred more. There has been much research on the use of music for emotional drivers. Certainly, music seems to be helpful in terms of reducing emotional effects and improving driving performance, but the specifics of music parameters should be further determined (Fakhrhosseini et al. 2014).

Finally, we envision that more auditory displays can be used specifically for electric vehicles and automated vehicles. Given that electric vehicles do not generate engine sound but some states in the US force them to have some noise for pedestrians, it would be of interest to conduct research on ambient sounds of electric vehicles: whether it should mimic engine sounds or make novel sounds. This sound will also be helpful for drivers in terms of providing situation awareness about vehicle states.

11.9 Tactile/Haptic Displays

For in-car interaction, there is considerable interest in the manipulation of visual, auditory, and tactile sensory modalities to improve human interaction with the external environment (e.g., Repperger et al. 2005). Tactile feedback—as a promising additional source of information—is still underused in vehicles today, but allows to access an important and often overlooked information channel. In driving, vision accounts for the majority of sensory input, but the sense of touch becomes a more common information source recently for some good reasons. Tactile feedback makes it possible, for instance, to find the radio dial in a car without looking, and to know when fingers are correctly placed on a "QWERTY" keyboard (because of the bumps on the "F" and "J"-keys) or other key panel. Numerous studies have shown that tactile feedback is useful to increase task performance, reduce mental load, improve on the perception of warning signals, etc. Considering the use of tactile sensations from the perspective of losing that sense provides interesting results. There is evidence that the total or partial loss of the sense of touch often cannot be adequately compensated for by the application of other sensory modalities and eventually results in a limited ability to perceive the environment or even to stand or move (Robles-De-La-Torre 2006, p. 24). As a consequence, it is assumed that a vehicular application, when providing improper haptic feedback, might impair driving performance in the same way as a major somesthetic loss would do. For example, it would be very difficult or even

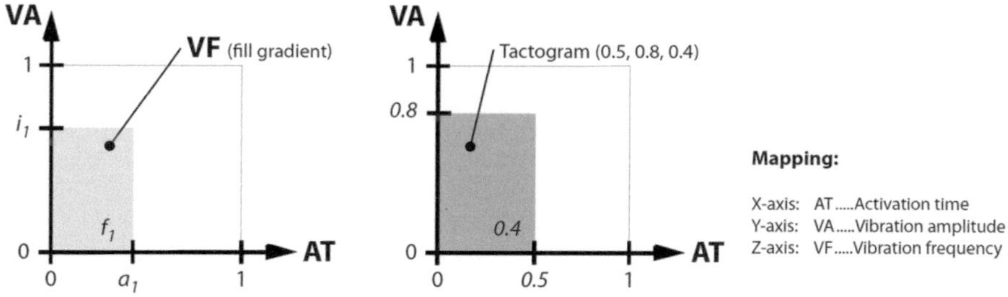

Fig. 11.7 A tactogram is unambiguously defined as a three-tuple. The variation parameters for a specific tactor element are activation time (AT), vibration amplitude (VA), and vibration frequency (VF). For local use, each dimension is scaled to the range [0.0.1] according to the physical limitations of the actual used tacile feedback system. (*Source* Riener 2010)

impossible to steer a car (using the brake pedal, throttle control, and/or clutch) with numb feet; it would likewise be demanding for a driver to steer the car when receiving invalid or improper tactile feedback on a particular driving activity.

That is why we suggest to define and use a tactile language/haptic alphabet for in-car use. Any tactile pattern applicable for the specific stimulation of a driver is defined in the "haptic alphabet." a pool of tactile signs or symbols ("Tactograms," Fig. 11.7). The definition of a haptic alphabet is not as easy as the definition of other alphabets, as the main variation parameters (activation time, frequency, intensity/amplitude) are continuous measures and the number of tactors to be used in an application might also be not constant. Nevertheless, the composition as well as the complexity of this alphabet should be oriented toward other, well-established touch-alphabets, such as the Braille alphabet, Tadoma, the Vibratese language, or even Fingerspelling and their attempts to identify, define, and classify characters or words on different levels of importance.

The term "Tactogram," though, is not new and has been used before in fields ranging from zoology, marine sensory biology to business or management (Riener 2009, p. 94). In the context of this work, a tactogram is understood as a three-tuple (duration, amplitude, frequency—which are similar to basic auditory variables) to characterize the activation of one tactile element. For vibration frequency, the frequency range is determined by the type of mechanoreceptors in the skin to be activated. In our studies, we used tactile stimuli to activate Pacinian corpuscles. Pacinian corpuscles are categorized as the fastest adapting class of mechanoreceptors, i.e., they adapt very quickly to stimuli and the effect of stimuli decays rapidly after onset (Wolfe et al. 2014). A problem of Pacinian afferents is that they discharge only once per stimulus application, hence they are not sensitive to constant pressure (Bicchi et al. 2003).

Using tactograms, it is easy to compose tactile melodies/patterns or a "vibration carpet" for multiple actuators as suggested in (Fig. 11.2). Here, each element is characterized as a 6-tuple (T, X, Y, AT, VA, VF). T corresponds to start time of activation, X and Y are the coordinates of the tactile actuator, AT, VA, and VF are used as before (Fig. 11.8).

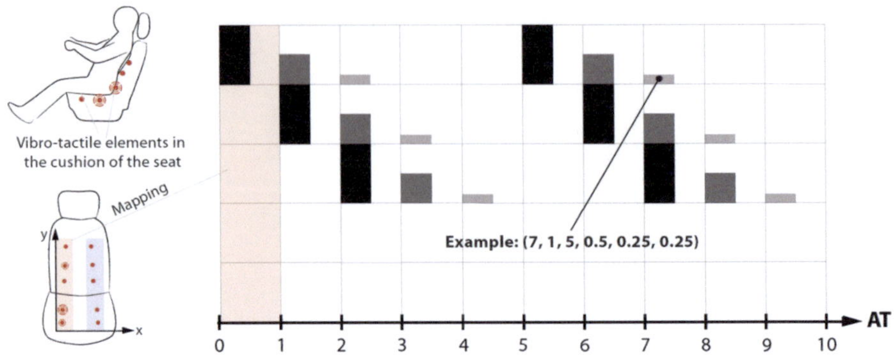

Fig. 11.8 "Melody" of a tactile stimulation system composed by units of 6-tuples. Shown is only the activation of the left strip of five actuators integrated into the car seat. (*Source* Riener 2010)

Depending on the type of information to convey (information, warning, hazard), tactile stimuli should be perceived comfortable/calming, attentive/arousing or even disturbing. A similar categorization was proposed by Matthews et al. (2004) for visual displays. Based on requirements for automotive applications and earlier work in the field, e.g., by Matthews et al. (2004), McCrickard et al. (2003) and Pousman and Stasko (2006), we have proposed to differentiate between four classes of tactile stimuli (a) Ignore, (b) Change blind, (c) Make aware, (d) Demand action, corresponding to releasing increased levels of attention (LOA).

"Ignore" represents a tactile pattern conveying unimportant information that will not require additional attention at the driver. This is a type of information that normally does not come from the tactile stimulation system, but from the environment. The class "Change blind" should be used if a notification requires no immediate action by the driver, but aims at evoking an action in the longer time. An example of this class could be an eco-driving system that informs the driver (in a non-attentive way) about current driving efficiency (Riener et al. 2010). The drivers might change their driving style (implicitly) to improve on driving efficiency, but no harm put on the drivers if they do not respond to the stimuli. Using the "Make aware" type of feedback requires immediate action by the driver, but, as before, would not cause serious danger if the reaction time is delayed. If a tactile signal, for example, informs the driver that the engine oil is at a low level, nothing dramatic will occur, if the driver continues to drive for some more kilometers (to stop at the next gas station). However, the tactile pattern needs to provoke a medium level of attention (LOA) at the drivers, so to make sure that they do not miss this information. The final class "Demand action" is used for feedback indicating high potential of risk and necessitates an immediate action by the drivers in order to prevent hazardous situations. Events of this notification level are critical enough to demand a driver's full attention, e.g., warning of a close-by pile-up. In this situation, the driver should immediately react and initiate emergency braking.

11.10 Conclusions

This chapter introduced the readers to the state-of-the-art challenges that automotive user interfaces researchers are facing terms sensor instrumentation of driving simulators and naturalistic driving environments for driver-in-the-loop research. The topic for this chapter became evident as a result of two workshops, of which contributions, summarized in the introduction, illustrate current trends in in-cabin sensing and monitoring, user modeling and assessment in simulator and on-road conditions. We provided then a compendium of techniques, tools, and protocols that researchers, especially, those starting their research in the field can select upon during the planning phase of their own studies. We produced a compiled guidance on quantitative and qualitative based best-known methods to be strategically considered in the study design phase and methodologies for carrying out user assessment and finally, we also provided extensive learning from qualitative methods and quantitative (visual, auditory, haptic, and tactile) mechanisms that we will further clarify in the following recommendations.

11.10.1 Recommendations

11.10.1.1 Behavioral and Qualitative Metrics

While thinking about the right methods, the factors time and (monetary) costs normally play an important role. Experiments using behavioral and qualitative metrics, such as usability tests are expensive as they rely on (many) test subjects and correct, accurate samples. Keep in mind that a lower number of study participants will inherently lead—as in any other kind of study—to less statistical reliability and validity and the outcome would be a lower scientific quality. Beside data collection, the analysis of qualitative data from user tests and interviews is even more time consuming. Online forms (e.g., LimeSurvey, Google Forms, or Typeform) could help to save time by automatically generating Excel sheets and providing basic preevaluation. Moreover, the choice of a standardized questionnaire like PANAS or an individualized survey or interview depends on the scope of your research project. If you want to have comparable results and you have a big number of test subjects, we recommend standardized methods. In case you have only a small sample, individualized interviews or surveys can reveal more detailed and interesting insights.

As already mentioned before, there are plenty of methods, which were all developed to collect data to help answering a certain (research) question. All of them have their eligibility. We recommend researchers—before actually thinking about a concrete study setting including several methods—to brood over following question: *What do I actually want to know?* The right method(s) to be chosen depend on the answer.

11.10.1.2 Visual Displays

Following the line of thought presented in the visual display section we recommend researchers to follow guidelines and principles established in automotive regulatory institutions as these are the strictest due to the visual nature of the driving task. Even HUD and novel reactive displays need to comply with eyes-off-the-road or eyes-off-focus requirements. We can foresee a relaxation of visual display requirements when vehicles operate at high automated levels (L3 and L4) but similar guidelines are being developed and we foresee will be imposed on this channel to govern transitions between levels for emergency takeover scenarios.

11.10.1.3 Auditory Displays

User experience research suggests that driving environments can benefit greatly from systems that apply auditory displays such as described sounds, voice and beeps for in-vehicle infotainment systems, warning, and speech-based application interactions with an intelligent in-vehicle agent. Auditory displays remain a strong choice for the most critical warning communications but in the path to autonomous systems, natural speech interfaces will play a critical role at allow information access for the in-cabin occupants. In the right conditions natural speech interfaces provide the lowest cognitive load and highest user satisfaction among different modalities. There are nonetheless, certain risks that still have to be addressed, as NVUIs do not report the same successful results for every subjects. Independently from typical speech recognitions issues, like slang terminology or strong local accents, users with long-time exposure to command-based voice applications need adaptation to natural speech input. Previous experience with intelligent agents, like Siri, Cortana, Alexa, or Google now experience-has a strong influence in the baseline mental model of the speech interaction.

Finally, designers must account for context-awareness in the implementation of their system interactions. In-cabin context-awareness can be applied following two approaches. On one hand, awareness focused on the user, where understanding his/her stress level, frustration or mood can determine the way the system reacts or presents information. On the other hand, context-awareness resulting from the interaction with the vehicle agent system with the vehicle's on-board computer or other electronic systems in the car, can help determine the amount of attention the user can procure and therefore modify the way information should be presented.

11.10.1.4 Tactile/Haptic Displays

In order to get deeper insights in the effect of various types and positions of tactile stimuli, over the past years we have undertaken several lab/field studies in different settings. Common findings are amongst others:

1. There is almost no limitation in the use of tactile feedback, i.e., any region of human body can be activated (finger/hand, wrist, arm, chest, back, thigh/shank, forehead, etc.). It is, however, important to consider spatial resolution of tactile receptors in the skin. A good reference is Weinstein (1968), who mapped back in 1968 the two-point touch threshold of the entire body (ranging from 2 mm at the fingertips to 45–50 mm at the back).
2. For classical haptic feedback, Pacinian corpuscles are best qualified as carriers. However, the human skin includes much more receptors that can also be employed for feedback similar to haptics. Beside seven kinds of mechanoreceptors (sensitive to pressure, vibration, and slip), the skin includes two different thermoreceptors (sensitive to changes in temperature), four kinds of nocioceptors (responsible for pain), and three types of proprioceptors (sensitive to position and movement) (Toney et al. 2003). Tactile stimuli are received from four types of cutaneous mechanoreceptors (Ruffini corpuscles, Merkel disks, Meissner corpuscles, and Pacinian corpuscles), each with its strengths and weaknesses. Depending on a specific application, it might be required to have closer look at other mechano- or other types of receptors to identify best-suited "sensor."
3. The higher the vibration amplitude (VA), the higher was the achieved level of attention (LOA) at the driver, i.e., amplitude control can be used to steer the level of attentiveness a signal is triggering. In this relationship, it should be mentioned that tactile perception is slightly different for male/female gender, but strongly depends on the age. As receptors in the skin die off with increasing age both perception and spatial resolution degrade over time. This fact needs to be considered when designing interfaces especially for young or old people, as one could easily miss the designated feedback class (according to the categorization introduced before).
4. A change in the vibration frequency is associated with an adjustment in the level of attention. For Pacinian corpuscles, best perception is at the nominal level of about 250 Hz (ranging between 80 and 1,000 Hz), perception (and with it LOA) decreases above or below this nominal frequency.
5. Changing the pulse-pause ratio directs to a higher level of attention when the ratio between pulse and pause increases (lower pause time) and vice versa, the longer the activation time, the higher the level of attention of the user (but it must be noted that human beings adapt to durable stimulation—and in that case the level of attention tends toward zero), and The rhythm of vibrations influences the generated level of attention (harmonic or disharmonic patterns cause lower or higher attention levels). Similarly as flashing light for the visual channel, tactile on-off pulses are perceived annoying when used for longer time, in particular when disharmonic, but can otherwise be used to very quickly direct attention to something. (This is also used with rumble strips on the centerline or shoulder of roads—driving over results in auditory/tactile notice and momentarily stimulates our perception system. According to a US FHWA report (Chappell et al. 2006), run-off-the-road (ROR) crashes on the New York State thruway were reduced 88 percent after rumble strips were installed and Virginia DOT reported a 52% reduction of ROR crashes on the state's interstate highway system).

References

Barrow, J. H., & Baldwin, C. L. (2009). Verbal-spatial cue conflict: implications for the design of collision-avoidance warning systems. In *Proceedings of the International Driving Symposium on Human Factors in Driver Assessment, Training and Vehicle Design*.

Belz, S. M., Robinson, G. S., & Casali, J. G. (1999). A new class of auditory warning signals for complex systems: Auditory icons. *Human Factors, 41*(4), 608–618.

Bicchi, A., Dente, D., & Scilingo, E. P. (2003). Haptic illusions induced by tactile flow. In *EuroHaptics Conference* (pp. 2412–2417).

Billman, G. E. (2013). The LF/HF ratio does not accurately measure cardiac sympathovagal balance. Frontiers in *Physiology*, *4*(26), 5.

Boverie, S., Giralt, A., Lequellec, J., & Hirl, A. (1998). Intelligent system for video monitoring of vehicle cockpit, SAE Technical Paper, 980613. doi:10.4271/980613.

Brooke, J. (1996). SUS-A quick and dirty usability scale. *Usability Evaluation in Industry, 189* (194), 4–7.

Chappell, D., et al. (2006). Priority, market-ready technologies and innovations: Rumble strips, US DOT, Federal Highway Administration, Technical report FHWA-HRT-06-048. http://www.fhwa.dot.gov/resourcecenter/teams/safety/saf_8RUM.pdf.

Chuang, L., & Bülthoff, H. (2015). Towards a better understanding of gaze behavior in the automobile. In A*djunct Proceedings of Automotive UI 2015, Workshop Practical Experiences in Measuring and Modeling Drivers and Driver-Vehicle Interactions* (p. 4).

Courage, C., & Baxter, K. (2005). *Understanding your users: A practical guide to user requirements: methods, tools, and techniques*. Gulf Professional Publishing.

DIN Deutsches Institut fur Normung. (2003). Road vehicles—Ergonomic aspects of transport information and control systems—Specifications and compliance procedures for invehicle visual presentation. Ref No: DIN EN ISO 15008:2003-10.

Fakhrhosseini, S. M., Landry. S., Tan, Y-Y., Bhattarai, S., & Jeon, M. (2014). If you're angry, turn the music on: Music can mitigate anger effects on driving performance. In *Proceedings of the 6th International Conference on Automotive User Interfaces and Vehicular Applications (AutomotiveUI'14)*, Seattle, WA, USA, September 17–19.

Fischer, E., Haines, R. F., & Price, T. A. (1980). Cognitive issues in head-up displays. NASA Technical Paper 1711, p. 32.

Gable, T. M., Raja, S. R., Samuels, D. P., & Walker, B. N. (2015). Exploring and evaluating the capabilities of Kinect v2 in a driving simulator environment. *Proceedings of AutoUI'15* (pp. 297–304). NY, USA: ACM.

Gaver, W. W. (1986). Auditory icons: Using sound in computer interfaces. *Human-Computer Interaction, 2*, 167–177.

Hammerschmidt, J., Tünnermann, R., & Hermann, T. (2014). EcoSonic: Auditory displays supporting fuel-efficient driving. In *Proceedings of the 8th Nordic Conference on Human-Computer Interaction: Fun, Fast, Foundational* (pp. 979–982). ACM.

Harris, H., & Nass, K. (2011). Emotion regulation for frustrating driving contexts. In *Proceedings of the ACM SIGCHI Conference on Human Factors in Computing Systems (CHI 2011)*. Vancouver, BC, Canada: ACM Press.

Hassenzahl, M., Schöbel, M., & Trautmann, T. (2008). How motivational orientation influences the evaluation and choice of hedonic and pragmatic interactive products: The role of regulatory focus. *Interacting with Computers, 20*(4), 473–479.

Hermann, T. (2008). Taxonomy and definitions for sonification and auditory display. In *Proceedings of the 14th International Conference on Auditory Display (ICAD2008)*. Paris, France.

Jeon, M., Park, J., Heo, U., & Yun, J. (2009). Enhanced turning point displays facilitate drivers' interaction with navigation devices. In *The 1st International Conference on Automotive User Interfaces and Vehicular Applications (AutomotiveUI2009)*, September 2009 (pp. 145–148). Essen, Germany: ACM Press.

Jeon, M., Davison, B., Wilson, J., Nees, M., & Walker, B. N. (2009). Enhanced auditory menu cues improve dual task performance and are preferred with in-vehicle technologies. In *The 1st International Conference on Automotive User Interfaces and Vehicular Applications (AutomotiveUI2009)*, September, 2009 (pp. 91–98). Essen, Germany: ACM Press.

Jeon, M., Walker, B. N., & Srivastava, A. (2012). "Spindex" (speech index) enhances menu navigation on touch screen devices with tapping, wheeling, and flicking gestures. *ACM Transactions on Computer-Human Interaction (TOCHI), 19*(2); *14*, 1–27.

Jeon, M., Walker, B. N., & Gable, T. M. (2015). The effects of social interactions with in-vehicle agents on a driver's angry level, driving performance, situation awareness, and perceived workload. *Applied Ergonomics, 50*, 185–199.

Joyce, J. D., & Cianciolo, M. J. (1967). Reactive displays: Improving man-machine graphical communication. In *Proceedings of the* November *14–16, 1967, Fall Joint Computer Conference (AFIPS'67 (Fall))* (pp. 713–721). New York, NY, USA: ACM. doi:10.1145/1465611.1465705.

Kern, D., & Schmidt, A. (2009). Design space for driver-based automotive user interfaces. In *Proceedings of the First International Conference on Automotive User Interfaces and Interactive Vehicular Applications (AutomotiveUI)*. Essen, Germany.

Kun, A. L., Boll, S., & Schmidt, A. (2016). Shifting gears: User interfaces in the age of autonomous driving. *IEEE Pervasive Computing, 15*(1), 32–38.

Lazar, J. (Ed.). (2007). Universal usability: Designing computer interfaces for diverse user populations. Wiley.

Lazar, J., Feng, J. H., & Hochheiser, H. (2010). *Research methods in human-computer interaction*. Wiley.

Lequellec, J.-M., & Lerasle, F. (2000). Car cockpit 3D reconstruction by a structured light sensor. In *Proceedings IEEE Intelligent Vehicles Symposium* (pp. 87–92).

Manseer, M., & Riener, A. (2014). Evaluation of driver stress while transiting road tunnels. In 6th *International Conference on Automotive User Interfaces and Interactive Vehicular Applications (AutomotiveUI'14)*, September 17–20 (p. 6). Seattle, WA, USA: ACM.

Matthews, T., Dey, A. K., Mankoff, J., Carter, S., & Rattenbury, T. (2004). A toolkit for managing user attention in peripheral displays. In *Proceedings of the 17th Annual ACM Symposium on User Interface Software and Technology (UIST'04)* (pp. 247–256). New York, NY, USA: ACM.

McCrickard, D. S., Chewar, C. M., Somervell, J. P., & Ndiwalana, A. (2003). A model for notification systems evaluation—assessing user goals for multitasking activity. *ACM Transactions on Computer-Human Interaction (TOCHI'03), 10*(4), 312–338.

Moldovan, L., & German-Sallo, Z. (2014). Wavelet transform based HRV analysis. *Procedia Technology, 12*(2014), 105–111.

Nass, C., Jonsson, I.-M., Harris, H., Reaves, B., Endo, J., Brave, S., & Takayama, L. (2005). Improving automotive safety by pairing driver emotion and car voice emotion. In *Proceedings of the SIGCHI Conference on Human Factors in Computing Systems (CHI05)*. Portland, Oregon, USA: ACM Press.

Nees, M. A., & Walker, B. N. (2011). Auditory displays for in-vehicle technologies. In P. DeLucia (Ed.), *Reviews of human factors and ergonomics* (Vol. 7, No. 1, pp. 58–99). Santa Monica, CA: Human Factors and Ergonomics Society.

Nees, M. A., Gable, T., Jeon, M., & Walker, B. N. (2014) Prototype auditory displays for a fuel efficiency driver interface. In *Proceedings of the 20th International Conference on Auditory Display (ICAD2014)*. NY, USA.

Pauzié, A. (2008). A method to assess the driver mental workload: The driving activity load index (DALI). *IET Intelligent Transport Systems, 2*(4), 315–322.

Pelaez, G., Garcia, F., de la Escalera, A., & Armingol, J. (2014). Driver monitoring based on low-cost 3-D sensors. *IEEE Transactions on Intelligent Transportation Systems, 15*(4), 1855–1860.

Pousman, Z., & Stasko, J. (2006). A taxonomy of ambient information systems: Four patterns of design. In Proceedings of the Working Conference on Advanced Visual Interfaces (AVI'06) (pp. 67–74). New York, NY, USA: ACM.

Repperger, D. W., Phillips, C. A., Berlin, J. E., Neidhard-Doll, A. T., & Haas, M. W. (2005). Human-machine haptic interface design using stochastic resonance methods. *IEEE Transactions on Systems, Man, and Cybernetics, Part A, 35*(4), 574–582.

Riener, A. (2009). Sensor-actuator supported implicit interaction in driver assistance systems, Ph. D. thesis, Johannes Kepler University Linz, Austria, 329 pp.

Riener, A. (2010). *Sensor-actuator supported implicit interaction in driver assistance systems* (p. 316). Vieweg+Teubner.

Riener, A. (2011). Sitting postures & electrocardiograms: A method for continuous and unobtrusive driver authentication. In I. Traore & A. A. E. Ahmed (Eds.), *Continuous authentication based on biometrics: Data, models, and metrics* (pp. 137–168). 701 E. Chocolate Ave. Hershey, PA 17033, USA: IGI Global.

Riener, A., & Anzengruber, B. (2012). "FaceLight"—potentials and drawbacks of thermal imaging to infer driver stress. In *4th International Conference on Automotive User Interfaces and Interactive Vehicular Applications (AutomotiveUI'12)*, October 17–19 (p. 8). Portsmouth, NH, USA: ACM.

Riener, A., & Ferscha, A. (2008). Supporting implicit human-to-vehicle interaction: Driver identification from sitting postures. In *International Symposium on Vehicular Computing Systems (ISVCS 2008)*. Dublin, Ireland: ACM Digital Library, Trinity College, July 22–24, 2008.

Riener, A., & Noldi, J. (2015). Cognitive load estimation in the car: Practical experience from lab and on-road tests. In *Adjunct Proceedings of Automotive UI 2015, Workshop Practical Experiences in Measuring and Modeling Drivers and Driver-Vehicle Interactions* (p. 4).

Riener, A., Aly, M., & Ferscha, A. (2009). Heart on the road: HRV analysis for monitoring a driver's affective state. In *1st International Conference on Automotive User Interfaces and Interactive Vehicular Applications (AutomotiveUI 2009)*, September 21–22 (p. 8). Essen, Germany: ACM.

Riener, A., Ferscha, A., Frech, P., Hackl, M., & Kaltenberger, M. (2010). Subliminal vibrotactile based notification of CO2 economy while driving. In *Proceedings AutomotiveUI 2010* (pp. 92–101). Pittsburgh, Pennsylvania, USA: ACM, November 11–12, 2010.

Robles-De-La-Torre, G. (2006). The importance of the sense of touch in virtual and real environments. *IEEE Multimedia, 13*(3), 24–30.

Russell, J. A., Weiss, A., & Mendelsohn, G. A. (1989). Affect grid: A single-item scale of pleasure and arousal. *Journal of Personality and Social Psychology*, *57*(3), 493.

Scholles, M. (2016). ECG monitoring of motorists without skin contact, Fraunhofer Group for Microelectronics, April 2016. http://www.mikroelektronik.fraunhofer.de/en/press-and-media/microelectronics-news/news/detail/News/ecg-monitoring-of-motorists-without-skin-contact-1943.html.

Stanton, N. A., Salmon, P. M., Rafferty, L. A., Walker, G. H., Baber, Ch., & Jenkins, D. P. (2013). *Human factors methods: A practical guide for engineering and design* (2nd ed.). Taylor & Francis Ltd.

Stevens, A., et al. (2002). Design guidelines for safety of in-vehicle information systems. *A Report by Transport Research Laboratory in the Department of Tansport, Local, Government and Regions*.

Tawari, S. M., & Trivedi, M. (2014). Continuous head movement estimator for driver assistance: Issues, algorithms, and on-road evaluations. *IEEE Transactions on Intelligent Transportation Systems, 15*(2), 818–830.

Toney, A., Dunne, L., Thomas, B. H., & Ashdown, S. P. (2003). A shoulder pad insert vibrotactile display. In *Seventh IEEE International Symposium on Wearable Computers (ISWC'03)* (pp. 35–44), White Plains, New York.

University of Florida. (2016). Drivers' motion depth database—DMDDB. Retrieved May 2nd 206 from http://research.dwi.ufl.edu/dmddb/.

Walker, B. N., Lindsay, J., Nance, A., Nakano, Y., Palladino, D. K., Dingler, T., et al. (2013). Spearcons (Speech-based earcons) improve navigation performance in advanced auditory menus. *Human Factors, 55*(1), 157–182.

Watson, D., Clark, L. A., & Tellegen, A. (1988). Development and validation of brief measures of positive and negative affect: the panas scales. *Journal of Personality and Social Psychology, 54*(6), 1063.

Weinstein, S. (1968). Intensive and extensive aspects of tactile sensitivity as a function of body part, sex, and laterality. In D. R. Kenshalo (Ed.), *The skin senses* (pp. 195–222). Springfield, IL: Charles C. Thomas.

Wintersberger, P, Riener, A., & Frison, A. K. (2016). Automated vehicle, male or female driver: Who'd You Prefer? Comparative analysis of passengers' mental conditions, emotional states and qualitative feedback. In *8th International Conference on Automotive User Interfaces and Interactive Vehicular Applications (AutomotiveUI'16)*, October 24–26 (p. 9). ACM: Ann Arbor, Michigan, USA. doi: http://dx.doi.org/10.1145/3003715.3005410

Wolfe, J. M., Kluender, K. R., & Levi, D. M. (2014). *Sensation and perception* (4th ed.). Sinauer Associates, Inc.

Wuhe, Z., Lei, Z., & Ning, D. (2014). Sensing driver awareness by combining fisheye camera and Kinect. In *Proceedings of SPIE 9276, Optical Metrology and Inspection for Industrial Applications III*, 927624, November 13, 2014.

Xu, L., & Fujimura, K. (2014). Real-time driver activity recognition with random forests. In *Proceedings of AutoUI'14* (8 p.). NY, USA: ACM.

Chapter 12
Towards Adaptive Ambient In-Vehicle Displays and Interactions: Insights and Design Guidelines from the *2015 AutomotiveUI* Dedicated Workshop

Andreas Löcken, Shadan Sadeghian Borojeni, Heiko Müller, Thomas M. Gable, Stefano Triberti, Cyriel Diels, Christiane Glatz, Ignacio Alvarez, Lewis Chuang and Susanne Boll

Abstract Informing a driver of a vehicle's changing state and environment is a major challenge that grows with the introduction of in-vehicle assistant and infotainment systems. Even in the age of automation, the human will need to be in the loop for monitoring, taking over control, or making decisions. In these cases, poorly designed systems could lead to needless attentional demands imparted on the driver, taking it away from the primary driving task. Existing systems are offering simple and often unspecific alerts, leaving the human with the demanding task of identifying, localizing, and understanding the problem. Ideally, such systems should communicate information in a way that conveys its relevance and urgency. Specifically, information useful to promote driver safety should be conveyed as effective calls for action, while information not pertaining to safety (therefore less important) should be conveyed in ways that do not jeopardize driver attention. Adaptive ambient displays and peripheral interactions have the potential to provide

A. Löcken (✉) · H. Müller · S. Boll
University of Oldenburg, Oldenburg, Germany
e-mail: andreas.loecken@uni-oldenburg.de

S. Sadeghian Borojeni
OFFIS – Institute for Information Technology, Oldenburg, Germany

T.M. Gable
Georgia Institute of Technology, Atlanta, USA

S. Triberti
Università Cattolica del Sacro Cuore, Milan, Italy

C. Diels
Coventry University, Coventry, UK

C. Glatz · L. Chuang
Max Planck Institute for Biological Cybernetics, Tübingen, Germany

I. Alvarez
Intel Labs, Hillsboro, OR, USA

G. Meixner and C. Müller (eds.), *Automotive User Interfaces*,
Human–Computer Interaction Series, DOI 10.1007/978-3-319-49448-7_12

superior solutions and could serve to unobtrusively present information, to shift the driver's attention according to changing task demands, or enable a driver to react without losing the focus on the primary task. In order to build a common understanding across researchers and practitioners from different fields, we held a "Workshop on Adaptive Ambient In-Vehicle Displays and Interactions" at the AutomotiveUI'15 conference. In this chapter, we discuss the outcomes of this workshop, provide examples of possible applications now or in the future and conclude with challenges in developing or using adaptive ambient interactions.

12.1 Introduction

For an effective human–machine interaction, the individual must be at least partly aware of the driving situation and the environment. Timely and user-appropriate presentation of information is one of the core challenges when designing safety critical or enjoyable interfaces. This remains a pressing challenge in the case of semi- or fully automated traffic. One of the reasons for this is that the effective allocation of the driver's attention depends on multiple factors, which include situation awareness, cognitive workload, and ever-changing task requirements.

According to Endsley (1995), the operator interface has a large influence on situation awareness and needs to be designed carefully. Cognitive workload is among the factors that affect situation awareness. Wickens' (2008) Multiple Resource Theory (MRT) suggests that some modalities might be more suitable for parallel perceptual and cognitive processing than others, depending on the available mental resources. It is therefore important to have a wide range of possible modalities for interactions with the vehicle. Although many interface concepts with multiple modalities have been proposed, ambient displays, defined as "aesthetically pleasing displays of information which sit on the periphery of a user's attention" (Mankoff et al. 2003), have not been investigated in detail in the automotive domain.

Discussions on peripheral interactions and adaptive displays already started as part of the workshop "Social, Natural, and Peripheral Interactions: Together and Separate" at Automotive UI'14 (Riener et al. 2014). In order to continue the discussions and achieve a comprehensive understanding of adaptive ambient displays and peripheral interactions, we held the "Workshop on Adaptive Ambient In-Vehicle Displays and Interactions" at AutomotiveUI'15. One of the main goals in the workshop was to arrive at comprehensive solutions for directing attention and communicating the right amount of information at the right time via adaptive User Interfaces. We were interested in particular, but not restricted to, ambient in-vehicle displays that make driving a safer and more enjoyable experience. For example, how should light displays be appropriately designed in order to indicate the timely need to perform a lane change maneuver without being unnecessarily disruptive (Löcken et al. 2015)? How can navigational support be given using vibro-tactile actuators (Asif and Boll 2010)? How can displays adapt to the driver's state? To adequately address such issues, it is necessary to engage the combined expertise from various fields.

The Adaptive Ambient In-Vehicle Displays and Interactions workshop hosted 40 participants across different academic disciplines (e.g., computer science, psychology, neuroscience) and industrial areas (e.g., design of infotainment UI, driving assistant systems). Our presenters interpreted the topic of the workshop from many different perspectives: touchless UI, strategies for task resumption and attention shifting, multi-modal approaches, locations for ambient displays and interaction in the vehicle as well as presented a literature survey of related and existing work. These works were presented at the beginning of our workshop. After the presentations, we discussed three topics in a World Café: (a) Displays which need focused attention, (b) Displays which work in the periphery of the attention, and (c) Peripheral interactions. We used submitted papers as a starting point to discuss the aforementioned topics. After a presentation of results from the World Café, we continued with a Group Work session, in which participants were divided into four groups to discuss: (1) Definitions of interaction with ambient displays, (2) potential applications today, (3) Potential applications in the future, and (4) grand challenges. The World-Café format is described in (The World Café Community Foundation 2016).

In both sessions, participants discussed and defined their understanding of adaptive and/or ambient interaction in the vehicle. We collected many ideas on technologies that spanned a plausible implementation outlook ranging from current time to "in 50 years." In this regard, we also identified grand challenges around driver involvement and shared control, not only with regards to technological implementation, but also as it pertains to perceived value and consumer acceptance.

In the following section, we define 'focal attention' and 'ambient attention,' based on our discussions, in order to be able to discuss interfaces with a need for focal attention and those which work in the periphery of the driver's attention. These definitions will be used throughout the chapter and help to have a common understanding of attention. The subsequent sections present a reflective summary of the presented workshop papers. Finally, we will provide examples for interaction concepts which could be realized now or in the future and discuss grand challenges for the design and implementation of such systems.

12.2 The Notion of Focal Attention and Ambient Attention

Recent explorations of attention in digital technology identify the gap between knowledge of attention shift and translating this knowledge into interactive systems and suggest that we should be "designing systems capable of reasoning about users' attention, and consequently deciding how best to disappear from or to gain and guide user's attention" (Roda 2011).

Two fundamental assumptions, which were also for example discussed by Pashler and Johnston (1998), were shared across our participants: (a) attention is a limited resource, (b) the allocation of attentional resources can be influenced.

In this section, we present the results of two discussions during the workshop. These discussions were intended to elicit operational and theoretical definitions of two aspects of attention as defined by Wickens (2002), as they would pertain to task performance: focal attention and ambient attention. In this context,

- **focal attention** is typically defined as the use of foveal vision resources for the tasks that necessitate fixation (e.g., reading).
- **ambient attention** refers to the ability to process information by peripheral vision (e.g., optic flow heading).

We did not restrict 'focal' and 'ambient' attention to vision and therefore needed to define them independent from modality. The results of the discussion established a common ground for subsequent discussions on how cues that are processed with ambient attention can, for example, result in effective and voluntary recruitment of focal attention. In the following, we elaborate on these two types of attention.

12.2.1 Focal Attention

Diverse perspectives were offered on what constitutes focal attention. These include psychological definitions that are constrained by empirical findings as well as operational definitions that are based on characterizations of the in-vehicle workspace.

12.2.1.1 Definitions

Physiologically speaking, focal attention is a part of vision, which can be described in terms of a distinct segregation of focal and peripheral processing (Leibowitz et al. 1984)—visual acuity in the foveal region (approximately 2° in the visual field's center) falls off exponentially and is reduced by a factor of 10, 20° in the visual periphery (Snowden et al. 2012). Thus, eye movements exist to align the fovea with critical aspects in the visual environment and, in doing so, apply focal attention to the relevant visual information. In this regard, focal attention in vision can be tracked with the use of eye-tracking technology and is an independent resource from other perceptual modalities, such as audition. This tracking of where foveal resources are directed to in the visual scene is typically referred to as overt attention and was agreed upon to be a suitable proxy for focal attention in the context of driving, which depends heavily on visual information.

Wicken's (2008) provides a framework that facilitates the discussion on parallel perceptual processing across different sensory channels. For example, drivers often feel capable of focusing their full attention on their driving tasks while listening to the radio. In operational terms, the MRT suggests that listening to the radio consumes peripheral auditory resources that are independent from vehicle handling that requires focal attention in the visual and manual responding domain.

We came to the general consensus that focal attention is a vital ability of the driver that ought to be taken into consideration in the development of in-vehicle systems. Tasks that employ focal attention can be reasonably assumed to be carried out more effectively and efficiently than those using ambient attention, thus resulting in better performance in less time. Conversely, the lack of focal attention on the driving task could result in critical misses, such as a failure to notice and respond to the sudden braking of the lead vehicle, or the failure to take over steering in an automated vehicle when required.

12.2.1.2 Directing Focal Attention

Peripheral cues (possibly in separate sensory modalities) can be designed to redirect overt or focal attention to critical aspects in the visual scene, whereby the inducement of appropriate gaze-movements can be treated as a validation of their implementation. For example, visual (or auditory cues) can be placed within the in-vehicle environment to 'capture' the driver's gaze, when deemed appropriate. Discussed scenarios included redirecting of a driver's gaze to either the road ahead when it has not been looked at after a given duration of time, or to unexpected events such as a fast-moving car that enters one's lane. For example, Pomarjanschi and colleagues (2013) used strategically placed LEDS in the driving cockpit to effectively guide eye movements towards appropriate locations at traffic intersections. It should be noted that covert shifts of attention can occur in the absence of overt eye movements (Wright and Ward 2008). Thus, shifts of attention cannot be entirely determined with the use of an eye-tracker.

In-vehicle cues for directing focal attention can take the form of either exogenous or endogenous cues. Exogenous cues are salient sensory events such as a bright light, while endogenous cues are learned through association (e.g., arrows). Exogenous cues can be preferred to endogenous cues as they are less likely to be ignored than endogenous cues. Furthermore, their effects are known to be less influenced by user states such as high workload (Jonides 1981; Müller and Rabbitt 1989). However, an exogenous cue that cannot be ignored by drivers could pose a problem in situations where drivers are committed to paying attention to the road ahead. Also, cues that are false alarms could result in safety risks or annoyance and thus bad user experience. Ambient exogenous cues could be an effective intermediate solution that could be less influenced by drivers' user state, compared to endogenous cues while allowing for drivers' to effectively ignore them when necessary.

We agreed that the control of focal attention should be guided by cues to support decision-making. In this regard, it is necessary for designers of in-vehicle ambient cues to be sensitive to where focal attention should be directed at which stage of a given vehicle-handling procedure, such as negotiating the lane-entry of an unexpected vehicle. In doing so, it is necessary to first determine why focal attention to the given situation was lacking in the first place. Many discussants noted that the introduction of automated driving is likely to increase drivers' lack of focal attention on driving-related tasks. As such, understanding deficits of focal attention to critical tasks in automated driving scenarios might be an especially timely research topic.

12.2.2 *Ambient Attention*

The discussion of ambient attention started with an attempt to define peripheral or ambient displays in the vehicle. We offer two definitions for ambient displays for further debate

1. Any modality of output that does NOT require immediate action and can change its state to communicate information.
2. Information in any modality that notifies the users to communicate with them, but cannot be manipulated.

We discussed different modalities such as visual, auditory, haptic, and olfactory and their advantages and disadvantages for ambient displays in the vehicle.

Visual cues have to be designed with great care as driving is already a task requiring much of driver's focal and ambient vision capabilities. Nevertheless, peripheral cues through LED lighting are an important research field.

Auditory cues were thought to be easily integrated into the vehicle's soundscape. Stereo qualities of auditory cues make them suitable for directional cueing of attention to certain situations the driver has to attend to. Both, stereo and mono auditory cues may help to decrease visual attention on the road. For more information regarding in-vehicle auditory displays see Nees and Walker (2011).

Haptic displays could be integrated into the parts of a vehicle where the driver is in contact with: steering wheel, seat, and seat belt. Our participants envisioned shape-changing materials for conveying, e.g. information on the road condition via the steering wheel.

Olfactory displays have not been researched much in the driving context. Smell is perceived quite individually and a driver might get used to a certain smell-scape. On the other hand, smell is well-suited for addressing visceral interactions.

When discussing these modalities, we also tried to imagine what information ambient displays could be well suited for. We identified a wide range of applications. In automated or semi-automated driving, peripheral cues can be used to keep the human in the loop, when he or she is not the active driver. Peripheral cues may also help speed up the takeover process from automatic to manual driving, by directing the driver's attention to the important aspects of the developing situation ahead. Contextual information on what is important and what is not will need to be inferred for each situation.

On the other hand, designers will need to take into account how often information to be conveyed via ambient displays occurs and how they can ensure that the cues 'remain ambient' to the driver without distracting him or her from the driving task. Our hypothesis is that ambient displays are well suited for probabilistic information. But will this extra information that is available but not really in focus help drivers stay in the loop? And what about external contextual information, e.g. an Internet of things scenario, where your fridge knows you are out of milk and the car knows that your route will take you by the supermarket? Would we want to be informed on this non-driving-related information at all? We see much research potential here.

12.3 Summary of Topics

The workshop participants were invited to submit position papers on the topic of adaptive ambient displays and interactions before the workshop and present it to the other participants. In the following, we summarize the seven relevant contributions.

12.3.1 Integrating a Touchless UI in the Automotive Environment

Toppan and Chiesa's contribution (2015), "Integrating a Touchless UI in the Automotive Environment", introduced touchless sensor devices as an opportunity to improve the interaction capabilities of infotainment systems in the car. Indeed, as this contribution points out, voice-based interaction is affected by complex implementation and cost-ineffectiveness (e.g. translation in different languages), and touch-based devices present possible risks for safety with the driver having to split attention between the interface and the road. On the contrary, the authors proposed that touchless sensor devices are able to recognize universal gestures the driver can perform with his/her hand without additional distraction if designed correctly. The authors used the Leap Motion which works through an infrared camera and red lens. It recognizes and detects movements of the hand/fingers positioned above it. Several gestures were implemented, such as swipe to left/right (a movement of the hand towards a direction) to navigate among menu pages; key tap (rotation of a finger down towards the palm, then returning to initial position) to perform a selection; anticlockwise rotation of the hand to come back at previous page; and others more complex, such as writing letters in the air with a finger in order to set destination on the map. Moreover, the system included audio feedback to confirm/redone selections, so that the driver's look was not needed for monitoring interaction; and a physical knob with which the driver could activate or deactivate the sensor in order to avoid giving commands to the infotainment system because of accidental gestures.

The authors performed pilot qualitative usability tests both in simulation environments and in the car. The results suggest that the easiest gestures were well performed by the testing subjects, while the more difficult ones presented some problems because of drivers' lack of precision and limited sensitivity of the sensors.

The report is overall convincing while showing the possible benefits of touchless control for infotainment systems in the car. Undoubtedly, such technologies still need systematic user tests in different contexts, real-life traffic included. Indeed, on the one hand, touchless control does not require the user to look away from the road if used in pair with auditory cues; but on the other hand, the cognitive resources demanded by touchless control partially overlap the ones demanded by driving (e.g. spatial working memory, manual response, motor control). According to Wickens' MRT (2008), such overlapping of resource demands for different tasks may present risks for attention efficiency. Future research should consider carefully the complexity of gestures these systems can support, in particular the use of audio within gestures, which is further covered in Gable et al. (2014).

With regards to this chapter, this work demonstrates how to remove the need for focal attention while interacting with the car using a touchless interface.

12.3.2 Take Me Where I Was: Assisting In-Vehicle Interruption Management with Peripheral Cues

Considering all the in-vehicle information systems such as navigation or entertainment and communication, driving nowadays has evolved into a multi-tasking occupation. The paper "Take Me Where I Was: Assisting In-vehicle Interruption Management with Peripheral Cues" (Sadeghian Borojeni 2015) proposed peripheral in-vehicle cues to aid task resumption, the idea being that recovering from an interruption of a task requires efforts and attentional resources. Thus, returning to the interrupted task increases mental workload by trying to reconstruct the previous state prior to interruption. For example, imagine a scenario where you are driving on a highway and talking to a passenger when your in-car navigation system informs you that you need to take the next exit. Once you are on the exit, you want to continue the conversation. Hence, interrupted tasks can be considered as prospective memory tasks where the goal of the interrupted task has to be retrieved from memory. Using a cue that is associated with the goal, operators can be primed to access the encoded intention more easily. More specifically, the paper proposes different approaches to design in-vehicle (light) cues facilitating task resumption. The proposed system presented cues before and after the interruption to first encode the intention and later to retrieve the encoded information again. In another approach, cues based on task completion should remind the user of an incomplete task while event based cues should remind him/her to take a certain action in order to complete the intended goal.

Driving happens in a dynamic context with short interaction times. If the resumption time for switching back to an interrupted secondary task is reduced, drivers will have more time to dedicate to driving and consequently will be less distracted. This raises the question of how to design a cue that will effectively prime the desired recall without being ambiguous. In a highly dynamic world where several things are co-occurring, the cue is prone to encode various events while it should trigger only one intention. Another interesting point that will require attention is how to design cues that trigger a specific memory. For example, we are involved in a conversation and before turning to a different task we are presented a cue. After the interruption is over, the cue is presented again to remind us of the conversation. While one might recall that they wanted to continue the conversation, it will be challenging to remind a person of where he/she left off and what he/she wanted to say next. Other challenges that future research will need to address when developing such systems is the prioritization of tasks and the use of modalities. For example, answering a phone call and therefore suspending the navigation demand until the phone call is over might not be desirable since we will subsequently have to drive back to where we had to turn to get to our destination. The other interesting topic that would need to be addressed is if the cue presentation should occur in the modality of the interrupted task or whether the cue should be presented using a sensory channel that at the moment is not occupied by other information.

Most relevant for this chapter is the idea of how to switch the focal attention back to an interrupted task.

12.3.3 AmbiCar: Ambient Light Patterns in the Car

"AmbiCar: Ambient Light Patterns in the Car" (Löcken 2015) aims at assisting driving performance when changing lanes by a peripheral light pattern. To avoid overloading foveal vision, peripheral cues with different levels of urgency were designed. These levels were based on the ones defined by Matthews et al. (2004). Löcken's designs range from slow moving light dot (level: change blind) to changing color (level: make aware) and red flashes (level: interrupt). It appears that a complex design of ambient light cues that seems intuitive when designing is hard to decode in the context of the actual driving task. The final design of the ambient light informed the driver that he or she is about to be overtaken by another car in the fast lane. The approaching car and its relative distance to oneself were displayed by an ambient light that moved towards the bottom of the A-pillar in order to display the distance. In addition, the brightness of the ambient light cue adapts according to the assumed need for attention. Even though reaction times and safety gaps improved significantly, participants' results in a driving simulator study did not report a subjective improvement of workload using variants of the NASA Task-Load Index (TLX).

Ambient lights may be a good way of enhancing situation awareness without directing attention away from the primary driving task. In the example of being overtaken by another vehicle one can keep the traffic ahead in focus while also being aware of vehicles within a collision trajectory if changing lanes. When implementing these systems in a car, it would be interesting to see how much of the visual resources are taken away from the primary task and whether they may cause a conflict with alerts that signal potential collisions, reduced headway, etc., and how operators are able to prioritize between the different information streams. While information can be presented visually, future research could address the implementation of other modalities in more urgent situations or in situations of high visual load when the driver needs to be alerted. This could potentially lead to a system that is informative as long as it is not urgent but turns into a warning when the situation becomes critical.

This work is relevant for this chapter, because it discusses how to develop an adaptive display for ambient attention.

12.3.4 Attention Enhancement During Steering Through Auditory Warning Signals

Drivers rely on the visual modality while driving. However, other modalities such as audition are not employed as heavily, and when they are, they are not applied as effectively as they could be. In the review paper "Attention Enhancement During Steering Through Auditory Warning Signals" (Glatz et al. 2015) the authors discussed this underuse of the auditory modality, pointing out why auditory signals might be a good way to convey warning signals to the driver, and pointing to considerations when employing the use of sound for warnings. The authors argued that while true multiple resource sharing does not take place such as that predicted by Wicken's MRT (2008), the auditory realm proves to be a good director of attention to important information. The authors also discussed the different types of auditory information that could be used for auditory warnings such as speech, natural, or other types of sounds and some pros and cons of each. Finally, they reviewed some literature on looming sounds and the effect of the changing location of a sound can have on attending mechanisms.

The document points out that auditory cues within the vehicle are underutilized. Within ambient displays, the use of the auditory modality could be approached in multiple ways such as small modulations of the currently played music, modulating the engine or road noise, or other types of small changes of sounds already present within the in-vehicle soundscape. Additional sounds can also be introduced to create ambient displays such as different types of sonification for data of interest, which could be approached in a naturalistic way, or through the introduction of new types of sounds within the cockpit. The use of auditory signals within the vehicle is still underutilized and researchers could take advantage of the modality at low cost when scaling due to the presence of speakers in almost all production vehicles. An in-depth view of the employment of the auditory modality within the vehicle can be found in Nees and Walker's paper titled "Auditory displays for in-vehicle technologies" (2011).

12.3.5 Approaching a Placement Strategy for Windshield Displays

Haeuslschmid and Shou (2015) describe an initial framework to inform the placement of driver information (e.g., media content, fuel level) in order to facilitate recognition and understanding in future large field of view Head Up Displays, referred to as Wind Shield Displays (WSD). The framework is based on the theory of Proxemics, the three zone model, the information context, and priority. For vehicle specific applications, the authors proposed four display zones: Private, Vehicular, Social and Public Display Zone.

- At a distance of 70–95 cm, the *"Private Display Zone"* is suggested to contain information related to the person and their social interactions such as phone content, including emails, messages, and photos.
- The *"Vehicular Display Zone"* (95–120 cm) would contain vehicle-related information including fuel level and indicators, for example.
- A *"Social Display Zone"* (120–360 cm) would be used for social interactions at a lower level of detail or lower personal relevance compared to the private display zone, and may include Tweets for example.
- Finally, the *"Public Display Zone"* (360 cm to infinity) is proposed to contain information related to the environment and can be displayed in a spatially registered (i.e. contact analogue) manner within this zone.

Information examples would include navigational hints, traffic lights, or vehicle headway. Moreover, the authors suggest that with the exception of the Public Display Zone, only one layer should be defined for all information which furthermore should be displayed at the outer borders to decrease physical strain. These terms refer to the idea that an increased virtual image distance may avoid accommodation-convergence issues leading to reduced visual strain and fatigue. The authors further recommend not to place information on the top part of the windshield, i.e. above the driver's focus point, and to avoid overlaying display areas on different depth levels. High priority information (e.g. crash warning) is recommend to be salient and presented in the central field of vision. Note that it is not quite clear from the text why this information should be presented in the middle of the windshield as opposed to around the line of sight of the occupant/driver which arguably would be superior in terms of target detection. Interactions which require visual tracking of the system's responses can be located in areas that allow the driver to fast-access and reading, as well as to gain visibility of the road ahead, in other words, near the line of sight. Ambient information (low priority information) should contain low level of graphical detail (simple symbols) to enable fast perception using peripheral vision if necessary. Finally, the authors suggest placing reading material near the line of sight given its high visual demand.

The paper is intended as a first step towards a theoretical framework for the design of wide field of view Head Up Displays and thus relevant for focal and ambient attention. Future research is expected to validate the recommendations provided by this paper.

12.3.6 Prototyping Adaptive Automotive UX: A Design Pedagogy Approach

Hendrie and colleagues discuss an experiential prototyping approach to developing future car experiences (2015). The approach was explored in a project involving graduate and undergraduate design students and representatives from the technology and automotive industry. Based on a human-centered design approach, the authors proposed a "design through making" approach to complement the more established "design thinking" approach. It is put forward as a pedagogical approach to allow students (and designers) to evaluate and reflect on users' interactions with technology, move between macro and micro issues, and explore and experience design constraints and opportunities. The processes are suggested to be facilitated by the use of prototyping requiring the designer to move from the speculative to the specific. The role of the design educator is to facilitate the design process and to teach basic tools and methods such as storyboarding, prototyping, and "Wizard-of-Oz" techniques. From a pedagogical perspective, and arguably by extension from an organizational perspective, the design process requires the development of a productive space where create thinking and making can co-exist. Second, when defining the project brief we recommend to not conceive it as consisting of a list of deliverables but instead focus on defining a project and explore its parameters. To this end, the design process is proposed to be split into phases with each phase producing a new brief for the next phase, and each phase involving a different group of designers.

- Phase 1, "Landscapes", involves the collecting and creation of inspirational materials from a wide number of sources.
- Phase 2, "The journey", augments the first phase by conducting in-car experiments to explore the user experience. Students created sketch vignettes articulating possible occupant experiences.
- Phase 3, "Interactions", students detailed and contextualized the vignettes and used illustrations and paper prototypes and conducted user trials.

Discussing the approach, the authors highlighted limitations. For example, there was a tendency of designers to treat insights and observations as structured, empirical data and of engineer to consider them anecdotes. See Chapter X[1] for more details on how to appropriately translate insights to engineering requirements. Second, there is a risk that participants may misunderstand prototypes and develop an incorrect or incomplete mental model which may diverge from the intended goal or meaning of the interaction. Finally, the authors discuss the level of fidelity required for prototypes and argue that the Minimum Viable Prototype (MVP) should not be treated as visually compelling, or technologically advanced product vision. Instead, as a targeted effort in user-centered design concepts to express and validate a design vision.

This work gives insights on how to design interactions and adaptive interfaces and was thus relevant for our discussions during the workshop.

[1]Reference to another chapter of this book, written by Ignacio Alvarez and colleagues, needs to be added here!

12.3.7 Visual Displays for Automated Driving: A Survey

Manca et al. (2015) provided a literature survey of visual displays used in automated driving. Indeed, emerging technologies that promote driving automation may bring collateral adverse effects, such as safety reduction related to system misuse, due to driver over/undertrust of the automation capabilities. Therefore, visual displays for automated driving can be very useful to provide the driver with continuous feedback of the automation devices performance, this way promoting the driver's situation awareness appropriate levels of trust. The preliminary literature search performed by the authors retrieved 23 documents on such displays. Six of them are described in the work, selected on the basis of system maturity and their clear aim in terms of situation awareness support. Although very diverse, the described displays provide a synthetic overview of interesting solutions. Some of them graphically represent the vehicle on the interface, for example from a bird's eye view (Alessandretti et al. 2014), highlighting danger zones when another vehicle or an obstacle approaches it (Müller et al. 2014); others employ Head Up Displays that represent digital information superimposed on the windshield (e.g. directions, takeover request, etc.); finally, some technologies are based on eye-catching LED lights positioned within the cabin, that vary in color or patterns (e.g. blinking) to attract driver's attention towards a potential threat on the road. Based on their literature survey, the authors proposed guidelines for future research or design, such as integrating the different solutions in more complex devices. Another interesting advice for designers regards the implementation of situation awareness inputs in mobile phones and tablets since, especially in highly automated vehicles, drivers may engage in activities other than driving, such as using their mobile devices.

The literature survey by Manca and colleagues, along with the discussed aspects, clearly shows that this particular field of automotive technologies begins to be sufficiently mature to permit critical overviews. Future studies should include systematic review methods in order to identify the best solutions to promote situation awareness in the automated vehicle, along with their specific characteristics in terms of driver experience.

Most of the presented works either needed focal attention or were designed to draw the attention of drivers and were thus relevant to the discussions in the workshop.

12.4 Interaction and Display Concepts

After the presentations, summarized in the previous section, our goal was to envision concepts for adaptive displays that target ambient attention or use peripheral interactions when driving nowadays and in the future. In the following, we first describe the general ideas and thoughts on concepts for adaptive displays and interactions. Thereafter, we will present examples for concepts.

12.4.1 Considerations for Adaptive Ambient Displays and Interactions

Before developing concrete design concepts for adaptive displays and interactions which do not need focal attention, we discussed requirements and expected characteristics of such systems that go beyond those of common interfaces for assistant systems.

12.4.1.1 Peripheral Interaction Versus Overlearned Behavior

First, we needed to define a line between peripheral interaction and overlearned behavior. Consider using the gears when driving in a right-hand drive vehicle (common in the UK) when you are used to driving a left-hand drive vehicle (common in the US). Being in the right-hand drive vehicle, suddenly shifting gears becomes a task that requires a lot of attention, while it seems effortless when you drive in a left-hand drive. The answer, whether this is overlearned behavior or peripheral interaction, was not found during the workshop but may have to be addressed repeatedly when designing for peripheral interaction in the vehicle.

12.4.1.2 Adaptive Interactions Need Intent Prediction

If it is possible to predict the driver's intention, the vehicle's interfaces could adapt accordingly to ease the execution of the intended interaction. This adaption must be done in a way such that the driver keeps the sense of agency. If through the adaptations of the interface the driver feels like he or she is losing control of what is going on, those interfaces will be rejected by the driver. On the other hand, subtle ways of guiding the driver's attention to the vehicle's features that help him or her to perform a task better might be embraced.

12.4.1.3 Adapt the Interface to Context and Driver's State

One of the prominent topics was context-sensitive information density adjustment.

Generally speaking, an effective system should only present as much information as the targeted user can effectively process. Presenting too much information, especially during the execution of a critical maneuver, could be detrimental. Discussions touched upon how the driver's state and capacity for information processing could be evaluated in the cockpit. Besides this, the scheduling and prioritization of information presentation was addressed. How should sub-systems negotiate, which system uses which display to show its information when and in which modality. Although novel methods exist to present extra information without overloading the driver (such as with tactile interfaces), the goal should always be not to take driver's competences away but rather to assist him or her in reaffirming his or her skills.

12.4.1.4 Personalization

Another topic was the personalization of the driving and vehicle experience. Ambient cues can make us "feel at home" in our car, which to some is the only sanctuary during a busy day. Bringing in personal mobile devices and integrating them into the vehicle is as important as setting up the vehicle to be as comfortable for the driver and passengers as possible. Design questions such as how to moderate the indoor experience and how to master personalization and customization with cues addressing all senses arose. One example is making use of the visceral properties of smells. The distinct smell of a new car might be intriguing to some but be rejected by others.

12.4.1.5 Shape-Changing Interfaces as Adaptive Ambient Displays

Shape-changing materials took a prominent role when thinking about designing ambient displays. Some suggestions concerned the steering wheel changing its haptics to represent road conditions or upcoming takeover requests for automated driving, other suggestions concerned the driver seat, e.g. the backrest changing to model the traffic situation behind one's vehicle.

12.4.1.6 Hierarchy for Tasks

We established a hierarchy where primary tasks include steering, lane-keeping, and headway maintenance; secondary tasks include route-finding, planning; tertiary tasks include telecommunications and social media consumption. This taxonomy is not dissimilar to the hierarchical levels *strategic*, *maneuvering* and *controlling* tasks as defined by Michon (Michon 1985). Primary tasks are on the *maneuvering* or *controlling* level, while secondary tasks are on the *strategic* level. However, since Michon's model was developed to describe the driving task only, tertiary tasks are not covered in his model.

12.4.2 Possible Applications Today

Automated driving will not become ubiquitous overnight. For the foreseeable future we will have a large fleet of manually driven vehicles which demand high attention and mental workload. Whereas the introduction of driving assistant systems has simplified many driving-related tasks and maneuvers, distraction from driving will nevertheless still lead to fatal accidents. "Notification displays tend to draw focal attention at key moments, while ambient displays typically avoid drawing focal attention and instead use divided attention" (Matthews et al. 2004). Hence, ambient displays can be designed to not distract drivers in critical situations. Also, in a condition like today's driving context, ambient displays can be used in in-vehicle infotainment systems. In the following sections, we will discuss possible applications of ambient displays today that were discussed in the workshop.

12.4.2.1 Blend It Inside Out!

One of the discussed applications for ambient displays is bringing the outside world into the car. Having in-vehicle adaptive ambient displays, drivers can benefit from getting more information about their outside environments such as vehicles around them and especially the ones located in their blind spots. Moreover, they can be informed about the climate and traffic changes or road conditions unobtrusively. For instance, using ambient light or tactile displays, we can inform drivers of traffic congestion or road conditions (e.g., bumpy/slippery road) without distracting their visual attention from the road.

Alternatively, augmented reality can be used to blend the inside and outside world. For example, having displays on car windows and windshield which augment information about navigating through a city. This information could include interesting route points and landscapes such as gas stations with low price rate or tourist attractions in a city.

12.4.2.2 What's My State?

Another application of an ambient display is to inform drivers of the vehicle state. This can be done by displaying information items which do not need to be monitored continuously (e.g., fuel or oil level, temperature, distance left to destination, etc.). For example, a well-designed ambient light display with a progress bar metaphor can be used to show the amount or progress of any quantitative data such as fuel level or time to destination needless of having drivers to focus on a numerical display to read the related digits.

12.4.2.3 Occluded View

A problem that drivers currently have to deal with is lack of awareness of the traffic ahead, in conditions where the front vehicle is a heavy duty vehicle (e.g., truck or bus) or the road is curvy. Other traffic participants or obstacles are hidden, thus in such conditions. Thus an ambient display could assist drivers by presenting relevant information, for example about traffic congestion or speed of approaching vehicles from the other side of the road.

12.4.2.4 Wake Up!

Today cars are able to detect drivers' fatigue. However, to our knowledge, most of them suffice with displaying a text or an icon (e.g., coffee cup) on the information cluster. It can be assumed that drivers with a level of fatigue which can be detected by the vehicle are prone to miss these alerts. We proposed not only to show the information but also to wake up the driver. Using ambient thermal or tactile displays on the steering wheel, or smell displays with awakening odors such as lemon or mint can make drivers aware of their fatigue level.

12.4.3 Possible Applications in the Future

The rise of driver assistance systems has simplified many tasks for drivers by reducing the number of their tasks (e.g., Adaptive Cruise Control (ACC) taking over longitudinal control). However, the drivers' role in the near future will not be eliminated and drivers have to keep a considerable level of situation awareness while driving. Despite the decrease in the number of tasks that the driver is responsible for, the amount of information to be monitored has not decreased. Keeping track of the information about the tasks delegated to automation and monitoring the state of the system while driving can be challenging. Even with full automation, drivers can be assisted in their new primary task, such as reading or time management, by using adaptive ambient interfaces.

12.4.3.1 What's Going on in the Platoon?

A good example of a near future situation is cooperative driving such as platooning. In this scenario, the longitudinal control of the vehicles in the platoon is automated, and emergency braking is just done by the platoon leader and communicated to other vehicles. Still, the amount of information to be processed by the driver is considerable. Information about ego vehicle state, vehicle relative to the platoon vehicles, and the whole platoon state are continuously presented. Moreover, other than keeping the control of the vehicle, the driver has to perform different maneuvers related to driving in a platoon (e.g., changing position in the platoon). All this information should be communicated to the driver in an unobtrusive way in order not to distract him/her from driving. The utilization of ambient displays could convey this information and the progress of the maneuvers without occluding drivers' visual attention.

12.4.3.2 What's Going on Around Me?

Another functionality of the ambient displays in the near future could be using them to communicate information about distance and speed of other vehicles and assist driving tasks such as overtaking. The system of Löcken (2015), which was presented earlier, already displayed the speed and distance of closing vehicles on the fast lane using ambient light. Pfromm and colleagues (2013) display other traffic participants using LEDs. Hence, it appears plausible that ambient light could be used to display speed and distance of other traffic participants to keep drivers of automated cars informed and, thus reduce takeover times.

12.4.3.3 What's the Car's Plan?

Ambient displays inside automated cars could be used to inform the passengers of the intention and processes of the automated car. Moreover, the information about the intention, state and next maneuvers of the automated vehicle might also be displayed outside the car to inform other traffic participants, such as drivers of other cars or pedestrians. Communication of such information, a big part of which is nowadays communicated by nonverbal and gesture communication of drivers and pedestrians could be handled by using ambient displays.

12.4.3.4 Don't Make Me Sick!

Ambient displays may also be used to display ego-motion to avoid occupants experiencing motion sickness (Diels and Bos 2015, 2016). In particular, when engaging in non-driving tasks using head down displays or rearward facing seating arrangements, ego-motion information received by our eyes may differ from that received from our organs of balance. Ambient displays have the potential to nullify such sensory conflicts.

12.4.3.5 My Very Own Vehicle

Ambient displays might also be used to recreate a customized interior, e.g., with a customized information cluster, preferred set of colors, ambient lighting, predefined priority of information items, its salience levels, as well as the used sounds or light patterns for notifications. If a driver needs to change cars, he or she can take the customized behavior of the ambient displays with him/her.

12.4.3.6 Entertain!

Using augmented reality, the car's windows can be turned into gaming displays for kids to be entertained during long trips, or show images of the surrounding environment in other seasons (e.g., displaying carnivals in November in Rio de Janeiro). These displays will need to adapt to the car's environment and passengers' attention.

12.5 Grand Challenges

The goal of this workshop was to foster a discussion of the adaptive and ambient in-vehicle displays and interfaces so as to complement the novel landscape that current technology might yield. We presented works of the workshop participants, discussed the difference of 'ambient attention' and 'focal attention' from our perspective and gave examples for possible application now and in the future.

We conclude this chapter with the various grand challenges that we came up with during the workshop. It was unsurprising that the discussion pivoted around automated driving and the novel experiences that this technology would bring to consumers. In this regard, adaptive in-vehicle technologies offer the opportunity to vary these novel experiences according to the estimated real-time needs of the driver, which could themselves vary in response to adaptive in-vehicle experiences.

12.5.1 The Subjective Value of Assistant Systems

It is first necessary to understand the value that is currently placed by drivers on their personal transport vehicles. The qualitative nature of such values include those that are functional (e.g., a reliable means to commute 20 km to work), psychological (e.g., manual gear transmission for driving involvement), or social (e.g., brand identification with one's lifestyle values). An accurate estimation of these values and how they map with consumer demographics would reveal the type of adaptive ambient in-vehicle technologies that might be desired. This is because such technologies hold the possibility of either eradicating or enhancing as well as introducing new frustrations, depending on the values that are currently attached to the driving experience. For example, drivers who value ease of handling over manual steering could be expected to especially appreciate an adaptive shift to autonomous vehicle handling during heavy traffic or when the vehicle senses high frustration in the driver. On the other hand, drivers who value autonomy could be increasingly frustrated with a system that steadily removes steering options, especially in difficult situations. There is unlikely to be a one-size-fits-all solution. Thus, it is necessary to understand consumer demands prior to the development of adaptive in-vehicle technologies, in order to identify the technologies that are most likely to be appreciated upon their realization. This constitutes a primary challenge to the development of ambient adaptive technology, which precedes any engineering challenges related to their implementation.

12.5.2 Possible Pitfalls

The perceived demand for adaptive ambient systems will vary immensely across individuals. Nonetheless, all discussants agreed that the following principles would characterize public acceptance of adaptive ambient in-vehicle technologies. It was generally agreed that drivers were likely to respond negatively to any technologies that appeared to: (a) manipulate their behavior; (b) subjugate their autonomy; (c) be less reliable than their believed performance. In other words, expectations of interactions with adaptive and ambient systems are likely to be similar as one's interactions with a fellow human driver or passenger. As we create ambient and affective interactions, we may run into an 'uncanny valley' similar to that reported in the computer graphics of human avatars (Mori et al. 2012). If this appears contrived, drivers might perceive that they are being manipulated. This could result in an effective system that is rejected for its perceived intrusiveness. For example, pleasant music that is played to adaptively lower driving stress, which is sensed non-intrusively by physiological sensors, could be perceived as being manipulative by some individuals. Unpleasant but highly effective warning cues could be rejected simply because consumers do not wish to introduce them into their environment. In the worst case scenario, a responsive system might be anthropomorphically perceived as a social agent with an annoying personality. Of course, such jarring experiences already manifest themselves when we travel with incompatible companions. The difference with ambient and adaptive technology is that such implementations could be perceived as undesirable features that contributed to the cost of the vehicle.

12.5.3 Understanding Consumer Motivation

Consumer ignorance is a strong challenge that innovations can expect. Adaptive and ambient technologies might be especially vulnerable to this challenge, specifically because they strive towards providing useful functions without intrusion. In other words, they are not noticeable except when they fail to accurately adapt to the needs of the user. This challenge already presents itself in consumers' perceptions of the automobile. Few consumers are aware of the complexities that underlie the effective functioning of their personal vehicles. Addressing the relevant complexities of a system can add value and allow consumers to discriminate between different vehicles with varying features. It will be a challenge to effectively communicate the design and user experience and their returns to the consumer, instead of using vague terms or concepts that consumers do not value or desire (e.g., focal attention, ambient cues, etc.). Such features ought to be characterized in terms of the novel and more pleasant experiences that they will support. For example, consumers readily perceive the benefits of technical innovations, such as cruise control and automatic gear change—such innovations are expressed in terms of the job that the consumer does not wish to perform, which the system may offers to take over. Why would a consumer wish to attend to something else besides driving? How can we convince the consumer that a given system will take over his/her responsibility effectively and reliably? Perhaps adaptive and ambient technology should be specifically designed to address these specific task scenarios that are explicitly desired, instead of a generic purpose of reducing the attentional demands of driving? Instead of ambient sensing and broad definitions of desired or undesired user state, it is essential to identify the source of the stress instead, and target it. Furthermore, brand identity creates the perceived value of a given technology and it should support a given brand identity. In this regard, adaptive and ambient in-vehicle technologies might effectively create further market diversification and, in doing so, cater to heterogeneous needs and wants.

12.5.4 Social Attachment Between Drivers and Their Vehicles

Could innovations be visceral, rather than purely technological? The social attachment between drivers and their automobiles is an aspect that could particularly matter to the adoption of adaptive and ambient technologies, especially if our interactions with them resemble those with social agents. Instead of task-segregation, wherein a given technology is delegated an undesirable task, perhaps the focus could be on creating a satisfying task-sharing environment instead. For some, the personal automobile is a personal and individualized space that is highly valued, with a characteristic feel that is removed with the use of shared public transport (e.g., taxis). Could adaptive and ambient technologies allow for effective on-demand customization of an in-vehicle environment? If so, this could lead to wider acceptance of car-sharing schemes, for which the emerging technology of self-driving cars are especially well-suited for. An example for efforts into this direction is the recent partnership of General Motors and Lyft (Fiergermann 2016). If physical vehicles are no longer identified with the individual consumer, could this connection be established between the user and a virtual agent that could be (re-)assigned to whichever vehicle is currently in use. If so, this could represent a dramatic shift in the automobile industry from the engineering of better vehicle handling to the development of a more personable ride experience that is sensitive to the desires of the individual.

12.5.5 Reliable Assessment of User States

Adaptive and ambient technologies will only gain acceptance if they are reliable and non-intrusive. While physical measurements, such as the estimation of lane markings and obstacle proximity, are increasingly robust, measuring user states in a non-intrusive fashion will continue to be a challenge for a long while. Technologies that track eye movements, physiological responses, and even brain activity can be effectively incorporated into an in-vehicle environment. Nonetheless, meaningful user states have to be established within the context of well-defined operational scenarios before it is even useful to determine if such states can be reliably inferred from such measurements. For example, while many might consider workload estimation as a worthwhile endeavor, it is unclear what is meant by the term 'workload' without a clear operationalization of the mental processes that might be involved given the context of a scenario or task.

12.6 Summary

The automobile space is heading towards a de-personalized future, which will be brought about by the technological realization of automated driving and social innovation of effective car-sharing. In other words, individuals will perceive the automobile purely as a mode of transport that, unlike public transport, offers customizable routes. Hence, it will be similar to our current use of taxis. The challenges that face such a future are not only technological in nature. They are also based on social acceptance, at both the individual level as well as a societal level. A successful implementation of adaptive ambient interfaces should seek, not only, to support driving performance in autonomous vehicles. It has the potential of providing an enhanced environment over existing in-vehicle experiences and, in doing so, allay the psychological concerns that could be introduced by a future of autonomous vehicles.

Acknowledgements We would like to acknowledge the authors of the submissions and those who attended the workshop for their participation. Further, we would thank the German Research Foundation (DFG) for financial support within project C03 of SFB/Transregio 161. Portions of this work were supported by a National Science Foundation Graduate Research Fellowship (DGE-1148903). This work was also partially supported by the funding initiative *Niedersächsisches Vorab* of the Volkswagen Foundation and the Ministry of Science and Culture of Lower Saxony as part of the *Interdisciplinary Research Center on Critical Systems Engineering for Socio-Technical Systems*.

References

Alessandretti, G., Amditis, A., Metzner, S., et al. (2014). InteractIVe—Final Report. Deliberable D1.9, version 1.2. http://www.interactive-ip.eu/index.dhtml/docs/interactIVe_SP1_201405-06v1.2-D19-Final_Report.pdf.

Asif, A., & Boll, S. (2010). Where to turn my car? Comparison of a tactile display and a conventional car navigation system under high load condition. In *Proceedings of the 2nd International Conference on Automotive User Interfaces and Interactive Vehicular Applications* (pp. 64–71). New York, NY, USA: ACM.

Diels, C., & Bos, J. E. (2015). User interface considerations to prevent self-driving carsickness (pp 14–19). ACM Press.

Diels, C., & Bos, J. E. (2016). Self-driving carsickness. *Transport* in the *21st Century*: *The Application* of *Human Factors* to *Future User Needs Part B*, *53*, 374–382. doi:10.1016/j.apergo.2015.09.009.

Endsley, M. R. (1995). Toward a theory of situation awareness in dynamic systems: Situation awareness. *Hum. Factors, 37*, 32–64. doi:10.1518/001872095779049543.

Fiergermann, S. (2016). You'll be able to call a self-driving car to pick you up, as Lyft and GM challenge Uber. In *Mashable*. http://mashable.com/2016/01/04/lyft-gm-self-driving-cars/#aCKfVe2Hvqq3. Accessed 23 Jun 2016.

Gable, T. M., May, K. R., & Walker, B. N. (2014). Applying popular usability heuristics to gesture interaction in the vehicle. In *Adjunct Proceedings of the 6th International Conference on Automotive User Interfaces and Interactive Vehicular Applications* (pp. 1–7). New York, NY, USA: ACM.

Glatz, C., Bülthoff, H. H., & Chuang, L. L. (2015). Attention enhancement during steering through auditory warning signals. In *Workshop on Adaptive Ambient In-Vehicle Displays and Interactions in AutomotiveUI'15*. http://www.auto-ui.org/15/p/workshops/5/glatz.pdf.

Haeuslschmid, R., & Shou, Y. (2015). Towards a placement strategy for windshield displays. In *Workshop on Adaptive Ambient In-Vehicle Displays and Interactions in AutomotiveUI'15*. http://www.auto-ui.org/15/p/workshops/5/haeuslschmid.pdf.

Hendrie, M., Alvarez, I., & Hooker, B. (2015). Prototyping adaptive automotive UX: A design pedagogy approach. In *Workshop on Adaptive Ambient In-Vehicle Displays and Interactions in AutomotiveUI'15*. http://www.auto-ui.org/15/p/workshops/5/hendrie.pdf.

Jonides, J. (1981). Voluntary versus automatic control over the mind's eye's movement. In *Attention and performance IX (Proceedings of the 9th International Symposium on Attention and Performance*, Cambridge, July 13–18, 1980). Erlbaum, pp 187–203.

Leibowitz, H. W., Shupert, C. L., & Post, R. B. (1984). The two modes of visual processing: Implications for spatial orientation. In *Peripheral vision horizon display (PVHD).*

Löcken, A. (2015). AmbiCar: Ambient light patterns in the car. In *Workshop on Adaptive Ambient In-Vehicle Displays and Interactions in AutomotiveUI'15*. http://www.auto-ui.org/15/p/workshops/5/loecken.pdf.

Löcken, A., Heuten, W., & Boll, S. (2015). Supporting lane change decisions with ambient light. In *Proceedings of the 7th International Conference on Automotive User Interfaces and Interactive Vehicular Applications* (pp. 204–211). New York, NY, USA: ACM.

Manca, L., de Winter, J. C. F., & Happee, R. (2015). Visual displays for automated driving: A survey. In *Workshop on Adaptive Ambient In-Vehicle Displays and Interactions in AutomotiveUI'15*. http://www.auto-ui.org/15/p/workshops/5/manca.pdf.

Mankoff, J., Dey, A. K., Hsieh, G., et al. (2003). Heuristic evaluation of ambient displays. In G. Cockton, & P. Korhonen (Eds.), *Proceedings of the Conference on Human Factors in Computing Systems—CHI'03* (p. 169). New York, USA: ACM Press.

Matthews, T., Dey, A. K., Mankoff, J., et al. (2004). A toolkit for managing user attention in peripheral displays. In *Proceedings of the 17th Annual ACM Symposium on User Interface Software and Technology—UIST'04* (pp. 247–256). New York, USA: ACM Press.

Michon, J. A. (1985). A critical view of driver behavior models: What do we know, what should we do? In L. Evans, & R. C. Schwing (Eds.), *Human behavior and traffic safety* (pp. 485–524). Springer, US.

Mori, M., MacDorman, K., & Kageki, N. (2012). The uncanny valley (from the field). *IEEE Robotics and Automation Magazine, 19*, 98–100. doi:10.1109/MRA.2012.2192811.

Müller, H., Löcken, A., Heuten, W., & Boll, S. (2014). Sparkle: An ambient light display for dynamic off-screen points of interest. In *Proceedings of the 8th Nordic Conference on Human-Computer Interaction: Fun, Fast, Foundational* (pp. 51–60). New York, NY, USA: ACM.

Müller, H. J., & Rabbitt, P. M. (1989). Reflexive and voluntary orienting of visual attention: Time course of activation and resistance to interruption. *Journal of Experimental Psychology: Human Perception and Performance, 15*, 315–330. doi:10.1037/0096-1523.15.2.315.

Nees, M. A., & Walker, B. N. (2011). Auditory displays for in-vehicle technologies. *Reviews of Human Factors and Ergonomics, 7*, 58–99. doi:10.1177/1557234X11410396.

Pashler, H., & Johnston, J. C. (1998). Attentional limitations in dual-task performance. In *Attention* (pp. 155–189). Hove, England: Psychology Press/Erlbaum (UK) Taylor & Francis.

Pfromm, M., Cieler, S., & Bruder, R. (2013). Driver assistance via optical information with spatial reference. In *16th International IEEE Conference on Intelligent Transportation Systems (ITSC 2013)* (pp. 2006–2011). IEEE.

Pomarjanschi, L., Dorr, M., Bex, P. J., & Barth, E. (2013). Simple gaze-contingent cues guide eye movements in a realistic driving simulator. In B. E. Rogowitz, T. N. Pappas, & H. de Ridder (Eds.), *Human vision and electronic imaging XVIII* (pp 865110-1–865110-8).

Riener, A., Pfleging, B., Jeon, M., et al. (2014). Social, natural, and peripheral interactions: Together and separate. In *Adjunct Proceedings of the 6th International Conference on Automotive User Interfaces and Interactive Vehicular Applications* (pp 1–6). New York, NY, USA: ACM.

Roda, C. (2011). Human attention and its implications for human–computer interaction. In C. Roda (Ed.), *Human attention in digital environments* (pp 11–62). Cambridge University Press.

Sadeghian Borojeni, S. (2015). Take me where I was: Assisting In-vehicle interruption management with peripheral cues. In *Workshop on Adaptive Ambient In-Vehicle Displays and Interactions in AutomotiveUI'15*. http://www.auto-ui.org/15/p/workshops/5/sadeghianborojeni.pdf.

Snowden, R., Snowden, R. J., Thompson, P., & Troscianko, T. (2012). *Basic vision: An introduction to visual perception*. OUP Oxford.

The World Café Community Foundation. (2016). The world cafe. In *World cafe*. http://www.theworldcafe.com/. Accessed 10 Jul 2016.

Toppan, R., & Chiesa, M. (2015). Integrating a touchless UI in the automotive environment. In *Workshop on Adaptive Ambient In-Vehicle Displays and Interactions in AutomotiveUI 15*. http://www.auto-ui.org/15/p/workshops/5/toppan.pdf.

Wickens, C. D. (2008). Multiple resources and mental workload. *Human Factors, 50*, 449–455. doi:10.1518/001872008X288394.

Wickens, C. D. (2002). Multiple resources and performance prediction. *Theoretical Issues in Ergonomics Science, 3*, 159–177.

Wright, R. D., & Ward, L. M. (2008). *Orienting of attention*. Oxford University Press.

Chapter 13
The Steering Wheel: A Design Space Exploration

Alexander Meschtscherjakov

Abstract The steering wheel is—besides pedals for acceleration and breaking—the most prominent interaction artifact between drivers and their vehicles. All cars have a steering wheel, which translates steering instructions from drivers to cars. "Eyes on the road and hands on the wheel!" is one of the most prominent paradigms in the automotive world. The driver should always have a grip of the steering wheel, making it also the most reachable area in the car for manual interaction. Automotive interaction designers have, rightly, used the area on and around the steering wheel to position interaction elements beyond steering. Today's cars are cluttered with buttons and switches to operate the car's information and entertainment system. New interaction modes, such as touch screens on the steering wheel or shape changing rims offer interaction designers new perspectives on utilizing the steering wheel, as a means for interaction with the vehicle. In this chapter, we describe the design space steering wheels offer for interaction beyond steering the vehicle. We collect and analyze various approaches from industry and academia on human-steering wheel interaction beyond traditional interaction and infer potentials and risks when utilizing such novel modalities in terms of interaction design. This analysis leads to a thorough discussion of the steering wheel interaction design space, resulting in related interaction design recommendations. Finally, we provide a look into the future when evermore advanced driving assistance systems pervade the car, eventually relieving the driver from the steering task with the emergence of autonomous vehicles.

A. Meschtscherjakov (✉)
Christian Doppler Laboratory "Contextual Interfaces", Center for Human-Computer Interaction, University of Salzburg, Salzburg, Austria
e-mail: alexander.meschtscherjakov@sbg.ac.at

G. Meixner and C. Müller (eds.), *Automotive User Interfaces*,
Human–Computer Interaction Series, DOI 10.1007/978-3-319-49448-7_13

13.1 Introduction

Drivers have to perform at least two things when driving a vehicle: they have to control the longitudinal and lateral movement of the vehicle. The longitudinal movement (i.e., accelerating and breaking) is achieved by pushing pedals with the feet. The lateral movement is achieved by turning tires to the left or to the right. In modern cars, this is achieved via a steering wheel positioned in front of the driver. The steering wheel is typically connected by one or more spokes to a steering column from which the driver input is transferred to the tires.

The steering wheel typically has the shape of a ring, which is grasped by the driver, and turned clockwise or counterclockwise. When the vehicle is moving forward, turning the steering wheel clockwise results in a movement to the right; a counterclockwise rotation results in a movement to the left. This mapping of the rotation of the steering wheel to the movement of the vehicle is intuitive for humans. Even toddlers discover this immediately when driving their toy karts. This mapping seems to be a natural one. Then again, the mapping is not as clear when driving backwards. Inexperienced drivers sometimes have a problem to estimate the trajectory of the car when driving in reverse.

It was not always the case that a car was directed with a steering wheel. In the nineteenth century, automobiles were directed with a reversed tiller, as they are used in boats to turn the rudder. In 1894, Alfred Vacheron was one of the first who used a steering wheel in an automobile race (Greathouse 2008). A steering wheel has the advantage that it could be handled easily and more precisely. Depending on the transmission ratio between the steering wheel and the tires, a major rotation movement of the steering wheel may result in a small movement of the vehicle. This allows steering the vehicle at a higher speed. Since then steering wheels have evolved both in their appearance and in their functionality.

Two of the most powerful transitions the steering wheel has undergone were usability and safety driven. In terms of usability, the introduction of hydraulic or electric power steering made it more convenient and easy to rotate the steering wheel, even while the car is standing still. The adjustable steering wheel targeted the ergonomic properties of the steering wheel. The tilt steering wheel[1] allows the driver to move the steering wheel up or down. The telescope steering wheel allows the driver to pull it closer or push it further away. The position of modern steering wheels may be adjusted to the ergonomic properties and preferences of individual drivers.

For safety reasons, all steering wheels in new cars must be equipped with an airbag. They inflate when the crash-sensors detect an impact of the vehicle. For interaction designers, there is always the constraint to find the right place and space for the airbag when designing new steering wheels. Another safety feature that has been integrated into the steering wheel was a button for the horn. Traditionally, it is

[1] http://automotivemileposts.com/autobrevity/tiltwheel.html (Accessed Sep 1 2016).

Fig. 13.1 The development of the steering wheel over time using the example of a Mercedes 190 SL, ca. 1960 (*left*) and a Mercedes E 220d, 2015 (*right*)

located in its center, although other approaches incorporate the horn into the spokes or the steering wheel rim.

Apart from these innovations, the steering wheel has increasingly become a place to position interaction possibilities for the driver with the vehicle, in addition to maneuvering the vehicle. Overtime, more and more interaction modes have been integrated into the steering wheel and it has evolved from a mere means to direct the vehicle to an interactive control center (see Fig. 13.1).

In this chapter, we are not interested in the steering property of a steering wheel or how it feels to steer the vehicle, which is influenced by a variety of factors such as resistance of the steering wheel, the immediacy of the steering, or the center-point-feeling. We are interested in how a steering wheel can be used as a *design space* for interaction designers to be able to incorporate all kinds of interaction approaches in the steering wheel. We will systematically identify the interaction properties of a steering wheel and derive conclusions for designers who want to use the steering wheel to provide a means for the driver to communicate with the vehicle. We will identify different properties of steering wheels and their implications on interaction design. We will outline recent approaches from industry and academia on how a steering wheel can be used for driver input and output. Finally, we will have a look into the potential future of the steering wheel.

13.2 Context Steering Wheel

In order to be successful in their design, interaction designers should always take the context into account in which the interactions occur. When designing interactions for a steering wheel, this is the automotive context. The automotive context is characterized, that a driver and passengers are inside a *moving vehicle*, with a surrounding that is constantly changing. More than that, drivers interact with other interfaces that are not part of the *primary task* of driving safely. Regarding the steering wheel itself, the *form factor* (e.g., its shape, size, and appearance) has an influence on the design space. We can identify different *interaction areas* on the steering wheel. In the following sections, we will discuss how these properties influence the steering wheel design space.

13.2.1 Moving Vehicle

One aspect when designing interactions for the automotive context is that the vehicle itself is moving. This movement results in a multitude of forces that are acting upon the person who is interacting with a system but also the interactive system itself. For the driver that means that both accelerating and decelerating forces, as well as centrifugal forces, influence the interaction with a system. Pushing a button or touching a screen at an exact position might be difficult, especially when the forces are often changing, as it is the case in a vehicle that moves through traffic.

Moreover, the structure of the road and the vibrations of the motor might be transferred to the steering wheel. For an interaction designer that means that interactive elements should be designed with a larger surface area to counteract the inaccuracy possibility of these factors. A designer should also design for error recovery, since some interaction errors might occur unintentionally due to the emerging forces. Above that, if a designer uses vibro-tactile feedback, this feedback should not conflict with natural vehicle vibrations.

Another aspect that should be taken into account is changing light conditions. The interface needs to be visible in different, fast changing light conditions but also not be distractive. For example, a bright display increases visibility in a bright environment but may lead to visual distraction when it is situated in a dark environment. That is why most car manufacturers, but also companies that produce route guidance systems, equip their interfaces with light sensors that change the display into a "night mode" during low light situations (e.g., at night or when entering a tunnel).

In addition, the sun might create a glare on a display on the steering wheel reducing readability. Visual elements, such as speedometer or tachometer, in the instrument cluster are often underneath a hood to protect them from direct insolation. This might not be easily achievable for the steering wheel. Therefore, designers need to make sure that readability is given under all light conditions.

Different sound sources might also interfere with some interaction approaches. On one hand, there are unavoidable driving noises from the motor, the tires, the wind, as well as other vehicles. On the other hand, many people like to listen to music or other audio inside the vehicle. An interaction designer has to take this into account when designing sound feedback or earcons for the automotive context. The noisy environment also has to be taken into account when using speech as an input modality. With respect to the steering wheel design space, it could be interesting to put microphones or loudspeakers and onto the steering wheel instead of using the onboard sound system. For a microphone, the distance between the driver and the steering wheel is rather short and, thus, interference might be reduced. When using a loudspeaker on the steering wheel, this might be used to provide the driver with spatial information of the sound source.

13.2.2 Primary Task

Apart from the moving vehicle, another context factor that has an influence on designing for the steering wheel is the primary task of the driver to always drive safely. The driver should keep their eyes on the road and hands on the steering wheel. This means that when an interaction designer offers a new interaction approach for the driver, this new approach must not distract the driver. Distraction can have different forms. In the automotive context, we distinguish between manual, visual, and cognitive distraction.

Regarding manual distraction, the steering wheel has some advantages over other options for interacting with the vehicle, such as touch screens or rotary knobs mounted in the center stack. When interaction elements are positioned near the steering wheel, the way hands move to and from the steering wheel is short. That is one reason why many car manufacturers have equipped their steering wheels with different interaction possibilities. Then again, the steering wheel itself is one of the crucial interfaces for driving safely. Thus, using the steering wheel for interactions other than steering has to take this into consideration. For example, even if an interaction designer places an interface on the steering wheel, this interface may only be used with one hand, since the other hand has to hold the steering wheel itself. Thus bi-manual interactions are only possible when simultaneously steering is possible. Currently, bi-manual interaction is used for handover procedures in autonomous vehicles. For example, the driver has to push two buttons on the steering wheel simultaneously with both thumbs in order to confirm the handover request from the autonomous vehicle.

Interfaces that require the driver to take their eyes off the road should be placed near the line of sight of the driver when looking at the road ahead. Both, the distance the eye has to move matters, as well as the directions of the eye movement are crucial. Vertical or horizontal eye movements are faster than diagonal ones. Thus, a glance at the steering wheel is typically faster than one to a visual item in the center stack. Nearest to the line of sight are head-up-displays, followed by items in the instrument cluster. Since the steering wheel typically is positioned so that the driver sees through the steering wheel to the instrument cluster, this area of the steering wheel might also be used for visual feedback, when the instrument cluster is not needed. The closer the direction of the glance is to the line of sight onto the road, the more likely it is for the driver to react to changes in the peripheral view. For the steering wheel design space, that means that the upper part of the steering wheel is a preferable place for visual information.

Regarding cognitive distraction, interaction with the steering wheel should be easy and effortless. More than that, interaction sequences should always be interruptible, meaning that the driver also should have the possibility to resume to a task and interaction should not be time critical. The driver should also have the possibility to undo interaction steps or exit an interaction procedure easily, at any time. To reduce mental effort, customization options should be provided.

13.2.3 Form Factor and Interaction Areas

So far, we have discussed the effects of the moving vehicle and the primary task of the driver on interacting with the steering wheel. Both aspects need to be taken into account for every interaction inside the vehicle (not only the steering wheel). What makes the steering wheel so special, is its form and the resulting interaction areas. They offer a huge variety of interaction potentials for a designer, but also some constraints.

The steering wheel is situated in front of the driver, making it one of the few objects inside the vehicle that combine good visibility with perfect reachability. Elements such as HUDs or instrument clusters might be even better in terms of visibility since they are even closer to the line of sight of the driver. But direct manipulation with these interfaces is not possible. Rotary knobs in the center stack, such as BMW iDrive, Audi MMI, or Mercedes Command, are relatively easy to reach but bear a risk of visual distraction. Touchscreens in the center stack have a good visibility but are not easy to be operated without changing sitting posture. The steering wheel seems to be a fitting interaction area for both input and output.

The steering wheel itself generally has a circular form, which can be easily grasped with both hands. Some vehicles have a more rectangle form such as the Austin Allegro[2] from the 1970s or modern Formula 1[3] steering wheels. The diameter of the steering wheel in standard factory model passenger cars is between 32 and 41 cm (Wolf 2009, p. 157). Trucks and busses often have a larger steering wheel; sport cars often have a smaller one. Steering wheels have at least one spoke (i.e., 1974 Citroën DS[4]) that connects the rim with a hub on the steering column.

One important aspect of a steering wheel is the fact that it is a rotating interaction element, which makes it very special in terms of interaction design. That means that the steering wheel might be in a rotated position when an interaction is started or even the whole interface might be rotating while interaction happens. From an input perspective, that makes blind interaction hard, since the driver might not exactly know where an interactive element is. From an output view, content rotation may be a problem. For an interaction designer, it is crucial to take the revolving nature of a steering wheel into account when designing interactions.

The design space of a steering wheel can be divided into three areas: the *front* of the steering wheel, the *back* of the steering wheel, and the steering wheel *rim* (see

---

[2]https://de.wikipedia.org/wiki/Austin_Allegro (Accessed Sep 1 2016).

[3]https://www.wired.com/2014/05/formula-1-steering-wheels/ (Accessed Sep 1 2016).

[4]https://en.wikipedia.org/wiki/Citro%C3%ABn_DS (Accessed Sep 1 2016).

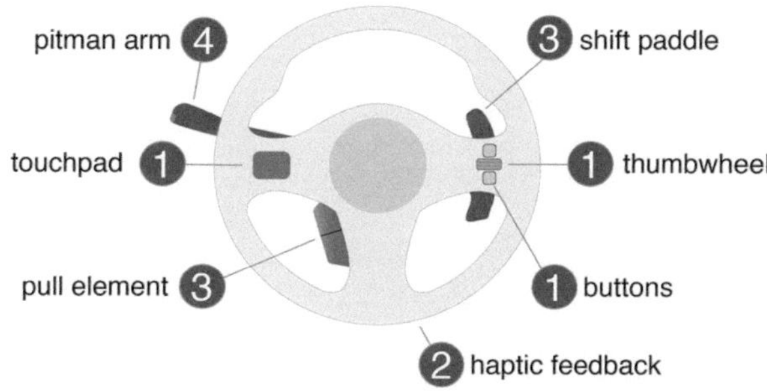

Fig. 13.2 Form factor and interaction areas. Interaction can take place on the steering wheel itself or in the area behind it. The steering wheel itself has three interaction areas: the front, the rim, and the back. At the front side of the steering wheel (*1*) different elements can be mounted such as buttons, thumbwheels, or touchpads. The steering wheel rim (*2*) for example can be used to provide haptic feedback. The back of the steering wheel (*3*) may contain shift paddles or other pull elements. Behind the steering wheel (*4*) often pitman arms are mounted. They do not rotate when the steering wheel is turned

Fig. 13.2). At the front side of the steering wheel various interaction elements can be mounted including buttons, thumbwheels, and touchpads. The rim of the steering wheel can be used to provide haptic feedback and may allow tab or swipe gestures. The back of the steering wheel often contains pull elements such as shift paddles. Apart from these three interaction areas *on the steering wheel*, designers can use the space *around* the steering wheel for interaction design. For example, behind the steering wheel often pitman arms are mounted. They are used for different purposes such as turn indicators, light switches, or windshield operators. Contrary to elements mounted on the steering wheel, these elements do not rotate with the steering wheel when turned. Other interaction approaches might use mid-air gestures around the steering wheel, for example, to change the volume. In this chapter, we will focus on the three interaction areas on the steering wheel itself.

13.3 Steering Wheel Design Space

The steering wheel spans a plane with its origin in its center (see Fig. 13.3). Angelini et al. (2014) have divided the steering wheel into 44 zones in which gestural interaction takes place on the steering wheel. These zones include segments of the steering wheel on the back of the rim, the front, on the outer part of the rim, on the inner part of the rim, and on the spokes. In a simulator study, they have shown that participants would perform gesture interaction at almost every one of the 44 zones, showing the potential of the whole steering wheel as an interaction space. Additionally, they have proposed a taxonomy of steering wheel interaction based on the type of interaction (i.e., tap, swipe, and gesture) and the part of the

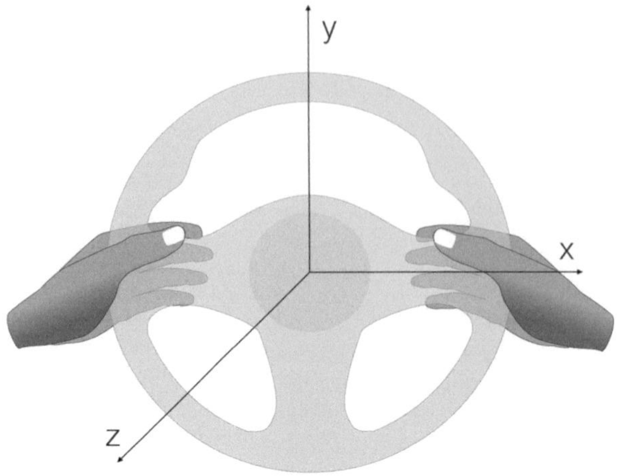

Fig. 13.3 The steering wheel lays in the *xy-plane*. Interaction may be done at the front, at the back or at the rim of the steering wheel. It might be interfering with the rotation of the steering wheel. Thus, gestures within the *xy-axis* might be more difficult than along the *z-axis* such as push for interactions on the steering wheel, pull for interactions behind the steering wheel, and tabs or flicks on the steering wheel rim

hand with which the interaction is conducted (thumb, index finger, whole hand, both hands).

We structure our design space discussion rather on the area of the steering wheel on which interaction takes place: the front of the steering wheel, the back of the steering wheel, and the steering wheel rim. We will describe interaction approaches from industry and academia to discuss the design space properties for each of these areas.

13.3.1 Front of Steering Wheel Interaction

In many modern vehicles, the front of the steering wheel is cluttered with interaction elements, making them to multifunction steering wheels. They are used as both input and output modalities. Input modalities include push buttons, switches, or thumbwheels. Push buttons or switches are common interactive elements on the steering wheel. They are often used for single functions (e.g., turn on speech recognition, set the cruise control speed, manipulate audio volume) but are also increasingly used as multifunction buttons (e.g., four-way buttons for menu selection in the instrument cluster).

For example, the BMW 7 series 2016 has 10 buttons, a rocker switch, and a thumb wheel (see Fig. 13.4). Interaction elements often have different functions depending on the context. Some are merely on/off buttons (e.g., cruise control), others have an incremental function though repeated pushes (e.g., volume control),

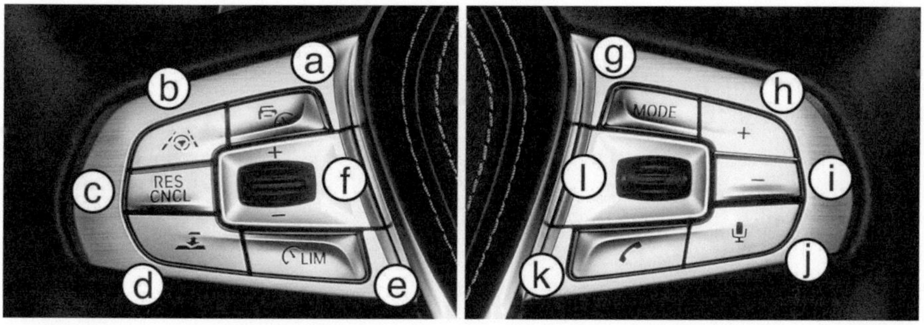

Fig. 13.4 Buttons, rocker switch, and thumbwheel on a BMW 740d, 2016. On the *left side*, five buttons are used for ADAS functions. Depending on the state of the vehicle, pressing a button has different meanings: (*a*) cruise control on/off, (*b*) steering and lane guidance assistant on/off, (*c*) pause cruise control or resume speed, (*d*) set desired distance by repeatedly pressing the button, (*e*) speed limiter on/off. The rocker switch (*f*) may be pressed to set the current speed as the target speed. Pressing the rocker switch up or down to a resistant point would increase or decrease the target speed by 1 km/h. Pressing it up or down past a resistant point would increase or decrease the target speed by 5 km/h. On the *right side*, five buttons and a thumb wheel are located. Buttons include (*g*) changing entertainment source, (*h*) increase volume, (*i*) decrease volume, (*j*) voice command activation, and (*k*) telephone pick up. The thumb wheel (*l*) may be turned to select elements from a list and pushed to confirm a selection

and yet again others are more complex. For example, the rocker switch (f) on the left side of the BMW 7 may be pressed, pushed up or down until a resistant point is met, or even pushed beyond that resistant point. Noteworthy, all elements are shaped and provide a haptic feedback, which allows for blind interaction.

For drivers, it is recommended that they grasp the steering wheel in a 9 o'clock position for the left hand and a 3 o'clock position for the right hand or even lower (e.g., left hand between 7 and 8 o'clock, and right hand between 4 and 5 o'clock as recommended by the NHTSA[5]). This prevents the airbag from pushing the driver's hands back towards the driver's body in case of a crash. In this position, elements placed on the steering wheel may be operated with the driver's thumbs. Thus, an interaction designer should make sure that elements may be reached with the thumbs while still grabbing the steering wheel. The advantage of this way of interaction is that the palm of the hand and the fingers may grasp the steering wheel while interacting and it is only the thumb, which presses buttons.

Another form of interaction is given when one of the hands has to relieve the steering wheel in order to be able to operate interactive elements on the wheel. Here, the advantage is not as big as if both hands are holding the steering wheel, but, still, the manual distraction (i.e., time that a hand is away from the steering wheel rim) is short due to the short distance it takes for the hand to switch between grabbing the steering wheel rim and interacting with a button on the steering wheel.

[5]http://www.nhtsa.gov/staticfiles/nti/teen-drivers/pdf/steeringtechniques.pdf (Accessed Sep 1 2016).

Interaction elements should be quickly reachable and, if possible, be operated without looking at the interactive element. This might be achieved by usage of larger buttons and haptic feedback on them so that the driver can feel the right button to be used.

When a vehicle is standing still (e.g., in a traffic jam), it might be even the case that a driver has the possibility to use both hands for interaction. In these cases, more advanced interaction concepts, such as (software) keyboards integrated in the steering wheel, might be useful. Bi-manual interaction on the steering wheel might also be used as a safety feature. For example, in the case of an autonomous vehicle, the handover procedure from the vehicle to the driver might be confirmed by the driver by pushing two buttons simultaneously: one with the left hand and one with the right hand. This ensures that handovers are only done on purpose and not accidentally.

Recent approaches use the entire front of a steering wheel for interaction. Pfeiffer et al. (2010) have proposed using multi-touch gestures on the steering wheel for operating infotainment systems. Döring et al. (2011) used that system to collect gestures drivers would use for 20 different infotainment commands, such as scroll in a list, volume up, or zoom into a map on the multi-touch steering wheel. Their results showed that drivers performed both single-hand gestures, as well as multi-hand gestures. They also reported that drivers preferred gestures compared to buttons on the steering wheel. In a follow-up study, they compared gesture input on the steering wheel with interaction with the center stack and found a significant reduction in gaze time and visual demand required for interaction with the multi-touch steering wheel. They also highlighted the flexibility for visualization and interaction on a multi-touch steering wheel.

The Austrian company Audio Mobil Electronik Gmbh went one step further, together with Takata AG, the global steering wheel manufacturer. They proposed the interactive Communication Steering wheel iCS. It combines a touchscreen with physical elements such as soft keys, buttons, and rotary knobs (see Fig. 13.5). In the upper middle part of the steering wheel, a touchscreen is located. It provides all relevant information while driving. For entering a destination, an on-screen keyboard may be visualized. On both sides soft keys are aligned in a way that they can be reached with the thumbs while steering. They provide physical shortcuts to interact with the information on the screen. Elements, that may be pulled and pushed, are located behind the steering wheel. They are used for light and windshield wiper control. Rotary knobs reachable from the front and the back are used for volume and menu control. Even the turn indicators are mounted on the steering wheel front. The iCS is designed to be the only interface between the driver and the vehicle and no extra visualization in an instrument cluster or center stack display is needed. Based on a similar approach, Osswald et al. (2011) have shown the feasibility of the touchscreen in terms of driver distraction, especially for low demand tasks, such as selecting an item from a list.

For text input, the iCS uses an on-screen keyboard. Other approaches, such as handwritten text input using fingers on a touchscreen, have also been suggested. For example, Kern et al. (2009) compared handwritten text on a touchscreen mounted

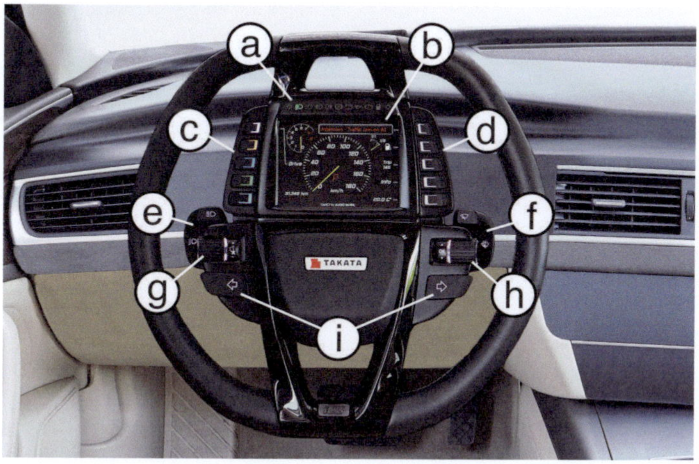

Fig. 13.5 The interactive communication steering wheel iCS, as developed by Audio Mobil and Takata. At the top indicator lamps (*a*) are located. The most prominent element is a touch screen (*b*) visualizing all relevant information. It may also provide an on-screen keyboard. On both sides soft keys (*c*, *d*) are aligned that dynamically match with functions on the touch screen providing haptic feedback. Behind the steering wheel pull/push elements for light (*e*) and windshield wiper (*f*) interaction are located. Two rotary knobs either operated form the front or from the back allow volume control (*g*) and menu selection (*h*). Turn indicators are implemented as buttons (i) on the steering wheel. No instrument cluster or center stack display is needed

on the steering wheel with similar text input on a touchscreen in the central stack. Their results show that steering wheel interaction led to less corrective actions and the remaining errors were significantly smaller (25% less). They also suggest visualizing character feedback in the dashboard rather than on the touchscreen itself.

Pfleging et al. (2012) have proposed a multimodal approach that combines simple speech commands and minimal touch input on the steering wheel. In a first step, speech is used as a selector and quantifier of the proposed interaction (e.g., "backseat windows, open"). In a second step, a gesture on a touchpad on the steering wheel (e.g., moving a finger up) allows for a fine-grained manipulation of the selected object. From an interaction point of view, this kind of multimodality is an interesting approach as it matches interaction modalities with the desired function: selecting an object from many is done with a voice command; precise adjustment is achieved by a generic gesture. Ulrich et al. (2013) have proposed a similar approach that uses a button-based mode selection for the left hand (e.g., chose radio) and contact-based gestures on a touchpad for the thumb of the right hand.

To date, none of these approaches have been introduced broadly on the market. An initial step in the direction of using gestures on the steering wheel was made by Mercedes. They included touch elements instead of 4-way buttons on the left and right spoke of their E 220d steering wheel (see Fig. 13.6). Gestures include up,

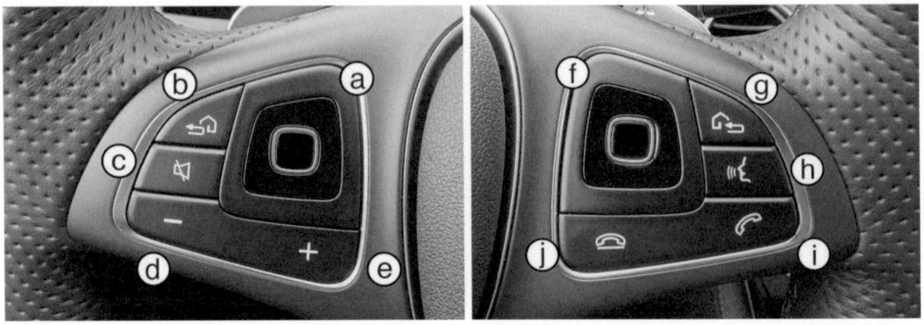

Fig. 13.6 Touch elements on the steering wheel of a Mercedes E 220d, 2016. With the *left touchpad* (*a*), the instrument cluster may be operated, with the *right touchpad* (*f*) the center stack display may be manipulated. Complex gestures such as entering a letter are not possible. Gestures include *up*, *down*, *left*, and *right*. It also includes two home/back buttons for the instrument cluster (*b*) and the center stack display (*g*). The *left* side is equipped with buttons for muting (*c*), decrease (*d*) and increase volume (*e*). The *right side* includes buttons to activate speech interaction (*h*), accept calls (*i*), and reject calls (*j*)

down, left, and right movements of the thumbs. They are mainly used to interact with the instrument cluster, the HUD, and the center stack display. Interestingly, on both sides a home/exit button is located: the left one for the instrument cluster, the right one of the center stack display.

Another approach is using force sensing as the switching mechanism without the need for physical displacement of a button as suggested by Diwischek and Lisseman (2015). They have combined mechanical switches and capacitive touchscreens to enable a broad range of different switching surfaces, materials, and geometries. A force detection system can not only predict touch interaction, but also quantify the touch location and its force and, thus, deliver tactile response while the operating finger is still mechanically coupled with the touch surface. In a user study, they compared two waveforms and four different frequencies as feedback. Results show that drivers preferred a pure single wavelength of 230 Hz over the other alternatives.

Regarding the revolving nature of a steering wheel, input might be difficult when it conflicts with the rotation of the steering wheel. For example, if there is a rotary knob on the steering wheel or a gesture takes place in the xy-plane of the steering wheel, it might conflict with the rotation of the steering wheel itself (see Fig. 13.7). Thus, interactions along the z-axis are preferable. Interface elements range from different kinds of buttons and rocker switches to thumb wheels. Often, these elements are positioned in a way that they can be operated with thumbs, even when the steering wheel is rotated (since hands move along with the steering wheel). This is only convenient a maximum of 90° rotation of the steering wheel in both directions. If the steering wheel is rotated 180°, interaction becomes complicated since, for example, right and left is switched.

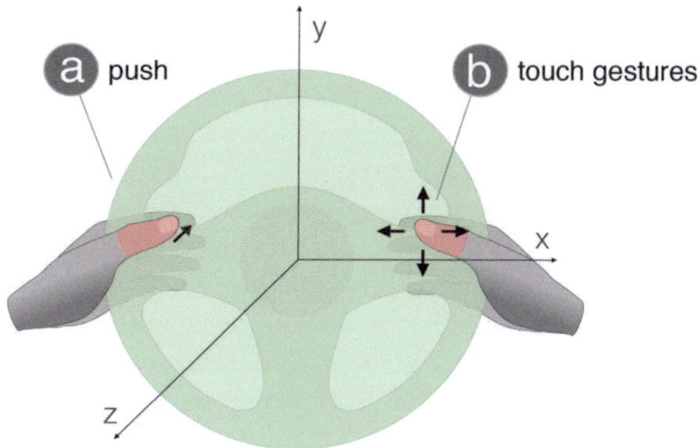

Fig. 13.7 When the hands are grabbing the steering wheel, interaction is done by the two thumbs. The most common approach is to use push (*a*) of buttons or other interactive elements along the y-axis. Other approaches use touch gestures (*b*) within the xy-axis which might be difficult to perform when the steering wheel is turning

This is not the case when the steering wheel has a fixed hub, such as the Citroën C4.[6] In this case, only the outer steering wheel rim is rotating, whereas the center of the steering wheel is fixed, which might have advantages when the airbag deflates. This makes it easier to locate input elements on the steering wheel even when it is rotated. On the other hand, thumb interaction is more difficult when the steering wheel in a rotated position. Then, thumbs are moved away from interaction elements and hard to reach even if the rotation angle is rather small.

A fixed hub has its advantages when we look at output modalities. If the steering wheel is used as a visual output modality showing text, numbers, or symbols, a severe problem for interaction design might be content rotation. When the steering wheel is turned, text might not be readable anymore or symbols might be misinterpreted. When positioning text on the steering wheel, the text might be either in the center of the steering wheel or relocated. If the text is displayed in the center of the steering wheel, content rotation can be compensated by turning the displayed information so that it is always aligned correctly. In case of a fixed hub, this alignment is not necessary.

However, there are at least two reasons not to place the display in the middle of the steering wheel. First, this is often the place where an airbag is positioned; second, the driver would have to look down rather far from the line of sight onto the road. Thus, a display should be positioned in the upper part of the steering wheel, where typically the instrument cluster is visible behind the steering wheel. When the information of the instrument cluster is visualized on the steering wheel, the instrument cluster itself may be negligible. In terms of display behavior, Wilfinger et al. (2013) have presented three approaches. The first approach, does not rotate the content of the display, thus when the steering wheel is turned content turns also. The second approach, labeled "Ferris Wheel[7]", aligned the content horizontally when the steering wheel was turned. The third condition, named "Sticky Edge",

[6]https://en.wikipedia.org/wiki/Citro%C3%ABn_C4 (Accessed Sep 1 2016).

[7]A similar approach has already been patented by Lahiff, as early as 1997 (Lahiff 1997).

followed the metaphor of a window through the steering wheel. Content virtually stayed at the same position and aligned horizontally so that the text stays in the same position from the driver's perspective. In a simulator study, glance and response time for all three conditions and a baseline condition (i.e., traditional) were compared. They found no significant differences in terms of visual cost between the conditions but more visual distraction for the baseline condition. It seems as if content rotation is manageable for interaction designers. This is also supported by basic research, which shows that the rotation of words below 60° has no negative effect on the legibility of words (Koriat and Norman 1985).

Another aspect of content visualization on the steering wheel is the fact that the content is closer to the eyes than it is when displayed in an instrument cluster, let alone a HUD. Although the distance between the eye and the steering wheel is enough to be readable the constant switch between the far away street and a near steering wheel can have a negative impact on eyestrain.

13.3.2 *Back of Steering Wheel Interaction*

The back of the steering wheel shares some properties with the steering wheel rim and the front of the steering wheel. Mainly, this is its ring-formed shape and rotary nature. Interaction is done while the driver is grabbing the steering wheel rim. In contrast to the front of the steering wheel, where thumb interaction is preferred at the back of the steering wheel, interaction is done with the fingers located behind the steering wheel. This supports bi-manual and multi-finger interaction (see Fig. 13.8).

Similar to bi-manual interaction on the front of the steering wheel, this form of interaction may be used at the back of the steering wheel when unintended interaction should be avoided (e.g., in autonomous vehicle handover situations). More interestingly, in terms of interaction design, is the possibility to use multi-finger

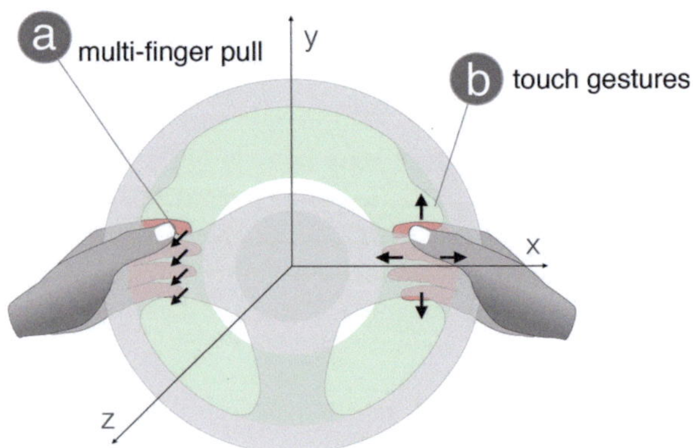

Fig. 13.8 Interaction with the back of the steering wheel is usually done with fingers located behind the steering wheel. Multi-finger pull gestures (*a*) seem to be promising, especially combined with chorded keys. Touch gestures (*b*) may be performed in the *xy-plane* but only in the area reachable by fingers

interaction in form of chorded keys. This is a concept that allows entering commands by simultaneously pressing a set of input elements (Engelbart and English 1968). Examples of such an input are chords played on a piano or the Braille alphabet for blind people (Braille 1829).

Such types of interaction on the back of a steering wheel were already suggested by Osswald et al. (2012). They prototyped a braille keyer with three buttons on the left and right rear side of a steering wheel and showed the feasibility of this concept in a user study. They describe three advantages of their approach: entering a text while driving with eyes on the road and hands on the wheel; the possibility to enter characters with a single interaction step (i.e., pressing a chord of keys); no need of hand–eye coordination as in touchscreen interaction. A major disadvantage of this approach is a cumbersome learning phase and a high cognitive load.

Murer et al. (2012) have proposed a slider as an interactive element on the back of a steering wheel. This approach does not allow chorded keys but gestures. Gestures are performed within the xy-axis of the steering wheel. One of their main findings was that the blind interaction that takes place behind the steering wheel needs an appropriate tactile feedback. They also compared buttons with touchscreen sliders. Buttons had the advantage of providing a clearer tactile feedback and, above that, allowed the fingers to rest on the button, whereas touching the slider element with a finger already may lead to some unintended interaction.

Meschtscherjakov et al. (2014) have discussed the back of the steering wheel design space in more detail. In terms of gestures in the xy-plane, they have argued that gestures that follow the rotation direction of the steering wheel are more likely to be confused with steering. Thus, radial gestures such as swipes from or to the origin of the xy-plane are preferable. They also suggested an interesting approach to enter characters by freehand gestures, e.g., with the index finger.

Another application for back of steering wheel interaction was proposed by Kuhn et al. (2013). They mounted three force-sensing resistors on the rear side of the steering wheel to allow driver authentication by means of tap sequences. Their results show the feasibility of the approach. Drivers could successfully enter reasonably long tap sequences (up to 8 taps). They suggest using at least five taps to protect authentication from shoulder-surfing attacks (i.e., direct observation techniques, such as looking over someone's shoulder to get passwords or PINs). They also found that binary input (tap vs. no tap) was superior to more sophisticated approaches, such as distinguishing between high pressure tabs and low pressure tabs. Regarding interaction design, this suggests that using the amount of pressure as input on the back of the steering wheel have to be used with caution.

13.3.3 Steering Wheel Rim Interaction

Interacting with the rim, or elements on the rim, of a steering wheel is different than interacting with elements on the front or the backside of the steering wheel in many

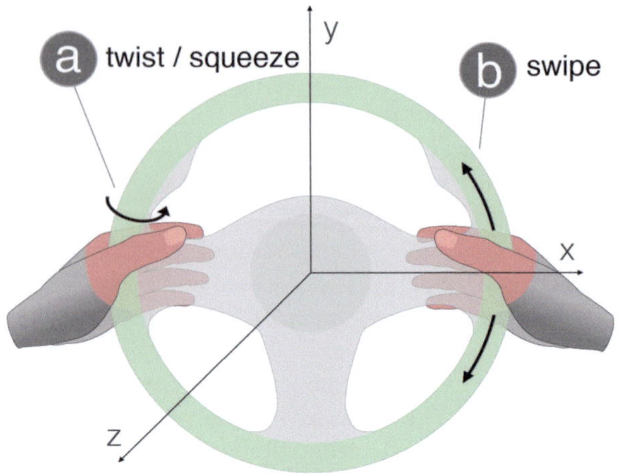

Fig. 13.9 Interaction with the steering wheel rim is manifold. Apart from tab gestures on top, the back, or the front it might be twisted or squeezed (*a*). Within the *xy-plane* and along the rim, swipe gestures may be performed (*b*)

aspects. The rim is always grasped by at least one hand. Thus, it has to be clear for both the interactive system and for the driver when an interaction with the system is intended and when it is a "natural" interaction with the steering wheel, i.e., steering. Interaction may include different forms of gestures such as click, tap, flick, stroke, and twist (Koyama et al. 2014) or even squeezing the rim (Angelini et al. 2013). Figure 13.9 illustrates the steering wheel rim interaction space.

Steering wheel rim interaction can either be located in certain areas of the steering wheel rim and/or the interaction could be specific. For example, the upper or lower part of the steering wheel could be equipped with sensors that allow interaction with these parts. Interaction can then be achieved by simply grabbing these areas of the steering wheel rim. Again, interaction designers do have to be cautious when implementing such features, as to not induce unwanted interactions eliciting annoyance of the driver.

Other approaches may require a certain interaction in these areas. For example, swipe gestures along the steering wheel rim might be used to increase or decrease volume. Conceptually, we can distinguish between interactions that are more likely to interfere with steering and those that are sufficient differing. Interaction along the steering wheel rim itself (i.e., within the xy-axis) might be misinterpreted more often than, for example, a twisting of the rim itself as if it was a motorcycle handle.

Angelini et al. (2013) have presented WheelSense, a system that allows four types of gestures on the right upper part of a steering wheel: tapping, dragging up, dragging down, and squeezing using Flexiforce sensors. Koyama et al. (2014) utilizing 120 infrared (IR) sensors embedded in the steering wheel and machine learning algorithms to detect flick, click, tap, stroke, and twist gestures.

Some approaches extend the steering wheel rim itself with interactive elements such as buttons or switches. They are located either at the front of the steering wheel rim or behind it. Then again, front of the steering wheel and back of the steering wheel interaction design guidelines can be applied. For example, Sandnes et al. (2008) suggest mounting three buttons on the inner side of the steering wheel rim at the 2 o'clock position. These three buttons may be pushed while driving with a combination of three fingers implementing a chorded key input. They also suggest using sequences of such chords based on visual mnemonics. In their case, three chords result in one input. For example, the letter L would be entered by pushing all three buttons in the first step, then pushing only the lowest button, and again pushing the lowest button in the final step of the sequence.

González et al. (2007) have mounted two small rectangular touchpads (2.8 by 3.2 cm) at the 2 o'clock and 10 o'clock positions of the steering wheel allowing thumb gesture interactions. They compared different forms of text entry, such as clutching through the alphabet (i.e., repeated strokes over the touchpads), dialing (i.e., circular movements to select characters), and gesture based character input. For the latter, they used the EdgeWrite (Wobbrock et al. 2004), a gestural text entry method where letters are entered by a single stroke on the touchpad. EdgeWrite was 20–50% faster than selection-based text entry methods. One drawback of this system is that it requires memorizing proprietary gestures for each character. Regarding interaction design, they struggled with the alignment of the touchpads. In their prototype, touchpads were tilted 45°; thus, scrolling through a horizontal list did not match with a horizontal movement of the thumb, but a wipe in a 45° angle when the steering wheel was in the home position. It also leads to the effect that gestures need to be executed twisted as well. This has to be taken in consideration when designing for input on the steering wheel rim.

From an output perspective, the steering wheel rim is utilized in many ways. Beruscha et al. (2011) distinguish between two types of tactile feedback for lane departure warning: synthetic steering wheel torque or vibration. For example, applying a steering torque jerk pointing towards the lane center informs the driver in which direction to steer to get back into the lane. Steering wheel oscillations (induced by subsequent steering wheel torque of equal extent in alternating direction) can inform the driver about the necessity of a steering reaction. Steering wheel vibration signals are either located at the left or right half of a steering wheel to warn the driver of an imminent lane departure.

The utilization of vibro-tactile feedback for drivers on a steering wheel has been proposed by Van Erp and Van Veen (2001). They suggest using vibro-tactile feedback for different forms of information, such as spatial information, warning signals, or communication. Some manufacturers use vibro-tactile feedback of the steering wheel to inform the driver about a lane change or as part of a lane departure warning system. Such vibro-tactile feedback has the advantage to be easily recognizable by the driver. However, it might also distract the driver, since, in some cases, a bumpy road might also cause a vibration of the steering wheel.

Berber-Solano and Giacomin (2005) have shown that the manipulation of steering wheel vibrational feedback can also be used to improve driver detection of the road surface. In terms of interaction design, this might bring the opportunity to enhance situational awareness for the driver. Modern cars are often designed to increase comfort and user experience by shielding drivers from environmental conditions, such as uneven roads or wind noise. This gives drivers the feeling of apparent safety but also bemuses the driver's senses and connection to the environment. Dynamic tactile feedback on the rim could be utilized to enhance the driver's feeling of driving. With the introduction of steer-by-wire, completely new forms of feedback could be implemented relieved from mechanical constraints.

Kern et al. (2009) used six vibration motors evenly distributed on the steering wheel rim under a layer of rubber to provide drivers with vibro-tactile representations of navigational information. They compared turn-by-turn navigation instructions provided via audio, visual, and vibro-tactile channels and a combination of those with each other. Best driving performances were reached by a combination of visual output and embedded vibration on the steering wheel rim.

Enriquez et al. (2001) have used inflatable pads in the rim to produce pulsations as feedback on the driver's hands. Pneumatic pumps are used to let the rim pulsate at varying frequencies. Their results show that the tactile feedback reduced reaction times and that the pulsation frequency could be used to convey information to the driver.

Shakeri et al. (2016) recently presented a haptic steering wheel with six solenoids embedded into the steering wheel. Three were mounted on each side of the steering wheel rim, so that they created three bumps in the median palmar region of each hand. Solenoids pins were covered with a latex shield and could be individually moved in or out, which results in, overall, 64 tactile feedback patterns. A driving simulator study showed that perception accuracy drops when more than three solenoids were reeled out and that haptic patterns mirrored symmetrically on both hands were perceived more accurately.

From a sensing perspective, the steering wheel rim can also be regarded as a driver behavior evaluation tool. For example, steering wheel reversal rates (i.e., frequency how often a steering wheel is moved in the opposite direction from the current direction of movement within a certain time frame, Savino 2009) can be used to assess driver performance. Steering wheel movement has been proven to monitor driver vigilance and drowsiness (Bergasa et al. 2006).

13.3.4 Design Space Properties

Based on the discussion in the last section, the following (Table 13.1) provides an overview of design space properties along with interaction design recommendations. They may serve as best practices for interaction designers, when conceiving novel interaction approaches on the steering wheel.

Table 13.1 Design space properties and related interaction design recommendations

Design space property	Interaction design recommendations
Forces and vibrations	• Make interactive elements large • Design vibro-tactile feedback distinguishable from natural vibrations • Design for error recovery
Changing light	• Adapt to changing light conditions • Design for backlighting • Design for glare of the sun • Use night visualization modes
Noise	• Adapt feedback to changing noise and sound conditions • Take noise into account when using speech input
Manual distraction	• Place interaction objects onto or near the steering wheel • Do not interfere with the steering task itself • Avoid-bi-manual interaction
Visual distraction	• Place visual information on the top of the steering wheel to reduce eye movement and allow peripheral detection • Avoid diagonal eye movements
Cognitive distraction	• Design interaction procedures to be interruptible • Provide a clear exit for each interaction step • Provide shortcuts and customization options
Front of steering wheel interaction	• Allow interactive elements on the steering wheel to be operated with thumbs • Allow one hand interactions to be made blindly by means of haptic elements • Interaction in the xy-plane have to be used with caution and must not interfere with the steering wheel rotation • Two-hand interaction are applicable when the vehicle is standing still or for safety reasons (e.g., takeover procedures in autonomous vehicles) • For text entry combine handwritten characters on a touchscreen with a dashboard visualization of the text • Use multimodal approaches function specific (e.g., speech as a selector, gesture for fine-grained manipulation) • Visualizations should be positioned as high as possible • Design for content rotation
Back of steering wheel interaction	• Multi-finger interaction and chorded keys are possible • Gesture interaction should be radial • Blind interaction needs appropriate feedback • Allow fingers to rest on the interactive elements without triggering an input • Pressure sensitive input should be used with caution
Steering wheel rim interaction	• Clear distinction between steering and interacting with the interactive steering wheel rim • Gestures may include flick, click, tap, stroke, twist or squeeze • Use different forms of tactile feedback (e.g., torque, vibration, pulsation, shape changes, etc.) • Haptic warnings should be clearly perceivable without being intrusive or disruptive • Consider rotation of the rim for input and output alignment • Use rim as sensor for driver physiological state • Utilize rim to convey unobtrusive information

13.4 Looking into the Future

If we look into the future of vehicle steering wheels, we can identify at least three developments that will open the design space for steering wheel interaction and pave the ground for completely new interaction possibilities with the steering wheel: steer-by-wire, the intelligent steering wheel, and steering wheels in autonomous vehicles.

13.4.1 Steer-by-Wire

To date, most steering wheels are connected mechanically to a steering column. In aviation, these mechanical connections have been replaced by electronic ones, also known as steer-by-wire. Steer-by-wire provides some advantages for interaction design. First, interaction designers would not be limited to the form factor of a wheel, but can imagine other shapes. One possibility is to exchange the steering wheel with a joystick. With a joystick it would not only be possible to steer the vehicle but also to accelerate and break. The two-dimensional nature of driving could be implemented by means of a two-dimensional interaction modality, which might have advantages from an ergonomic perspective (Bubb et al. 2015, p. 334).

Apart from the form factor, new functionalities may be implemented on the new steering device. Novel forms of driver assistant systems may be developed and new ways of interaction designs may be possible. Anand et al. (2013) have shown that fast drivers adapt their driving behavior to different levels of steering wheel force feedback regardless of cognitive efforts in secondary tasks. They suggest the design for a personalized force feedback of the steering wheel based on the driver's preferences.

For interaction designers, steer-by-wire will open completely new possibilities to design steering wheels and ways how to interact with them. Steering wheel shapes and their esthetic appeal may be freely chosen. The place where the steering wheel is positioned inside a vehicle may be altered. Also, new forms of interactions with the steering wheel will be possible.

13.4.2 Intelligent Steering Wheel

Another trend will be the intelligent steering wheel. With the rise of evermore inexpensive and exact sensors, steering wheels will be equipped with more and more sensors and actuators. On one hand, these sensors will make it possible to evaluate driver behavior in real time. On the other hand, physiological data from the driver will be used to diagnose driver state. In both cases, the steering wheel will act accordingly by either interventions in maneuvering the vehicle or by providing appropriate feedback for the driver.

For example, Assuncao et al. (2015) have shown that driving event identification is possible by correlating steering wheel movements with lane deviations in online applications. They have shown that their technique identifies dangerous lane departure events with a 91.42% precision and a low false positive rate. Ibragimova et al. (2015) have proposed a smart steering wheel cover. This design idea contains telephone and music controls on the inner side of the rim, as well as vibro-tactile and visual feedback (green LEDs embedded in the whole rim). In their approach, driver behavior data (e.g., acceleration and breaking) is collected by a smartphone and sent to the steering wheel that provides visual feedback, for example, about fuel economy. Baronti et al. (2009) have proposed a concept that integrates 16 low-cost sensors in the rim for measuring grip force and hand position on a steering wheel in order to detect driver's fatigue. Creating a chain of different kinds of sensor units can serve as platform for multiple driver behavior and driver state measurements.

Intelligent steering wheels will offer interaction designers new possibilities of personalization and context awareness. Interaction designers will have access to physiological data from the driver, as well as information from the environment and other vehicles. This will allow interaction designers to aggregate and analyze data from different sources in real time and provide them to the driver. The way how drivers will interact with this data will be a challenge for interaction designers.

13.4.3 Autonomous Vehicles

Finally, the rise of autonomous vehicles will change how we interact with the steering wheel in the future tremendously. It will make interaction with the steering wheel simpler, but also more complex. Depending on the autonomy level of the vehicle, the driver has different tasks to complete. With the introduction of new advanced driving assistant systems (ADAS), new functions have to be integrated in the steering wheel. Most OEMs include new buttons and switches to allow the driver to operate such systems on the steering wheel. If the car is capable of performing the driving task autonomously, there are two possibilities for the driver; either the driver has to constantly monitor the car and its behavior and be capable to intervene at any time; or the driver is free to carry out other activities and has to be ready to take over control after advanced warning. In both cases, the driver has to be able to interact with the steering wheel when the vehicle is driven manually. Additionally, interfaces will have to be designed to allow such takeover procedures and inform the driver about vehicle status and intentions. This will make interaction design more complex.

If we look further into the future and assume that fully autonomous vehicles are on the road, interaction with the steering wheel might become very simple, since the vehicle is driving on its own. Then, the steering wheel might be retractable and only available in case of an emergency. Some approaches, like the Google car,[8] were even presented without any steering wheel at all.

[8]https://www.google.com/selfdrivingcar/ (Accessed Sep 1 2016).

13.5 Conclusion

In this chapter, we have investigated the design space of steering wheels from an interaction design perspective. We have concentrated on interactions with the interactive elements on the steering wheel itself, distinguishing between the front and back of the steering wheel, as well as, interacting with the steering wheel rim. Of course, interacting with elements situated at these areas is not mutual exclusive but may overlap. Henceforth, some properties of the design space are applicable regardless of the interaction space.

We have shown that both the driving context, as well as the fact that the driver always has a primary task, have a major influence on the interaction design on the steering wheel. For most steering wheels, the form factor and the resulting interaction areas are settled. Its circular form spans an interaction plane, which is rotating when the driver is steering. This makes interacting with elements mounted on the steering wheel complicated and the rotation of interaction elements and visual content needs to be taken into account. Since the driver should always have their eyes (and also the mind) on the road, blind and intuitive interaction supports a good interaction design.

Furthermore, at least one of the driver's hands should have a solid grip on the steering wheel. With two hands on the steering wheel, the back of the steering wheel offers potentials for multi-finger and chorded interaction. The front of the steering wheel is most often operated with the thumbs. Newer approaches suggest touchscreens on the front of the steering that enable rich visualizations and different kinds of touch gestures. The steering wheel rim offers a multitude of interaction potentials. Different forms of interaction gestures have been proposed, such as tabbing, swiping, flickering, twisting, or even squeezing. The rim also has been used as a feedback modality, for example, through vibro-tactile or shape changing approaches. Moreover, since there is always at least one hand on the steering wheel, it may be used as a sensor device. On one hand, explicit steering behavior may be monitored. On the other hand, implicit measures, such as skin resistance, may be used to derive information about the driver state and behavior, which can also be used to allow a better interaction design.

For the future interaction design on the steering wheel, we mainly see three areas that have a big influence. First, steer-by-wire will allow completely new ways of interaction design on the steering wheel. When the steering wheel is mechanically decoupled from the steering itself, it allows completely new forms of steering wheels. They might be replaced by joysticks or other forms of interactive objects. Second, we are facing the development of evermore-intelligent steering wheels. Intelligent steering wheels will be capable of sensing driving behavior and driver status. Together with context sensors and all forms of x2car communication they will be able to inform the driver by providing complex information in new ways. Finally, the raise of autonomous vehicles will make interaction with future steering wheels both, more complex and also simpler. Interaction will be more complex since driver will want to use their time to interact with the steering wheel not only for maneuvering the vehicle, but also for operating infotainment systems and interacting with their mobile workplace. It will become simpler, since the car will take over many of the primary tasks a driver has to conduct today. For interaction designers, this will lead to new freedoms but also new forms of responsibility.

Acknowledgements I thank Dr. Arno Laminger for providing the photographs and Thomas Grah, MA for the graphic arts shown in this chapter. The financial support by the Austrian Federal Ministry of Science, Research and Economy and the National Foundation for Research, Technology and Development and AUDIO MOBIL Elektronik GmbH is gratefully acknowledged (Christian Doppler Laboratory for "Contextual Interfaces").

References

Anand, S., Terken, J., & Hogema, J. (2013). The effect of cognitive load on adaptation to differences in steering wheel force feedback level. In *Proceedings of the 5th International Conference on Automotive User Interfaces and Interactive Vehicular Applications* (pp. 158–164). ACM.

Angelini, L., et al. (2013). Wheelsense: Enabling tangible gestures on the steering wheel for in-car natural interaction. In *International Conference on Human-Computer Interaction* (pp. 531–540). Berlin: Springer.

Angelini, L., et al. (2014). Gesturing on the steering wheel: A user-elicited taxonomy. In *Proceedings of the 6th International Conference on Automotive User Interfaces and Interactive Vehicular Applications, New York, NY, USA, 2014* (pp. 31:1–31:8). ACM.

Assuncao, A. N., de Paula, F. O., & Oliveira, R. A. (2015). Methodology to events identification in vehicles using statistical process control on steering wheel data. In *Proceedings of the 13th ACM International Symposium on Mobility Management and Wireless Access* (pp. 1–4). ACM.

Baronti, F., Lenzi, F., Roncella, R., & Saletti, R. (2009). Distributed sensor for steering wheel grip force measurement in driver fatigue detection. In *Proceedings of the Conference on Design, Automation and Test in Europe, Leuven, Belgium* (pp. 894–897). European Design and Automation Association.

Berber-Solano, T. P., & Giacomin, J. A. (2005). Perception enhancement system for automotive steering. In *Proceedings of the 2005 Annual Conference on European Association of Cognitive Ergonomics* (pp. 147–153). University of Athens.

Bergasa, L. M., Nuevo, J., Sotelo, M. A., Barea, R., & Lopez, M. E. (2006). Real-time system for monitoring driver vigilance. *IEEE Transactions on Intelligent Transportation Systems, 7*(1), 63–77.

Beruscha, F., Augsburg, K., & Manstetten, D. (2011). Haptic warning signals at the steering wheel: A literature survey regarding lane departure warning systems. In Bubb, H., Bengler, K., Grünen, R. E., & Vollrath, M. (Eds.) Haptics-e (Vol. 4, No. 5, pp. 1–6). Automobilergonomie. Springer.

Braille, L. (1829). *Procedure for writing words, music and plain song using dots for the use of the blind and made available to them*. Paris: Royal Institution of Blind Youth.

Bubb, H., Bengler, K., Breuninger, J., Gold, C., & Helmbrecht, M. (2015). Systemergonomie des Fahrzeugs. In *Automobilergonomie* (pp. 259–344). Springer.

Diwischek, L., & Lisseman, J. (2015). Tactile feedback for virtual automotive steering wheel switches. In *Proceedings of the 7th International Conference on Automotive User Interfaces and Interactive Vehicular Applications* (pp. 31–38). ACM.

Döring, T., et al. (2011). A gestural interaction on the steering wheel: reducing the visual demand. In *Proceedings of the Annual Conference on Human Factors in Computing Systems* (pp. 483–492). New York: ACM.

Engelbart, D., & English, B. (1968). A research center for augmenting human intellect. In *Proceedings of the 1968 Fall Joint Computer Conference* (Vol. 33, pp. 395–41). San Francisco, CA.

Enriquez, M., Afonin, O., Yager, B., & Maclean, K. (2001). A pneumatic tactile alerting system for the driving environment. In *Proceedings of the 2001 Workshop on Perceptive User Interfaces* (pp. 1–7). New York: ACM Press.

González, I. E., Wobbrock, J. O., Chau, D. H., Faulring, A., & Myers, B. A. (2007). Eyes on the road, hands on the wheel: Thumb-based interaction techniques for input on steering wheels. In *Proceedings of Graphics Interface 2007* (pp. 95–102). New York: ACM Press.

Greathouse, J. (2008). Reinvent the wheel—A nonstandard look at standards. Retrieved Sep 1, 2016 from http://johngreathouse.com/wheel/.

Ibragimova, E., Mueller, N., Vermeeren, A., & Vink, P. (2015). The smart steering wheel cover: Motivating safe and efficient driving. In *Proceedings of the 33rd Annual ACM Conference Extended Abstracts on Human Factors in Computing Systems* (pp. 169–169). ACM.

Kern, D., Schmidt, A., Arnsmann, J., Appelmann, T., Pararasasegaran, N., & Piepiera, B. (2009). Writing to your car: Handwritten text input while driving. In *Proceedings of the 27th International Conference Extended Abstracts on Human Factors in Computing Systems* (pp. 4705–4710). New York: ACM.

Kern, D., Marshall, P., Hornecker, E., Rogers, Y., & Schmidt, A. (2009). Enhancing navigation information with tactile output embedded into the steering wheel. In *International Conference on Pervasive Computing* (pp. 42–58). Berlin: Springer.

Koriat, A., & Norman, J. (1985). Reading rotated words. *Journal of Experimental Psychology: Human Perception and Performance, 11*(4), 490.

Koyama, S., et al. (2014). Multi-touch steering wheel for in-car tertiary applications using infrared sensors. In *Proceedings of the 5th Augmented Human International Conference* (p. 5). ACM.

Kun, A. L., Royer, T., & Leone, A. (2013). Using tap sequences to authenticate drivers. In *Proceedings of the 5th International Conference on Automotive User Interfaces and Interactive Vehicular Applications* (pp. 228–231). New York: ACM.

Lahiff, J. (1997). Vehicle information display on steering wheel surface. Nov. 25 1997. US Patent 5, 691–695.

Meschtscherjakov, A., Wilfinger, D., Murer, M., Osswald, S., & Tscheligi, M. (2014). Hands-on-the-wheel: Exploring the design space on the back side of a steering wheel. In *European Conference on Ambient Intelligence* (pp. 299–314). Springer International Publishing.

Murer, M., Wilfinger, D., Meschtscherjakov, A., Osswald, S., & Tscheligi, M. (2012). Exploring the back of the steering wheel: Text input with hands on the wheel and eyes on the road. In *Proceedings of the 4th International Conference on Automotive User Interfaces and Interactive Vehicular Applications*. ACM.

Osswald, S., et al. (2012). Back of the steering wheel interaction: The car braille keyer. In *International Joint Conference on Ambient Intelligence*.

Osswald, S., Meschtscherjakov, A., Wilfinger, D., & Tscheligi, M. (2011). Steering wheel-based interaction: Potential reductions in driver distraction. In *International Joint Conference on Ambient Intelligence*.

Pfeiffer, M., Kern, D., Schöning, J., Döring, T., Krüger, A., & Schmidt, A. (2010). A multi-touch enabled steering wheel: Exploring the design space. In *Proceedings of the 28th of the International Conference Extended Abstracts on Human Factors in Computing Systems* (pp. 3355–3360). New York: ACM.

Pfleging, B., Schneegass, S., & Schmidt, A. (2012). Multimodal interaction in the car: Combining speech and gestures on the steering wheel. In *Proceedings of the 4th International Conference on Automotive User Interfaces and Interactive Vehicular Applications* (pp. 155–162). New York: ACM.

Sandnes, F. E., Huang, Y. P., & Huang, Y. M. (2008). An eyes-free in-car user interface interaction style based on visual and textual mnemonics, chording and speech. In *Proceedings of the 2008*

International Conference on Multimedia and Ubiquitous Engineering (pp. 342–347). Washington, DC: IEEE Computer Society.

Savino, M. R. (2009). Standardized names and definitions for driving performance measures. *Dissertation*, Tufts University.

Shakeri, G., Brewster, S. A., Williamson, J., & Ng, A. (2016). Evaluating haptic feedback on a steering wheel in a simulated driving scenario. In *Proceedings of the 2016 Conference Extended Abstracts on Human Factors in Computing Systems* (pp. 1744–1751). ACM.

Ulrich, T. A., Spielman, Z., Holmberg, J., Hoover, C., Sanders, N., Gohil, K., & Werner, S. (2013). Playing charades with your car–The potential of free-form and contact-based gestural interfaces for human vehicle interaction. In *Proceedings of the Human Factors and Ergonomics Society Annual Meeting* (Vol. 57, No. 1, pp. 1643–1647). SAGE Publications.

Van Erp, J., & Van Veen, H. (2001). Vibro-tactile information presentation in automobiles. In *Proceedings of Eurohaptics* (Vol. 2001, pp. 99–104). DTIC Document.

Wilfinger, D., Murer, M., Osswald, S., Meschtscherjakov, A., & Tscheligi, M. (2013). The wheels are turning: Content rotation on steering wheel displays. In *Proceedings of the SIGCHI Conference on Human Factors in Computing Systems*. (pp. 1809–1812). New York: ACM.

Wobbrock, J. O., Myers, B. A., Aung, H. H., & LoPresti, E. F. (2004). Text entry from power wheelchairs: EdgeWrite for joysticks and touchpads. In *ACM SIGACCESS Accessibility and Computing* (Vols. 77–78 pp. 110–117). ACM.

Wolf, H. (2009). Ergonomische Untersuchung des Lenkgefühls an Personenkraftwagen. *Dissertation*, Technische Universität München.

Part IV
Tools, Methods and Processes

Chapter 14
The Insight–Prototype–Product Cycle Best Practices and Processes to Iteratively Advance In-Vehicle Interactive Experiences Development

Ignacio Alvarez, Adam Jordan, Juliana Knopf, Darrell LeBlanc, Laura Rumbel and Alexandra Zafiroglu

Abstract In-vehicle experiences are made up mainly of mundane small moments, repeated practices, and taken-for-granted decisions that make up daily experiences in and around private passenger vehicles. Understanding what those experiences are for drivers around the world presents an opportunity for designing novel interactive experiences, technologies, and user interfaces for vehicles. In this chapter, we present a set of tools, methodologies, and practices that will help reader create a holistic design space for future mobility. Transitioning between ethnography, insights, prototyping, experience design, and requirements decomposition is a challenging task even for experienced UX professionals. This chapter provides guidance in this matter with practical examples.

I. Alvarez (✉) · A. Jordan · J. Knopf · D. LeBlanc · L. Rumbel · A. Zafiroglu
Intel Corporation, Santa Clara, CA, USA
e-mail: ignacio.j.alvarez@intel.com

A. Jordan
e-mail: adam.jordan@intel.com

J. Knopf
e-mail: juliana.l.knopf@intel.com

D. LeBlanc
e-mail: darrell.leblanc@intel.com

L. Rumbel
e-mail: laura.m.rumbel@intel.com

A. Zafiroglu
e-mail: alexandra.c.zafiroglu@intel.com

G. Meixner and C. Müller (eds.), *Automotive User Interfaces*,
Human–Computer Interaction Series, DOI 10.1007/978-3-319-49448-7_14

14.1 Creating a Holistic Design Space for Future Mobility

For user experience designers, the interior of a vehicle is a unique canvas for interaction creativity. The vehicle cockpit is an immersive environment in which interactions become second nature for users. But it is also a very complex design space where regulatory requirements coalesce with human factor ergonomics, functional safety engineering, aesthetics, and savvy user technology adoption.

Balancing these factors in a successful manner to stay on the vanguard of innovation requires sensitivity, skills, tools, and good practices. In this chapter, we present some of the best known methods that Intel has developed over several years of research and development of novel in-vehicle experiences. The first section introduces the reader to ethnographic processes Intel has used to orient automotive user experience designers to current experiences, priorities, and challenges of private car drivers and passengers. The second section presents a tool for iterative solution design of automotive experiences. The third section guides the reader to the study planning and user assessment practices needed to objectively prove, advance, and analyze in-vehicle experience concepts. Finally, we illustrate the process through which prototype concepts are converted into product requirements and the conditions that guide technology adoption and marketability.

14.2 The Insight Capture Process

In 2010, a team of user experience researchers in Intel Labs started a set of research projects on automobility that over the course of two years included field research in eight countries and a wealth of types of research data: audio recording of interviews, video recordings of drives, detailed inventories of the contents of cars from Singapore to Munich to Sao Paolo, GPS logs of daily car journeys, and logs of smart phone use during these trips. What was remarkable about these projects was lesser individual types of data collected, than how they were combined to address our deceptively simple primary research question, namely: *What is a car?* The question grounded all of our inquiries, and allowed us to make visible, and set aside, many of the assumptions that we make about the nature and experiences of owning and using a vehicle, based on our own experiences as owners and drivers. What did we really know about how others experience a car? What type of object is a private passenger vehicle? How is it incorporated into the daily lives of middle-class households around the world? How do cars enable particular types of mobility? What characterizes the diversity and the commonality of automobility globally?

Such foundational research allowed us to begin to imagine, design, and test multiple possible automobility futures grounded in the realities of existing transportation systems rather than in our own flights of fancy. Futures are, after all, not wholly of our own making. To paraphrase Marx, we make our history out of conditions that already exist and are close at hand (Marx 2008). To identify opportunity areas for designing experiences, technologies, and automotive user interfaces, we chose to study and take inspiration from the mundane small moments, repeated practices, and taken-for-granted decisions that make up people's daily experiences with cars.

In this section, we describe two rounds of foundational ethnographic research on automobility, detailing research methods and reporting methods that encouraged reframing, discovery, engagement, and interaction with findings by our colleagues and business partners. We also detail how we framed research findings to be *useful* rather than merely *interesting*, and provided materials to partner with designers, prototyping teams, and product engineering teams to translate opportunity areas identified in research into experience definition, rapid prototyping, and product generation. Our focus here is on research and report methods, rather than on specific findings presented elsewhere (Bell 2011; Zafiroglu et al. 2011, 2012).

14.2.1 Exploratory Automobility Research

Before planning and executing field research, we reviewed existing social science literature on the history and social and cultural experiences of automobility in major world markets. We thus began our field research with an appreciation of how, during the course of the twentieth century, automobility transformed how we live and created a new global transportation system, realized in many local, market-specific variations. Cars are much more than simply vehicles that transport us from one location to another. They are complex cultural and socio-technical artifacts that dramatically shape our experiences of mobility, identity, geopolitics, built environment, and social relationships.

As we planned field research, we also took into account that we were on the cusp of a new transformation of our transportation systems. Intelligence, slowly building in our vehicles and automobile ecosystem in the forms of traffic signal controllers, inductive loop detection, ramp metering, RFID, license plate recognition, SCATS, highway advisory radio, contraflow, speed cameras, personal navigation devices, ATMS, crowdsourced traffic apps, was beginning to transform cars, our roadways and surrounding legal, risk, monitoring, enforcement, and planning systems. Computation, advanced sensing, big data and analytics, and connected services across vehicles, infrastructure, smart devices, and the cloud enabling smart cars, autonomous vehicles and intelligent transportation systems are shaping the twenty-first century.

14.2.2 Phase 1: Car Turn Outs

We began small-scale research on private passenger cars in 2010. Our methods and timelines for research represent a fairly typical investment strategy for user experience research. We first conducted a phase of low-intensity research to determine if there was enough value in the space to justify a larger investment of resources for more in-depth research. Though our initial investment in exploring "*What is a car?*" was limited, we were committed to multi-sited ethnography (Marcus 1995) of car ownership and use as we wanted to understand how the twenty-first century transformation would be both globally reaching as well as locally realized.

We started with a small number of interviews in five countries: The United States (California and Oregon), Great Britain (Greater London area), Australia (Melbourne and New South Wales), Singapore, Malaysia (Penang), and China (Hangzhou and Beijing). Field sites were initially opportunistic, based on locations we were already conducting unrelated research or had other occasion to visit. Over time, sites were intentionally chosen as locations we believed might offer a substantially different or unique glimpse of vehicle experiences than we had yet to encounter. Singapore, with its particular incentives and laws around car ownership and use was one such site. Rural New South Wales was another. Three sets of questions structured our research:

- What are the lived experiences of owning, using, maintaining, parking, and caring for a car?
- What are the characteristics of local automobility ecosystems? How do people learn to drive? Access roads? Buy insurance and use it? How are road rules enforced?
- What technologies are present and used in cars? What other objects do people carry with them in cars? How can these objects illuminate what cars are and how they are used?

All research participants were visited at their homes and in the presence of the primary private car they owned or to which they had regular access; almost all were revisited approximately 1–2 weeks after the initial interview. The overall research protocol consisted of four activities; interviews, ride-alongs, video diaries, and car inventories. In 2.5–3 h broad, semi-structured ethnographic interviews, we asked participants about daily routines, mobility practices, and personal and household histories with private cars, including their current vehicle. During the course of the interview, we generally moved from the home to the location of the participant's car, and into their car. In approximately half of the interviews, we also "rode along" with the participants during a daily trip such as picking up a dependent from school or running an errand. This questioning was supplemented by an assignment to complete a series of self-recorded video interviews, on which they reflected on a recently completed trip. We reviewed the video responses with participants during the second visit and asked follow-up questions that were not covered in the initial interview.

Influenced by archeology, we used the objects people carry with them in their cars as a means of understanding how cars are used and what they mean to people. We playfully named this car inventory activity "carchaeology" (Zafiroglu et al. 2011). We brought a water-resistant fabric shower curtain liner and a folding step stool to each research interview and had participants systematically unpack all of the objects in the interior, in the trunk and rear bed of their vehicle, including both "brought in" (easily removable) and "added-on" objects (permanent or semipermanent). We cataloged each object, noted its location and asked the car owner a series of open-ended questions, including—but not limited to:

- What: *What is it? What does it signify to the owners? What is its value? What is its use in surrounding automobility practices?*
- Who: *Who originally brought it in? Who in the past, present and future uses it, has ownership or use rights to it? To whom does it belong?*
- When: *When was it added to the car? When is it used? When was it last seen? When will it move from the car (if ever?)*
- Why: *Why was it added to the car? Why is it still in the car? Why is this an important part of their car experience, or why is it not?*

Each object was placed in turn on the shower curtain liner and when the car contents were fully present, the entire assemblage was photographed. We then worked with the participant to repack the car, noting which objects were not replaced, but either moved to the trash or to another location.

What became clear is that the objects people have in their cars are an essential part of their car experiences. They are rich sources of information about important activities, routines, and social relationships that take place in the car. Cataloging and asking about them was a useful way to help participants remember and relate stories about what happens on a daily basis in their cars, as well as unusual events. They also served as useful probes to help participants talk about how they think about their car, take care of it, and manage a range of obligations and activities involving the vehicle.

From a research and design perspective, this methodology jump-started our thinking about what cars currently do for people, what they mean to a range of stakeholders, and what they might do and mean in the future. Inspired by these belongings, we created a series of design questions around the possibilities enabled by cars as components in intelligent transportation networks.

What we did not fully understand yet through this round of research were the small moments that made up lived experiences of spending time in with and around cars, and the ephemerality of movement. For example, it was difficult to understand how smart phones were used in cars. When asked directly, research participants could not answer exactly how, when, or why they used their phones (Zafiroglu and Healey 2011). Their responses were vague based more on the details of their phone using being so mundane, as to not be non-memorable, rather than on a fear of telling us that they had broken a traffic rule. They literally could not recall when, where, and why they were using their phones.

14.2.3 Phase 2: Local Experiences of Automobility

In the next car research project, *Local Experience of Automobility* (LEAM), we narrowed our focus to three of the world's four largest car markets; Brazil, Germany, and China. Each market offered opportunities to explore automobility practices in contexts beyond US-based audiences. They also offered significant contrasts in the history of private car ownership, the development of road systems and automobile manufacturing.

We continued applying the semi-structured ethnographic interviews, ride-alongs, video reports, carchaeology, and follow-up in-home interviews. But this time, we added a number of data sets made possible by in-vehicle sensors, GPS tracking devices and smart phone tracking software. During the first interview, we installed sound pressure sensors, accelerometers, light sensors, and temperature sensors in the cabins of participant's cars. We also installed a single passive recording GPS in each car, and an application monitor on each participant's Android smart phone. We collected data through these sensing and data recording instruments for one month of routine car and phone use.

At the end of the data collection period in each country, we collected the sensors and GPS devices and started an intensive three-week period of data review, analysis, and preparation of materials to review with participants during follow-up interviews. GPS and smart phone application monitoring data sets were time-stamped aligned, GPS stops hand-labeled, smart phone data verified, and the "clean data set" used to produce representations of journeys in Google Earth (Fig. 14.1) that showed travel dates, times, paths, speeds, and use of mobile phone during, and around, each journey. We revisited research participants in their homes,

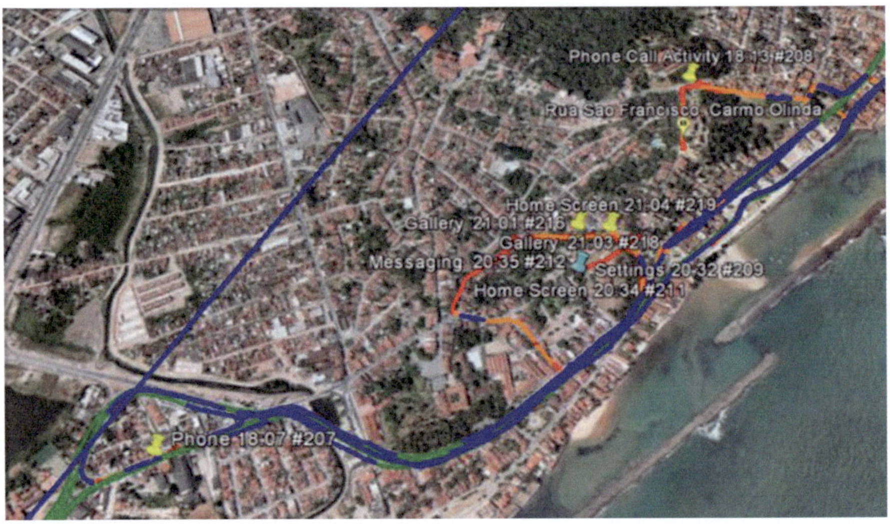

Fig. 14.1 Example map of participant's journey

this time with a laptop projector and large folding screen rather than a folding stepstool and shower curtain liner. Using the Google Earth representations, participants, and researchers co-explored mobility patterns. As we have argued (Zafiroglu et al. 2012), contrasting the remembered drives participants shared during the initial interviews, with the recorded drives from the clean data set allowed us to more deeply explore the experiences of car use than either data set alone. Additionally, the Google Earth representations were useful memory-jogs for participants, similar to the questioning about objects in cars during the earlier carchaeology sessions.

14.2.4 Connecting Insight to Design and Production Teams

This type of in-depth qualitative user experience research can be quite powerful when presented in the right way. In the wrong way, it is easily dismissed as anecdote, and given the damning evaluation of being interesting. The unspoken second half of such evaluation is that the work is not *useful*, meaning that the stakeholder does not know how it directly impacts her work or the broader business opportunity being explored. We sought ways to have stakeholders viscerally connect with the research findings, and to get them to "think with" the insights, using insights as a tool for *them* to go farther, rather than as a static set of data to digest and either accept or reject.

We used two techniques to accomplish this. First, in presentations we shifted away from a set linear narrative of: methods, data, and results, to a model based on directly engaging automotive design and engineering teams with a curated set of research findings that could be explored through a digital portal (Fig. 14.2).

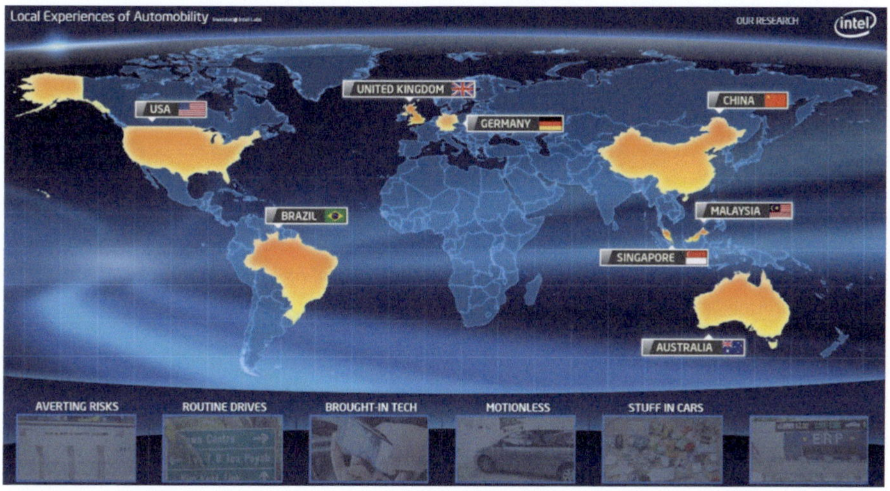

Fig. 14.2 Research portal for local experiences of automobility

Fig. 14.3 Carchaeology artifact simulacra

Second, we created tangible objects for stakeholders to explore, allowing them to discover patterns by themselves and ask the research team questions. These include fleece blankets printed with life-size photographs of carchaeology blankets and life-size plastic simulacra of objects found in cars that could be picked up, and had pithy research insights and design provocations printed on the reverse side (Fig. 14.3). We often places these simulacra onto fabric-covered conference room walls using Velcro. Stakeholders could choose which material to engage with. This gave them the opportunity to guide conversations to the topics they found most compelling and spur their creativity (Zafiroglu et al. 2013; Singer 2014).

These materials and the particular presentation methods were used in the iterative solution design process described in the following section where a mixed team of user experience researchers, ethnographers, automotive user interface designers and engineers followed an agile iterative prototyping process to create automotive user experiences.

14.3 Creational Prototyping

Innovation is an experimentation process that should be intrinsically iterative and generative. In order to explore design solutions to the real-world problems identified in our foundational research, we created a driving simulator platform whose main purpose is to unleash creativity and enable design thinking practices through making. In this section, we describe this automotive user experience prototyping tool, named Skyline, and describe the methodologies for creational prototyping from low to high fidelity user experiences.

14.3.1 Skyline Architecture

Skyline is a prototyping platform developed in Intel Labs and targeted to support user experience researchers in designing, implementing, and evaluating in-vehicle concept experiences. Skyline enables iterative development of in-vehicle user interactions integrated into configurable driving scenarios. As opposed to other driving simulator platforms, Skyline is not focused on providing training for driver's behavior or measuring driving performance, but rather prototyping and answering design hypothesis through user assessment in an effortless manner. The platform was presented to the public for the first time at the 7th International Conference on Automotive User Interfaces and Interactive Vehicular Applications (Alvarez et al. 2015).

Skyline is composed of modular and flexible hardware and software components. The physical configuration, see Fig. 14.4, resembles a minimalist vehicle cockpit where materials and finish are chosen to provide a generic in-cabin feeling. This neutral environment facilitates user experience assessment. The mainframe dashboard enables flexible placement of vehicle controls (steering wheel, gear shift and pedals) as well as dynamic placement of a varying number of in-vehicle displays and sensors. The dashboard sections that are not occupied by screens are covered with snap-in panels to provide a uniform surface.

The display components are stand-alone computer electronic devices (tablets) connected via Wi-Fi to the Skyline wireless local area network. All display devices, including participant brought in personal devices such as smartphones are integrated this way. Each of the displays is identified upon connection. Finally, a computer connecting to the driving environment display (TV or projectors) is the centralized server that hosts both the human–machine interaction server and the driving simulator software. In-cabin sensors such as driving controls, pressure sensors, microphones, or cameras can be wired directly to this computer. A flexible number of leather vehicle seats complete the standard deployment (Fig. 14.5).

Skyline software platform is a real-time run-time environment that uses open-source web technologies to integrate display and interaction interfaces with the virtual world on which the user will drive. Figure 14.2 presents a high-level diagram

Fig. 14.4 Skyline driving simulator

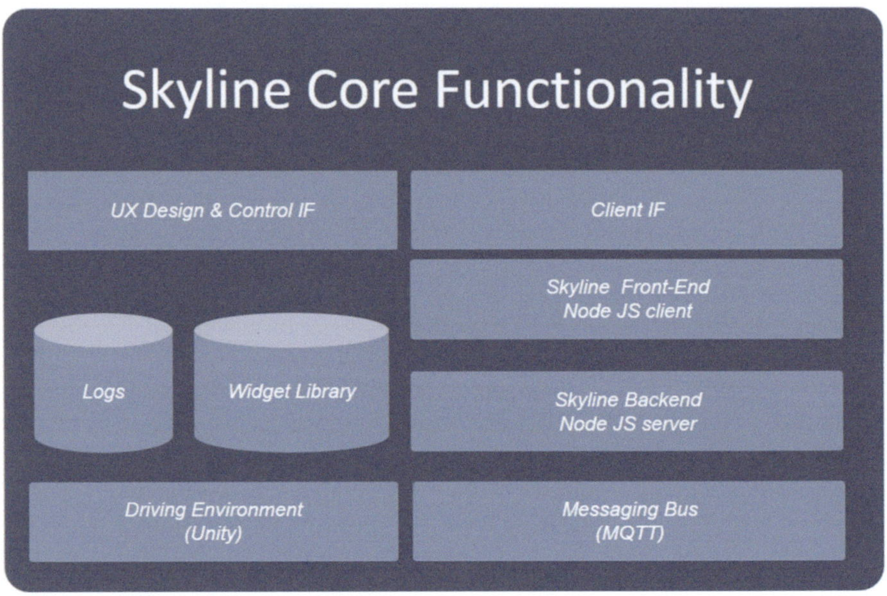

Fig. 14.5 Skyline software architecture

of the software components. The virtual driving environment, also known as Skyline world was developed using the popular game engine Unity (Unity 2015). NodeJS provides the main back-end and front-end components. Node.js is an open-source, cross-platform environment for web applications with real-time requirements (Joyent 2015). Node.js applications are natively developed in JavaScript, and can be executed in various operating systems through a web browser. The messaging bus between applications is based on Message Queue Telemetry Transport (MQTT). MQTT follows a publish-subscribe model, where a message broker is responsible for distributing information to interested clients based on the message topic (MQTT. org 2015). The front-end graphical user interface (GUI) elements are self-contained JavaScript components that can be programmed to respond to MQTT messages. We name these GUI components “widgets”. Widgets are developed following the UI standards advocated by React (Gackenheimer 2015) and communicate with the Skyline backend via Socket.io library (Socket.io 2015).

Widgets can be displayed in any of the available Skyline screens upon UX designer configuration. Skyline offers a library of widgets that can be used and configured to recreate any in-vehicle interaction scenario. Simple widgets show static images, or play audio files, but complex ones can be iteratively built into a complete navigation system with buttons, map displays, and text to speech. This widget library is the core of Skyline’s UX development and Skyline includes a suite of design and control applications to help prototypers use these assets in a driving simulator environment.

14.3.2 Skyline Principles for Automotive Interaction Design

While the previous section provided an overview of Skyline components and functionality, this section illustrates how Skyline design pillars can be leveraged to provide a platform that accelerates innovation and research findings. Skyline’s design was conceived to as a flexible development platform for in-vehicle user experiences that enabled reuse of software components. Skyline was therefore focused to address driving simulators limitations by making sure six fundamental attributes were satisfied, namely adaptive, flexible, participatory, predictive, accessible, and rapid development.

Adaptive

Skyline is designed to help understand and define user goals and preferences and then rapidly build solutions in response to those user needs. Adaptation can be done at multiple levels, from the reconfiguration of the display components to the modification of information placement using the display zones. Within the widget component itself, Skyline allows customization of message notification or even graphical elements. For example, prior to running a user assessment we required the participant to email us a picture of themselves. We then used this image and their name as an avatar to personalize certain interfaces throughout the user experience. While the first effect of this personalization is usually surprising to the participant it makes the experience more realistic. Having a system address the users by name made it easier to collect feedback that would apply to their daily lives. Developing a system that is more adaptive to the individual needs of the user means that we could better simulate the ownership experience and the participant could evaluate the probes through a longitudinal context.

Flexible

Skyline offers researchers the ability to support multiple driving experiences from a single platform, from driver-only to driver–passenger interactions to a complete family environment with back seat experiences. This requirement inspired the modular approach to in-vehicle displays and the software-defined execution environment. Rather than having access to a driving simulator with fixed screens, we envisioned the complete in-vehicle interior to be a virtual display space that could to be explored according to the research goal. With Skyline, researchers have the opportunity to integrate large displays into the vehicle cockpit and divide display areas within the screen, or rather place a myriad of smaller displays that behave as independent displays or as displays zones for a virtualized display covering multiple devices. This encourages new use case development, rapid onsite ideation, and real-time curation and placement of content for interaction and probes.

Participatory

The recent boom of autonomous driving related research is a sound indicator that vehicles are on a trajectory to become fully connected mobile systems. Skyline was engineered to address a participatory need and the access to cloud services was natively integrated as it was architected to support web technologies. It is fairly simple to develop widgets that can connect to Cloud Services via public APIs. This process allows researchers to create widgets that connect to a social network accounts and pull or push data. Since the development is native to web technologies there is a limited learning curve to native operating systems. Access to cloud services can be done directly using rest interfaces. Similarly, widgets can push data to any online service allowing the prototyper to customize integration with any big data storage system which reduces the system requirements of the physical configuration of Skyline and allows sharing of raw data sources. Researchers can thus operate in distribute environments using Git repositories to update their local Skyline libraries and exchange configuration files, logs or access sensor data dumps across the globe.

Predictive

Skyline was built to facilitate the understanding of user needs, patterns, and preferences. There are two ways in which Skyline can appear predictive to the end user. The first one is to utilize a Wizard of Oz (WOZ) setup (Dahlbaeck et al. 1993) where the UX researcher can gate the interaction of the participant by triggering or pushing transitions between states from a control interface. This can mimic UI interactions such as buttons, gesture recognitions, or verbal to test user's reactions to an interaction concept without the hurdle to fully develop the systems.

The second way to appear predictive is the use of trained user models to control the interaction logic of the widgets. In longitudinal studies where users go through several interactions with the system, initial data captures can be applied to generate user models that will govern interaction of the visual element. Certain sensor platforms such as Intel® RealSense™ are uniquely built for such behavioral models development and application (Intel Corporation 2015).

Accessible

Skyline was developed using open-source web technologies and standards. We purposely chose widespread technologies that had a strong user community, such as MQTT and NodeJS, with a hope to lower learning curves and allow researchers to have easy access to the core components of Skyline. The platform is built to provide users with simple tools that allow them to run a journey experience and modify it to create their own interactions. But we understand that users will most likely quickly jump into creation of new widgets to fulfill specific research purposes and hopefully share these components back with a growing community of Skyline developers/users. By developing simplified menus to build scenarios, there is no longer a development divide that separates UX researchers from developers. The platform is accessible to anyone on the team.

Rapid Development

At the core of Skyline principles is the ability to rapidly iterate design concepts. Swift scenario creation and hypothesis evaluation is the most valuable attribute that Skyline can bring to a team of experienced UX researchers and developers. Skyline enables quickly setting up a driving experience, building the desired interaction and testing it with users in a realistic driving environment, whether it is in the form of a low fidelity WOZ setup or a high fidelity experience. The modular nature of Skyline also allows for between-subjects experiment modifications for exploratory research. After a first round, user feedback for information positioning can be automatically updated and a second experiment can be run with the same participant.

14.3.3 Iterative Development Cycles in Skyline

Figure 14.6 summarizes the process of planning, executing, and iterating rapid development cycles in Skyline. This section corresponds to the use of the tool and does not consider the user study planning or user assessment methodologies and processes when evaluating in-vehicle experiences. These are explained in the following section.

Using the using the Journey Builder a researcher is able to lay out an HMI experiment updating the placement and appearance of widgets in the Skyline HMI library without need of coding or software development skills. The researcher can focus on instrumenting the adequate control level required for her/his experiment. At the beginning of the research program researchers might wish to quickly architect researcher-controlled scenarios while later on the experience can be tightly integrated with the user-driven interaction. Once a “journey” is build it can be saved for execution. The Journey Runner provides a web interface to launch and control a running driving experiment. The tool is accessible from any device connected to the Skyline network including handheld devices such as smartphones, giving the

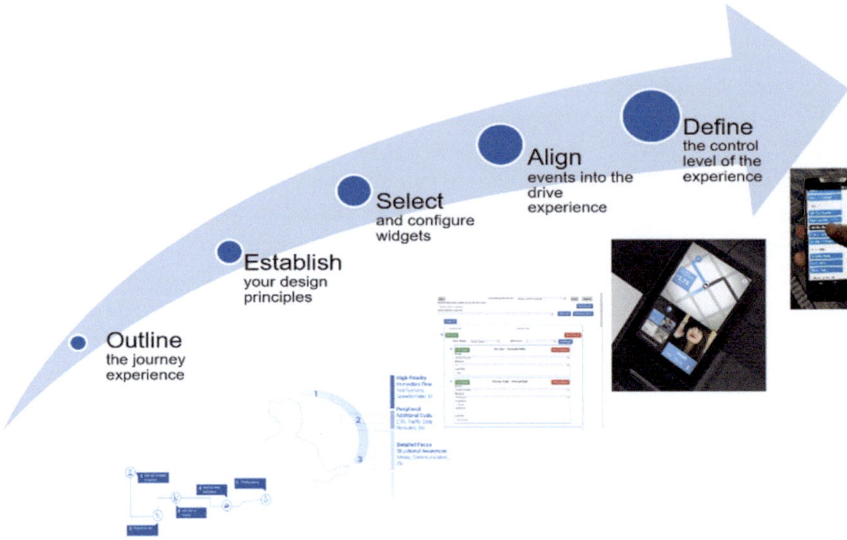

Fig. 14.6 Design process in Skyline

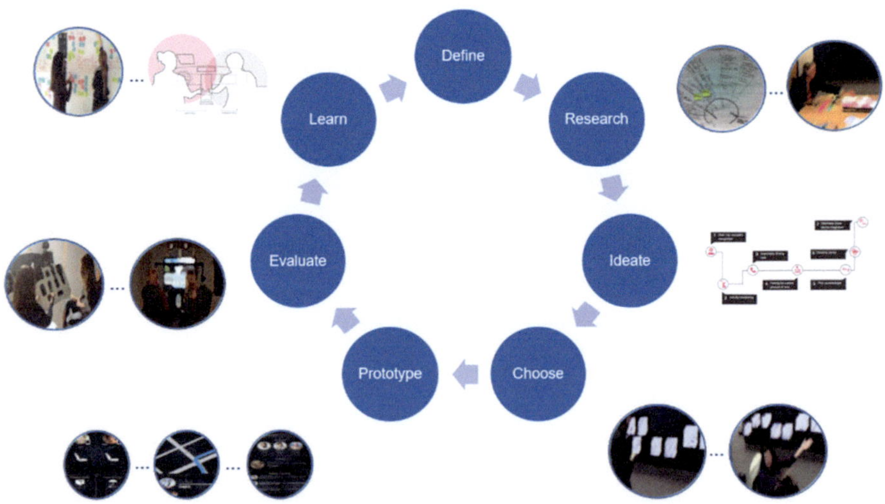

Fig. 14.7 The design making iterative process

researcher the flexibility to guide the user experiment in the simulator while discretely controlling the experience flow. The Journey Runner allows the selection of stored journeys and provides control over the HMI states. During a controlled experiment, the researcher can trigger execution of events using this interface. Every interaction is also recorded in a log file which can be stored and reviewed during the data analysis process. As the data is reviewed and discussed, the process becomes a circular self-optimizing loop that we like to call design making, as described in Fig. 14.7. We outlined the principles of automotive design making theory in Hendrie et al. (2015).

14.4 Assessing Automotive User Experiences

In this section, we review the entire user assessment process including planning, user definition, recruiting, assessment, data collection, and analysis of qualitative and quantitative data. We provide guidance through each step to help the reader understand how to effectively extract value propositions from user assessment in Skyline.

14.4.1 Research Goals

The first step in running any user assessment is determining the research goals and creating an actionable test plan. To do this, an understanding of what question(s) are we trying to better understand is needed, as well as what is it we need to know. An example of a research question can range from "Where do users prefer a text alert notification on the HUD?" to "In what interactions do users want the system to act on their behalf versus in what interactions do the users want control?"

Once these key research questions have been determined, a test plan can now be created, using Skyline as the assessment tool. If probing on the question of automotive agency, it follows that designing a methodology to probe specifically on the value of trust could surface data that could lead to new areas of discovery. A key question to ask in this process might be, how can the user experience push the boundaries on human control versus automotive agency?

14.4.2 Methodology

Ideally, the methodology used to refine the research question enough to run a use experience assessment using Skyline follows a "winnowing" process that leverages several rounds of foundational research. This foundational research is usually aimed at either better identifying and characterizing the targeted end user group, or a process of refining the research question itself. A sample research "flow" in this

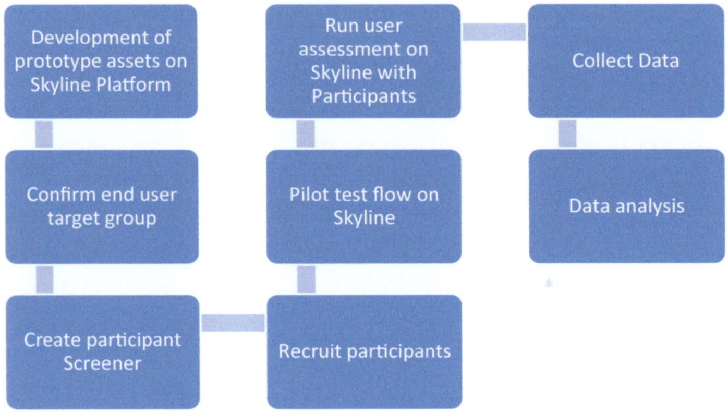

Fig. 14.8 Research planning phases

vein starts with Ethnographic pathfinding, is followed by conceptualization, then A/B testing and finally optimization. Skyline is best utilized during the optimization portion of this process flow given the iterative nature of refinement associated with user feedback gathered from user assessment testing. The test plan for a round of rapid prototyping leveraging Skyline generally follows the steps showed in Fig. 14.8.

14.4.3 The Recruiting Process

The recruit for any study is of course dependent on research goals and the research question, but should always be representative of the targeted end user group. The targeted end user group itself may come from several sources: a customer or client, previous persona development work, or market segmentation research.

The end user group may include characteristics from all three, one, or even none, if no previous research has been conducted. It should be able to be characterized by different factors that can be screened for, such as identified key socioeconomic factors, demographic data (such as age range, income level, race, etc.). Some of these factors may include key persona characteristics; for instance: Stay-at-home mothers. If no previous research has been conducted to help identify the targeted end user group, it should be seriously considered at this juncture. The goal in this step is to create a screening document which can be used as a questionnaire with potential participants to make sure that they are a good fit for the study, and who will help effectively answer the research question. If participants are not screened in this manner, it increases the risk that the data will not be robust or conclusive. Once participants have been screened and targeted end users have been identified as participants, prep can start on Skyline itself to prepare for the study.

14.4.3.1 From Research Goals to an Actionable Test Plan

Once participants have been selected and dates have been confirmed to run the research assessment on Skyline, there are number of things that are best known methods to make the study flow as effectively as possible. These things can be started in the preparation phase when asking the participants for feedback. For example, it is often good to ask participants to send in a photo of themselves so that the Skyline experience being tested can be personalized for them. This means that the test participant will see an avatar or photo of themselves in particular interactions within the HMI. This better immerses the participant in the test experience, reducing the facilitator to explain interactions and contributing to the overall fidelity of the test environment. It is also good to ask users to bring their own commonly used devices (such as a phone or wearable.) The participant's phone can be connected to Skyline so when alerts or texts are received in this simulated driving environment, the participant is receiving them on their own device. This is an added layer of immersive experience than if this was a proxy device handed to the user so when probed they can answer from a place of reflection rather than abstraction.

As the test scenario is prepared to be loaded into the Skyline virtual world, there are several things to keep in mind as visual assets are developed. Skyline is a medium to high fidelity simulated driving environment and not the exact experience as participants will find they have driving a vehicle in the wild. To account for that gap, there are key learnings acquired for better outcomes for general test scenarios. For example, make sure to use large, easily identifiable graphics and text. Keep text as limited as possible, and place any graphics or pop-up cues in upper quadrants of screens so that they are easily visible to the driver. Bright colors also make a difference, as light conditions in the lab environments will vary. Skyline tends to have a highly stimulating visual cockpit so critical cues (such as those emergency alerts) should have both audio and visual alerts that mute any other audio playing (such as the radio) for phase one concept tests so users can clearly hear and react to stimuli. It is also helpful to designate specific areas of the Skyline device components for "driving" alerts versus "other" notifications. This is meant to delineate what alerts are focused on driving critical information versus other notifications such as texts. These best known methods have been confirmed with multiple rounds of user research using Skyline, and should serve as a baseline design guide for any scenarios to be tested.

While following up with participants for personal assets, there are a number of best known methods to keep in mind while creating your Skyline scenario and research protocol. For instance, take into account time needed to allow the participant to acclimate to driving on Skyline (vs a real car.) This process is called a test run and can be described as allowing the test participant time to familiarize themselves with the simulated driving environment. This is done by the facilitator allowing them the opportunity to take a "test drive" through a modified environment, time using the steering wheel, brake, and accelerator all while taking in the multiple screens within the cockpit itself. After much data analysis we find a test run in the simulator environment to be very helpful, allowing the participant to overcome the learning curve, so that when the actual test scenario is being played, the participant is able to better participate rather than trying to drive. As a part of this "test drive", an overview of Skyline should be conducted, such as location of the pedals, what the HMI is, what exactly a HUD (Head's up display) is. The time needed to familiarize participants with Skyline should be built into your study script.

14.4.4 The Facilitator Role

The role of the facilitator is extremely important when running a prototyping study on Skyline. After all, the facilitator is the person with whom the participant will have the most direct interaction, so how they act during the simulation is important. They are the ones who have the ability to make the experience as natural as possible and as true to real life experience in a vehicle as can be had.

That being said, there are several things the facilitator can do to make the experience easier and more realistic for the participant to enable a less "directed" experience. He or she can perform typical passenger tasks during the session, for instance, and allow the participant to place their personal devices where they normally would in their own vehicle (dependent on test constraints, of course.) They can also direct the participant to perform normal typical driving tasks (such as clicking their seatbelt, checking for their phone, and starting the "car") to reinforce the sense of driving reality during the simulation. Allowing the participant to perform their typical "while driving" behaviors (such as checking their phone) during the simulation also further helps create a sense of driving reality.

14.4.5 Stimuli Versus Probes

During the simulation itself, the facilitator's use of stimuli and probes is extremely important in eliciting important contextual data from participant's responses. When appropriate, it is best for the facilitator to interject verbally during driving to prepare participants for upcoming stimuli, and then prep participants to stop at natural pauses in the virtual environment. It is also helpful to build in time to accommodate discussion after longer and more complicated usages (such as when multiple events occur close together.) Incorporating normal driving tasks and directions (such as "Take a left at the light") helps reinforce the role of the participant as the driver and increases their engagement with the simulation, despite the distraction of constantly stopping to answer queries. The facilitator can also query for quantitative feedback while the driver is driving, and then follow up during a pause in the simulation for deeper qualitative questioning. Use of a remote to delay or start stimuli in the simulated environment is also useful in giving the facilitator more time to probe when necessary, and also keeps the participant from accidentally tripping code to trigger starting another test stimuli too early.

14.4.6 Holistic Data Analysis for Use Case Requirements

There is an anonymous quote that states "UX is the radical notion that if you are designing for someone, you might actually want to learn something about them". There are two well-known and accepted general research methods for this learning process, and they are qualitative and quantitative based. Qualitative research has been described as the "development of concepts which help us to understand social concepts in naturalistic settings, giving due emphasis to the meanings, experiences and views of the participants" (Pope and Mays BMJ 1995; 311:42–45). It is subjective in nature. A few typical data collection methods for this type of research are interviews, observations and focus group discussions. On the other hand, quantitative research is used to measure a research question by generating a numerical data set or data that can be transformed into useable statistics. It is objective in nature. Typical quantitative research methods tend to be more structured than qualitative, and can be used to surface patterns in data. These methods include online surveys, analytics, and systematic observation. The data from both quantitative and qualitative methods is intrinsic to the method by which it was collected, so data analysis methods will also be native to the research type. Quantitative analysis will focus on statistical trends and patterns, while qualitative analysis can be approached using inductive methods, an emergent framework to initially find similar patterns and then associate them with relationships, or deductive methods, using the research questions to group the data and look for similarities and differences. If this iterative process is mature, these findings offer the researcher an opportunity to document this data into a UX format commonly referred to as use case requirements. A use case is a description of the needs a group has for a system as well as documenting how the system will meet those needs. Use cases can be developed at different stages of the product development lifecycle, depending on if you are using an agile or a waterfall development process. As the use case prototypes are evaluated and the user assessments provide data supporting or denying the initial design hypothesis, the researchers can then start documenting specifics on the usage requirements. We have found that capturing both quantitative and qualitative metrics help in the discussion and prioritization of use cases in the product development phase as we illustrate in the following section.

14.5 From Use Case Requirements to Product Development

Now that you have learned how to prototype and assess in-vehicle concepts, you are equipped to bring amazing ideas to life and evaluate them. In this section, we describe what to do when too many ideas and solutions seem promising. How to listen to user preferences, prioritize use cases and understand the challenges for integration of new technologies/use cases into an existing platform. When it comes to automotive, requirements are some of the most stringent in the industry. Weighting the cost of bringing the technology to market, answering the right questions, and demonstrating promising results that justify investment in novel ideas will facilitate streamlining your new interactive experiences into the market.

Consumers expect their computing devices to offer fresh and dazzling new functionality every time they upgrade their device or sign up for a new phone plan. Automotive engineers and designers are under pressure to incorporate new technologies and usages into the vehicle on development cycles that are closer to iPhone release cycles than traditional car platform cycles. As technology plays an increasingly important role in car buying decisions, engineers welcome any help in identifying technologies and usages that are ready for the car.

Finding usages that have broad appeal and can be delivered to market in a cost effective way is a very difficult challenge. Car platform development cycles are often several years long, which makes it difficult to pick technologies and usages that would not be on the showroom floor anytime soon. On the other hand, the technology will be in your driveway for a decade or more. Finally, the average car has many millions of lines of existing code. With these challenges, how can automotive manufacturers and their subcontractors integrate complex new technologies and deliver compelling usages to market in a safe and secure manner? How can you turn emerging technologies into concrete usages that customers are willing to pay for?

14.5.1 Exploring Technology via Usages

To help engineers deploy the right solutions, we built a driving simulator that supports modular interfaces for HMI's, messages, and data. Emerging technologies can be integrated with HTML interfaces to speed prototyping and experimentation. We know that cars of the future will have more powerful graphics capabilities to support multiple displays; have multiple microphones for enhanced voice recognition and interaction capabilities; cameras that watch and sense occupants; high speed to connections to the cloud and other cars; and many other forms of passive and active sensors in and around the car. Driver assist technologies will augment or act on behalf of the driver. Turning technology into usages and then into hardware and software ready for high volume manufacturing is at the heart of the issue.

The Skyline prototyping platform proved to be attractive to engineers because it allowed high fidelity user experiences development incorporating new technologies without writing lots of complex code. It also allowed experiences to be moved from one screen in the car to another screen (e.g., from the heads up display to the instrument cluster or the center console screen) easily. Along the way, customers gave valuable feedback on what they liked as well as what they did not like. For example, having the car display information from your calendar is nice, but what if you do not want your passengers to know your itinerary for the day? This insight allowed the team to explore multiple HMI options and use technologies to determine who is in the car and have that influence how and when information is displayed. In some cases, understanding what customers clearly did not want in the car was as valuable as what they did want. Removing options early in the prototype phase is much cheaper than finding it out later in the process.

14.5.2 Defining a Path to Product

From the design exploration work we have completed with Skyline, we have established an overall process for moving from experiential prototyping to product and solution development. This process starts with understanding the larger business and technical context for the concepts, interactions, and experiences that are being prototyped. Several questions can help define this context, such as: What are the business goals and objectives for product or solution? How will the experiences be brought to market? What technical capabilities are required to enable the experiences? What value is being provided to end users? And what problems are the set of interactions and experiences ultimately solving?

Establishing a business and technical context provides a foundation to guide the process of translating the usages that have been prototyped with Skyline to specific use cases that we discovered in the foundational ethnographic phase. With Skyline we are able to rapidly explore a range of different in-vehicle experiences, and investigate several different design approaches for those experiences. However, having a specific set of business objectives and hypotheses about technical implementation allows us to narrow down the large possibility space that Skyline affords, and establish metrics for prioritizing the key experiences and usages. Sometimes these requirements come from specific geographic markets or target audiences identified in the market study. Having this foundation also enables us to assess other inputs that feed into the product definition process, such as specific customer requirements, and decide if those inputs are aligned with the overall user value propositions.

Once a business and technical foundation is established, we are able to take the experiences, scenarios, and interface concepts from the Skyline prototype and define a set of specific, high-priority product use cases. These use cases describe the primary interactions that people will have within the vehicle, as well as the functionality that the vehicle will need to provide to accommodate those interactions. The primary use cases also reflect the key user value propositions of the product; a disconnect between the use cases and value propositions indicates that the use case development process is not yet complete and requires further iterations.

After the primary use cases are defined, they are decomposed into technical requirements. These requirements may encompass many different parts of the overall system, including interior displays, cameras, sensors, communication or connectivity technologies, or functionality provided by a local server or by the cloud. Requirements may point to technologies currently available or still under development; requirements may also point to differing levels of compute workload.

With a complete set of use cases and technical requirements, specific product implementations can then be evaluated.

14.5.3 Bringing Real Technology into Skyline

Intel has developed many technologies that at first glance have promising automotive application, but without applied usages do not have enough value to warrant the cost of putting them in the car. That is—they appear interesting, but without compelling usages that customers can grasp quickly, the OEM is never going to incur the cost across the full product line. One example is cameras inside the cabin. Are they obtrusive? Do they feel like "big brother" or do they provide enough value to offset initial misgivings? Intel originally developed the RealSense depth sensing camera for PC's and tablets. Could it be applied to vehicles? Why? How? Can enough usages be found to make it worth the cost of adding it to every vehicle in a product line? By incorporating RealSense into Skyline, we not only enabled auto manufacturers to explore usages, we enabled them to show many uses for the same camera(s), which amortizes the cost of the cameras across many usages. Interestingly, different OEM's have found different applications for the cameras—customers will not only see cameras pointed within the car of the future, they will see a diverse range of new offerings that will be both intuitive and desirable.

14.5.4 OEM's Want to Be Inspired

Car companies are setting up labs in Silicon Valley to explore how technology can inspire future generations of vehicles. Engineers who would otherwise have worked at Google and Amazon are now working on cars. These teams want to be inspired by technology and explore options quickly (fail early, fail often). Using tools like Skyline, they are evaluating Intel technologies and see them in action. In turn, this has made it easier to convince the purchasing department that the cost of integrating the technology will pay off via new customer options and packages on the dealer's showroom floor.

Sometimes, the inspiration does not directly lead to a new product, but gives insights into where the market is headed. We integrated smartphone wireless charging into Skyline in spite of the fact that no smartphone vendor has made this a standard feature of the phone… yet. Showing car OEM's how a phone could be charged in the car without wires allows them to think about where it should be stored (a binnacle?), which could help reduce unsafe driving habits by stashing the phone out of easy reach of the driver—all while improving Bluetooth connectivity and allow for applications like Pandora to work wirelessly in the car. In this example, a technology allowed for exploration of a usage that spans in many directions.

An additional challenge for the car company is to find a way to move the technology and usage from the lab to their manufacturing line. This often takes a significant amount of time.

14.5.5 The Power of Bringing Experiences to Life

In this chapter, we have described how Skyline can be a powerful tool for sketching out interactions and interface concepts, for representing complete experiences that be put in front of users for assessment and usability, and for illustrating functionality that can be decomposed into use cases and technical requirements. However, in our product-focused collaborations with partners and OEMs, we have also seen the power of simply bringing experiences to life.

When usages are made tangible in the Skyline prototyping environment, engineers and management can directly interact with product functionality before committing resources and budget to actually developing it. Experiencing the usage can help with the decision to bring the solution to market and translate it to a real product. Bringing the usages to life in Skyline also provides a more holistic view of how each of the individual usages connect to form a complete experience. This total experience in Skyline can help partners look beyond incremental improvements, and instead see greater opportunities that tie together multiple features and functionality. A total experience can also serve to align internal teams around a single, unified design direction. Finally, seeing and experiencing vehicle functionality first hand can establish greater empathy for the end user, since usages do not simply exist as a line on a spreadsheet or as an item in a requirements document, but instead as functionality that real people will use, and derive value from.

14.6 Conclusion

As we reach the end of the chapter, we hope this journey on which we introduced a holistic set of practices, methods, and processes to discover and define the future of automobility has been useful to both experienced automotive user experience developers and to those who are taking their first steps into this field. With the advent of commercially available autonomous vehicles around the corner, this is certainly an exciting time to imagine what in-vehicle interactions are possible. At Intel we believe it is our mission to bring smart, connected devices to every person on Earth. This requires that we drive technology revolutions and share the tools and platforms that can fuel innovation. From the archeology-influenced ethnographic field research and the data-based insight extraction to the design making principles that guide our iterative prototyping work and the analysis and prioritization of amazing ideas intro products, we hope that the proposed practices are adopted and refined across industry and academia and we look forward to advocating standardized practices that bring us forward.

References

Alvarez, I., Rumbel, L., & Adams, R. (2015). Skyline: A rapid prototyping driving simulator for user experience. *Proceedings of the 7th International Conference on Automotive User Interfaces and Interactive Vehicular Applications—AutomotiveUI 15* (pp. 101–108), Nottingham, UK. ACM.

Bell, G. (2011). Unpacking cars: Doing anthropology at Intel. Anthronotes v. 32, no 2 32.

Dahlbäck, N., Jönsson, A., & Ahrenberg, L. (1993). Wizard of Oz studies. *Proceedings of the 1st International Conference on Intelligent User Interfaces—IUI 93* (pp. 193–200). New York, USA: ACM.

Gackenheimer, G. (2015). What is react? Introduction to react (pp. 1–20). Apress, Springer Science Business Media.

Hendrie, M., Hooker, B., & Alvarez, I. (2015). Prototyping adaptive automotive UX: A design pedagogy approach. *Extended Proceedings of the 7th International Conference on Automotive User Interfaces and Interactive Vehicular Applications—AutomotiveUI 15.* Retrieved September 3, 2015, from http://www.auto-ui.org/15/p/workshops/5/hendrie.pdf.

Intel Corporation. (2015). Intel Developer Zone: RealSense. Retrieved November 15, 2015, from https://software.intel.com/en-us/realsense/home.

Joyent, Inc. (2015). Nodejs.org. Retrieved February 27, 2015, from http://nodejs.org/.

Marcus, G. (1995). Ethnography in/of the World system: The emergence of multi-sited Ethnography. *Annual Review of Anthropology, 24*, 95–117.

Marx, K. (2008) The 18th Brumaire of Louis Bonaparte. New York: Cosimo.

MQTT.org. (2015). Retrieved February 27, 2015, from http://mqtt.org/.

Singer, N. (2014). Intel's sharp-eyed social scientist. New York Times, Retrieved February 14, 2014 from permalink: http://nyti.ms/1fpyssn.

Socket.io. (2015). Socket.IO. Retrieved February 27, 2015, from Socket.IO: http://socket.io/.

Unity Technologies. (2015). Unity3D. Retrieved February 24, 2015, from http://unity3d.com/.

Zafiroglu, A., Bell, G., & Healey, J. (2011) (C)archaeology: Car turn outs and automobility. *Adjunct Proceedings of Automotive UI Auto UI '11*, Salzburg, Austria, Nov 30–Dec 2 (pp. 49–50) ACM.

Zafiroglu, A. & Healey, J. (2011) Unmuting Smart Phones in Cars: getting phones to speak to us in ways their owners can't. *Adjunct Proceedings of Automotive UI AutoUI '11*, Salzburg, Austria, Nov. 30–Dec 2 (pp. 180–182). ACM.

Zafiroglu, A., Healey, J., & Plowman, T. (2012) Navigating to multiple local transportation futures: Cross-Interrogating remembered and recorded drives. *Proceedings of Automotive UI '12.*

Zafiroglu, A., Healey, J., & Plowman, T. (2013). Step on, pick up, tap through: Moving findings beyond PowerPoint. In *Ethnographic Praxis in Industry Conference Proceedings* (Vol. 2013, No. 1, pp. 410–410).

Chapter 15
Virtual Reality Driving Simulator Based on Head-Mounted Displays

Quinate Chioma Ihemedu-Steinke, Rainer Erbach, Prashanth Halady, Gerrit Meixner and Michael Weber

Abstract This chapter presents the development of an innovative and interactive 3D virtual reality driving simulator based on head-mounted displays, which gives the driver a near-realistic driving experience for the development and evaluation of future automotive HMI concepts. The project explores the potentials and implementation of virtual reality in the automotive sector for the analysis of new HMI concepts and safety functions in the automotive sector. Special emphasis is laid on hazardous situations which are ethically not possible to evaluate on a real road at the early stage of the concept, when the risk involved for both the driver and the prototype, for example driver distraction and autonomous vehicle studies is not yet ascertained. The 3D virtual reality approach was meant to overcome some of the limitations of conventional 3D driving simulators, such as lack of total immersion and intuitive reaction of the test driver, necessary for an effective analysis of a particular driving situation. The sense of presence offered by virtual reality is essential for the research and evaluation of safety functions, since appropriate and reliable solutions are only possible when the problem associated with a particular traffic situation is well understood. The focus was on the following aspects: 3D modeling, correct simulation of vehicle and traffic models, and integration of a motion platform to give the feel of a real car and control devices and finally,

---

Q.C. Ihemedu-Steinke (✉) · R. Erbach · P. Halady
Center of Competence HMI, Robert Bosch GmbH, Leonberg, Germany
e-mail: quinatechioma.ihemedu-steinke@de.bosch.com
quinate.ihemedu-steinke@uni-ulm.de

R. Erbach
e-mail: rainer.erbach@de.bosch.com

P. Halady
e-mail: halady.prashanth@de.bosch.com

Q.C. Ihemedu-Steinke · M. Weber
University of Ulm, Ulm, Germany
e-mail: michael.weber@uni-ulm.de

Q.C. Ihemedu-Steinke · G. Meixner
UniTyLab, Heilbronn University, Heilbronn, Germany
e-mail: gerrit.meixner@hs-heilbronn.de

G. Meixner and C. Müller (eds.), *Automotive User Interfaces*,
Human–Computer Interaction Series, DOI 10.1007/978-3-319-49448-7_15

head-up display use cases. Finally, the solutions to eliminate simulation sickness were reviewed and implemented. A prototype was developed which displays dynamic head-up-display features.

15.1 Introduction

How would a driver react if faced with a hazardous situation? When is it best to alert the driver about unforeseen danger? What is the best way to interact with the driver in order to achieve the desired reaction? Is it possible to develop a near-realistic driving simulator? The above questions can be answered by analyzing the driver's behavior within a driving situation taking into account other drivers, the road conditions, and the car dynamics. This is important since human error is one of the principal factors that lead to road accidents, and is attributed to increased mental workload, mostly caused by distractions such as operating devices or eating while driving. The most effective way to understand how drivers master situations which could lead to an accident would be to place them in that same situation. Placing drivers into a real driving situation in order to evaluate their behavior is too dangerous whereas testing environments such as crash test environments are very time and cost intensive. Therefore, driving simulators are commonly used for research purposes in monitoring the driver's behavior and for design and evaluation of new Human–Machine Interaction (HMI) concepts. Mostly Two-Dimensional-Driving Simulation (2D DS) environments are used in the automotive industry and for automotive research. These conventional driving simulators often lack the immersion of the driver into the driving scenario, and hence do not reveal the natural reaction and interaction required to understand the drivers' behavior.

Emerging technologies in the field of Virtual Reality (VR) from the area of consumer electronics and increasing processing power offer potentials for new highly immersive Driving Simulation (DS) concepts, for example in the development and evaluation of new Advanced Driver Assistance Systems (ADAS).

Most of the leading automotive manufacturers use VR in various phases of their development process for example; Ford, for interior and exterior design (Ford 2015; Howard 2014), Audi, for car configuration (Robarts 2015), Renault for research and development (Renault 2014), Lexus for test driving of virtual prototypes (Eedelstein 2014) and BMW for car development (BMW 2003), just to name but a few.

VR DS with a Head-Mounted Display (HMD) such as the Oculus Rift or the HTC Vive provides a different DS perspective to conventional DSs used in the development of automotive products. Users feel present in the computer-simulated environment due to the wide field of view, stereoscopic 3D effect and extremely low latency offered by novel HMDs. This leads to the sense of presence which means that the test driver is physically and mentally involved in the driving experience for a natural reaction and interaction with the system. Therefore, it is possible to collect and analyze data from complex and critical driving situations in a controlled environment without endangering the life of the driver or destroy the prototype.

In order to realize a driving simulator with a HMD, the device has to be integrated into the control loop of the DS. In most cases, a driving model is implemented, which consists of integrated traffic scenes and real driver input hardware like gas pedal, brake pedal, and steering wheel. A VR driving simulator as basis for future automotive HMI applications is expected to solve the problem and challenges ADAS researchers' face of scientifically proving the reliability and safety of HMI concepts. This is due to lack of proper analysis at the early stage of development before the physical prototype is developed. The virtual prototype enables fast prototyping and early analysis of a concept to unforeseen circumstances without waiting for the physical prototype. The quality of the product and hence, a better user experience is also expected to improve considerably. This is because it is a user-oriented development where considerable number of variants can be shown, modified, and assessed at a very early phase of the process, thereby reducing the time and costs required for the overall process (Volkswagen 2015).

15.1.1 Motivation

The level of immersion experienced in real time using VR devices such as head-mounted displays due to the stereoscopic three-dimensional (3D) effect and the wide field of view is expected to enhance development of driving safety functions. Therefore, a DS with integrated HMD would facilitate rapid prototyping and introduction of new HMI-features since all changes could be made directly on the virtual prototype until expected result is achieved. This would serve as basis for a physical prototype. This is very important in the automotive industry with high level of competition, need for cost-reduction, and globalization, and most especially as vehicles are getting highly connected and interactive giving room to more driver distraction concerns.

Another benefit of 3D DS development is flexible, user-oriented adaptation of functions and HMI components in the context of a recursive, user-centered development for an enhanced user experience. Implementation of new ADAS, for example, an adaptive-cruise-control (ACC) or adaptive collision-control could be rapidly and flexibly evaluated using a complete VR approach with regard to usability and acceptance without the use of real hardware which are usually very expensive. Some VR DS systems are composed of only the Logitech steering wheel and pedals, 3D models, sound systems, and a HMD, the surrounding environment are all virtual objects. DS in a 3D environment with space depth and reliable distances offers the user an immersive opportunity to drive in an environment almost similar to a real driving vehicle. This environment is safe and risk free for the test driver and the prototype.

Furthermore, best tools for developing and interacting with 3D applications could be discovered through reviews and benchmarks. Likewise, solutions to tackle the unpleasant effects of simulation sickness could be disclosed, implemented, and evaluated. Finally, through the integration of a real car and a motion platform to simulate motion cues, user tests of new concepts could be performed in an ergonomic environment similar to reality, the limitations of the automotive concept could be disclosed, and effective users' feedbacks collected for future work in eliminating these limitations on the physical prototype and subsequently, the actual product.

15.1.2 Overview

Section 15.1 introduced this work, describing the motivation and a brief description of the purpose of a VR DS. Section 15.2 discusses the state of the art of the key technologies in a VRDS. A historical background, characteristics, and applicability of VR and DS in various industrial sectors is described in detail. This chapter is finalized by a short summary of both technologies.

Section 15.3 describes the key points in the realization of a VR DS, the concept, tools, and implementation of a prototype. This chapter also provides a detailed description of the steps required for the development of the simulator. These steps include; VR Simulator concept, modeling of high resolution 3D objects, prototype development, and interaction concepts. This is followed by the selection of an appropriate motion platform for the simulation of motion cues.

Section 15.4 illustrates the shortcomings and limitations of simulated environments. The limitations of systems using VR and DS technologies are discussed and suggested solutions stated.

Finally, Sect. 15.5 presents the conclusion of this work. This section also states the challenges and limitations faced by automotive components suppliers regarding reliability of the developed applications and also the limitations connected with the use of 2D DS, and why VR technology could be a better solution.

15.2 State of the Art

15.2.1 Virtual Reality

VR as it is mostly called is a 3D multi-sensory highly interactive artificial environment which gives the user the sense of spatial presence through total immersion (VirtualReality 2009; Ni et al. 2006). The feeling of immersion could arise as a result of the user feeling isolated from the real environment but being part of the simulated environment, and being able to interact and manipulate virtual objects in a natural way which provides the illusion of moving in the virtual environment. Depending on the level of immersion experienced, the sense of spatial presence could be achieved through multiple sensory channels interaction and stereoscopic 3D effect. The sense of presence in the virtual world is reached, when the user perceives the virtual world as real and believes to be in a different location rather than the actual real environment.

VR is an old technology that originated from research on 3D interactive graphics and vehicle simulation in the 1960s and 1970s. The term VR is associated with the following terms; computer-generated graphics, virtual environment, stereoscopic 3D, real-time tracking, highly interactive, and multiple sensory channels. The virtual environment could be real for example the simulation of a real flight cockpit for training pilots, or an imaginary flight cockpit for gaming.

VR is classified into various categories based on the level of immersion it offers. These ranges from Non-immersive for example desktop visualization of 3D objects, Augmented VR for example Head-up Display (HuD) to Immersive VR achieved with the help of a HMD, for example, HTC Vive or a CAVE (Cave automatic virtual environment). The CAVE, which was discovered in 1992, is a projection-based system which uses 3D images projected on the walls and floor of a cube-like room to create an immersive experience. This technology immerses users into the virtual world with scenes projected on every corner of the room and surround sound, they could also navigate freely with the help of 3D glasses. Therefore, the CAVE provides a high level of immersion without disconnecting the user from the real world and, it enables multiple users to share the experience and move and interact freely in the virtual world. However, when compared to HMDs, it is very expensive to set up and requires an entire room.

HMDs such as Oculus Rift, Samsung Gear VR, and HTC Vive are currently the most used VR devices on the market (Fig. 15.1). Unlike the CAVE which involves a room and projectors, HMDs are affordable, easy to transport, use small display screens worn close to the user's eyes and move with the viewer, and provide complete immersion in the virtual world by disconnecting viewers from their real environment, therefore providing a totally immersive VR experience.

For a fully immersive experience in the virtual environment, extra control devices are necessary for an optimal interaction with virtual objects. These devices aid to navigate freely within the virtual world. For example, the *Virtualizer* (Cyberith 2014) aids to move freely (walk, jump, sit, and run) in the simulated world, tracking the position of head, hand, or the entire body. Conventional control devices like Mouse, Trackball, and Joystick are not very effective with HMDs because the eyes are completely covered and users especially non-gamers, find it difficult to locate the position of the navigation buttons. Data Gloves with inbuilt sensors on fingers track the user's hand and finger motions and could be used to manipulate virtual objects. For driving simulators, Logitech steering wheels and pedals with force feedback are widely used to enable a realistic driving experience. Other tracking devices such as the Kinect, are used to scan and create user's avatar and this is mapped to the user's motion in order to improve the level of immersion experienced (Aitpayev and Gaber 2012).

15.2.1.1 Historical Background

This subchapter illustrates some of the VR highlights of the past such as the inventions of two VR pioneers namely; Morton Heilig who invented the Sensorama

Fig. 15.1 Oculus Rift and HTC Vive virtual reality head-mounted display

machine in 1957 and Ivan Sutherland, who invented the Ultimate Display headset in 1965. Stated in this chapter are also some of the reasons why VR was not widely used in the past decades though the potentials it offers in various industrial sectors were very obvious then, until the rebirth of VR in 2012 by Palmer Lucky who founded the Oculus Rift.

Morton Heilig, a cinematographer, who had the vision of an interactive 3D in the 1950s, invented the multi-sensory motorcycle simulator known as the Sensorama in 1957. The simulator enabled users to watch 3D films, which was later patented in 1962 (Morton 1962). The Sensorama combined all sensory channels by generating visual scenes of the city driven, engine and city sounds for auditory stimulation, smell of exhaust and food and seat vibration in its interaction concept in order to give its users the illusion of being part of the virtual environment. Apart from the generated vibrations of the seat, the machine also generated wind for an immersive experience of the visually perceived scenes. Heilig understood in his research of "The Cinema of the future" that in order to recreate reality, all sensory channels that influence human perception of the reality has to be taken into consideration and not just sight and sound (ArtMuseem 2000). This is actually what makes VR so complex to achieve. Though it is difficult to say for sure who invented VR, most papers refer Morton Heilig as the *father of VR* because of his contribution in insuring a near-realistic virtual experience through his multi-sensory interaction concept. The Sensorama was however not a big commercial success but laid down a blueprint for future 3D interaction concepts (Fig. 15.2).

Another great invention of the 1960s in the area of VR is the Ultimate display HMD by Ivan Sutherland in 1965. The Ultimate display was part of the PhD dissertation concept of Ivan Sutherland in 1963. The concept was later developed to the Sword of Damocles, which is claimed by some researchers to be the world first HMD to track the head in real time. Just like the Sensorama, it is an interactive device which generated acoustic, olfactory, taste, and tactile stimulation. Sutherland on his research on the world of 3D graphics described his headset as a window in the artificial environment. He described the screen as a window through which one

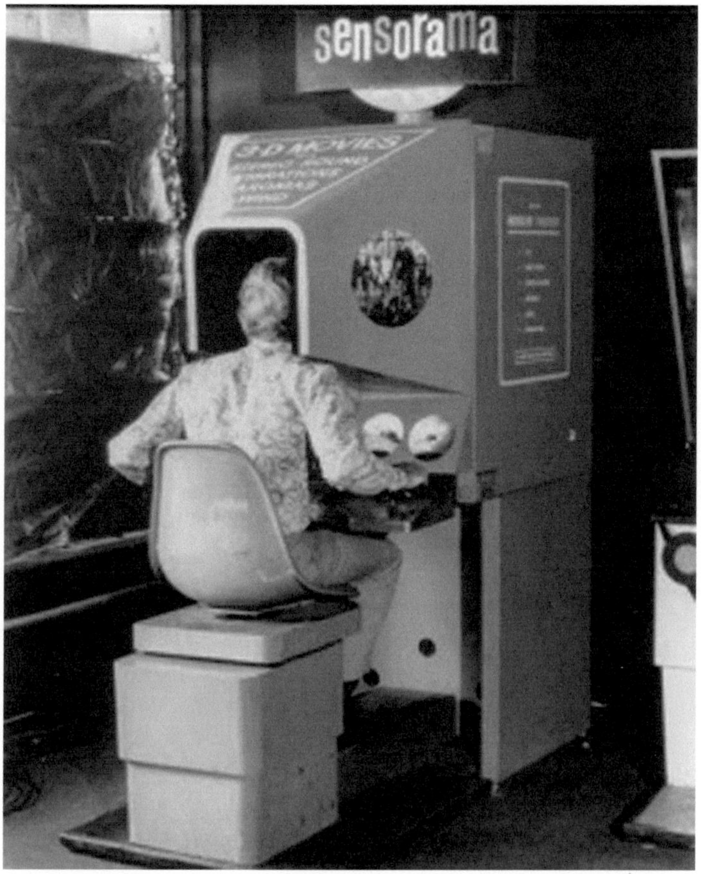

Fig. 15.2 The Sensorama motorcycle (Morton 1962)

sees a virtual world but, the challenge still remains to make that world look real, act real, sound real, and feel real. This challenge as he rightfully quoted then has not changed. Though the Sensorama and The Sword of Damocles devices achieved the ultimate goal of VR by placing the users in a virtual environment, the Sword of Damocles goes further to track the user's head movement which was correctly mapped to the stereo view in real time (Fig. 15.3).

Other devices and applications followed such as the Visually Coupled Airborne Systems Simulator (Kocian 1977) which is a visual system which comprises of a helmet with a tiny television tube and imaging optics meant to resolve visual presentation problems of flight simulators by imposing HuD contents in the pilot's view. However, due to software and hardware limitations, side effects such as simulation sickness, high cost of setting up a VR labs, lack of accurate head-tracking, computers with limited processing power for HMDs in the past decades, VR was not successful and not widely applied. Nevertheless, this marked

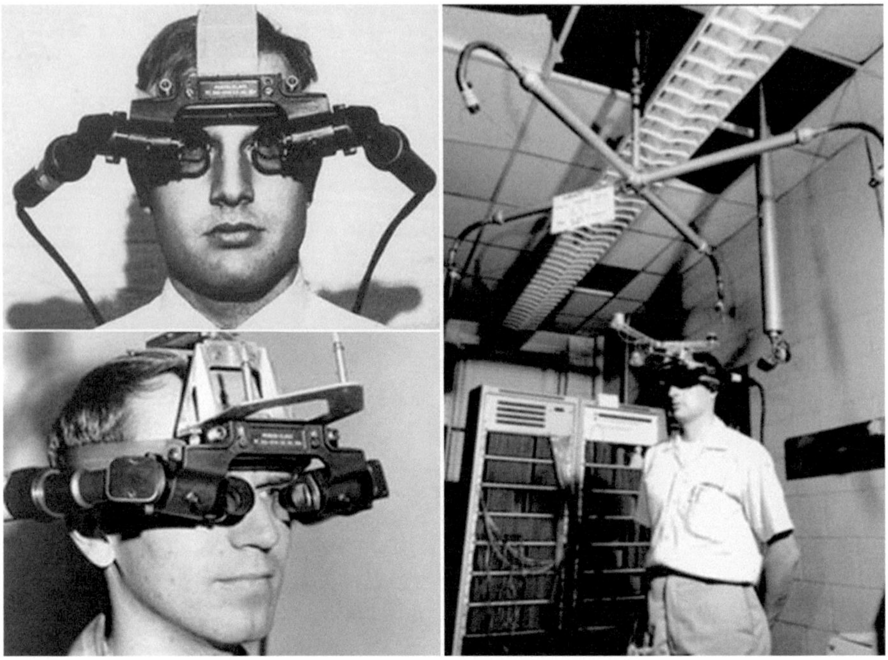

Fig. 15.3 The Sword of Damocles HMD (Sutherland 1965)

the beginning of an era which would change the way users interact with their systems and basis for the development of better virtual devices such as research into 3D interactive concepts with virtual objects through stimulation of all sensory organs and just like Sutherland rightfully stated, making it feel real, act real, sound real, and look real for the users of virtual environments. Morton Heilig and Ivan Sutherland also greatly inspired many works in the area of VR, highly interactive computer graphics, and HMI concepts.

15.2.1.2 Industrial Application of Virtual Reality

The psychological effects VR devices have on its users by creating a sense of presence in the virtual environment and at the same time, disconnecting them from their real surroundings could provide a natural form of interaction and thus, enhanced HMI concepts. For example, a test driver could be confronted with a dangerous traffic situation and reacts with panic although the environment is safe and absolutely risk free. This in turn, could create new forms of HMI concepts applicable and profitable to several industrial sectors. Though at the mention of VR, most people associate it directly with gaming only; VR could be applied in various industrial sectors from the design phase to marketing and maintenance of products and services. VR enables 3D visualization for a better understanding of concepts and how they interact. A virtual prototype of a new concept could be developed and tested on its expected functionality. This enables the discovery of discrepancies or safety challenges at the early stage of development.

That is especially essential for a sector like the automotive industry and other sectors that carefully need to avoid safety critical situations or situations with significant distraction of the driver from his primary task. Hence, the offered solutions need to be thoroughly evaluated before being introduced into a product and a proper evaluation is performed on a real environment. This allows a 3D perception of structures and provides new and risk-free methods of evaluation of critical traffic situations in the automotive industry without necessarily taking users to the road especially at the early stage of the development. It also reduces the number of prototypes developed because the prototype is not exposed to any physical damage and all changes could be performed on the virtual prototype.

Most of the leading automotive manufacturers use VR in various phases in their development process (refer to Table 15.1).

VR could be applied to develop virtual architectural models of houses before they are built in order to foresee damages and prepare adequate preventive measures. Virtual prototypes enable potential customers to step into the computer-generated building and interact with their future house before it is built instead of just watching a non-interactive image. This has the advantage that users can visualize, explore, and create a better understanding of the house they had in mind, therefore achieving a better customer cooperation and satisfaction.

Table 15.1 Few automotive manufacturers who apply the VR technology

Automotive manufacturers using virtual reality		
Manufacturer	Application area	Benefits
Audi	Customer vehicle configuration	• Enhanced user experience • Virtual showroom saves space
BMW	Vehicle development-virtual prototype and car designing	Fast prototyping—Saves time and costs
Chevrolet	Advertisement and sales	Virtual showroom-immersive driving experience
Fiat Chrysler	Advertisement and Sales	Explore a car before it is built
Ford Motor	• Vehicle interior and exterior design • Autonomous vehicle technology	• Design better and safer cars • Rapid prototyping
Lexus	Virtual prototyping	Test cars before they are built
Mercedes Benz	Virtual test drive—Marketing	Test cars before they are released
Nissan	Virtual test drive	Product awareness
Porsche	Customer entertainment	Product awareness
Renault	Research and development	Saves time and costs
Toyota	Driver distraction campaign	Creates awareness to safe driving
Volvo	Virtual test drive—Marketing	Product awareness

In the medical field, VR is mostly used for training and therapy. An example is the training of medical students to perform virtual surgery and get acquainted with the practice before the real surgery is performed on real patients. The stereoscopic 3D and high resolution displayed by HMDs presents detailed near-eye resolution of medical imaging diagnostics. Through the simulation of the entire physiology of the human body and organs, doctors could easily diagnose problems and chose the appropriate therapy faster. Therapists may also use the technology to treat people who are afraid of heights and needles with the help of a virtual application and virtual needle therapy. The surgeon using a HMD may have a complete simulated view of the surgery and detect new and easier ways to perform the surgery. For medical students learning how to operate, the best way would be to start with virtual patients. With the use of self-avatar, stroke patients could be stimulated to move disabled body parts by seeing their body do the movements in the virtual world.

In the Air Force, VR is applied for training of complex tasks and techniques without putting the users in any kind of risk. A full simulation to replicate a real-life situation, for example a simulated aircraft with all elements like in a real cockpit and place pilots in that position in order to improve their skills in a controlled environment. Table 15.2 shows some industrial application areas of VR.

Table 15.2 Some application areas of virtual reality

Industrial sector	Application area
Automotive	• Vehicle designing • Immersive virtual driving tests • Marketing and sales • Collaborative engineering • Evaluation of concept and performance targets
Healthcare	• Training and education • Diagnosis and treatment • Surgery simulation • Phobia treatment • Various therapies
Construction	• Virtual architectural design • Virtual buildings • Marketing and sales
Packaging	• Design and virtual prototypes
Entertainment	• 3D gaming • Virtual gallery and museums
Business	• Training employees • Virtual tours • 3D Product representation • Advertisement • Virtual meeting
Military	• Posttraumatic therapy • Training

15.2.1.3 Summary

VR is a real-time simulation that offers a high-end user interface that tricks the brain with the illusion of handling in the computer-generated world rather than operating from outside the virtual world. The main objective of VR is to create a real or imaginary world and objects to represent real-life situations, which are easy to operate and understand by the user for example, driving simulators to represent complex traffic scenarios. Furthermore, VR devices like HMDs with its real-time position tracking ability, stereoscopic 3D, near-eye resolution, and ultra-low latency provide a natural form of interaction within the virtual environment due to the multi-sensory channels. The good news is, most of the obstacles that hindered the advancement of VR have been resolved or its resolution in progress, and the devices are very much affordable ranging from 99 US Dollars for Samsung Gear VR, 599 US Dollars for the Oculus Rift to 799 US Dollar for the HTC Vive. Those who cannot afford the renowned devices mentioned above, could still experience full immersion using the ultra-low-cost Google cardboard and many other cardboards affordable from 5 US Dollars onwards.

As amazing and captivating as VR may seem, it still has a long way to go before its full integration into industrial processes is achieved. Creating an accurate replica of reality required to fool the human brain demands high computing power with appropriate programming. However, the experience and potential it offers, shows that it is worth the trouble since users get so captivated and spend a lot of time marveling at the virtual environment, which in turn would create awareness about the product or service in question. It is recommended to make the experience as interesting as possible by creating near-realistic scenes for experiences meant to replicate real-life products or services. The time spent in the virtual world should be limited because not every user is comfortable with HMDs and simulation sickness could occur at any stage during the experience. With the rapid emergence of new technologies such as better graphic cards, computer processing power, 3D modeling tools, it is expected that content building, which is still a major setback and other limitations would be easily dealt with in the future. VR should complement and not replace development processes or physical prototypes.

15.2.2 Driving Simulation

DS is a representation of driving scenarios and its complexity in a computer-simulated environment. It is used for research, development of new products, future enhancement for a better product quality, product verification with regards to reliability, and system robustness. It is composed of several highly performing subsystems for example, visual assets, motion system, interaction tools, and car dynamics which are integrated to reproduce near-realistic traffic models in real time. According to a research carried out by the Center of accident of the Monash University in Australia on the application of driving simulators in the prediction of changes in real-world crash risk, driving simulators are important and essential for pre-evaluation of interventions not yet tested (Rubin-Brown et al. 2009).

New automotive HMI components that are often derived from CE products are normally meant to be tested in real vehicles, on a real road, and with a real driver. To test a new component for its reliability and safety for the road could be very costly and risky and it should be repeatable. For this reason, driving simulation just like any simulation system is used to represent a real-world model of such concepts for example product or process model. The models are recreated on a computer to better understand, analyze, design, and easily manipulate virtual prototypes of complex systems mostly developed in Matlab/Simulink (Robinson 2004). Most of these simulations are running on computers with vehicle models implemented in a close loop. The use of these models in place of real vehicles and physical prototypes of automotive components, saves time, work, reduces cost, and avoids unforeseen hazardous situations which could arise from an unproven technology. In order to achieve a realistic driving experience in a virtual environment, it is important to have a realistic description of all factors involved in a driving scenario for example traffic situation, the road and driver behavior, when creating a simulation model for a vehicle (Gühman et al. 2012). Likewise, it is crucial to simulate speed and acceleration correctly when assessing ADAS for a realistic and correct simulation of vehicle behavior (Kemeny 2014).

The two main types of driving simulators which are widely used for research purposes are **HIL** (Hardware in the Loop), which is a form of real-time simulation mostly used in testing complex embedded systems with integrated real test components for example an electronic control unit (ECU), and **SIL** (Software in the Loop), which consists of only software model in a close loop. During a **HIL** simulation, a mathematical model of the process normally referred as the **Plant** such as a vehicle model or an engine is simulated and integrated with the real device to be tested. The test object is manipulated to accept the simulated model as a real object. Since the real environment where the hardware is supposed to run is a simulated model, a possible damage to the plant such as car engine is avoidable and therefore, reduces cost and risk (Halvorsen).

15.2.2.1 Choosing the Right Driving Simulator

The choice of a driving simulator system depends on the purpose and project for which the simulator is required. Driving simulators range from very simple and compact systems to highly sophisticated and multi-million US Dollars systems such as the Dome. The advanced and highly immersive Dome driving simulator engulfs a real vehicle, offers a visual field of up to 360° and has high performance motion simulators of 6–9 degrees of freedom (DOF). The Dome is so far the most sophisticated and advanced driving simulator meant for research and engineering purposes.

The next category of driving simulators are the so-called full-scale driving simulators which use a real vehicle or a semi-vehicle in order to improve comfort or ergonomics by presenting the user with a real vehicle components and dashboard.

Fig. 15.4 D-BOX actuators (*left*) and AMS motion platform (*right*)

Fig. 15.5 Audi A8 driving simulator with D-BOX actuators

In order to reflect the behavior of the real vehicle, a motion platform is integrated, for example, the D-BOX motion simulators which move the entire car while it reproduces in real time the motion cues which are correctly mapped to the visually perceived motion to the driver (D-BOX 2016) (Figs. 15.4 and 15.5).

The last type of driving simulator to be considered in this chapter is the compact driving simulator. An example of this type of driving simulator system is the A3 from the Atomic Motion System shown in Fig. 15.6. Compact driving simulators are low-cost simulators which could accommodate extra screens attached direct to the system or on the wall. Steering wheels and pedals such as the Logitech G27

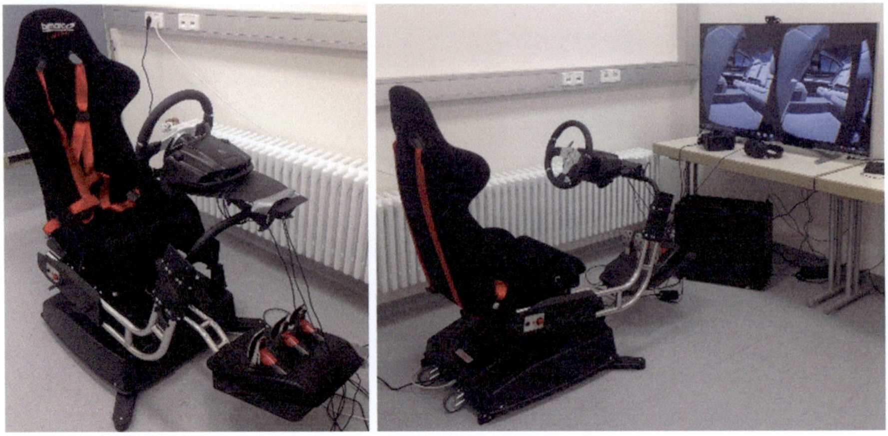

Fig. 15.6 Atomic motion system compact driving simulator

with force feedback could be easily integrated. Compact as the name states, it is easy to transport for instance, to fares and for customer demonstration. Though most come with integrated motion simulators, this could be easily integrated depending on the compatibility to third party products.

15.3 Realization of the Virtual Reality Driving Simulation

15.3.1 Concept and Architectural Design

The idea of integrating VR to a DS is meant to speed up the development process of automotive components through early evaluation with an interactive virtual prototype. These virtual prototypes drag the users into the system thus enabling an immersive and natural interaction with a product yet to be built. This will in turn reduce the number of iterations through reducing the number of physical prototypes developed since all changes could be made directly on the virtual prototype (Coates et al. 2002). The most important aspect of the immersive quality provided by a VR DS is the feeling of being part of the simulated environment which encourages the acceptance of this artificial environment as real. This provides room for a realistic testing of critical safety systems at the early stage of development when the concept is still not yet ascertained. The acceptance of the virtual environment as real has the advantage that the feedback provided by test drivers relates very closely to a realistic driving experience on a real road. This feedback could be effectively used as a basis for the development of the physical prototype and thus, saving production costs and time.

In order to develop a virtual driving simulator which could be effectively integrated in the development cycle of automotive components, specific requirements, and steps have to be considered and effectively implemented. These involve the selection of VR-specific tools such as simulator software with VR devices integration possibility for example 3D engines such as Unity3D and Unreal Engine 4. Most important is the correct simulation of the virtual environment contents such as 3D models of cars, landscapes, roads, and the vehicle control model. The visual quality of the environment determines if the artificial environment is accepted as real or not since the appearance of pixels as a result of low resolution definitely does not promote realism. Therefore choosing appropriate tools for VR systems plays a major role especially when it involves complex systems like a driving simulator with high performance meant for the evaluation of safety functions.

The process of selecting the tools should be done through benchmarking of the performance of the tools into consideration and recommendation of VR device manufacturers such as Oculus VR for systems using Oculus Rift (Oculus 2016). Though the use of renowned driving simulation software such as SILAB would be the most appropriate because of the many features and ready-made virtual environment which has passed the test of time, 3D simulation tools compatible with head-mounted displays are favored because most renowned head-mounted displays support Unity3D and Unreal Engine 4. It is however recommended to consider the integration concept and limitation factors of all components especially Third-Party tools, for example input and output devices such as Logitech steering wheels and pedals when considering 3D engines because not all have the ready plugins. Unity3D has support for most Third-Party tools.

The process of driving coupled with the different driving maneuvers in order to perform the needed driving tasks, yield different types of motion cues which are essential for the driver's effective and intuitive maneuver of the traffic situation. The visual cues are not enough when driving since the motion felt is needed in order to connect the driver's body to the vehicle mechanics. This makes it a must to reproduce the most important visual motion cues by selecting a high performance motion platform in order to simulate the required motion cues. The integration of the chosen motion platform poses a big challenge because a 1:1 scaling of the motion implemented and felt is necessary. For example, the driver should be able to differentiate about a vibration felt on a smooth surface and that felt on a rough road surface in order to react appropriately. Therefore, it is essential to reproduce the motion accurately since research has proven that no motion at all is better than bad motion. This is because if the motion does not represent the visually perceived motion, it could result in the wrong maneuver and also simulation sickness especially in a vigorously moving environment like driving simulators.

The concept of a VR DS is centered on the Human actor. The Human actor wears the head-mounted display, controls all input devices, and experiences various feedback based on the inputs. During driving, the input devices send signals to the simulation model. The simulation model accepts this signal and calculates the corresponding feedback. It then provides an adequate response to the output devices. This response goes back to the Human actor. The system works continuously in a close loop until the application stops. This simple concept encourages

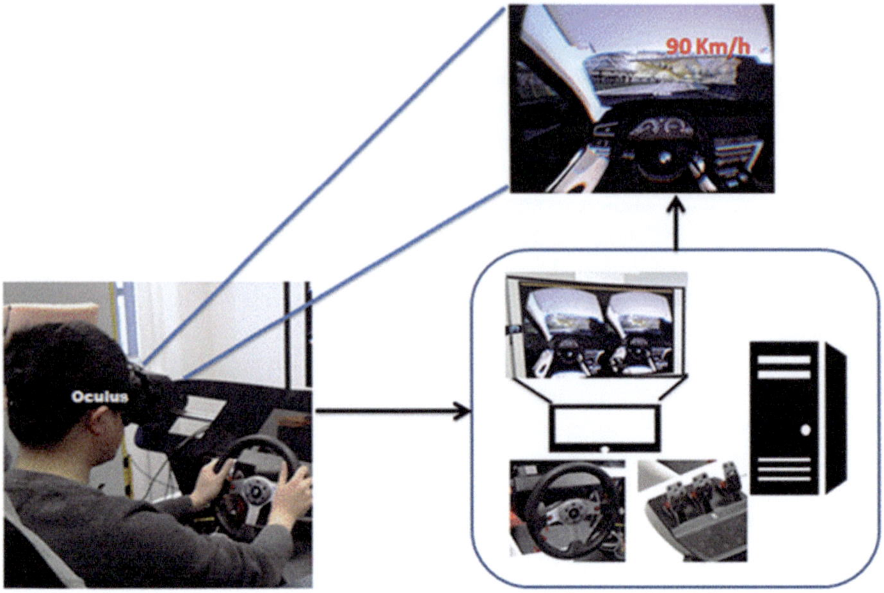

Fig. 15.7 A VR driving simulation centered on the Human actor

early validation of the components and system as a whole. Test drivers could experience real-time simulation of new safety products behavior and enables a near-realistic interaction. This would reduce the development time since discrepancies are detected and readily taken care of at the early stage of development. The development of a virtual driving simulation should be model-based and should support easy integration and reusability of components which could arise as a result of changed requirements. Figure 15.7 above illustrates this concept.

15.3.2 Implementation

The implementation of the VR DS at the *Center of Competence HMI, Robert Bosch GmbH in Leonberg* in cooperation with the *UniTyLab of the University of Applied Sciences in Heilbronn* is based on two driving simulation concepts, namely the *Compact* and the *Full Scale* driving simulators. While the Full Scale composed of a semi-Audi A8 with full integrated components and real car cabin, and a D-BOX motion platform, the compact simulator consisted of the Atomic Motion Systems A3 (AMS A3). This work will however focus on the implementation of the AMS A3 Compact driving simulation system because the full-scale simulator system is still under development. Figures 15.5 and 15.6 show the semi-Audi A8 full scale and the AMS A3 driving simulator mock up.

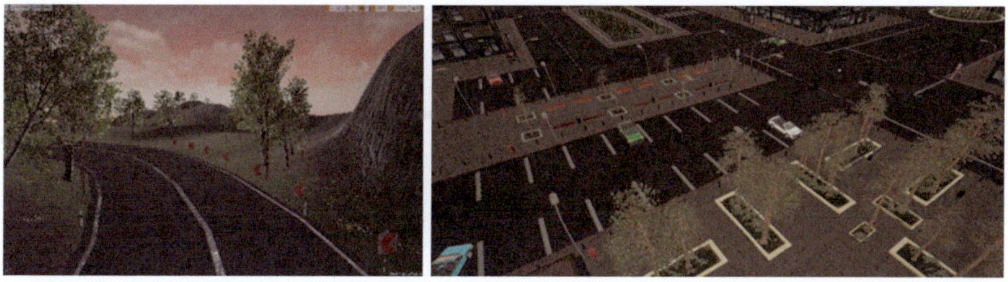

Fig. 15.8 3D model of a typical German country road (*left*) and traffic model (*right*)

In order to develop a complex VR application such as a driving simulator, a high-end computer with a powerful graphic card is recommended. The central processing unit (CPU) used for the prototype development is the Intel Core i7-5960X which has an eight core processor unit and an Asus X99 mainboard which enables the addition of multiple graphic cards. Two GeForce GTX 1080 graphic cards are selected for an optimal graphical performance since this is very essential for VR application. Other components such as the RAM and disk drives are selected based on the application to be developed and benchmark.

Both Unity3D and Unreal Engine 4 are considered for the implementation of the simulator software. Based on the benchmark carried out on a previous work, Unity3D was considered because of its large user community for easy problem tackling and availability of most third-party plugins. Unreal engine 4 on the other hand, provides a higher and better visualization performance of the virtual environment for a near-eye visualization and also the Blueprint technology for non-developers. The implementation also involves the modeling of the 3D assets for the virtual environment such as roads, traffic signs, landscapes, and many more assets that make up a realistic driving environment (Fig. 15.8).

The prototype development consists of two phases; the first phase consists of the simulation of the traffic environment and a car model, taking into consideration an accurate simulation of the vehicle dynamics. This phase also involves the integration of control devices such as Logitech steering wheel, shifter and pedals, and a 3D audio for the surrounding and engine sound. The second phase consists of the integration of the motion system to the simulator software and the implementation of dynamic head-up display features for various use cases which was actually one of the main objectives of this simulator concept. Head-up displays enable the driver on a real road to focus on the road and therefore avoiding distractions since all the necessary features are projected on the windscreen. The easiest implementation of head-up display could be done as a standard floating GUI text. This is however not enough for complex and dynamic features. Unreal engine 4 on the other hand with its Blueprint technology provides a faster and better solution for head-up display contents. Figure 15.9 shows prototype without head-up display and with head-up display.

Finally, an appropriate haptic feedback device for hand-finger tracking and when necessary, with upper or full body tracking so that drivers could handle freely and intuitively like he would in a real car and also interact with clickable interfaces like

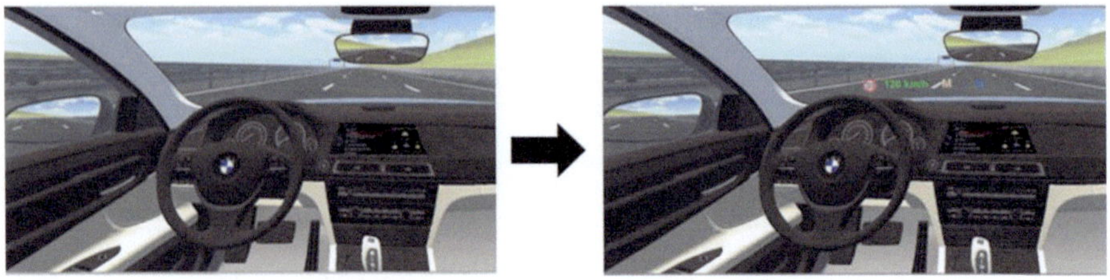

Fig. 15.9 Wrong (*left*) and correct (*right*) display of a BMW Dashboard

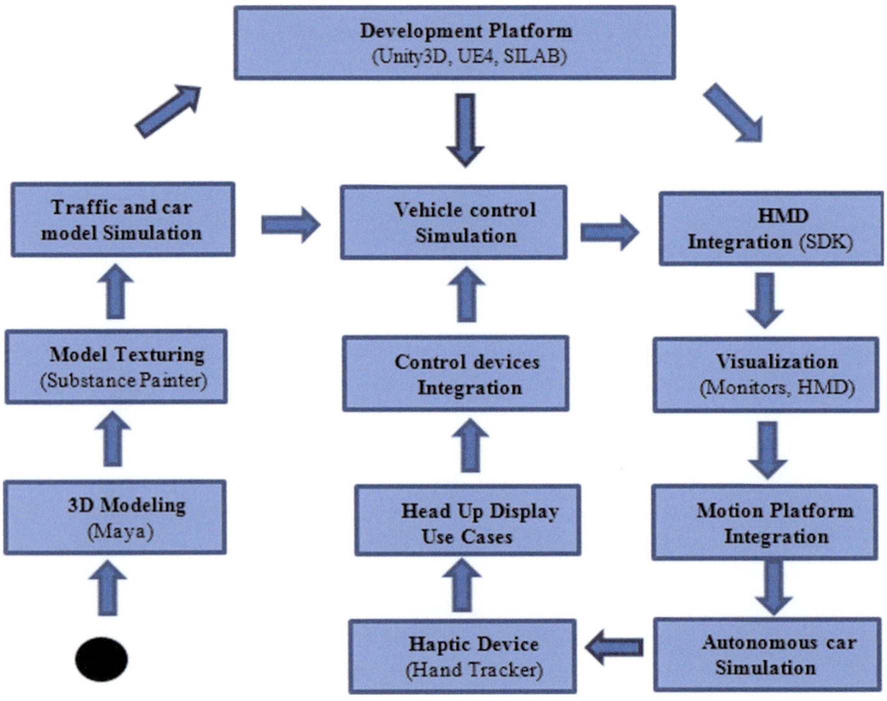

Fig. 15.10 The workflow of the simulator implementation

in a real car infotainment system. Considered for this simulator is the *Manus VR* Glove which tracks the user's hand movements in real time, with the help of sensors inside the glove (MANUS VR 2016). Figure 15.10 shows the most important steps and functions to be considered when implementing a VR DS. In addition to the steps illustrated below, the choice of a VR—ready computer with high-end processing power and graphics is very important.

15.3.2.1 Driving Scripts

The driving script is the main controller in this driving simulator. Due to the fact that 3D Engines offer only wheel colliders and simple Physics, it is therefore very hard to get a realistic driving model. This is not enough to develop a realistic driving simulator with all the complexity involved in a driving scenario. In order to implement a realistic car model, the driving script was implemented in JavaScript and C# in Unity3D which consists of transformer object for the steering, wheel colliders for the Front Left and Front Right wheels, vehicle dynamic parameters, for example, engine maximum Torque and controller functions.

The scripts consists of a Start function, where the center of mass of the rigid body is initialized, an Update function, which is constantly reading the inputs from the driver and transforming the resulting movements to the wheels and steering of the moving vehicle and a graphical user interface function, which updates the display values in the car cockpit. Figure 15.11 shows the driving script control loop.

Finally the driving script is integrated into the Unity3D VR DS project. During this integration, the Front Right and Front Left wheel colliders of the car are mapped to the respective wheel collider variables of the driving script as well as the steering transform object.

15.3.2.2 Control Devices

In order to achieve a fully immersive experience and give the driver the illusion of presence in the driving virtual environment, a near-realistic interaction with virtual objects is essential. These devices aid the user to move freely (touch, feel, walk, sit, run, and jump) and manipulate objects like in real-life. Conventional control devices like Mouse, Trackball, and Joystick are not very effective with head-mounted displays because the eyes are completely covered and for

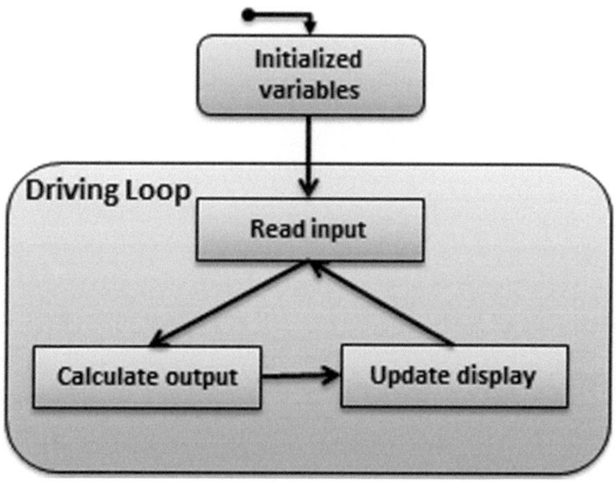

Fig. 15.11 Driving script control loop

Fig. 15.12 MANUS VR Glove for hand tracking

non-gamers, might be difficult to locate the various keys or buttons needed for specific navigations. Driving simulators third-party hardware such as Logitech G27 steering wheel and pedals with its force feedback transmission used to control the direction and speed of movement provides an optimal interaction when accurately implemented for a realistic driving impression. The control flow between the Logitech device and the 3D vehicle model simulation is realized in the 3D engines with the help of Logitech Plugins and Assets. The system could however be configured in a way that, if the Logitech is not connected, the system automatically starts using the keyboard as the driver input. Also very essential for a better interaction and user experience are hand and finger tracking devices such as Data Gloves with inbuilt sensors for tracking the test driver's hand and finger motions in real time and also to manipulate virtual objects such as clickable interfaces (Fig. 15.12).

15.3.2.3 Three-Dimensional Modeling and Animation

Three-dimensional models are a vital component in the development of a VR DS. The 3D models should consist of 3D stereoscopic display information reflecting the objects in the real world. Third-party assets could be used except the needed models are not available or not good enough for the required purpose. Unity3D asset store and Unreal engine 4 offer a good number of models for easy integration. Most 3D engines support .fbx and .obj assets. These 3D models were downloaded from the Unity Asset store. Most models were retextured using the Substance painter texturing software. Most assets come with very high polynomial numbers which could slow down rendering and also affect the performance of the implemented software negatively, some parts of the asset for example the car model could be removed by a 3D-Artist and any resulting issues could be easily fixed (Fig. 15.13). It is very essential to add AI vehicles and animated pedestrians in order to give the illusion of a congested traffic model.

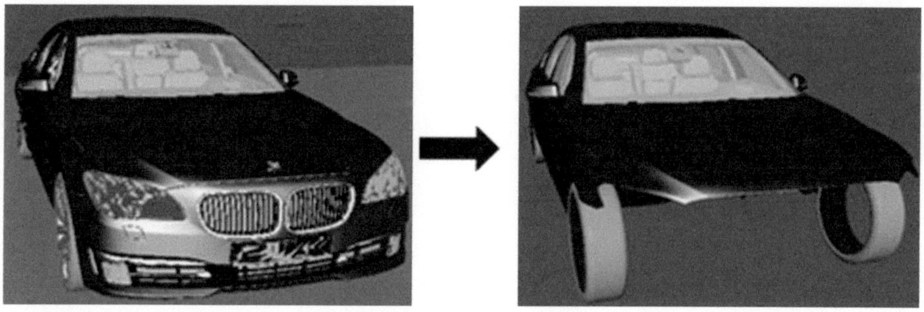

Fig. 15.13 Model reduction of parts not visible by the driver

15.3.2.4 Visualization Components

This consists of the head-mounted display, desktop monitor, or any other display monitors. Most important is the choice of a high-end graphic card and a renowned head-mounted display like the Oculus rift consumer version and HTC Vive. The visualization components interact with the simulation through the head-mounted display SDK which provides the integration package with Unity3D or Unreal engine 4, Library containing all the Prefabs, and plugins necessary to create and run a VR application. The main prefabs are Player Controller and Camera Rig. The camera controller is positioned in front of the driver, so that the actor wearing the device can be following the vehicle movement. The tracking sensor of the device is placed in a suitable position in order to accurately track the position of the device.

15.4 Challenges

It is quite complicated to replicate a real traffic situation because of the complexity of traffic conditions. It is even more complicated to develop a VR DS which the human brain has to accept as real. Though VR is not a new technology, its application is associated with a lot of limitations which must be overcome in order to explore the benefits of the technology. Few of these limitations associated with VRDS are; the lack of accurate motion cues which correlates the vehicle mechanic to the human body, lack of haptic feedback when the driver operates interfaces that should give him a clue of the next action without necessarily looking at the button, low resolution which does not correlate with reality because of the appearance of pixels, a traffic environment which does not correlate to reality, and the inconsistency in the speed driven and the speed felt.

All factors that influence the driving conditions must be taken into consideration when implementing a VRDS to be applied on a real product development. For example, the behavior of the vehicle must be consistent with the driver's expectations if the driver is expected to accept the environment as real. The punishment for not achieving the aim, which is to accurately fake reality, could result to simulation sickness. Simulation sickness which is a form of motion sickness, experienced by a person exposed to simulated environments.

Simulation sickness could be caused by sensory conflict as a result of a conflict between the visual perception of the simulated motion and the motion felt (Karl et al. 2013). People exposed to simulated environment like VR and DS could experience general discomfort which could influence the user's mental and physical state during evaluation sessions on a simulated environment. Therefore, it is relevant to find a solution on how to eliminate or reduce these symptoms to minimum. Almost all the limitations of VR and DS induce simulation sickness. For example, the lack of a motion platform or inappropriate motion cues could lead to simulation sickness because of conflicting information perceived visually and felt by the inner ear. While some researchers argue that little motion, appropriate or not is better than none, other studies suggest that no motion is better than a badly simulated motion. Dr. Kemeny however presented a motion cueing algorithm in his research titled *Motion cueing algorithms for a real-time automobile driving simulator*, which takes into consideration driver's attention and the road information in order to reduce false motion cues presented to the driver (Kemeny 2012). It is therefore essential to integrate motion simulators to generate the visually perceived motion in the form of vibration and tilting which gives the feel of a real car. The vibration is synchronized with simulated engine revolution per minute (RPM) for an improved perception of the virtual speed, while tilting offers a vestibular stimulation to suppress simulator sickness during acceleration, braking, and steering or cornering (Weinberg and Harsham 2009).

High latency which reflects a delay between an action such as pressing on the gas pedal and the response such as the car moving accordingly in the visually perceived world poses a big problem associated with all VR applications. While some platforms offer accurate Vehicle Physics for a near-realistic vehicle simulation model, it still remains *near-realistic* and the discrepancy could be felt by an attentive and well-experienced driver. This could limit the level of immersion experienced. Another big challenge is the integration of audio systems that reproduces the correct vehicle sound and surrounding traffic sounds. These are just few of the problems accounted during the development of a VR DS prototype. Every DS is different and is faced with task or system-specific problems which could be solved accordingly.

15.5 Conclusion

As vehicles become more complex and connected, there is need to implement solutions that avoid and reduce any form of distraction to the driver. DS provides a computer-generated environment for various research purposes in the automobile industry. The application of DS is preferable to real-life experiments on the road because it is cost-efficient and enables rapid prototyping but most especially, it is a controlled environment and totally risk free for the users. This facilitates rapid development of virtual products for evaluation of new concepts before the physical product is produced. Emerging technologies in the field of VR from the area of consumer electronics and increasing processing power offer potentials for new highly immersive evaluation concepts through the combination of a VR device to a driving simulator. The integration of a HMD to a DS is to simulate a realistic environment of imaginary or real objects which is easy to operate and accepted by the user as real. VR and DS use the same technology to render user interaction and the sense of presence in the virtual environment and are highly interactive computer-generated graphics, which offer a certain level of immersion in the virtual environment and in real time (Kemeny 2014). Because of the above-mentioned reasons, both technologies could easily be combined to develop a highly immersive VR driving simulator framework, for rapid prototyping of HMI concepts using highly performing HMD or the CAVE for multiple user experience.

It is very essential to implement the following factors correctly in order to have the full benefits of a VR DS. First, the modeling of high-quality visual assets which represent real vehicle and traffic models is of paramount importance. Second, congested traffic models which consist of all components that make up a real traffic situation such as artificial intelligent (AI) cars with drivers, traffic lights and signs, landscape, pedestrians have to be correctly simulated and animated where necessary. Since the aim of VR is to impose realism to the human brain, there is need for a natural form of interaction such as the use of an instrumented glove with haptic feedback for hand and finger tracking, high-resolution graphic quality for a pixel-free visualization, correct engine, and surrounding sounds. Most important is the 1:1 mapping of the motions seen and felt because the motion cues enables the driver to have the right reaction, for example, slow down when driving on a rough surface. The correct implementation of the factors mentioned above due to the combination of high-end visualization, real-time computer simulation and a multi-channel interaction would also help to reduce the occurrence of simulation sickness.

Virtual environments are highly captivating at a first try but, in order to remain captivated into this illusion of realism created by VR, the virtual environment has to be properly simulated to reflect the corresponding real world it is meant to represent. If the technology is applied to evaluate a product, the product must be simulated in its correct size and behavior, a high graphical representation and a near-realistic interaction concept established. This also involves the addition of appropriate virtual objects that make up the simulated product such as animated pedestrians and traffic lights into a driving simulator in order to represent a real traffic environment.

A VR DS would enable easy evaluation of emerging technologies still under research such as the autonomous and highly connected vehicles. For example, unforeseen issues which could arise from the behavior of the human driver and the autopilot could be investigated with regards to autonomous vehicles. Although DS and VR are two technologies are highly associated with simulation sickness as mentioned above, these technologies are however very important in the evaluation of new products. VR device manufacturers and researchers are working on possible solutions to reduce or completely eliminate these limitations. Many solutions have been studied and suggested by researchers such as reduction of the field of view, addition of visual assets, integration of motion platforms, high-end graphic cards and many more. Most of these solutions are application-specific. While complex applications like a DS might consider the correct implementation of a traffic model, vehicle control model, and most of the solutions mentioned above, applications which are only meant for vehicle interior design might focus on the graphical performance of the system.

The challenges of creating a VR application with a high graphical performance and an intuitive interaction concept which is convincing and acceptable to the human brain as real for total immersion in the artificial world are very high and should not be overlooked. However, as VR evolves, graphic cards and CPUs are getting more powerful by the day, and HMD manufacturers are also improving the optics of their devices because it plays a major role on the visuals. Controls devices are also getting better and cheaper, for example the MANUS VR glove for just 250 US Dollars. VR DS could effectively be integrated at the early stage of development to complement conventional 2D DSs and real-road evaluations in the development and evaluation of future automotive HMI concepts.

References

Aitpayev, K., & Gaber, J. (2012). Creation of 3D human avatar using kinect.

ArtMuseem. (2000). *Pioneers*. Von http://www.w2vr.com/timeline/Heilig.html abgerufen.

BMW. (2003, May). (B. Group, Herausgeber) Von Virtual reality in car production: http://goo.gl/xe42M7 abgerufen.

Coates, N., Ehrette, M., et al. (2002). Head-mounted display in driving simulation applications in cards.

Cyberith. (2014). *Cyberith*. Von http://cyberith.com/ abgerufen.

D-BOX. (2016). *D-BOX*. Von http://www.d-box.com/corporate-page/about-d-box/ abgerufen.

Eedelstein, S. (2014, October 03). *DigitalTrends*. Von http://tinyurl.com/ovx2w7e abgerufen.

Ford. (2015). (F. Media, Herausgeber) Von Virtual_Lab: http://goo.gl/AsFREq abgerufen.

Gühman, C., Riese, J., & Wolter, T.-M. (2012). *Simulation und Test für die Automobilelektronik IV*. Berlin: Expert-Verlag.

Heilig. (1962). Von http://www.mortonheilig.com/InventorVR.html abgerufen.

Howard, B. (2014, June). Von ExtremeTech: http://goo.gl/IrbMz0 abgerufen.

Karl, I., Berg, G., Rüger, F., & Färber, B. (2013, January 23). Driving behavior and simulator sickness while driving the vehicle in the loop: Validation of longitudinal driving behavior. Neubiberg: *IEEE Intelligent Transportation Systems Magazine*.

Kemeny, A. (2014). From driving simulation to virtual reality. In *VRIC '14 Proceedings of the 2014 Virtual Reality International Conference*. New York: ACM.

Kocian, D. F. (1977). *A visually-coupled airborne systems simulator (VCASS)—An approach to visual simulation*. Ohio.

MANUS VR. (2016). Von https://manus-vr.com/ abgerufen.

Ni, T., Schmidt, G. S., Staadt, O. G., Livingston, M. A., Ball, R., & May, R. (2006). A survey of large high-resolution display technologies, techniques, and applications. In *Proceedings of the IEEE Conference on Virtual Reality* (pp. 223–236).
Oculus. (2016). Von https://www.oculus.com/en-us/oculus-ready-pcs/ abgerufen.
Renault. (2014, March 26). *Photo-realistic digital mock-up design review in a five 4Kx4K sided CAVE*. VR Center and Immersive Simulation, Renault.
Robarts, S. (2015, January 15). *gizmag*. Von http://www.gizmag.com/audi-vr/35620/ abgerufen.
Robinson, S. (2004). *Simulation: The practice of model development and use*. Chichester: Wiley.
Rubin-Brown, C. M., Williamson, A., et al. (2009). *Can driving simulators be used to predict changes in real-world crash risk?*. Australien: Monash University.
Sutherland, I. (1965). *The ultimate display*. New York.
VirtualReality. (2009). *Virtual reality: What is virtual reality*. Von http://goo.gl/oFQ1Wf abgerufen.
Volkswagen. (2015). Von Virtual_Technology: http://goo.gl/UyhdGj abgerufen.
Weinberg, G., & Harsham, B. (2009). *Developing a low-cost driving simulator for the evaluation of in-vehicle technologies*. Essen: ACM 978-1-60558-571-0/09/009.

Bibliography

Arjona, J. T., & Menéndez, J. M. (2005). *Virtual reality devices in driving simulators: state of the art and ongoing developments at U.P.M.* Universidad Politécnica de Madrid, E. T. S. Ingenieros de Telecomunicación.
Arkviz. (2013, July). *Turbosquid: BMW 7 series Long 2013*. Von http://goo.gl/aVaZXI abgerufen.
ARM. (2015). Von ARM: Connected_Community: http://goo.gl/51hAxD abgerufen.
AutomotiveWorld. (2015). Von Bosch: Facts about driver assistance. http://goo.gl/CjBwgu abgerufen.
Bauer, H. (1998). *Fahrsichereitssysteme* (2 Ausg.) (R. B. GmbH, Hrsg.) Stuttgart, Deutschland.
Baur, M., Fullerton, M., & Busch, F. (2010). Realizing an effective and flexible ITS evaluation strategy through modular and multi-scaled traffic simulation. *IEEE Intelligent Transportation Systems Magazine*, *2*(3).
BoschMobilitySolution. (2015). Von Driver Assistance Systems: http://goo.gl/eOcFlq abgerufen.
Breuer, J. J., & Kaeding, W. (2006). Contributions of driving simulators to enhance real world safety. *Driving Simulation Conference, Asia Pacific*. Tsukuba.
CarBodyDesign. (2015). Von Designers at work: Virtual reality: http://goo.gl/oeAUYE abgerufen.
CarMaker, I. (2015). Von http://ipg.de/simulationsolutions/carmaker/ abgerufen.
Choi, S., & Cheung, H. (2012). *Virtual prototyping for rapid product development, modeling and simulation in engineering*, Catalin Alexandru (Ed.) (Bde. ISBN: 978-953-51-0012-6.). (InTech., Hrsg.) Available From: http://goo.gl/IocEIz. Von http://goo.gl/IocEIz abgerufen.
Daimler. (2010). Von Driving simulator in Sindelfingen: Investment in cutting-edge technologies: http://goo.gl/3XTlZN abgerufen.
Davis, S., Nesbitt, K., & Nalivaiko, E. (2014). *A systematic review of cybersickness*. New York, USA.
De Winter, J., Van Leuween, P., & Happee, P. (2012). *Advantages and disadvantages of driving simulators*.
DLR (Hrsg.). (2015). (DLR Institute of Transportation systems) Von VR-Lab and HMI-Lab: http://goo.gl/bjmguJ abgerufen.
dSPACE. (2015). Von One-step solution for HIL simulation: https://www.dspace.com/en/inc/home.cfm abgerufen.
ElectricTV. (2012). *Oculus rift's virtual reality HMD unreal game integration teaser*. Von http://goo.gl/Anzvdq abgerufen.

ETAS. (2015). Von Virtual_Vehicle_LabTruck: http://goo.gl/tVBIvS abgerufen.
FurryBall 4.8 GPU Benchmark. (2014). Von http://goo.gl/9NMSwA abgerufen.
Gaudiosi, J. (2015, September 23). *Fortune*. Von http://fortune.com/2015/09/23/ford-virtual-reality/ abgerufen.
gizmag. (2010). Von Lexus unveils 'world's most advanced' driving simulator: http://goo.gl/wA31JQ abgerufen.
Google. (kein Datum). Von http://goo.gl/eurfYc abgerufen.
Singh, H., Barbour, B. M., & Cox, D. J. (2011). Driving rehabilitation as delivered by driving simulation. *Handbook of Driving Simulation for Engineering, Medicine and Psychology*, 1–12. https://viragesimulation.com/
Halvorsen, H.-P. (kein Datum). *Hardware-in-the-loop simulation*. Porsgrunn, Norway.
Heise (Hrsg.). (2014, March 20). Von VR-Brillen: Sony Morpheus und Oculus Rift 2 im Vergleich: http://www.heise.de/ct/artikel/VR-Brillen-Sony-Morpheus-und-Oculus-Rift-2-im-Vergleich-2150859.html abgerufen.
Hoeberichts, Fons; Niewold, Frans. (2013, October 22). Moog driving simulator speeds Dallara's race car development and enhances driver training.
iGroup. (2008). Von http://tinyurl.com/oll6d7z abgerufen.
Illinois, U. o. (Hrsg.). (2013). *Projection-based CAVE*. Von http://goo.gl/9oz2hV abgerufen.
IPG. (2014). Von IPG simulation solutions: http://ipg.de/en/simulationsolutions/ abgerufen.
James, C., Kearney, J., & Papelis, Y. (1996). Driving simulation: Challenges for VR technology. *Computer Graphics and Applications, IEEE, 16*(5), 16–20.
Kemeny, A. (2012). Motion cueing algorithm for a real-time automobile driving simulator. *Driving simulation Conference*. Paris, France.
Koenig, S. (2015). *Virtual reality driving simulation*. Von http://www.virtualgamelab.com/ abgerufen.
Krueger, H.-P. (2005). *SILAB—A task-oriented driving simulation*. Orlando.
Kunochan. (2007). *A brief history of (virtual) reality*. Von http://goo.gl/QSskXA abgerufen.
Leap Motion. (2015). Von Unity 3D: Leap motion core assets: http://goo.gl/dZ4lTo abgerufen.
lfs. (2015). Von https://www.lfs.net/ abgerufen.
Logitech Gaming SDK. (2014, October). Von http://goo.gl/Ku4GNV abgerufen.
Mathworks, T. (2015). Von http://www.mathworks.com/index.html?s_tid=gn_loc_drop abgerufen.
Mathworks, T. (2015). Von Non Real-time Simulink Car Model: http://goo.gl/GFhVw0 abgerufen.
Maxwell Driving School. (2015). Von http://tinyurl.com/l6oalsx abgerufen.
Mayden, A. (2014, November). *Digitaltutors: Unreal engine 4 vs. unity: Which game engine is best for you*. Von http://goo.gl/C4jViF abgerufen.
Mixamo. (2014, January). *Male character pack*. Von http://goo.gl/LvA26p abgerufen.
mortonheilig. (kein Datum). Von http://www.mortonheilig.com/InventorVR.html abgerufen.
Muhanna, M. A. (2014). *Virtual reality and the CAVE: Taxonomy, interaction challenges and research directions*. Amman, Jordan: King Hussein School for Computing Sciences, Princess Sumaya University for Technology.
Nelson, F., & Yam, M. (2014, April 30). *The pioneers speak*. Von The Past, Present, and Future of VR and AR: http://www.tomshardware.com/reviews/ar-vr-technology-discussion,3811-2.html abgerufen.
Nelson, F., & Yam, M. (2014, April 30). *The pioneers speak (Oculus Rift)*. Von The past, present, and future of VR and AR: http://www.tomshardware.com/reviews/ar-vr-technology-discussion,3811-4.html abgerufen.
Nelson, F., & Yam, M. (2014, April 30). *The pioneers speak (Sony Morpheus)*. Von The Past, Present, and Future of VR and AR http://www.tomshardware.com/reviews/ar-vr-technology-discussion,3811-4.html abgerufen.
Ni, T., Schmidt, G. S., et al. (2006). A survey of large high-resolution display technologies, techniques, and applications. In *Proceedings of the IEEE Conference on Virtual Reality*, (pp. 223–236).

Nordmark, S., Jansson, H., & Palmkvist, a. H. (2004). The new VTI driving simulator. Multipurpose moving base with high performance linear motion. In *Driving Simulation Conference 2004*. Paris.
Oculus (Hrsg.). (2015). Von https://www.oculus.com/dk2/ abgerufen.
Olarte, O. (2011, April). *Alertdriving*. Von http://goo.gl/p8Mn7K abgerufen.
OpenSceneGraph. (2015). (OpenSceneGraph, Herausgeber) Von http://www.openscenegraph.org/ abgerufen.
Rademacher, M. H. (2014). *Virtual Reality in der Produktentwicklung*. Ilmenau: Springer Vieweg.
Reddit. (2014, July). *Buying graphics cards for DK2*. Von http://goo.gl/aiGyfA abgerufen.
Richardson, L., & Shaw, K. (2009). *Traffic safety in the European Union*. Von http://www.unc.edu/euce/eusa2009/papers/shaw_03F.pdf abgerufen.
Salvador, B., Fernandez, M., & Perez, M. (1996). Virtual reality for driving simulation. *Communications of the ACM, 39*(5), 72–76.
Samsung. (2014). *Gear VR design*. Von http://www.samsung.com/global/microsite/gearvr/gearvr_design.html abgerufen.
Samsung. (2014). *Gear VR specifications*. Von http://www.samsung.com/global/microsite/gearvr/gearvr_specs.html abgerufen.
Schick, B., Henning, J., Wurster, U., & Klein-Ridder, B. (2008, April). Simulation methods to evaluate and verify functions, quality and safety of driver assistance systems in the continuous MIL, SIL, and HIL process. *3. Tagung Aktive Sicherheit durch Fahrerassistenz*.
sensoriumresearch. (2012). Von http://goo.gl/v924xo abgerufen.
SiLab, W. (Hrsg.). (2014). Von http://www.wivw.de/en/silab abgerufen.
Slob, J. (2008). *State-of-the art driving simulators, a literature survey*. Eindhoven: Control Systems Technology Group, Eindhoven University of Technology, Department of Mechanical Engineering.
Smaragda, C., Michael, D., Gregoriades, A., & Pampaka, M. (2013). Design of a 3D interactive simulator for driver behavior analysis. CA.
Sommer, M. (2012, December). *Car tutorial*. Von http://goo.gl/DYZqg8 abgerufen.
Tesis-DYNAware. (2015). Von Vehicle Dynamics Simulation: http://www.tesis-dynaware.com/en.html abgerufen.
TopTenReviews. (2015). *Graphics cards review and comparisons*. Von http://goo.gl/LD63o8 abgerufen.
Unity Answers. (2015). Von Unity Community: http://goo.gl/2lGtDl abgerufen.
Unity Insider. (2014). Von Deutschland führendes Unity Netzwerk: http://goo.gl/btTXNf abgerufen.
Unity Manual. (2015). Von Unity documentation: http://goo.gl/KKSnws abgerufen.
Unity3D. (2015). Von http://unity3d.com/unity abgerufen.
Unity3D. (2015). Von Industry leading 2D and 3D power: http://unity3d.com/unity/2d-3d abgerufen.
Unreal. (2014). Von https://www.unrealengine.com/what-is-unreal-engine-4 abgerufen.
Unreal. (2014, November). Von Unreal vs unity: http://goo.gl/u1tnBQ abgerufen.
Visbox. (2015). *vixbox*. Von http://www.visbox.com/products/cave/ abgerufen.
vrealities. (2013). *vrealities*. Von https://www.vrealities.com/products/data-gloves abgerufen.
vrvana. (2014). Von http://www.vrvana.com abgerufen.
Vuzix. (2014). Von Wrap 1200DX-VR: http://www.vuzix.com/consumer/products_wrap_1200dxvr abgerufen.
Wikipadia: Unreal_Engine (Hrsg.). (2015, January). Von http://en.wikipedia.org/wiki/Unreal_Engine abgerufen.
Wikipedia. (2008, October). Von http://tinyurl.com/34mjxf abgerufen.
Wikipedia. (2013). Von http://preview.tinyurl.com/pjrgebu abgerufen.
Wikipedia. (2016). Von https://en.wikipedia.org/wiki/Cave_automatic_virtual_environment abgerufen.
Wikipedia: Driving_Simulator. (2014, July). Von http://en.wikipedia.org/wiki/Driving_simulator abgerufen.

Wikipedia: Unity (Hrsg.). (2015, January). Von http://en.wikipedia.org/wiki/Unity_(game_engine) abgerufen.

wtvox. (2014, July). *wtvox*. Von http://goo.gl/3yoTmA abgerufen.

Yoshimoto, K., & Suetomi, T. (2006). The history of research and development of driving simulators in Japan. In *Driving Simulation Conference, Asia Pacific*. Tsukuba.

Chapter 16
Methods to Validate Automotive User Interfaces Within Immersive Driving Environments

Diana Reich, Christian Buchholz and Rainer Stark

Abstract To ensure safety and usability of automotive user interfaces, prospective validations during early prototyping stages are important, especially when developing innovative human-cockpit interactions (HCI). Real car driving studies are difficult to control, manipulate, replicate, and standardize. Additionally, compared to other study designs, they are also time consuming and expensive. One economizing approach is the implementation of immersive driving environments in simulator studies to provide users a more realistic awareness of the situation. Using simulator test environments puts the question of driving simulator validity forward, meaning the extent to which results generated in simulated environments can be transferred to real world environments. Thus, in this chapter the 'Immersive model-based HCI validation method', which was developed by the authors, will be introduced. First, the state of the art of driving simulators will be analyzed. For this, the authors defined the degree of fidelity based on the used elements. Next, findings of a series of driving simulator tests will be presented, which investigate the influence of immersive parameters in driving environments. Visual and auditory immersive parameters were used to analyze the validity of driving simulator environments, as well as different technologies (HMD, holobench, PC). Different levels of immersion (from low to high fidelity) were tested to examine this methodology. Thus, main intention was to demonstrate the generalizability and transferability of the 'Immersive model-based HCI validation method' for different use cases. Objective and subjective data show advantages regarding the situational awareness and perception for highly immersive driving environments while interacting with a navigation system.

D. Reich (✉) · C. Buchholz · R. Stark
Technische Universität Berlin, Berlin, Germany
e-mail: diana.reich@tu-berlin.de

G. Meixner and C. Müller (eds.), *Automotive User Interfaces*,
Human–Computer Interaction Series, DOI 10.1007/978-3-319-49448-7_16

16.1 Introduction

Infotainment systems, such as navigation systems, MP3, CD, or DVD players, belong to the standard equipment of a modern car cockpit and consist of so-called automotive user interfaces. These automotive user interfaces are controlled by different human-cockpit interactions (HCI), e.g., a spin controller in the center console, keystrokes, or touch screens. However, studies reported negative user validations regarding usability, user satisfaction and allocation of attention while driving and interacting simultaneously (e.g., Ablassmeier 2009; Lansdown 2001). Various cognitive processes (e.g., situational and risk perception) are involved when drivers interact with automotive user interfaces via different modalities (Jahn et al. 2005). A challenging question in this context is: What should innovative automotive user interfaces offer to allow an effective and safe HCI while driving? To answer this question, user research and prospective validation of automotive user interfaces during early prototyping stages are necessary (Mayhew 1999). Ideally, innovative automotive user interfaces increase driving comfort and ensure the safety of the people in the car and other road users. To test these requirements, reliable and valid validation methods have to be developed. This is the aim of this research by introducing the 'Immersive model-based HCI validation method' (see Fig. 16.1).

Prior to validating data of automotive user research, it has to be taken into account that one fairly common study design is the dual task design, in which a driver has two tasks: the first one is the (actual) driving task and the secondary one

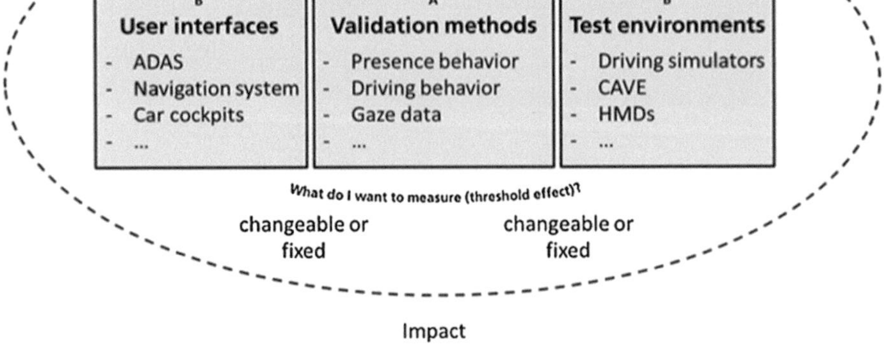

Fig. 16.1 Proprietary development of the 'Immersive model-based HCI validation method'

usually involves interacting with the automotive user interface (e.g., navigation system). Numerous methods are available to validate these dual task scenarios, (see (Breuer et al. 2003) for an overview). These methods significantly differ in terms of time, cost and test environments. Potential test environments can be divided into two types based on the realness of the test conditions and the possibilities of manipulations of variables (Zoeller 2015). These two types are laboratory (driving simulators) and field (real car) studies (Bruder et al. 2007). Depending on the aim of the validation, a specific test environment can be more or less advantageous. Driving simulator studies are becoming more important and already are a viable alternative to field studies (Green 2005; Pinto et al. 2008; Bella 2008). Currently, dual task scenarios are mainly evaluated using driving simulators, which can be more or less realistic compared to real world driving (Schindler et al. 2013). Knappe et al. (Knappe et al. 2006) wrote that driving simulator studies are characterized by high internal validity, but lack ecological validity due to reduced reality and presence experience of users. Ecological and external validity is highest for real car driving experiments, but these are only possible after the early prototyping stage has been completed. Furthermore, real car driving studies are difficult to standardize (low internal validity), more time consuming and expensive.

Therefore, one economizing suggestion could be implementing immersive driving environments for prospective human-cockpit interaction (HCI) validations. The more immersive the (driving) environment, the more realistically it will be perceived by users, who might perceive themselves as more present in the (simulated) situation (Moreno and Mayer 2002). In this context, immersion signifies the experience of submersion applied to a computerized representation or simulation. "Immersion is a description of a technology, and describes the extent to which the computer displays are capable of delivering an inclusive, extensive, surrounding, and vivid illusion of reality to the senses of a human participant" (Slater and Wilbur 1997). The success of immersing the user by presenting an (driving) environment depends on many factors, e.g., visibility, surround sound and interactive user-input (Nechvatal 2009). With highly realistic (and therefore immersive) test environments, more reliable and valid findings overall are expected. The aim of this research is to combine advantages of driving simulator experiments (e.g., repeatability) with the positive aspects of immersion (e.g., presence experience) to increase ecological and external validity of HCI validations in driving simulators. To do so, the state of the art of driving simulators will be described and classified based on their ability to reproduce the drivers' perception of an every day's journey to feel physically real to the simulator user. This ability is called fidelity level of the driving simulator (Roza 2004), which lays the ground for immersion. After that, research investigating the influence of a continuum between high and non-immersive experimental test environment setups is discussed. Next, based on this research, three selected, representative simulator studies and setups were chosen to compare their immersion degree using the standardized Lane Change Task (LCT; (ISO 2008)).

In order to better classify and modularize different aspects of an experimental setup (the validity of used methods, settings and devices), the 'Immersive model-based HCI validation method' will be taken into account (see Fig. 16.1). The authors developed a methodology which consists of three parts. The validation methods are in the center (A), with the user interfaces (B) and the test environments (B) on either side. The general logical rules are 'A is equivalent to B' and 'A if and only if B'. The validation methods include all dependent variables such as presence behavior, driving behavior, gaze data, subjective data, objective data and physiological data. The user interfaces can be, in example, an ADAS, a navigation system or car cockpits, and the test environments can cover everything from HMDs, over driving simulators to CAVEs. The validation methods are fixed (the same in every experimental condition), whereas interfaces or environments can be either fixed or can be varied depending on the validation question or which effect is to be tested (e.g., 'What do you want to measure?'). In a user experience study testing the usability of a user interface such as a navigation system, the test environment is usually fixed, and the interface (e.g., the navigation system) will be varied. One typical question for a fixed environment and a variable interface is 'Which performance of the system is best?'. If the research question is 'How does presence experience change?', it makes sense to use the same user interface and change the test environment (e.g., HMDs vs. CAVE).

16.2 Simulator Classification and Fidelity Levels

Current driving simulators are computer supported tools for investigating the driver-car interaction in simulated traffic conditions (Hucho 2005). Therefore, a broad range of different hard- and software setup exists. They range from 'minimal setup simulators', which have standard PC hardware, 2D PC screen and gaming steering wheel where drivers sit on desk chairs to 'maximum level simulators', which are multi-axis motion platforms with 360° stereo visualization having the ability to carry full passenger cars accommodating the driver. Due to the wide range of technical complexity, big efforts and costs of running a complex driving simulator, an important question needs to be asked: What is the appropriate frame for the technical equipment? For answering this question a target area for simulator usage should be defined first.

Following Bruder et al. (Zoeller 2015), the 'every day journey' defines the upper end of the target area based on the degree of realism of a test drive. This 'every day journey' can be separated into the following three parts: (1) track and environment the driver and his or her car drive through (2) the car and its geometric and functional properties, and (3) the human driver's perception of all of them.

Following Negele (2007), the sensory input of sight, noise and touch are relevant for driving situations. Neglecting the sense of smell and taste can be justified, because the primary requirements of the HCI in traffic are the primary requirements of the HCI in traffic as they are the perception of the immediate driving status and the anticipation of upcoming situations. As it investigates static simulation environments, this paper focuses on the two senses of sight and noise. Negele (2007) describes these senses as 'distant senses' that enable a temporal and spatial forecast, as it is required for anticipation of upcoming situations (Negele 2007).

Drive Cycles, such as the Artemis Drive cycle (Nicolas 2013) can be used for describing the track and environment of journeys. The Artemis Drive Cycle is based on a statistical study in Europe that aimed to gather data about everyday car use in different European environments describing the car speed profile over a cycle time and distance. Besides its speed profile, the track is described by track dimensions such as track width or the corner radius of the tracks the car drives through. Those dimensions are specified by legal directives and guidelines such as the German RAA (2008). Furthermore, the street surface has specific characteristics, for instance roughness or surface irregularities. The environment is characterized by interaction either with other traffic participants like other cars, bikes, or pedestrians, with traffic management elements such as traffic lights or road signs or with different ambient light conditions. The car is described by geometric and functional properties. The primary geometric properties affecting the sensory input of sight are the passenger compartment design, especially the front windscreen, which is limited by the A-pillars on its left and right side, the car roof and the instrument board. The primary functional properties affecting the acoustical input are the engine sound and the interaction between tire and track, which both could be affected by the structural characteristics of the car such as acoustic insulation.

Subsequently, the two senses of sight (visual perception) and noise (acoustic perception) will be described in detail and analyzed in the context of the everyday journey. The focus will lie on requirement needs in a driving simulator.

16.2.1 Visual Perception

The human sense of sight is determined by the capabilities of the human eye and the ability of the human brain to process the gathered information. Thus, the sense of sight depends on the functional properties of the single eye, like the visual acuity and the visual field as well as the ability of the human visual system to recognize colors, brightness, speed and distance. The visual acuity describes the ability of an eye to recognize details within an angle of 0.01° (Bubb et al. 2015). This equals a value of 0.6 arc minutes or in other words, the ability to recognize details of an object, which are 1,5 mm in size from a distance of 5 m. As it is shown in Fig. 16.2 (left diagram), the visual acuity is not constant over the eye's visual field but decreases sharply from the so-called functional vision field (FVF) of about 15° around outward direction. Traffic regulations describe this sharp field as "distance vision field" and define a square area on the windscreen that, in the case of a mirror damage, must not be repaired due to the importance of a clear and uninterrupted view. Figure 16.3 (right picture) shows this area which is defined to be horizontally 145 mm to the left and right from the center of the steering wheel and vertically limited by the passing field of the windshield wiper.

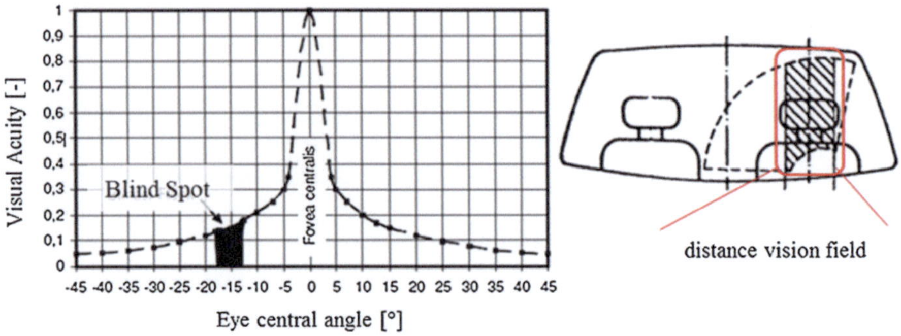

Fig. 16.2 Visual acuity and distance vision field

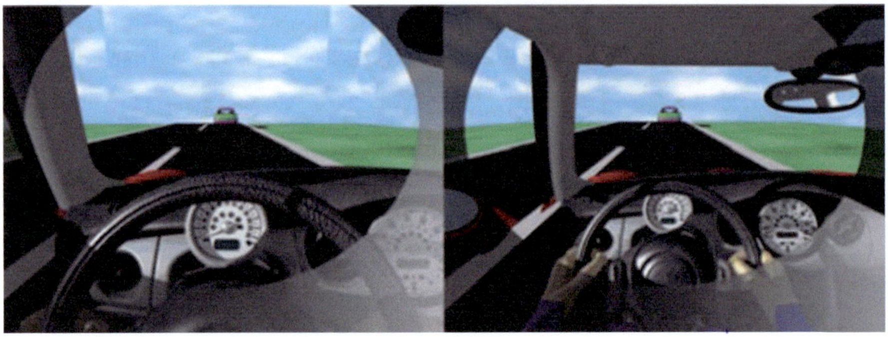

Fig. 16.3 Small and medium sized driver visual field

As objects in the distance of the functional visual field have a 100 percent chance to be perceived, objects in the peripheral vision, which provides more than 99% of the vision field, have a lower chance of being noticed (Bubb et al. 2015). Nevertheless, this field is highly important for perception as information in the peripheral field is used to initiate eye and head movements. The maximum visual field of view of a single eye is about 150° wide. The resulting maximum of the two-eyed visual field is about 180° to 200° and can be extended to 220° with an eye rotation (Negele 2007). With a head rotation, it can be extended to up to 320° (Bubb et al. 2015). Being limited by car geometry, especially by the A-pillar, traffic regulations describe a horizontal field of view of 140° to be sufficient for participation in road traffic (Negele 2007). Nevertheless, (Negele 2007) suggests using a field of view from 180° to 220° to cover the full frontal field of view. A visual field of view of 360° is not required by the Artemis drive cycle as no parking or backward driving takes place. Nevertheless, it could make sense to implement it due to simulator use. The required vertical field of view is limited by the car's front windscreen, a range from 40° to 60° is assumed to be sufficient for all cases (Negele 2007). On the left side, Fig. 16.3 shows the visuals field of a small woman and on the right the visual field of a medium sized man (Remlinger 2013).

Negele (2007) describes a visual acuity of 1 arc minute (0,017°) required to enable the readability of road signs whose sizes are regulated by traffic regulations. It is described that a visual acuity of 2–3 arc minutes is sufficient for driving cases without relevance of realistic traffic signs and visual acuity of more than 6 arc minutes (0.1°) as to be inappropriate for driving. Visual acuity depends on both, the brightness and color of light. Bubb et al. (2015) describe a maximum acuity to be achieved at a brightness of about 10.000 cd/m^2. This equals light conditions occurring in summer at noon. A higher brightness of about 1.000.000 cd/m^2 and more leads to driver blinding effects. As the brightness of car headlights range from 1 to 100 cd/m^2, a brightness range from 1 to 10 cd/m^2 can be determined as a lower boarder for brightness. The human visual system can differentiate between about 7 million color stimulus specifications (Bubb et al. 2015). Until 2011, the German traffic regulation FeV restricted the drivers licence for drivers with restricted red color perception (Deutsche Ophthalmologische Gesellschaft e.V. 2013). Thus a minimum color stimulus of 16 colors that include red color can be assumed to be appropriate for driving. Another limit is given by the fact that objects, which are perceived in direct outward direction with an angular velocity lower than 2 m/s, are no longer perceived as moving (Bubb et al. 2015). The perceived object's angular velocity rises with a rising eye angle and can be described by a circular boundary line over the visual field. This limit is important for the perception of transverse dynamic characteristics of the car, particularly the over- or understeer. Figure 16.4 shows exemplary circular boundary lines for the car speed of 50 km/h (blue), 80 km/h (red) and 120 km/h (green).

In laboratory testing, it has been found that the human eye can distinguish up to 500 flashes of white light per second which equals a frame rate of 500 Hz (Davis et al. 2015). The frame rate required for driving was investigated in driving simulator studies. The required rate for a fluent representation is more than 60 Hz, whereas frame rates of less than 30 Hz are the lower border (Negele 2007).

The perception of distances is very important for the anticipatory and compensatory car management and is determined by two different optical mechanisms (Negele 2007). The depth perception, which is the so-called 'lateral disparity' (the

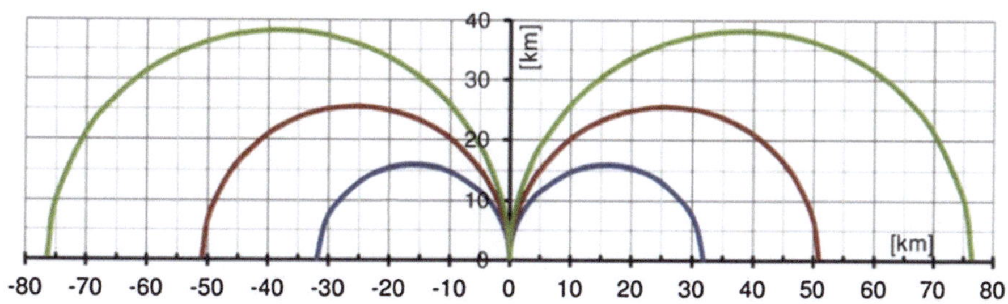

Fig. 16.4 Circular boundary lines for different car speed

perception of two slightly different images produced in both eyes) is also-called 'simultaneous parallax' or 'stereo vision'. The simultaneous parallax for depth perception is relevant in the field up to five (maximum ten) meters. At longer viewing distances, the stereo vision is not important anymore for the perception of depth.

The second mechanism is called accommodation and describes the adjustment of the eye to the distance of the viewed object by changing the curvature of the lens (Negele 2007). As a result, that distance and also the area in front of and behind this distance, respectively, can be perceived as sharp. Depending on the kind of object, which is pictured at an accommodation level, a distance impression arises. In humans this range is approximately 1.2 cm at a viewing distance of 20 cm, even 15 m at 5 m and already infinitely from 10 m. This fact is relevant in interpreting the vision system for selecting the distance of the image medium.

16.2.2 Acoustic Perception

The sense of hearing is the second remote sense of humans and allows a perception when the source of a sound is either concealed or invisible (Bubb et al. 2015). On the road the differences in sensitivity of the hearing at various frequencies and all the types of occurring sounds, such as the engine sound, squealing tires or acoustic signals in general, are relevant. The acoustic information in the car covers a very large area of the frequency spectrum. It starts at low-frequency vibrations like a wheel imbalance (about 20 Hz) and ends with the high-frequency sound of a screeching brake (20.000 Hz) or a blocked tire. Thus, this frequency range is required to enable the perception of all kinds of car characteristics. As there are several more or less important driving information over the whole frequency range, gaps in this range are assumed to be affecting the drive realism a lot.

Sound allows the interpretation of its cause and the location of its source (Bubb et al. 2015). Even a time difference of several milliseconds between the impact of the sound waves at the left and right tympanic membrane is sufficient to estimate the location of the source of the sound. Research reveals that a localization accuracy of up to 3° and an appropriate accuracy of about 15 to 20° are possible (Skluzacek 2012).

16.2.3 Perception Target Field

From processed analysis of human perception in a "every day journey", target parameters will be extracted and collected in this subchapter. Aiming to compare the state of the art of drive simulators regarding their fidelity level requires the definition of the term 'fidelity'. Furthermore, the authors will define a validation scale as a comparison criterion.

Roza (2004) provides an overview of fidelity theories, and states that qualifying and quantifying the level of fidelity is 'an area in which there exist many incomplete, inconsistent, and widely scattered views, concepts and approaches'. Different expressions are encountered in literature such as physical fidelity, objective fidelity, perceptual fidelity, behavioral fidelity, functional fidelity, attribute fidelity, abstract fidelity, psychological fidelity, and concrete fidelity. This paper follows the resulting assumptions of Roza (2004) and defines fidelity 'as realism or faithfulness of the simulation in the broadest sense'.

To evaluate the fidelity of a target field parameter, a three-stage valuation key will be applied to the respective parameter (see Fig. 16.5). Based on the analysis, the fidelity degree of a target field parameter can be evaluated to be not sufficient (red), partly sufficient (yellow), or fully sufficient (green). Figure 16.5 furthermore describes target field for visual perception with its target parameters visual field, visual acuity, brightness, color stimulus, frame rate and stereo vision.

Figure 16.6 describes target field for acoustic perception with its target parameters frequency range and localization range and accuracy.

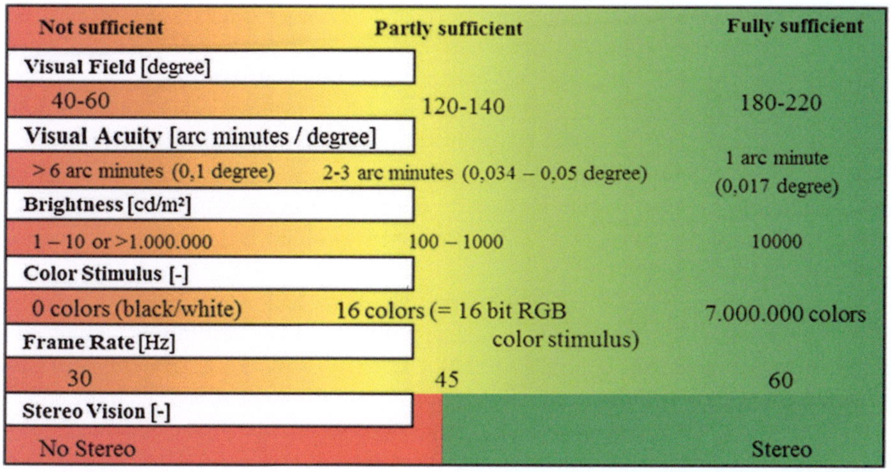

Fig. 16.5 Target field visual perception

Fig. 16.6 Target field acoustic perception

16.2.4 Subchapter: State of the Art of Driving Simulator Elements

1. **Visual Simulation**

The latest visual simulations in driving simulators use two different technologies for image generation: Electronic visual displays—screens—and projections systems consisting of projectors in combination with projector screens.

With the cathode ray tube (CRT) technology having been the state of the art of screen technology for multiple decades, liquid crystal displays (LCD) mostly replaced them in recent years (Elze and Tanner 2012). Being based on polarizable liquid crystal material with LED light sources in the back, the LCD Screen technology enables the design of very flat—and thus comparably light—image generating devices. As the commercial end-customer market is a huge driver of LCD Screen technology, driving simulators primarily use commercial LCD Screens as well (Dierig 2009).

Thus, the end-customer LCD market could be taken as an indicator for technical data on the current state of screen technology for driving simulators.

For driving simulations, projectors from end-customer markets are usually not used, due to the technological limits such as—in comparable price ranges–worse image resolution, color fidelity, and contrast and much worse brightness (2013). Using LCD technology connected with different light sources, such as laser light, can lead to brighter results than LED light sources used in end-customer projectors. Other disadvantages of end-customer market projectors are the installation efforts and space requirements, which are usually higher than those of screens. Thus, high end laser light based projection systems as used in cinemas are more common in driving simulator applications.

Following Negele's (2007) validation, the main technical criteria for image generation technology are the available screen size, image resolution, image generation speed, brightness, contrast and color fidelity. Table 16.1 gives an overview on technical data currently available for both generations of image technologies.

Depending on the image generation speed of the respective image generator, both, LCD Screens and Projectors can be used to generate stereo pictures. With stereo use of LCD Screens being very common for end users, validations show stereo application in today's driving simulators only in combination with projectors. To enable stereoscopic view in this setup, the most common technology is the use of polarization filters, e.g., by special polarization glasses to separate the single picture from the projector into two different pictures with offset for left and right user eye. To enable the required image generation speed, projectors with 120 Hz are used (Negele 2007).

A further stereo application is the head-mounted display (HMD). It consists out of a portable device with two LCD Screens for each left and right user eye. Limited in size, technical data of HMD are generally lower than from non-portable LCD Screens. Table 16.2 shows the technical data a HMDs currently used in drive simulator applications, the Oculus Rift (Hardware et al. 2016) (see also Fig. 16.7).

Table 16.1 Technical data for LCD screen (left) and projection systems (right)

LCD screen		Projection system	
Screen size (usual) [m] (maximum) [m]	Image resolution (usual) [-] (maximum) [-]	Screen size (maximum) [m]	Image resolution (maximum) [-]
600 mm × *400* mm *2000* mm × *1000* mm	*2048* × *1536 Pixel* *5120* × *2880 Pixel*	*Distance depending – Example:* *5* m × *2,8* m	*4096* × *2160 Pixel*
Frame rate (maximum)	Brightness (maximum) [cd/m^2]	Frame rate (maximum) [Hz]	Brightness (maximum) [cd/m^2]
60	*400*	*60*	*Distant depending:* *≈200*
Contrast (maximum) [-]	Color Fidelity (maximum) [%]	Contrast (maximum) [-]	Color fidelity (maximum) [%]
4000:1	*99%*	*500:1 (Cinema projector)*	*Distant and ambient condition depending*
Stereo vision [-]		Stereo vision [-]	
No		*Yes*	

Table 16.2 Technical data for HMD (Oculus Rift)

Visual field [degree]	Image resolution [-]
110 deg × *60 deg*	*2160* × *1200 Pixel*
Frame rate [Hz]	Brightness [cd/m^2]
45 Hz	*200*
Contrast [-]	Color Fidelity [%]
Up to 6000:1	*80%*
Stereo vision [-]	
Yes	

Fig. 16.7 Exemplary HMD in Use

To evaluate the fidelity level of visual simulation systems, the physical setup has to be taken into account as it has a lasting effect on the visual field, visual acuity, brightness and the color stimulus. The visual field depends on the screen size and the distance to the driver (Negele 2007). To achieve accommodation effects, (Negele 2007) suggests a minimum distance of 2.5–3 m between the screen and the user. Full accommodation is given at a distance of more than 5 m.

Furthermore, it has to be taken into account that the use of more than one LCD screen leads to vertical cuts in the displayed picture as LCD screens have a visible edge. Following Negele (Negele 2007), this is a major advantage of the projector technology which is able to create seamless pictures. In driving simulator application practice three LCD screens are used side by side at the most. The color stimulus is affected by the technical aspects of contrast and color fidelity as well as the distance between the user and screen. The following typical visualization setups used in driving simulators will be evaluated regarding their fidelity:

- **Setup 1:** Single LCD screen, about 0.8 m distance to the driver. This setup is commonly used in basic driving simulator configurations.
- **Setup 2**: Three LCD screens side by side, 2.5–3 m distance to the driver. This setup is commonly used in high dynamic driving simulators with the screen fixed onto the dynamic platform.
- **Setup 3:** Single projection wall with a distance of 2.5–3 m to the driver. This setup is commonly used in driving simulator configurations with a horizontal visual field of about 140–160°.
- **Setup 4**: Multi projection wall with more than 5 m distance to the driver. This setup is commonly used in fixed screen driving simulator configurations with a horizontal visual field wider than 180°.
- **Setup 5**: Head-Mounted Display.

On Fig. 16.8 upper left, the fidelity validation for the visual field is shown. Due to their limited size, the LCD screen based setups 1 and 2 cover the outward direction but are not able to cover big angles. With the position flexibility between projector and screen, the projector based setups 3 and 4 are able to cover bigger surfaces for the cost of reduced visual acuity. This interdependency is shown at the upper right sub-figure of Fig. 16.8, as the setup 4, which is characterized by a higher distance between projector and screen, is worse in visual acuity. Furthermore, the sub-figure shows the LCD screens with their higher resolution delivers better results in visual acuity when using comparable same technical setups.

HMDs enable the use of LCD screens with their advantages in image resolution —and thus visual acuity—connected with a wide visual field of up to 360°. As exemplary described in the technical description of Oculus Rift, today's HMDs have less image resolution capabilities than high end LCD screens. The comparison in brightness is shown by the middle left sub-figure. As the LCD based setups 1 and 2 as well as the HMD based setup 3 show good results in brightness, (Negele 2007) describes this to be a major disadvantage of projector technology compared to LCD. Compared to the target field, brightness of all visualization technology is much

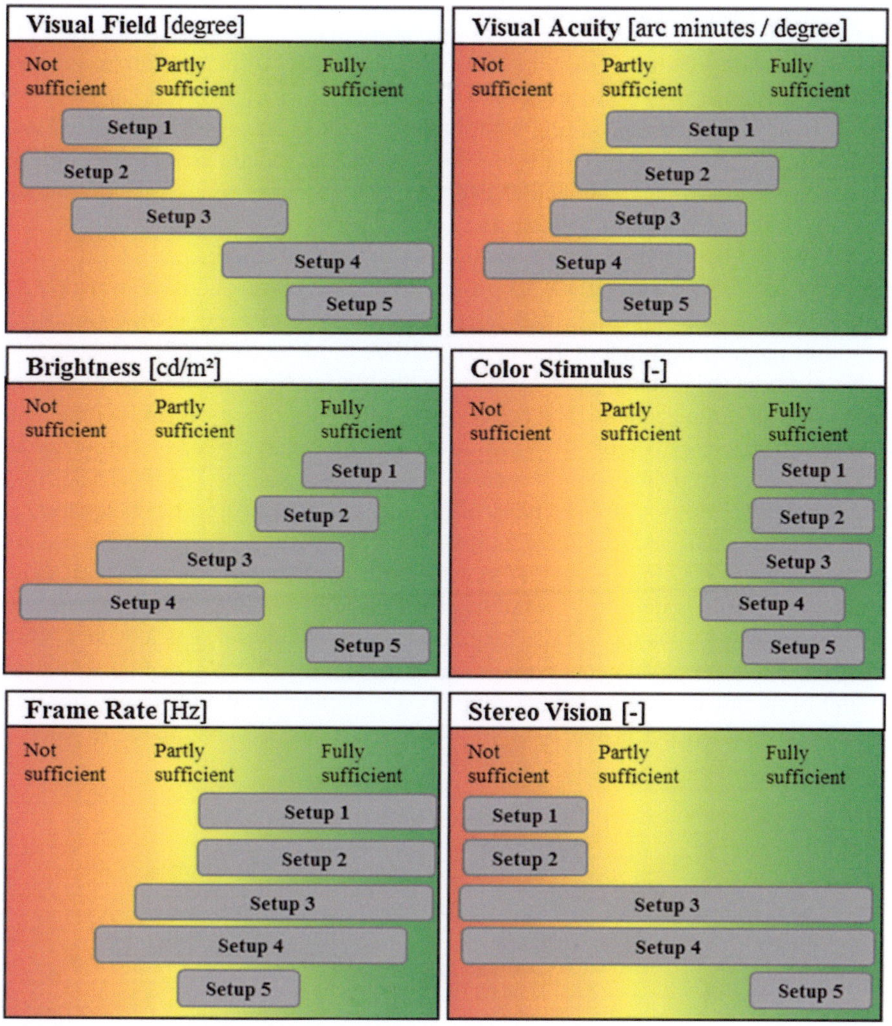

Fig. 16.8 Fidelity validation for state of the art visualization technology

lower. The best of LCD screens reach a brightness of about 400 cd/m^2. Color stimulus of all today's visual technologies is fully sufficient with small disadvantages of projector based compared to LCD based technologies.

As it is shown at the lower left of Fig. 16.8, the frame rate of LCD based technology is generally better compared to projector technology. Current HMDs are comparable to lower end LCD screens with their lower frame rate capabilities. Stereo vision perception can be implemented by usage of projection based technology as well as HMDs.

2. **Acoustic Simulation**

For generating audio signals, loudspeakers are used. A loudspeaker is a device containing one or more electroacoustic transducers which convert an electrical audio signal into a corresponding sound (Ballou 2013; Talbot-Smith 2013). There are three different types of loudspeakers, depending on the frequency range at which they are working: The woofer is designed to produce low-frequency sounds, typically from 60 Hz up to 250 Hz, the squawker produces mid-range frequencies from 250 to 2.000 Hz and the tweeter high audio frequencies, typically from around 2.000 to 20.000 Hz (Stern 1960). The speakers for the normal consumer market are usually full-range speakers, which have a good low and a good high-frequency response, ranging from about 100 to 10.000 Hz.

To evaluate the fidelity level of acoustic simulation systems, the physical setup has to be taken into account, as it has a lasting effect on both localization range and localization accuracy. Figure 16.9 shows two representative setups used in typical driving simulator applications (Negele 2007): The left side of Fig. 16.10 shows a 5.1 Dolby Surround setup. It consists of five full-range speakers (black squares) L, C, R, SBL and SBR and two woofers (blue squares) SL and SR. This setup enables the localization of sound sources. A stereo setup consisting of two black full-range speakers L and R is depicted on the right side of Fig. 16.9.

Figure 16.10 shows the fidelity range covered by these two typical setups. Both setups enable a broad frequency range by being limited in high frequencies (setup 1) and high and low frequencies (setup 2). Depending on the exact position and speaker quality, setup 1 is able to reproduce a nearly full localization range whereas setup 2 does not enable sound source localization.

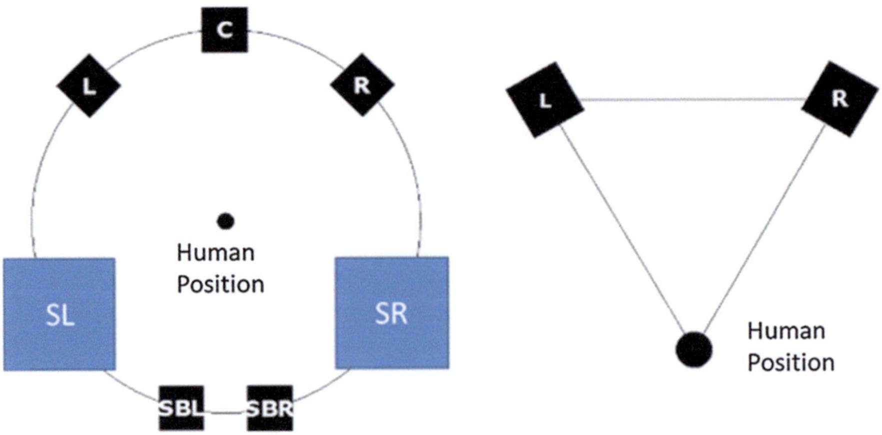

Fig. 16.9 Visualization technology setup for Fidelity validation—Setup 1: Dolby Surround (*left*) and Setup 2: Stereo (*right*)

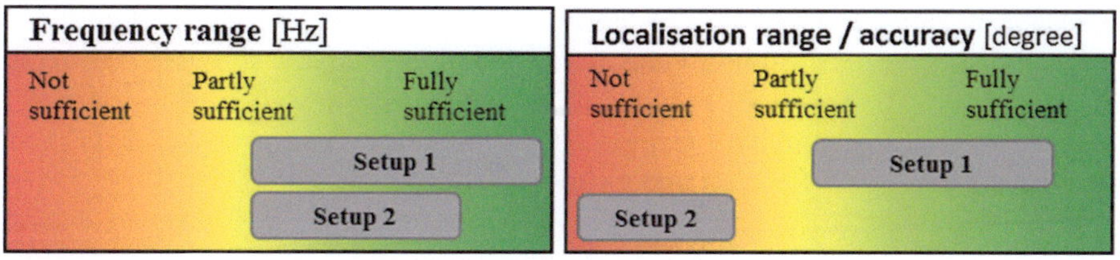

Fig. 16.10 Fidelity validation for state of the art acoustic technology

16.3 Empirical Driving Simulator Studies Within Immersive Environments

As shown in the previous parts, the technical fidelity can be determined quite well. But in order to configure driving simulation environments as realistic as possible, human perception also has to be taken into account (Roza 2004). What has yet to be examined is the relationship of fidelity and immersion in the context of drivers in the simulator. Sensory signals, namely visual and acoustic input, could increase the immersion and presence in driving simulations. The subchapter of this paper focused on these two parameters.

In order to analyze the influence of visual and acoustic parameter, a series of empirical driving simulator studies were conducted. The experimental designs and setups of these studies were developed using the 'Immersive model-based HCI validation method' (see Fig. 16.1). The authors chose the second approach for the first study. The user interface (a navigation system) was fixed and immersive parameters of the test environment were varied. A driving simulator and a holobench were used which were varied to either being highly immersive (stereo vision, car sound), intermediate immersive (stereo vision, without car sound or 2D vision, with car sound) or low immersive (2D vision, without car sound). Study 3 followed the same approach, but other test environments were validated. Here, the highly immersive test environment was characterized by a HMD (stereo vision with car sound) and the low immersive test environment by a PC setup (2D vision, without car sound). A third possibility, which is also described by the 'Immersive model-based HCI validation method', is shown in Study 2. Here, the HCI user interfaces and the test environment were varied. The test environment was either highly immersive (stereo vision, car sound) or low immersive (2D vision, without car sound). Subject had to interact with each of the three HCIs: touch control, spin control, free-hand gestures.

To achieve more comparability between the three driving simulator studies, the authors used the same use case and primary user task for them. One common task incorporated in the studies was the standardized Lane Change Task (LCT; (Mattes 2003)), which represents a highly validated and ISO-normed tool for measuring driver distraction while driving and performing secondary tasks (ISO 2008). The subject has to drive on a three-lane road and execute several lane change maneuvers indicated by signs on both sides of the road. The speed is constant at 60 km/h (37 mph). To analyze the results of the LCT using the analysis software, the *mean deviation* was extracted from the LCT. This reflects the deviation of the subjects' driven course from a normative model and gives information about driver distraction, e.g., caused by the secondary task.

In the following, a selection of description and findings of a series of driving simulator tests within immersive driving environments will be presented. Thus, main intention is to demonstrate the generalizability and transferability of the 'Immersive model-based HCI validation method' to different use cases.

16.3.1 Driving Simulator Study 1: Influence of Visual and Acoustic Parameters Within Immersive Environments

The experiment was implemented as a completely crossed 2 × 2 within-subjects design to investigate the influence of immersive visual and auditory parameters (Reich et al. 2014). The independent variables were the visual representation of the driving environment (2D versus stereo) and, the auditory representation of driving sound (no sound versus car sound). Thus, there was one *low immersive* condition without any immersive parameter and one *highly immersive* condition with two immersive parameters. Objective and subjective data were captured as dependent variables. The collected objective data included the driving parameters of the LCT for determining the driving performance for each condition, eye-tracking data, and the total scores of the SAGAT (Endsley 1995) questions for measuring situational awareness. Subjective dependent variables were assessed with four different questionnaires, namely the NASA-TLX for subjective mental workload (Hart and Staveland 1988), the perceived driving performance (PDP), the questionnaire for Presence and Immersive tendency (PIT) in virtual realities (Scheuchenpflug 2001) and the simulator sickness questionnaire (SSQ, (Kennedy et al. 1993)). With the PDP, the subjects assessed their own driving performance, and the PIT measured the subjects' perceived quality of experience of virtual reality.

The sample was composed of 20 participants with an average age of 28.65 years (SD = 5.95). The driving environment was presented on the Barco TAN Holobench, a projection table consisting of two orthogonal projection surfaces (each 1.80 m × 1.10 m). For this study, only the vertical screen was used to display the driving environment (three-lane road, signs, background). An active shutter stereo technology allows a stereoscopic (stereo) representation of the scene. Shutter glasses worn by the subject synchronously block the respective eye by polarization of the Liquid Crystal Display (LCD) glasses. Consequently, the image displayed is synchronized with left and right eye image, similarly to the screen with a frequency of 100 Hz. Thus, each eye perceives 50 frames per second, which creates a stereo view for the subject by merging the two images. The synchronization is generated

Visual Simulation – Holobench, distance 2m to Driver		
Visual Field [degree]	**Visual Acuity** [arc minutes / degree]	**Brightness** [cd/m²]
≈ 50	2-3 arc minutes	50
Color Stimulus [-]	**Frame Rate** [Hz]	**Stereo Vision** [-]
> 100.000	30	Stereo
Acoustic Simulation – PC Loudspeaker with Subwoofer, in front of driver		
Frequency range [Hz]	**Localisation range / accuracy** [degree]	
48 - 20000	≈ 150degree source, 15-20degree localisation range	

Fig. 16.11 Fidelity level of visual and acoustic simulation system

by the computer rendering the LCT. The signal is transmitted to the shutter glasses via an infrared transmitter. The change between monoscopic (2D) and stereoscopic (stereo) view can be evoked by switching the holobench to active stereo mode. Therefore, the shutter glasses need to be turned simultaneously turned on/off.

The second immersive parameter was the car sound to increase the immersion of the driving environment. The driving sound was generated automatically by starting the LCT and was reproduced by speakers connected to the LCT computer and holobench. Switching between car sound/no car sound conditions was done by turning on and off the speakers. Figure 16.11 shows the fidelity level for the experimental setup of visual and acoustic simulation system.

16.3.1.1 Principal Finding

The scores of the PIT questionnaire in virtual environments were compared across the conditions using a within-subjects repeated measure ANOVA with Greenhouse-Geisser corrections. The questionnaire consists of three dimensions: (*a*) *spatial presence,* (*b*) *quality of the interface and* (*c*) *involvement.* Figure 16.12 shows the differences between the immersive conditions regarding the three dimensions of the PIT questionnaire.

The ANOVA revealed a statistically significant difference in ratings of *spatial presence* among the four immersive conditions, $F(3,\ 57) = 3.13$, $p = < 0.05$, $n^2 part = 0.141$. Significant pairwise comparisons regarding *spatial presence* were found between the two extreme immersive conditions (2D; no sound versus stereo; car sound), $F(1,\ 19) = 5.71$, $p = < 0.05$, $n^2\text{part} = 0.231$. Corresponding planned t-tests showed that spatial presence was rated significantly lower in the low immersive condition, 2D no sound ($M = 2.92$; $SD = 0.84$) compared to the high immersive condition, stereo car sound ($M = 3.32$; $SD = 0.83$), $t(19) = -2.39$,

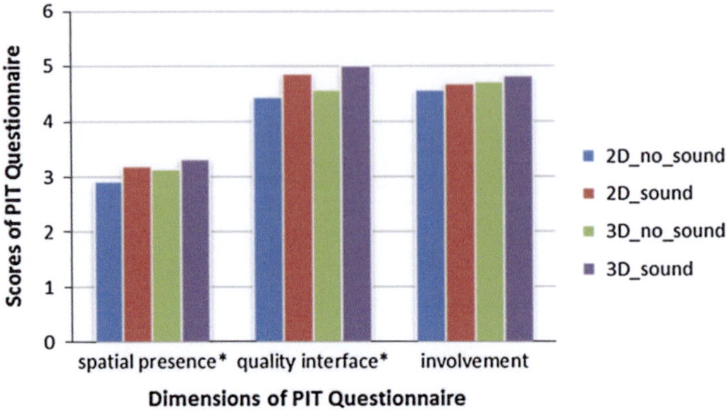

Fig. 16.12 PIT scores by immersive conditions

$p = 0.027$. The ANOVA also revealed a statistically significant difference in the ratings of *quality of interface* among the four immersive conditions, $F(3, 57) = 4.17$, $p = < 0.05$, $n^2 part = 0.180$. Significant pairwise comparisons regarding *quality of interface* were found between the two extreme immersive conditions (2D; without car sound versus stereo; car sound), $F(1, 19) = 7.63$, $p = < 0.05$, $n^2 part = 0.287$. Corresponding planned t-tests showed that *quality of interface* was rated significantly lower in the low immersive condition, 2D no sound (M = 4.44; SD = 0.71) compared to the high immersive condition, stereo car sound (M = 5.0; SD = 0.66), $t(19) = -2.90$, $p = 0.009$. No significant differences were present in the *involvement* data. In Fig. 16.12 it can be seen that the overall involvement rating for each immersive condition showed the highest scores for the most immersive condition (M = 4.82; SD = 0.93) and the lowest scores for the least immersive one (M = 4.57; SD = 0.86).

16.3.2 Driving Simulator Study 2: Influence of Immersive Test Environments on HCI

The second experiment was designed as a completely crossed 2 × 3 within-subjects design, with the aim to investigate the influence of immersive environments and three interaction modalities (Reich and Stark 2015). The *level of immersion* of the driving environment (low versus high) was the first independent variable. Here, visual representation (2D versus stereo view) and auditory parameters (no sound versus car sound) were varied. There are no immersive parameters in the *low immersive* (2D view; no sound), and the *high immersive* condition has two (stereo view; car sound). The second independent variable was *interaction modality* for a human-computer interface (HCI). Entries into a navigation system could made using a (1) touch controller, a (2) spin controller, or (3) free-hand gestures. Twenty participants (male = 12; female = 8) with an average age of 29.7 years (SD = 10.4) took part in this study.

The primary user task was again the Lane Change Task which was displayed on the holobench connected to the driving simulator, too. The same fidelity level for the experimental setup of visual and acoustic simulation system was presented as in study 1 (see Fig. 16.11 for technical details). The secondary task consisted of entering addresses into a navigation system, which were announced by the study manager (country/city/street).

16.3.2.1 Principal Findings

A repeated measure ANOVA with a Greenhouse-Geisser correction was performed for each AOI on (1) fixation count total, (2) fixation count total [%], (3) fixation time total [ms], and 4) dwell time total [%]. The AOIs of interest were the driving environment (holobench) and the navigation system interface. Significant differences between immersive conditions (low; high) were found for touch modality. Here, most glance parameters became significant regarding AOI 'navigation system interface' (Reich and Stark 2014). The ANOVA for *fixation count total* showed significant differences between low and high immersive environments, $F(1, 15) = 5.9$, $p = < 0.05$, $\eta^2 = 0.282$. Figure 16.13 shows exemplary that driving in high immersive conditions leads to fewer visual fixations on the navigation system interface (M = 121; SD = 50) compared to the low immersive conditions (M = 100; SD = 45). The ANOVA for *fixation count total [%]* showed significant differences between low and high immersive environments, $F(1, 15) = 6.3$, $p = < 0.05$, $\eta^2 = 0.295$ and fewer visual fixations (in %) on the navigation system interface (M = 8.3; SD = 4.2) compared to low immersive condition (M = 9.9; SD = 4.9). The ANOVA for *fixation time total [ms]* showed also significant differences between low and high immersive environments, $F(1, 15) = 6.8$, $p = < 0.05$, $\eta^2 = 0.312$. Driving in high immersive conditions leads to less visual

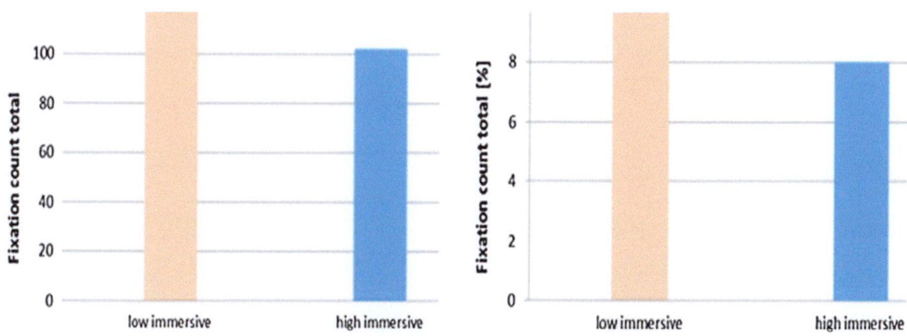

Fig. 16.13 Glance data for touch modality by immersive environments

fixation time on the navigation system interface (M = 14790; SD = 8705), compared to the low immersive condition (M = 18870; SD = 9361).

16.3.3 Driving Simulator Study 3: Influence of Immersive Test Environments on Situation Awareness

The experiment was implemented as a completely crossed 2 × 2 within-subjects design ($N = 20$). The first independent variable, *mode of visualization,* was varied as non-immersive and highly immersive (Reich 2015). Here, the highly immersive mode of visualization (Oculus Rift) was characterized by stereoscopic 3D view, stereo car sound, 360° head tracking, and an approximately 110° field of view. The non-immersive mode of visualization (PC) was characterized by 2D view, stereo car sound, no head tracking, and a 30° field of view. Figure 16.14 shows the fidelity level for both experimental setups of visual and acoustic simulation system. The second independent variable was type of stimuli (driving-relevant versus driving-irrelevant). The user task was again the LCT, displayed on the Oculus Rift. The LCT was rebuilt with the programmable open source driving simulator OpenDS (Math et al. 2013). The software is programmed entirely in Java and is based on the JMonkeyEngine framework, which is a video game engine. For additional details, see the home page of OpenDS (http://www.opends.eu).

The situation awareness was measured by SAGAT (Endsley 1995) during each LCT condition. At varying, unpredictable moments during the simulation, the participants were asked to stop four times per run. The screen (driving environment) was turned off and the participants were asked to answer situational questions, e.g., 'What did you see at the roadside?' 'What was the color of the object?' or 'What is your actual lane positon?' After answering the questions the screen was turned back on and the subject completed the LCT. The answers could only be right or wrong, which determined the corresponding values of 1 or 0 as results. To indicate the level of situation awareness the number of right answers was counted (the possible result was a score from 0 to 8). The stimuli relevant for driving were: (1) deer, (2) baby buggy, (3) person, and (4) road sign. The driving-irrelevant stimuli were: (1) triangle, (2) circle, (3) square, and (4) cone.

Visual Simulation Setup 1 - MacBook 15 Zoll / distance 0,6m to Driver			Visual Simulation Setup 2 – Oculus Rift		
Visual Field [degree]	Visual Acuity [arc minutes / degree]	Brightness [cd/m²]	Visual Field [degree]	Visual Acuity [arc minutes / degree]	Brightness [cd/m²]
≈ 30	1 arc minute	389	110	1 arc minute	200
Color Stimulus [-]	Frame Rate [Hz]	Stereo Vision [-]	Color Stimulus [-]	Frame Rate [Hz]	Stereo Vision [-]
> 5.000.000	60	No Stereo	> 1.000.000	45	Stereo

Acoustic Simulation – PC Loudspeaker with Subwoofer, in front of driver	
Frequency range [Hz]	Localisation range / accuracy [degree]
48 - 20000	~ 150degree source, 15-20degree localisation range

Fig. 16.14 Fidelity level of visual and acoustic simulation system for PC condition and HMD

16.3.3.1 Findings

The scores of the SAGAT data were compared across conditions using a repeated measures 2 × 2 ANOVA with a Greenhouse-Geisser correction. The two main factors were mode of visualization and type of stimuli (see Fig. 16.15). A significant main effect for mode of visualization was found, $F(1, 18) = 7.23$, $p = 0.015$, $\eta^2 = 0.287$. Another significant main effect for type of stimuli was found, $F(1, 18) = 19.00$, $p = 0.000$, $\eta^2 = 0.514$. Furthermore, there was a significant interaction between the type of visualization and stimuli, $F(1, 18) = 14.57$, $p = 0.001$, $\eta^2 = 0.447$. Figure 16.15 shows that participants answered SAGAT questions best when performing the LCT in non-immersive PC conditions and irrelevant stimuli ($M = 7.5$; $SD = 0.82$).

The scores for presence and immersive tendency scores were weighted and compared across conditions using paired t-tests with a significance level of 0.05. The t-tests showed significant differences between PC and Oculus Rift conditions, $t(16) = -2.18$, $p = 0.045$ for spatial presence ratings. They were highest when performing the LCT in the highly immersive Oculus Rift conditions ($M = 3.19$; $SD = 0.46$) compared to the non-immersive standard PC conditions ($M = 2.86$; $SD = 0.76$). The other two dimensions, quality of the interface and involvement, showed the same descriptive statistics. However they did not become significant.

The scores of the NASA-TLX questionnaire were weighted and compared across conditions using a repeated measures 2 × 2 ANOVA with a Greenhouse-Geisser correction. The two main factors were *mode of visualization* and *type of stimuli* (see Fig. 16.16). There was a significant main effect for *mode of visualization*, $F(1, 17) = 18.23$, $p = 0.001$, $\eta^2 = 0.518$. Cognitive workload was lower in non-immersive PC conditions ($M = 34.48$; $SD = 19.44$), compared to highly immersive conditions ($M = 46.91$; $SD = 21.21$). Neither a significant main effect found for *type of stimuli* nor an interaction between both factors was found.

Weighted SSQ scores, which were compared across conditions using a repeated measured ANOVA with a Greenhouse-Geisser correction. The ANOVA showed

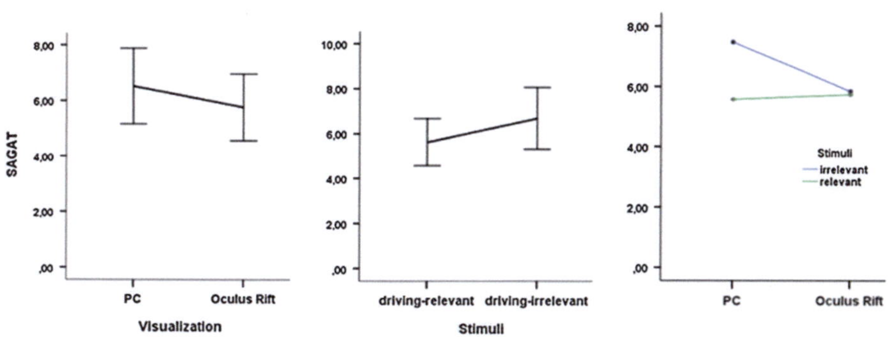

Fig. 16.15 *Left* Main effect for mode of visualization, *middle* Main effect for type of stimuli, and *right* Interactions for situation awareness ratings

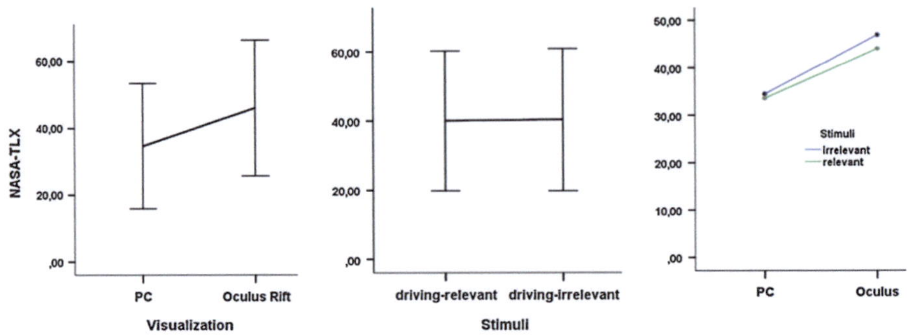

Fig. 16.16 *Left* Main effect for mode of visualization, *middle* Main effect for type of stimuli, and *right* Interactions for cognitive workload ratings

significant differences between the pre, PC and Oculus Rift conditions, $F(2, 38) = 4.96$, $p = <0.05$, $\eta^2 = 0.207$. Simulator sickness symptoms were lowest when participants arrived (pre) ($M = 18.35$; $SD = 2.15$) and remained almost the same for non-immersive PC conditions ($M = 18.45$; $SD = 2.65$) compared to high immersive conditions ($M = 20.90$; $SD = 3.92$), where simulator sickness symptoms were significantly higher.

16.3.4 Discussion

The empirical section discusses research comparing the influence of non-immersive and highly immersive driving environments while performing the LCT. Here, the 'Immersive model-based HCI validation method' was used to generate the experimental designs. Findings showed that performing the LCT in these different immersive environments leads to significant differences regarding the subjective and objective data. Subjective user ratings of spatial presence experience were significantly higher when performing the LCT in highly immersive (stereo view, car sound; e.g., Oculus Rift) conditions, compared to the non-immersive ones (2D view, no sound; e.g., PC setup). This result confirms Roza's (2004) recommendations to include stereo vision and car sounds into a driving simulation to render it more realistic. Because of the increased immersion, the authors conclude that in more immersive environments, participants felt more aware of the driving situation and therefore anticipated risks better. These results demonstrate that driving simulators can offer sufficient fidelity (Kappé and van Emmerik 2005) and mitigate Vlakveld's (2005) concern of simulator environments of being not rich enough to increase the situation awareness or risk perception.

The second experiment investigated the influence of immersion (low; high) in driving environments while interacting with a navigation system in three different HCI modalities (touch; spin controller; free-hand gestures) in order to validate the third approach of the 'Immersive model-based HCI validation method'. Differences regarding interaction modalities in this study are not diagnostically conclusive, because the development stages are not comparable yet (i.e., technical implementation, applicability, product reliability). The results showed significant differences in gaze data for touch interactions regarding immersive conditions. Thus, driving and interacting in highly immersive condition led to less visual attention on the automotive user interface (navigation system), suggesting that immersive environments seem to lead to higher situation and risk awareness and therefore to less visual attention on the secondary task.

Simulator sickness remains a bit problematic, since it was significantly lower in non-immersive conditions. Some participants also reported feeling awkward when they were wearing a closed HMD, which does not allow them to see their own hands on the steering wheel while driving a (virtual) car. It might therefore be more beneficial to perform the driving task in an immersive environment which enables drivers to see their own hands on the wheel, e.g., a CAVE or a powerwall. Future research could examine how to reduce simulator sickness by trying to match the inputs from the simulator and other sensory inputs more closely. To increase immersion, one should think about additionally implementing haptic and vestibular input, which could intensify the presence and immersion of the 'every day journey' (Roza 2004; Zoeller 2015). However, it is important to provide an optimal mix of the ratios of the different immersion parameters to create driving simulator experience which is as close to a real drive as possible (see Negele 2007).

16.4 Overall Conclusion and Outlook

Immersion experience is not equal to the fidelity level (De Winter 2007). It is unclear whether the inherent limited fidelity of a driving simulator undermines its effectiveness, or what kinds of tasks would be most affected. Vlakveld (2005) questioned whether driving simulators produce virtual environments that are rich and varied enough for the acquisition of higher order skills such as situation awareness and risk perception. But no ideal technology is available yet that is able to do all. Immersion depends on an intelligent mixture, which is not necessarily the best-rated one. Thus, empirical user tests are required. For which the 'Immersive model-based HCI validation method' could be a helpful scheme to construct experimental designs and setups.

To sum up, objective and subjective data lead us to conclude that higher immersive driving environments (stereo view; car sound) are useful for the automotive context. Immersive environments not only increase the situation awareness and therefore reduce visual attention on the secondary task but they also increase presence and immersive tendency (reality awareness) for driving tasks compared to conventional driving simulator environments.

Future work will implement the LCT in driving simulators with higher fidelity degrees using up to 360° 3D projection systems or dolby surround sound systems.

Furthermore, testing variants will be added that enable driver-simulator interaction using further senses. A broad range of tactile human-car interfaces as different steering wheels or pedal systems will be used as well as vestibular interaction systems as multi-axis motion platforms. These additional implementations in the setups will make it possible to also analyze other immersive influencing factors (e.g., haptic, tactile or kinesthetic feedback), which may affect presence experience positively, too.

The last step will be a real car driving experiment, which will provide data for a comparison with all conducted driving simulator studies. Such an investigation will allow to further check the validity of the 'Immersive model-based HCI validation method'.

References

Ablassmeier, M. (2009). Multimodales, kontextadaptives Informationsmanagement im Automobil. Doctoral Thesis. Technische Universität München.

Ballou, G. (2013). *Handbook for sound engineers* (4th ed., p. 597). Taylor and Francis. ISBN: 1136122532.

Bella, F. (2008). Driving simulator for speed research on two-lane rural roads. *Accident Analysis and Prevention, 40*(3), 1078–1087.

Breuer, J., Bengler, K., Heinrich, C. & Reichelt, W. (2003). Development of advanced driver attention metrics (ADAM). In H. Strasser, K. Kluth, H. Rausch, & H. Bubb (Eds.). Quality of Work and Products in Enterprises of the Future. Proceedings of the 50th—Anniversary Conference of the GfA and the XVII Annual ISOES Conference in Munich, 2003 (pp. 37–40). Stuttgart: Ergonomia.

Bruder, R., Abendroth, B. & Landau, K. (2007). Zum Nutzen von Fahrversuchen für die Gestaltung. In R. Bruder & H. Winner (Hrsg.) (Eds.), Wie objektiv sind Fahrversuche? Darmstädter Kolloquium Mensch & Fahrzeug, S. 79–93, Stuttgart: Ergonomia.

Bubb, H., Bengler, K., Grünen, R. E., & Vollrath, M. (2015). *Automobilergonomie*. Wiesbaden: Springer Verlag.

Davis, J., Hsieh, Y.-H., & Lee, H.-C. (2015). Humans perceive flicker artifacts at 500 Hz. In *Scientific Reports* 5, Article Number 7861. doi:10.1038/srep07861.

Deutsche Ophthalmologische Gesellschaft e.V. (2013). Fahreignungsbegutachtung für den Straßenverkehr, 6. Auflage, München.

De Winter (2007). Driving simulator fidelity and training effectiveness. In P. A. Wieringa1, J. Dankelman1, M. Mulder2, M. M. van Paassen2 & S. de Groot1 (Eds.), BioMechanical Engineering Department Faculty of Mechanical, Maritime and Materials Engineering Delft University of Technology 2628 CD Delft, The Netherlands J.C.F.deWinter@TUDelft.NL.

Dierig, C. & Jüngling T. (2009). Der Boom der LCD-Fernseher geht zu Ende, WeltN24 GmbH, Abruf 08.01.2009.

Elze, T., & Tanner, T. G. (2012). Temporal properties of liquid crystal displays: Implications for vision science experiments. *PLoS ONE, 7*(9), e44048. doi:10.1371/journal.pone.0044048.

Endsley, M. R. (1995). Measurement of situation awareness in dynamic systems. *Human Factors, 37*(1), 65–84.

German RAA (2008). Forschungsgesellschaft für Straßen- und Verkehrswesen e.V., Richtlinien für die Anlage von Autobahnen (RAA), FGSV-Verlag, Köln, 2008. http://www.forschungsinformationssystem.de/servlet/is/275112/.

Green, P. (2005, November 30–Dezember 2). How driving simulator data quality can be improved. In *Proceedings of the Driving Simulation Conference DSC Orlando*.
Hart, S., & Staveland, L. (1988). Development of NASA-TLX (Task Load Index): Results of empirical and theoretical research. In P. Hancock & N. Meshkati (Eds.), *Human mental workload* (pp. 139–183). Amsterdam: North Holland.
Hucho, W. H. (2005). Aerodynamik des Automobils – Strömungsmechanik, Wärmetechnik, Fahrdynamik, Komfort. 5. Auflage. Wiesbaden: Vieweg.
ISO, (2008). Road cars—Ergonomic aspects of transport information and control systems—Simulated lane change test to assess driver distraction. ISO/TC 22/SC 13 N WG8 N579, DIS 26022.
Jahn, G., Oehme, A., Krems, J. F. & Gelau C. (2005). Peripheral detection as a workload measure in driving: Effects of traffic complexity and route guidance system use in a driving study. *Transportation Research Part F-Traffic Psychology and Behaviour* 8(3), 255–275.
Kappé, B. & van Emmerik, M.L (2005). *The use of driving simulators for initial driver training and testing*. Soesterberg, Netherlands.
Kennedy, R. S., Lane, N. E., Berbaum, K. S., & Lilienthal, M. G. (1993). Simulator Sickness Questionnaire (SSQ): An enhanced method for quantifying simulator sickness. *International Journal of Aviation Psychology, 3*(3), 203–220.
Knappe, G., Keinath, A. & Meinecke, C. (2006). Empfehlungen für die Bestimmung der Spurhaltegüte im Kontext der Fahrsimulation. In M. R. K. Baumann, S. Leuchter, & L. Urbas (Hrsg.) (Eds.), MMI-Interaktiv, Aufmerksamkeit und Situation Awareness im Fahrzeug (Vol. 11, pp. 3–13). ISSN: 1439-7854.
Lansdown, T. C. (2001). Causes, measures and effects of driver visual workload. In P. A. Hancock & P. Desmond (Eds.), *Stress, Workload, and Fatigue*. Lawrence Erlbaum Associates.
Math, R., Mahr, A., Moniri, M. M., & Müller, C. (2013). OpenDS: A New Open-source Driving Simulator for Research, GMM-Fachbericht-AmE 2013.
Mattes, S. (2003). The lane change task as a tool for driver distraction validation. In H. Strasser, H. Rausch, & H. Bubb (Eds.), *Quality of work and products in enterprises of the future*. Stuttgart: Ergonomia Verlag.
Mayhew, D. J. (1999). *The usability engineering lifecycle*. San Francisco: Morgan-Kaufmann.
Moreno, R., & Mayer, R. E. (2002). Learning science in virtual reality multimedia environments: Role of methods and media. *Journal of Educational Psychology, 94*(3), 598–610.
Nechvatal, J. (2009). *Immersive Ideals / Critical Distances* (pp. 48–60). LAP Lambert Academic Publishing.
Negele, H. J. (2007). Anwendungsgerechte Konzipierung von Fahrsimulatoren für die Fahrzeugentwicklung, Dissertation, TU München.
Nicolas, R. (2013). The different driving cycles. http://www.car-engineer.com/the-different-driving-cycles/. Abruf 01.06.2016.
PC Games Hardware, Oculus Rift: Release, Preis, Spezifikationen, Systemanforderungen; pcgameshardware.de/Oculus-Rift-Hardware, Zugriff 08.07.2016.
Pinto, M., Cavallo, V., & Ohlmann, T. (2008). The development of driving simulators: Toward a multisensory solution. *Le travail humain, 71*(1), 62–95.
Reich, D. (2015, September 1–2). Comparison of immersive and non-immersive driving task environments. In *International Conference and Exhibition on Automobile Engineering*. Valencia, Spain.
Reich, D. & Stark, R. (2014, September 17–19). 'To See or not to See?' how do eye movements change within immersive driving environments. In *Adjunct Proceedings of the 6th International Conference on Automotive User Interfaces and Interactive Vehicular Applications (AutomotiveUI,14)*. Seattle, WA, USA. http://dx.doi.org/10.1145/2667239.2667268.
Reich, D. & Stark, R. (2015, January 5–8). The influence of immersive driving environments on human-cockpit validations. In *Hawai'i International Conference on System Sciences*. Kauai.

Reich, D., Wittkugel, H. J., & Stark, R. (2014). How do immersive driving environments influence user performances and experiences? In T. Ahram, W. Karwowski & T. Marek (Eds.), *Proceedings of the 5th International Conference on Applied Human Factors and Ergonomics AHFE 2014*.

Remlinger, W. (2013). Analyse von Sichtverdeckungseffekte im Fahrzeug. Dissertation, Technische Universität München.

Roza, Z. C. (2004). Simulation Fidelity Theory and Practice: A Unified Approach to Defining, Specifying and Measuring the Realism of Simulations. Doctoral thesis: Delft University of Technology, Faculty of Aerospace Engineering, Delft.

Scheuchenpflug, R. (2001). Measuring presence in virtual environments. In M. J. Smith, G. Salvendy, & M. R. Kasdorf (Eds.), *HCI International 2001* (pp. 56–58). New Orleans: HCI International.

Schindler, J., Kelsch J., Heesen, M., Dziennus, M., Temme G. & Baumann, M. (2013). A Collaborative Approach for the Preparation of Co-operative Multi-User Driving Scenarios. 10. Berliner Werkstatt Mensch-Maschine-Systeme, 10.–12. Okt. 2013, Berlin, Germany.

Skluzacek, S. (2012). Angular Resolution of Human Sound Localization, Thesis, Winsconsin, USA.

Slater, M., & Wilbur, S. (1997). A framework for immersive virtual environments (five): Speculations on the role of presence in virtual environments. *Presence: Teleoperators and Virtual Environments, 6*(6), 603–616.

Stern, L. (1960). Personalized stereo fidelity. In *Popular mechanics part II*.

Talbot-Smith, M. (2013). *Audio engineer's reference book* (p. 2.52). CRC Press. ISBN: 1136119744.

Vlakfeld, W. P. (2005). *The use of simulators in basic driver training*. Leidschendam, Netherlands.

Zoeller, M. (2015). Analyse des Einflusses ausgewählter Gestaltungsparameter einer Fahrsimulation auf die Fahrerverhaltensvalidität. Dissertation. Technischen Universität Darmstadt.

Part V
Applications

Chapter 17
User Experience with Increasing Levels of Vehicle Automation: Overview of the Challenges and Opportunities as Vehicles Progress from Partial to High Automation

Patrice Reilhac, Katharina Hottelart, Frederik Diederichs and Christopher Nowakowski

Abstract The long awaited arrival of automated driving technology has the automotive industry perched on the precipice of radical change when it comes to the design of vehicle interiors and user experience. Recently, much thinking and many vehicle concepts have been devoted to demonstrating how vehicle interiors might change when vehicles reach full automation, where a human driver is neither required nor in some cases, even allowed to control the vehicle. However, looking more near term across all global market segments, we will likely see an increasing number of vehicles with widely varying automation capabilities emerging simultaneously. Any system short of full automation will still require driver control in some set of situations, and some fully automated vehicles will still allow driver control when desired. While it is unlikely that the basic seating arrangement, steering wheel, and pedals will be radically altered in this emerging segment of partial to highly automated vehicles, it is quite clear that the overall user experience during automated driving will need to evolve. Drivers will not be content to hold the steering wheel and stare at the road waiting for what may be a very infrequent request to take-over driving. The chapter presents the research conducted to develop the Valeo Mobius® Intuitive Driving solution for providing an embedded digital experience, even in lower levels of automation, and all while still promoting both shorter transition response times and better transition quality when emergency situations call for a transition from automated to manual control.

P. Reilhac (✉) · K. Hottelart · C. Nowakowski
Valeo, Paris, France
e-mail: patrice.reilhac@valeo.com

F. Diederichs (✉)
Fraunhofer Institut IAO, Stuttgart, Germany
e-mail: frederik.diederichs@iao.fraunhofer.de

G. Meixner and C. Müller (eds.), *Automotive User Interfaces*,
Human–Computer Interaction Series, DOI 10.1007/978-3-319-49448-7_17

17.1 Introduction

17.1.1 Driving Automation Past

The arrival of automated driving has been long awaited since the concept was first introduced at the 1939 New York World's Fair's Norman Bel Geddes Futurama exhibit, sponsored by General Motors. In his book, Magic Motorways (Geddes 1940), Norman Bel Geddes laid out a vision of a high speed, automated future for the automobile, where crashes and traffic jams would be a thing of the past, and travel time was freed up for leisure activities. Geddes was optimistic that such a future could be built quickly and called for major infrastructure investments to redesign and expand the highways system with an eye towards increased safety and efficiency.

In contrast to his optimistic time frame, Geddes was pessimistic on the capabilities of, and necessity for, humans drivers, and even today, one of the most common reasons cited for pushing for of driving automation technology is the current toll in human lives and economic loss resulting from crashes. It has long been held that driver error is the largest contributing factor in up to 95% of crashes. This statistic was originally reported as 93% by Treat, Tumbas, McDonald et al. (1977), and later was estimated at 95% as reported by the National Highway Transportation Safety Administration (NHTSA 2008). However, this statistic is not necessarily surprising given that drivers are currently responsible for almost all vehicle sensing, planning, decision making, and control. Since beginning of the decade, the National Center for Statistics and Analysis (2015) has estimated traffic fatalities in the U.S. have hovered at just under 33,000 per year, at a rate of around 1.1 fatalities per 100 million Vehicle Miles Traveled (VMT). Furthermore, NHTSA estimated that the annual economic impact of traffic crashes in the U.S. reached $871B (Blincoe et al. 2014), but the safety argument is not the only thing driving automation research and development.

Eliminating the human as a driver enables commercial economic gains in the form of labor reduction when it comes to taxis, ridesharing, and the delivery of goods. It also allows for a complete rethinking of both vehicle usage and design. And recently, much thinking and many vehicle concepts have been devoted to demonstrating how vehicle interiors might change once vehicles reach full automation, where a human driver is neither required nor, in some cases, even allowed to control the vehicle. Volvo's Concept 26 (2016) envisions seating adjustments that allow the driver to back away from the steering wheel and completely disengage from the driving task. Similarly, the Mercedes F 105 concept (Mercedes-Benz 2015) allows the driver and front passenger to completely swivel their seats to rear-facing during automated driving. Finally, IDEO proposed even more radical driverless automation concepts such as Cody, the driverless mobile delivery platform, and WorkOnWheels, a mobile conference pod allowing a team to work around a central conference table while traveling to their next meeting (Stinson 2014).

Over the years, there have been many vehicle automation research programs including the GM Firebird II concept vehicle (1956), the U.S. Department of Transportation sponsored National Automated Highway Systems Consortium (NAHSC) demonstration (1997), the U.S. Defense Advanced Research Projects Agency (DARPA) Grand Challenge (2004, 2005) and Urban Challenge (2007), and countless university research projects. In Europe, the first groundbreaking automated driving results were achieved in the PROMETHEUS project (1987–1995), but more recent EU automated driving projects include CityMobil and CityMobil2 (2006–2016), Have-IT (2008–2011), SARTRE (2009–2012), interactIVe (2010–2013), COMPANION (2013–2016), and AdaptIVe (2014–2017).

Despite the great interest and effort, 75 years later, it seems that automated driving technology has been proverbially just over the horizon, waiting for a confluence of sensor technology and computing power advances to catch up with the driverless vision. As we enter the second decade of the twenty-first century, the technology has finally caught up, and the inevitable move towards automated driving is forcing the automotive industry to rethink the car as we know it. It is clear that the future will include radical change when it comes to the design of vehicle interiors once vehicles are capable of fully automated driving, but novel seating arrangements constitute only one aspect of the new automated driving user experience. If one thing is certain, even with only partial automation, drivers will not be content with maintaining today's status quo.

17.1.2 Driving Automation Present

17.1.2.1 The Case for a Future Including Partial Automation

Several similar taxonomies have been proposed for driving automation systems, but SAE J3016 (2014) defined perhaps the most comprehensive scale including six levels starting from Level 0, completely manual driving, and progressing through Level 5, full automation. SAE Level 2 systems or partial automation systems still require the driver to pay attention to the roadway and system operation while being prepared to intervene at a moment's notice. Such systems range from the Mercedes Traffic Jam Assist introduced in 2014 to the Tesla Model S Autopilot introduced in 2015. SAE Level 3 or conditional automation systems would allow the driver to disengage from the driving task, but requires that the driver remain ready to reengage when given a reasonable lead time. SAE Level 4 or high automation systems also allow the driver to disengage from the driving task, but even at this level, it is possible that the system could encounter conditions that cannot be handled by the automation without driver intervention. Thus, any system short of full automation may still require driver control in some set of situations, and it is possible that some fully automated vehicles will still allow driver control when desired.

While there are plenty of compelling arguments for leaping directly to full automation, there are also a number of arguments as to why the driverless revolution will probably not be as abrupt as envisioned by some. First, as described above, driving automation systems are already being introduced on vehicles as the technology becomes available, so we are already seeing SAE Level 2 systems on the road. Given that the average age of a vehicle on the road is 11.5 years in the U.S. (Naughton 2015), even a relatively rapid progression from partial to full automation into the marketplace means that partial, conditional, and high automation systems will survive on the roads for decades.

Second, cost is always a major factor in the uptake of new technology. If full automation ends up being significantly more expensive than partial automation, some market segments may show a preference for the “good enough” that can be more affordably achieved with lower levels of automation. Thus, especially looking across all global market segments, cost concerns could provide a strong counter-argument against leaping directly to full automation. Cost constraints could push the market towards a future where widely varying automation capabilities emerge simultaneously.

Finally, trust in full automation may take some time to build. While some manufacturers might feel that the technology for full automation without a driver is imminently ready for deployment, there has been some skepticism expressed by both the public and various regulatory agencies. The California Department of Motor Vehicles (California DMV 2015) proposed regulations that would prohibit the deployment of completely driverless automated vehicles, at least for the first three years of any deployment program. In the public workshops surrounding this proposal, members of the public and some public advocacy groups expressed a general distrust of both the technology and the manufacturers, citing the record number of U.S. automotive safety recalls in 2014. Both public demonstration projects and a steady progression from partial to full automation might be paths that could be used to bolster public trust in automated driving systems.

17.1.2.2 In Automation We Trust?

On the one hand and as mentioned earlier, skepticism has been expressed in the public policy debates surrounding future rules and regulations for automated driving systems, and studies have suggested that trust in full automation may take some time to build, especially when it comes to convincing individuals to purchase an automated vehicle. In a series of surveys conducted over the years following the recent boom in automated vehicle technology advancements, announcements, and news media coverage, Schoettle and Sivak (2014a, b, 2015, and 2016) found a high amount of concern over automation technology was still being expressed in all of the surveyed countries, U.S., U.K., Australia, Japan, India, and China. Across countries, less than 15% expressed absolutely no concerns over automation technology, and up to 33% said that they simply would not ride in a completely self-driving vehicle. Most of the concerns expressed related to safety, reliability, and security, but legal and public policy issues such automated vehicle liability were also expressed. Furthermore, at least in the U.S., the overwhelming preference, almost 80%, was for manual or partial automation, rather than completely self-driving vehicles.

On the other hand, the issue of trust in automated driving systems is complex and multifaceted. What drivers will say, what drivers will do, and what drivers really want to do are often not completely aligned. Once drivers experience an automated driving system, even an imperfect one, complacency and boredom quickly set in, and it becomes clear that automated driving is going to require an entirely new user experience. A study comparing SAE Level 1 and 2 driving automation systems (Llaneras et al. 2013) found that even when the drivers were told that the driving automation system required them to constantly monitoring of the roadway, 50% of the drivers quickly abandoned the monotonous monitoring task and began to engage in smartphone tasks such as reading, texting, or emailing. After all, one of the goals of automation has always been to free us from tedious tasks and allow us to engage in the kinds of tasks that we'd rather been doing. As the automated driving systems improve, drivers will not be content to hold the steering wheel and stare at the road waiting for what may be a very infrequent request to take-over driving. Yet, until the systems are near perfect, the greatest challenge for partial, conditional, and high automation will be creating that new user experience while still maintaining safety when a transition of control is required.

17.1.2.3 Focus on Transitions of Control

Perhaps the largest human factors challenge in the field of automated driving is the question of how to create and guarantee a seamless transition of control from the automation to the human driver across the diversity of potential scenarios that might require the driver to take-over control. While there has already been quite a bit of research on automated driving transitions of control, the issue is complex, and the results obtained from each study are highly nuanced due to the study design and the metrics chosen for analysis. Some studies consider hands-on-the wheel or foot-on-the-brake as the take-over-response, while others consider the start or end of the avoidance maneuver as the take-over-response. In some studies, the driver is paying attention to the road, while in others, some form of non-driving task or distraction is presented. It is important when looking across studies, to consider that distraction engagement level, time pressure, HMI design, and the complexity of the required take-over maneuver will all play a role in both the timing and quality of the control transition.

As an example, Blanco et al. (2015) noted that when there was little time pressure, completing non-driving tasks was usually given priority over regaining the vehicle control. In a test track automated driving study, drivers were given up to 50 s to respond to an automation take-over request, and the drivers took a mean of 17 s to finish what they were doing before attempting to regain control from the system. Conversely, when imminent haptic alerts were presented for lane drift or automated braking occurred when an obstacle was detected, the mean take-over time, as measured by driver placing his or her hands on the steering wheel, ranged from 1.3 to 2.4 s. However, the study also noted that when cautionary visual alerts were presented alone, these alerts were often completely missed by the drivers.

Along a similar thread, Gold, Damböck, Lorenz, and Bengler (2013) examined driver take-over time in a freeway driving simulator study and showed that the quality of the driver response was dependent on the time that could be allocated to that response. The study used a lead vehicle cut-out scenario to reveal a stopped vehicle in the travel lane while the drivers were engaged in an in-vehicle task displayed in the center console. The automation system prompted the drivers at either 5 or 7 s of Time-To-Collision (TTC). When given 7 s to respond, 100% of the drivers simply performed a lane change, but when given only 5 s to respond, about 20% of the drivers brought the vehicle to a stop in the lane before attempting a lane change. A follow-on study using a similar scenario, Radlmayr et al. (2014), further concluded that both the demands of the secondary task and the demands of the surrounding traffic can influence the driver take-over time. In this study, the mean driver take-over response times ranged from 2 to 3 s, but upper end of the response times were almost double the mean, ranging as high as 6.5 s.

While the above studies looked at in-vehicle, center console distractions, a more likely scenario is that the driver of an automated vehicle will engage in a secondary task using their own mobile device, rather than something presented on the vehicle's center console. Melcher, Rauh, Diederichs, et al. (2015) conducted a driving simulator study examining mobile phone usage at the time of the take-over request and various HMI strategies, including presenting the alert on the mobile phone. However, presenting the take-over request on the mobile phone did not dramatically decrease the response time. Mean response times were still on the order of 3.4–3.8 s, and the longest response times were in the range of 6.7 s.

There are two troubling trends that should be noted across studies. First, take-over response times are highly variable across subjects, trials, and the specific take-over scenarios. Each of the studies described above included outlier trials where the take-over response time was easily double the mean response time. Second, in some studies listed above, trials with infinite take-over response times were found and often discarded because the drivers simply never responded to automation take-over request. From a design perspective, neither of these outcomes is ideal to ensure safe and successful transitions of control with lower levels of automation. Essentially, SAE Automation Levels 2 and 3 still rely up on the driver to serve as a backup to the driving automation system. Thus, if the expected range of the driver take-over time doubles, then the vehicle sensing range would also presumably double, and we will need to concede that some situations where the driver does not respond will still lead to crashes.

17.1.3 Driving Automation Future: A New User Experience

As we enter a future where driving automation becomes commonplace, two key questions need answering: what will drivers want to do during the automated drive and what will an appropriate human machine interface (HMI) should look like. Certainly when we reach full automation, the concept of a driver could completely disappear, but in the near future, the spectrum of driving automation systems will likely include a mix of vehicles with varying levels of automation. During this interim, anything short of full automation will still require driver control in some set of situations, and for this emerging set of partial to highly automated vehicles; it is unlikely that the basic seating arrangement, steering wheel, and pedals will be radically altered. However, it is quite clear that the overall user experience during automated driving will need to evolve. Drivers will not be content to hold the steering wheel and stare at the road waiting for what may be a very infrequent request to take-over driving. In fact, research has shown that drivers will most likely disengage quite quickly from such a monotonous monitoring task, and who can blame them?

Humans have never been good at maintaining vigilance in supervisory tasks, and the promise of automation has always been to free us from this sort of tedium. As we move from partial to high automation (SAE Levels 2 through 4), the interactions and communication between the driver and vehicle become extremely important. Drivers will desire to hand the driving task over to the car and so that they can engage in other activities, but the car is also depending on the driver to be able to resume driving as quickly as possible when needed. The transitions must be both intuitive and seamless, or the automated driving user experience will seem clumsy and error prone. The automated driving system must be able to communicate when it is OK for the driver to engage in other activities versus when it is time to pay attention to the road because something abnormal has been detected or simply because there is low confidence in the performance of one of the sensors. The chapter presents the research conducted to develop and test the Valeo Mobius® Intuitive Driving solution for providing the driver with an embedded digital experience, even in lower levels of automation, all while still promoting both shorter transition response times and better transition quality when emergency situations call for a transition from automated to manual control.

17.2 Valeo Mobius® Intuitive Driving Concept

17.2.1 What Is Intuitive?

The paradox of driving automation systems that fall short of full automation is that these systems will take the driver out of the loop, but then expect the driver to regain both situational awareness and control of the vehicle when the automation encounters a situation that it cannot handle or any sort of equipment failure. Rather than replacing the user, some near term automated driving systems will rely on the user as a central component in the system. However, it is naïve to think that the driver will be content with passively monitoring the automated driving system without distraction and without a steady decline in vigilance. Thus, starting from the assumption that the user will continue to play a critical role in an automated driving system, the vision at Valeo is simple. Automated driving should be "Intuitive Driving." We need to go beyond smart sensors and perception and control algorithms, and we need to consider a user-centered approach in designing a new user experience for automated driving. On the road from SAE Level 0 automation, where clearly "I Drive", to SAE Level 5 automation, where clearly "It Drives", there exists a less defined middle ground encompassing SAE Levels 2 through 4, where we must create a collaboration between the human driver and that automation that can be conceptualized as "We Drive" (Reilhac et al. 2015). Much like the rider and horse analogy (Flemisch et al. 2003), the user experience during automated driving should be natural, connected, and intuitive, but the question is, what does that mean? What does it mean to be intuitive in the context of driving?

In a study conducted in 2012 in Germany, France, China and the US we asked people what would be an intuitive driving experience for them? In the abstract, the answer can be summed up by saying that intuitive flows from establishing a connection between the natural world and technological world. Intuitive technology correctly understands our goals and anticipates our needs. We also asked what intuitive automotive features might look like? Surprisingly, we found out that the end-users already perceive some of today's driving assistance systems as very intuitive. For example, automatic lighting, rain-sensing wiper systems, and Autonomous Emergency Braking (AEB) were all cited as current technologies that were highly intuitive.

What people need and want is a natural driving experience where the car's driving assistance systems support the driver's goals with its predictions, anticipations, and reactions. Just as the driver (or a perfect human driver) would do him-/herself. The overall objective is to eliminate physical and cognitive friction and to support the driver in a natural, easy to understand way without surprises—a natural process. As driving automation systems evolve along the continuum from "I Drive" to "It Drives", the user experience should also evolve to allow the driver more and more free time as would intuitively be expected.

We also asked what people want to do during the free time that they will gain thanks to automated driving. The top answer coming up was the usage of their smart devices to read or send text messages and e-mails, or simply to surf the internet. To verify this, one only needs to observe behavior at any train or metro station, because it is quite evident that what most people are doing while being transported is using their smartphones. While our study was conducted in 2012, more recent studies (Stokes 2015) have also concluded that mobile device use ranks high among the top desired activities as shown in Fig. 17.1. In fact, many of the respondents confessed that they already used their smartphones while driving, even without an automated driving system and even knowing that this behavior can be very dangerous.

Essentially, what the driver of future wants is to use their smartphones safely and effectively, while still retaining a locus of control over the vehicle. This actually makes a lot of sense. If the model of collaboration is "We Drive" than an intuitive automated driving system must both keep the driver in the loop and allow the driver

Fig. 17.1 What do drivers want to do while driving an automated vehicle?

to safely engage in the desired secondary task activities. This was further highlighted when drivers were asked about preferred display locations. Although most research studies have tended to look at secondary tasks and distraction presented in the center console, the drivers we surveyed reported that their preferred display for the smartphone functions would be directly in front of them, allowing them to easily keep an eye on the road ahead. It seems that intuitive is synonymous with a seamless integration of not only all information relevant to the act of driving, but also a seamless integration of the personal computing and connectivity power already available in their smartphones.

17.2.2 Valeo's User-Centered Design Process

Based on these findings in the "Intuitive Driving Study" we conducted in 2012, we started our user-centered innovation process using Design Thinking with the goal of developing a HMI concept that provides the user with the best possible user experience during automated driving. The Design Thinking process focuses on the

Fig. 17.2 The four phases of the design thinking process

four phases shown in Fig. 17.2: understand, ideate, prototype, and test. At the core of this process is a bias towards action and creation: by rapidly creating and testing, you can continue to learn and improve upon your initial ideas. In the following section we will describe the different phases of the design thinking process that ultimately lead us to the Valeo Mobius® intuitive driving concept.

Understand

The first step of the Design Thinking process is to understand the user in order to clearly define the problem. We need to fully understand both the current experiences and the desired experiences of the users for whom we are designing the product. While many user-centered design processes start with understanding and the problem, the Design Thinking process differs because it emphasizes the experiences and emotions of the users. If you study just the tasks and the problems encountered by the users, then you might risk simply optimizing the current solution, rather than thinking about solutions that might be outside-of-the-box. Understanding requires observation, questioning, and ultimately, immersing yourself in the user's world with the goal of creating a user's point of view that we want to address with our product.

As mentioned earlier, we started the process of understanding the user's about intuitive HMIs with our first survey on the topic in 2012. But this was just the start of the research that we would conduct on the topic. With the possibilities of automated driving, this first survey simply showed that new expectations concerning communication and entertainment were rising in the minds of the likely users of these new automated vehicles, but this was by no means the end of the story. While people might be excited to gain some free time during their commutes, the issue of trust in automation is also a looming concern.

In more than 60 in-depth interviews conducted in Germany, China and France in 2014 Valeo found out that the drivers' needs during automated driving are not limited to communication or entertainment. The HMI, with its displays and indicators, was still considered necessary to build trust in the system and reassure the drivers that the system was working correctly and safely. The surveyed drivers wanted transparency about what the car is detecting and predictability about what the car is going to do next so that they can easily understand what the car is doing and why. Building a shared sense of situational awareness and transparent decision making is very important for trust building. People are not yet readily willing to hand over full responsibility for their safety to a machine. To build trust, the car needs a carefully thought-out HMI concept which can reassure the driver when needed and permit the driver to concentrate on other tasks when it's appropriate.

It is also worth noting that many of the participants did presume that they might need less information from the vehicle after an initial adaptation phase. As positive experiences boost trust, the need for information is progressively reduced, but in general, these drivers never wanted to be left out of the driving loop. They always wanted to keep an eye on all information relevant to driving, like speed, GPS, traffic information, etc. (Reilhac et al. 2016).

In addition to these opinions and attitudes, Valeo also gathered data on the drivers' physical abilities and limitations that could play an important role for automated vehicles. Driving tests focused on topics like situational awareness and motion sickness in order to evaluate the effects of different HMI concepts (Diels et al. 2016). (See Fig. 17.3.) Since the tests were simply focused on perception and well being, the tests were conducted using a Wizard of Oz protocol. The subject was seated as a passenger in a normal car driven by another driver to simulate the effect of being a passenger in an automated vehicle. The test included 30 participants who were asked to read various news articles on different display locations throughout the driving cockpit. The study showed that situational awareness could be increased by 30% during automated driving simply by using a higher positioning of the display on which the driver was reading an article. This adjustment enhanced the driver's perception of the environment by keeping his gaze closer to the road. Simultaneously, we found that the higher display position reduced the feelings of motion sickness by 50%.

Having analyzed the different research and tests that we have done, we learnt that from the user point of view the HMI is on the one hand a very important instrument to create trust and to make people feel safe in an automated car and on the other hand it is the medium in the car where we can create a real added value for the end-user. These insights allow us to enter the next phase of our process.

Ideate and Prototype

The second and third steps in the Design Thinking process are to ideate and prototype, based on the analyses conducted and insights gained during the understanding phase. These two steps are generally performed by a multidisciplinary team in a creative workshop setting. The workshop is broken up into design sprints, where the team first ideates or brainstorms a very wide set of possible solutions, and

Fig. 17.3 Testing the effect of a higher positioned instrument cluster

then precedes to rapidly prototype the highest ranked solutions. This phase was when the ideas of intuitive driving started to come together and take the physical form that would become the Valeo Mobius® Intuitive Driving solution. In this phase we transformed our ideas into solid physical forms to be able to interact with them, experience them, and learn from them in order to make the next prototypes even better. In Fig. 17.4 you can see some of our early prototypes.

One thing that you will quickly notice is that the Design Thinking process emphases creating "quick and dirty" prototypes, and that's just what our interdisciplinary team did. In fact, we prototyped a wide range of ideas starting in cardboard, spare parts, and anything else that we had lying around the lab. Some ideas were ruled out quickly, while others were merged, combined, and given the group OK to move on to the next iteration. This process allowed us to experience the ideas and generate new ideas at the same time, and we continued the iteration process until we came up with one very concrete concept that showed enough promise to justify the investment of more time and resources to prototype further and start testing.

Test

The fourth and final step of the Design Thinking process is to test the prototypes, but the process of understanding, ideating, prototyping, and testing is both iterative and continuous. In an ideal iterative design process, we would not only test the product ourselves, but we would also involve colleagues that do not work on the topic and, most importantly, we would involve potential end-users at various stages of the process. The testing phase should be used to generate more observations and feedback, which allows us to better understand the users and the problems, allowing us to ideate more concepts, refine our prototypes, and enter another phase of testing.

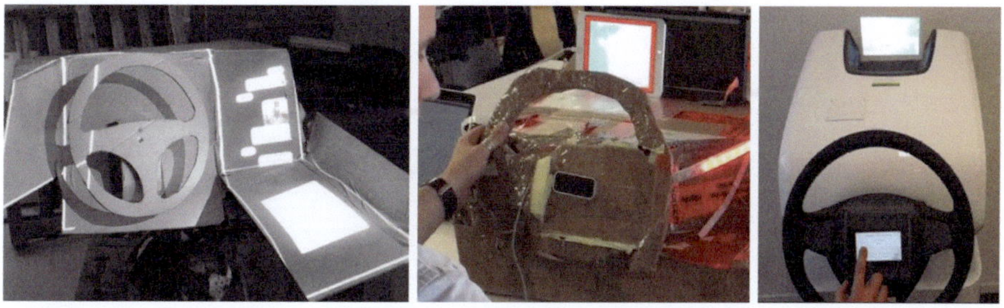

Fig. 17.4 The first prototypes created during the ideation and prototyping workshops

The more that you learn from and about your users, the more you can refine your original point of view to ensure that the solution that you are designing is the correct solution for the user's needs.

17.3 User-Testing of the Valeo Mobius® Intuitive Driving Concept

17.3.1 Description of the Valeo Mobius® Intuitive Driving Concept

The Valeo Mobius® Intuitive Driving Concept combines three major innovations to showcase our vision of an embedded digital experience centered around the automated driving system HMI. (See Fig. 17.5). First, the traditional fixed gauge instrument cluster will be replaced by a configurable digital instrument cluster using a 12" TFT display. The move to a digital instrument cluster would allow for the mirroring of a driver's smartphone or smart device content during automated driving, while still allowing enough space in the periphery of the display to provide the driver with any essential information to support maintaining a sense of situational awareness over the automated driving system.

Two keys to this concept are the instrument cluster location and the actual mirroring of the smartphone in the instrument cluster display. The instrument cluster display location should be placed above the steering wheel, as close as possible to the direct line of sight, in order to allow peripheral perception of traffic and frequent glance changes between road and display (Feron and Nicolas 2012). This location supports situational awareness and counteracts motion sickness, which can occur during automated driving when the driver is engaged in a secondary task (Diels and Bos 2015). The configurable display could then present a classical instrument cluster when in manual driving mode, but during automated driving mode, it could be used to show your smartphone screen content.

Second, the instrument cluster display and smartphone content would be controlled through a new generation of reconfigurable steering wheels switches,

Fig. 17.5 The Valeo Mobius® initial concept

consisting of a 1.8" touch screens on each side of the steering wheel, similar in size to a smartwatch. This configuration allows the driver to interact with their smartphone without taking their hands off the steering wheel, allowing for a much quicker and safer take-over response when driving needs to transition from the automation to manual control.

Finally, the third component of the concept is the inclusion of a driver monitoring system. The driver monitoring system would consist of a forward facing camera and hands-on detection on the steering wheel. Take-over alert times could be adjusted based on whether or not the driver is currently attentive, and rather than making the driver press a button to disengage the system or hold the wheel for five seconds while the system counts down, the driver monitoring system will be simply intuitively hand over control when the driver is ready and receptive.

In early 2015, Valeo launched the first tests with the prototype to evaluate usability and acceptance. The study included driver clinics and in situ interviews, where subjects were interviewed during the testing. End-users perceived the overall Valeo Mobius® concept as very appealing and attractive. The control through the touch screen steering wheel switches was rated as both intuitive and innovative, and the control provided over the full mobile device content during automated driving was a real highlight. During the simulated transition phases drivers felt well guided and supported by the system.

What was striking in this study was that the majority of the study participants were not aware of the added value in safety that this concept could offer. People saw the added value in the use of communication and entertainment media in automated mode, but some participants questioned why they could not just use their handheld smartphones directly, rather than operating them via the Valeo Mobius® controls. The potential danger of handheld phone usage during automated driving and the Valeo Mobius® concept's ability to improve safety would need to be demonstrated and tested in follow-on studies.

17.3.2 Valeo Mobius® Driving Simulator Study

17.3.2.1 Driving Simulator Study Method

A second study was conducted in April 2015 in cooperation with the Fraunhofer Institute IAO in Stuttgart. The main goal of the evaluation was to compare the Valeo Mobius® concept with a handheld smartphone (Samsung Note 4) in an automated driving scenario that might involve unexpected driver Take-Over Requests (TORs) such as those that might be required in SAE Level 2, 3, or even 4 automation.

The study's baseline condition simulated the ideal SAE Level 2 automation condition, where the driver is performing constant supervisory control by being attentive with both hands on wheel and both eyes on road. For the two test conditions, we selected a non-driving related reading task, with engagement either through a smartphone or through the Valeo Mobius® prototype. The text was either presented on the handheld smartphone or on the 12.3 inch instrument cluster display above the steering wheel. Text scrolling was either performed via the Valeo Mobius® press-touch-displays on the steering wheel or by swiping the screen of the Android phone. Hence, in the Valeo Mobius® condition, the driver's hands remained on the steering wheel, while in the smartphone condition the driver's hands held the phone and the driver's head was often directed downwards as shown in Fig. 17.6. Consequently the smartphone condition should result in total cognitive, visual, and manual distraction, while the Valeo Mobius® condition should only result in cognitive distraction and partial visual distraction (Ganzhorn et al. 2013).

The study was conducted in a stationary chassis driving simulator located at Fraunhofer IAO vehicle interaction lab, a platform suited for studies on driver state detection and infotainment control, where vehicle dynamics have relatively little impact (Diederichs et al. 2014, 2015). Although the simulator was on a stationary chassis, it did contain a motion platform allowing small movements such as braking jerks. The Silab 4.0 driving simulation software was projected with a 180° forward field of view, combined with projection screens for the rearview mirrors, and the

Distraction:	SAE Level 2 Baseline hands on wheel	SAE Level 3 using Valeo Mobius™	SAE Level 3 using Smartphone
Manual	–	–	✔
Visual	–	o	✔
Cognitive	–	✔	✔
Head Position			
Hand Position			

Fig. 17.6 Study test conditions

Valeo Mobius® cockpit was fully integrated with the driving simulator's adaptive multimodal dashboard (Sulzmann et al. 2013).

The driving scenario consisted of a busy highway scenario with three lanes in each direction, but traffic only allowed for an automated driving speed of 60 km/h. The subject vehicle typically drove in the middle lane following a truck in order to hinder the driver's view of the road ahead. For most of the experiment, the vehicle drove in automated mode, allowing the subjects to engage in one of the three experimental conditions. Several times during the experiment, two of the three lanes of the traffic suddenly stopped. The third lane, either the right or left lane, continued driving at the original speed, and a gap in traffic was timed such that it would allow an attentive driver sufficient time and space to manually change lanes and overtake the braking truck in the center lane such as is shown in Fig. 17.7. The simulated automated driving system was not able to perform an automatic lane changes, and thus, could only apply an emergency braking behind the truck.

As shown in Fig. 17.8, approximately 3400 ms before the automated driving system initiated automated emergency braking (AEB), a take-over request was issued to the driver instructing the driver to take-over driving and perform a lane change. The take-over request was composed of a multimodal warning, applying elements from a study by Melcher et al. (2015). The warning consisted of an acoustic gong and a flashing red LED array in the windshield, while simultaneously the Valeo Mobius® display switched to manual mode.

During the experiment, the abrupt traffic stop followed by a driver take-over request occurred five times. The first time that traffic came to an abrupt stop was considered a surprise take-over request, and thus could only be analyzed as a

Fig. 17.7 Traffic situation in the moment of TOR. The *left* lane allows manual lane change

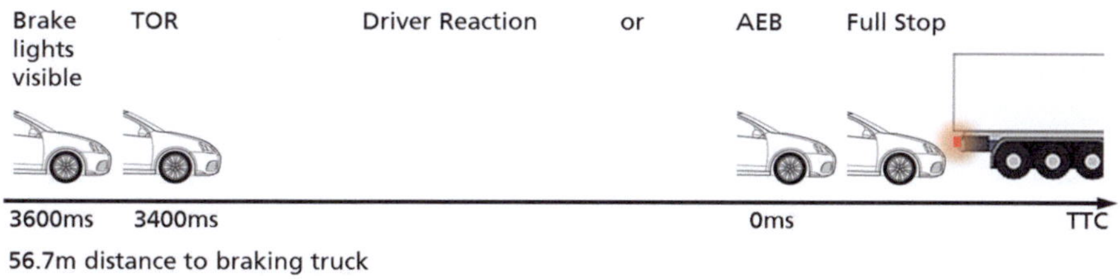

Fig. 17.8 Timing of the take-over situation at 60 km/h with 56.7 m headway to braking truck

between-subjects effect with respect to the baseline, Valeo Mobius®, or smartphone distraction method. Subsequent take-over requests were considered learned, and each driver was given at least one take-over request in each of the experimental conditions, baseline, Valeo Mobius®, and smartphone, allowing for a within-subjects analysis across these three conditions.

17.3.2.2 Driving Simulator Study Results

In total, the data of 20 female and 22 male participants (age mean 36.5 years, range 23–71 years) was analyzed. All participants held a driver license for at least 2 years, had a minimum annual mileage of 5000 km per year, and all used a smartphone or tablet at least twice a week. The dependent variables measured included the reaction times during the surprise take-over requests, steering quality during the manual lane-change maneuver, and driver's subjective assessment of the overall user experience when comparing the Valeo Mobius® concept against a handheld Android smartphone while in automated driving mode.

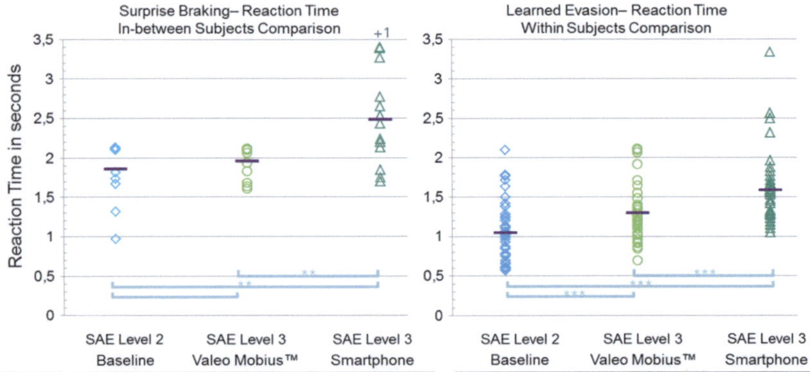

	SAE Level 2 Baseline	SAE Level 3 Valeo Mobius™	SAE Level 3 Smartphone	SAE Level 2 Baseline	SAE Level 3 Valeo Mobius™	SAE Level 3 Smartphone
Mean	**1.86**	**1.96**	**2.48**	**1.05**	**1.30**	**1.59**
Max.	2.13	2.12	3.40	2.10	2.12	3.33
Min.	0.97	1.60	1.70	0.57	0.70	1.05
Variance	0.13	0.04	0.32	0.14	0.12	0.19
Age range	23 - 51	23 - 51	23 - 59	23 – 71		
Gender ratio	6 f / 6 m	8 f / 8 m	6 f / 7 m	20 f / 22 m		
N	12	16	13 +1*	42		

— Mean *** P < 0.001 ** P < 0.05 * P < 0.1 n.s.

Fig. 17.9 Reaction times for the surprise braking event and the learned evasion maneuvers

Figure 17.9 depicts the analysis of reaction times for the surprise and learned take-over request reaction times. It should be noted that in the surprise take-over request comparison, one participant in the smartphone condition did not react within the allotted 3.4 s, and thus was excluded from analysis. The preferred response for all participants was to brake after receiving the take-over request. The overall effect of the condition in an unrelated one-way ANOVA was significant ($F_{(2,38)} = 8.68$, $p = 0.001$, $\eta^2 = 0.314$). When Bonferroni adjustment was made for the number of post hoc comparisons, *t*-tests revealed significant differences in means between both baseline and smartphone conditions ($p = 0.014$) and Valeo Mobius® and smartphone conditions ($p = 0.025$). There was no significant difference between Valeo Mobius® and baseline conditions ($p = 0.402$). Because of a possible violation of homogeneity of variances and of normality due to outliers in the data subsets, the respective non parametric tests were conducted, replying the same results regarding testing decision.

In the learned evasion condition, all participants managed to evade correctly to the left lane. A one-way repeated-measures ANOVA revealed a significant overall effect of condition ($F_{(2,82)} = 34.664$, $p < 0.001$, partial $\eta^2 = 0.458$). Bonferroni adjusted *t*-tests revealed significant differences in means for all post hoc comparisons (each $p < 0.001$).

For the learned evasion condition, the steering maneuver was also measured and analyzed. Both the extremes of the steering wheel angle and the standard deviation of the steering wheel angle were measured between the take-over request and a point 300 m past the initial lane crossing were analyzed for 36 of the participants. In Fig. 17.9, the assessed steering quality indices are described on the left side. On the right side the results for absolute value of steering wheel angle extreme values are depicted. A one-way repeated-measures ANOVA revealed a significant overall effect of condition ($F_{(2,70)} = 9,352$, $p < 0.001$, partial $\eta^2 = 0.211$). Bonferroni adjusted post hoc *t*-test revealed a significant difference in means between baseline and smartphone conditions ($p < 0.001$), a marginally significant difference between Valeo Mobius® and smartphone conditions ($p = 0.070$) and no significant difference between baseline and Valeo Mobius® conditions ($p = 0.210$).

In Fig. 17.10, the results of the analysis of steering wheel standard deviation analysis are depicted. For the standard deviation of steering wheel angle until lane crossing, a one-way repeated-measures ANOVA showed a significant effect of condition ($F_{(2,70)} = 7.171$, $p = 0.001$, partial $\eta^2 = 0.170$). Bonferroni adjusted post hoc *t*-test revealed a significant difference in means between baseline and smartphone conditions ($p = 0.003$), a marginally significant difference between Valeo Mobius® and smartphone conditions ($p = 0.097$) and no significant difference between baseline and Valeo Mobius® conditions ($p = 0.328$). Since Mauchly's test of sphericity was significant ($x^2 = 16.594$, df = 2, $p < 0.001$), Greenhouse-Geisser correction was applied to the one-way repeated-measures ANOVA for the standard deviation of steering wheel angle between lane crossing and 300 m after TOR, indicating a significant effect of condition ($F_{(1.443, 50.498)} = 12,992$, $p < 0.001$, partial $\eta^2 = 0.271$). For Bonferroni adjusted post hoc *t*-tests significant differences in means between both baseline and smartphone conditions ($p < 0.001$) and Valeo Mobius® and smartphone conditions ($p = 0.036$) were found. There was a marginally significant difference between Valeo Mobius® and baseline conditions ($p = 0.071$) (Fig. 17.11).

Finally, the results for usability and user experience rating are shown in Fig. 17.12. The analysis compared the Valeo Mobius® concept and smartphone for reading and scrolling during automated driving with system-initiated take-over requests possible. While no significant differences regarding subjective Usability between Valeo Mobius® and smartphone could be found on the standardized System Usability Scale (Brooke 1986), when asked which system participants would choose for non-driving-related tasks while driving an automated vehicle, a clear preference for the Valeo Mobius® system can be seen, especially along the axis of safety. A more detailed analysis of the user experience evaluation can be found in (Diederichs et al. 2015), including qualitative statements. Valeo Mobius® was rated significantly more convenient and more comfortable to use while driving automated, whereas smartphone was rated significantly more often to be distracting. Furthermore participants reported a better connection to the traffic situation as main advantage for Valeo Mobius®.

17.3.2.3 Driving Simulator Study Conclusions

The results showed how a well integrated HMI, such as the Valeo Mobius® concept, can support non-driving-related tasks during automated driving, while still

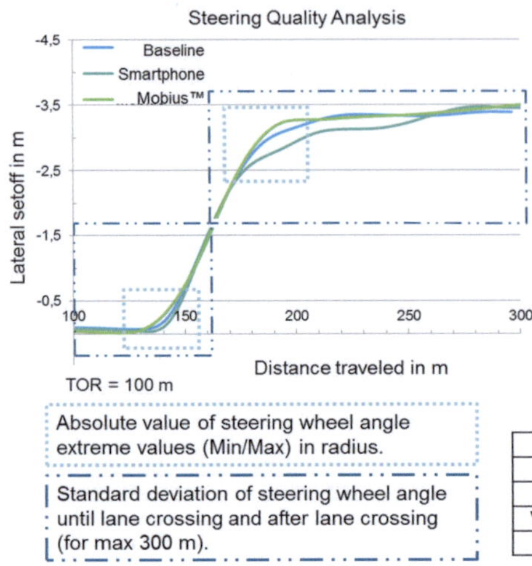

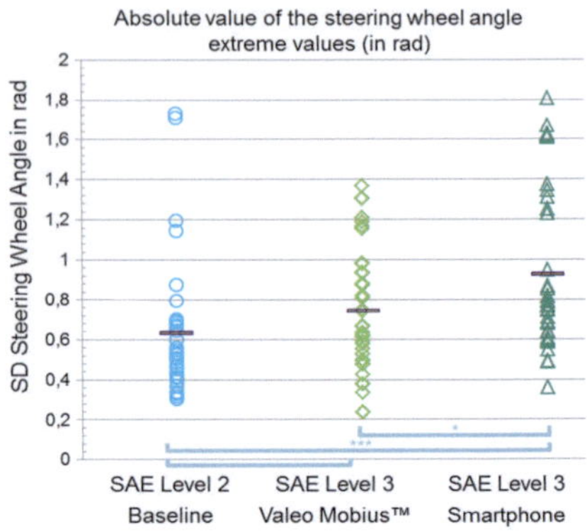

	SAE Level 2 Baseline	SAE Level 3 Valeo Mobius™	SAE Level 3 Smartphone
Mean	**0.63**	**0.74**	**0.93**
Max.	1.73	1.38	1.81
Min.	0.31	0.24	0.36
Variance	0.11	0.1	0.15
N	36	36	36

Fig. 17.10 On the *left side* the measurements of steering quality are exemplary depicted for one participant. On the *right side* the absolute Values of steering wheel angle extreme values for each condition are depicted

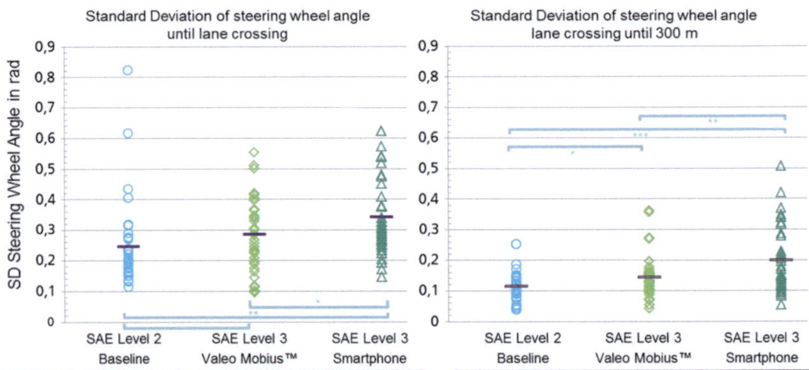

	SAE Level 2 Baseline	SAE Level 3 Valeo Mobius™	SAE Level 3 Smartphone	SAE Level 2 Baseline	SAE Level 3 Valeo Mobius™	SAE Level 3 Smartphone
Mean	**0.25**	**0.29**	**0.34**	**0.12**	**0.15**	**0.20**
Max.	0.82	0.55	0.62	0.26	0.36	0.51
Min.	0.12	0.10	0.15	0.03	0.04	0.06
Variance	0.02	0.01	0.02	0.00	0.01	0.01
N	36	36	36	36	36	36

Mean; *** P < 0.001; ** P < 0.05; * P < 0.1; n.s.

Fig. 17.11 Standard deviation of steering wheel angle

maintaining safety during the take-over scenario. As shown in Fig. 17.13, during an unexpected take-over situation, the smartphone users needed significantly more time to react than either an alert driver or a driver using the integrated Valeo Mobius® HMI concept. All the participants in the Valeo Mobius® condition showed a self-initiated braking reaction within 2.2 s, while over 75% of the drivers using the

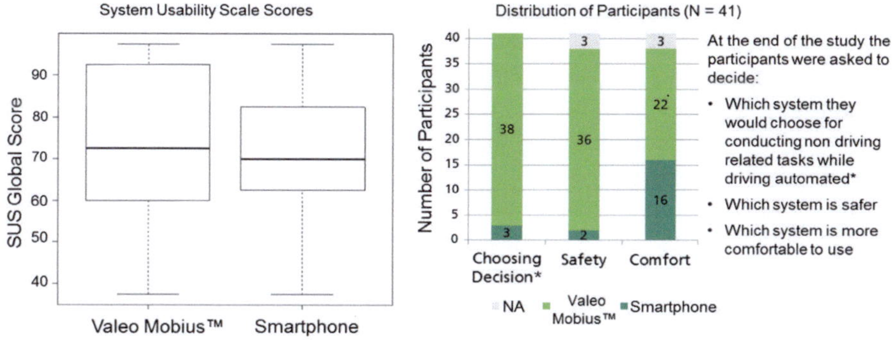

Fig. 17.12 System usability scale and UX evaluation

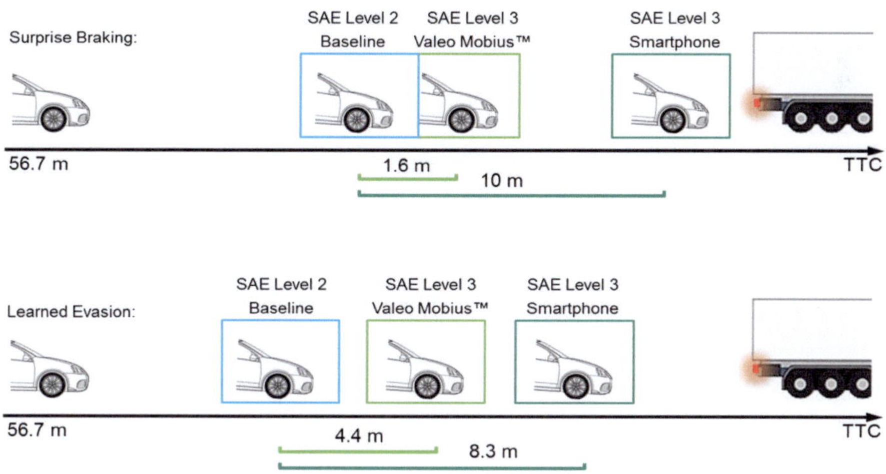

Fig. 17.13 Differences in stopping distance between the conditions

handheld smartphone took longer than 2.2 s, and one participant did not even react before the AEB kicked in.

In the learned evasion maneuver (also shown in Fig. 17.13), the differences were less pronounced, but there was still a significant difference in both response time and response quality between the integrated Valeo Mobius® HMI and the handheld smartphone conditions. The steering maneuvers with the integrated concept were less abrupt and stabilized quicker after the lane change. However, it should be noted that in this scenario, the quality of the steering maneuver is also related to the response time since a slower response time would require a more abrupt steering maneuver to make the lane change without hitting the stopped lead vehicle.

While we found no significant difference in response time between the baseline attentive driver and the Valeo Mobius® distraction condition for the learned maneuver, the baseline condition fostered, on average, a 320 ms quicker reaction time than the Valeo Mobius®. This was readily explainable given the scenario design. In the simulation, the brake lights became visible 3.6 s before the AEB, while the TOR was emitted at 3.4 s before AEB. Thus, the attentive driver had an advantage of at least 200–300 ms over either of the distraction conditions. The advantage was less pronounced in the surprise condition because driver was not expecting traffic to suddenly stop.

Furthermore, we may ask why smartphone users react 500 ms slower than Valeo Mobius® users in surprising take-overs, and 300 ms slower in learned take-overs. The smartphone users need to move eyes and the head in order to focus the traffic. In addition, some smartphone users placed the phone somewhere else before moving the hands to the steering wheel. Others kept grabbing the phone and steered with one hand, a possible explanation for reduced steering quality in this condition. The difference between surprising and learned take-overs can be attributed to the development and training of a strategy how to solve this challenge.

Overall, the results support the hypothesis that an integrated HMI concept like Valeo Mobius® has the potential to disarm the tradeoff dilemma between taking the driver out of the loop during non-driving related tasks, whilst leaving him with full accountability for intervening when the automated driving system requests it. The concept enables the driver to accomplish highly demanding non-driving related tasks while still maintaining the ability to take over manual control within a significantly shorter time than when using a handheld smartphone.

17.4 Conclusions

17.4.1 Applicability of Valeo Mobius® to Different Levels of Automation

While there remains a strong need for further research—especially on the question of exactly which tasks can safely be performed with an integrated HMI such as the Valeo Mobius® Intuitive Driving concept, the initial results of our development and testing suggest that this concept might be highly applicable to SAE Automation Levels 2 through 4. The advantage of an integrated HMI over a handheld smartphone is both clear and especially relevant for short take-over times, in the range of 2 to 5 s. As the allotted take-over time increases and approaches 10 s, the warning strategy simply becomes less relevant, as noted in Melcher et al. (2015). However, allowing for a driver take-over time in the range of 10 s is probably unreasonable for any driving automation system less than SAE Level 4, where the system would be fully capable of bringing the vehicle to minimal risk condition even without driver intervention. For a highway automation system traveling at 70 mph (110 kph), the minimum sensing range alone would need to exceed 300 m to detect objects with a time horizon of at least 10 s, and even if the sensing was perfect, accurately predicting all vehicle movements over the next 10 s is likely to be an impossible system constraint, especially for lower levels of driving automation.

Taking the levels of automation on a case-by-case basis, SAE Level 2 systems are the most challenging because, by definition, they require constant supervisory control by the driver and could require driver intervention at a moment's notice either with or without warning from the system. These systems essentially contain minimal sensor redundancy and limited sensor reliability, and they simply do not know what the sensor suite might be missing. As an example, the first crash of a SAE Level 2 system resulting in a fatality (Tesla Motors 2016) cited a number of extenuating circumstances such as a bright white truck on a bright background which may have fooled the camera and a precrash configuration where the radar was looking under the truck. However, most of the time, Level 2 systems will control the vehicle just fine, lulling the driver into a false sense of security. While clearly more research would need to be conducted before proclaiming the concept sound for SAE Level 2 automation, an integrated HMI concept like the Valeo Mobius® could allow the driver to engage in non-driving related tasks when the sensor confidence is high, while directing the driver's attention away from non-driving related tasks when the sensor confidence is low. At a minimum, the system would certainly be no less safe than drivers choosing to engage in handheld smartphone use.

SAE Level 3 systems are also challenging, but for a different set of reasons. SAE Level 3 automation should be capable of driving the vehicle in most normal circumstances, but may still require driver intervention in the event of a system failure or when driving out of systems operational domain. In either case, unlike a SAE Level 2 system, a Level 3 system must be able to detect, predict, and request a driver intervention with a reasonable amount of time for the driver to actually intervene. An integrated HMI concept like the Valeo Mobius® system is essentially perfect for SAE Level 3 automation because it allows drivers to both disengage from the driving task, while still promoting the ability to reengage in the driving task almost as fast as if the driver was already attentive to the road. More research is certainly needed on exactly how to best maintain driver situation awareness when heavily involved in non-driving related tasks, and more research could be needed on the potential cognitive tunneling effects that could be seen as the automated driving proceeds successfully for longer and longer periods of time between intervention requests.

Finally, SAE Level 4 systems, by definition, will be capable of bringing the vehicle to a minimal risk condition, even without driver intervention, and thus, these systems could be essentially driverless. However, it is likely that some SAE Level 4 systems will still allow the driver to engage or intervene in the driving task, especially when we consider the fact that the SAE Levels represent a continuum, rather than necessarily discrete levels. The road from SAE Level 3 to 4 might end up being a more gradual progression than a simple leap of technology. As an example, an automated driving system might not be able to navigate every single construction zone, but if the system is capable of recognizing situations that it might not be able to handle and bring the vehicle to a safe stop if the driver does not effectively intervene, then the system would qualify as a SAE Level 4 system. Continuing with our example, let's suppose that the automated driving system can handle most construction zones, unless there is a flagger directing traffic. An integrated HMI like the Valeo Mobius® system could still be beneficial, even if not necessary, by promoting a faster driver intervention than if the driver was using a handheld smartphone.

17.4.2 Automated Driving Future User Experience

As we enter a future where driving automation becomes commonplace, there's no doubt that drastic changes to vehicle interior layouts are on the horizon, especially once vehicles reach full automation, where a human driver is neither required or even allowed to control the vehicle. However, we must also recognize that even in the nearer term, where systems just shy of full automation will still require driver control in some set of situations, the automated driving user experience must also change and evolve. Drivers will not be content to hold the steering wheel and stare at the road waiting for what may be a very infrequent request to take-over driving. We must move beyond the simple interface design questions regarding buttons, switches, and displays, and we must start considering the overall user experience desired by future automated vehicle drivers.

It is quite clear that in today's society, most people lead lives digitally connected through their smartphones and other computing devices, and given a 30-second lull in any conversation, many, if not most, people will be reaching for their smartphone to fill the time. It is naïve to think that the drivers of automation will not do the same, whether or not they have read or clicked "I Agree" to a EULA that might have instructed them otherwise.

The automated driving user experience of the future must account for the increasing digitalization of life and provide an intuitive solution that allows the drivers to engage in non-driving tasks while both maintaining a level of situational awareness and promoting safe, efficient, and effective transfers of control when needed. This chapter presents Valeo's vision of such a future, along with the research conducted during our development of the Valeo Mobius® Intuitive Driving solution.

References

Blanco, M., Atwood, J., Vasquez, H. M., Trimble, T. E., Fitchett, V. L., & Radlbeck, J. (2015). *Human factors evaluation of level 2 and level 3 automated driving concepts* (Report No. DOT HS 812 182). Washington, DC: National Highway Traffic Safety Administration.

Blincoe, L. J., Miller, T. R., Zaloshnja, E., & Lawrence, B. A. (2014). *The economic and societal impact of motor vehicle crashes, 2010* (Report No. DOT HS 812 013). Washington, DC: National Highway Traffic Safety Administration.

Brooke, J. (1986). *System Usability Scale*. Reading, England: Digital Equipment Corporation.

California DMV. (2015, December 16). *Proposed Express Terms, Title 13, Division 1, Chapter 1, Article 3.7—Autonomous Vehicles. Sacramento, CA: California Department of Motor Vehicles*. https://www.dmv.ca.gov/portal/wcm/connect/ed6f78fe-fe38-4100-b5c21656f555e841/AVExpressTerms.pdf?MOD=AJPERES. Accessed 27 June 2016.

Diederichs, F., Pöhler, G., & Spath, D. (2014). Driving maneuver prediction based on driver behavior observation. In *Proceedings of the 5th International Conference on Applied Human Factors and Ergonomics,* Kraków, Poland, July 19–23.

Diederichs, F., Bischoff, S., Widlroither, H., Reilhac, P., Hottelart, K., & Moizard, J. (2015). New HMI concept for an intuitive automated driving experience and enhanced transitions. In *Proceedings from the 7th Conference on Automotive User Interfaces and Interactive Vehicular Interactions*, Nottingham, UK, Sept. 1–3.

Diederichs, F., Schüttke, T., & Spath, D. (2015). Driver intention algorithm for pedestrian protection and automated emergency braking systems. In *Proceedings of IEEE 18th International Conference on Intelligent Transportation Systems*—(ITSC 2015), Canary Islands, Spain, Sept. 15–18.

Diels, C., & Bos J. (2015). Self-driving carsickness. In *Special Issue of Applied Ergonomics—Transport in the 21st Century: The Application of Human Factors to Future User Needs* (In Press).

Diels, C., Bos J., Reilhac, P., & Hottelart, K. (2016). Motion sickness in automated vehicles: the elephant in the room. In G. Meyer & S. Beiker (Ed.), *Road Vehicle Automation 2*. Lecture Notes in Mobility. Springer International Publishing Switzerland.

Feron, S., & Nicolas, L. (2012). Eyes close to the road: the perfect restitution? Ergonomic and technical perspectives on two solutions: HUD and top mounted instrument cluster. In *IQPC 3rd International Conference on Automotive Cockpit HMI*, Bonn, Germany, Sept. 24–26.

Flemisch F. O., Adams C. A., Conway S. R., Goodrich K. H., Palmer M. T., & Schutte P. C. (2003). The h-metaphor as a guideline for vehicle automation and interaction, NASA/TM-2003-212672.

Ganzhorn, M., Diederichs, F., & Höfer, M. (2013). Eine bewerterbasierte Ablenkungsskala (BABS) zur objektiven Beurteilung von Unaufmerksamkeit. In: *Der Fahrer im 21. Jahrhundert.* Düsseldorf: VDI-Verlag. pp 133–146.

Geddes, N. B. (1940). *Magic Motorways*. New York, NY: Random House.

Gold, C., Damböck, D., Lorenz, L., & Bengler, K. (2013). "Take over!" How long does it take to get the driver back into the loop? In *Proceedings of the Human Factors and Ergonomics Society 57th Annual Meeting*, San Diego, CA, Sept. 30–Oct. 4.

Llaneras, R. E., Salinger, J., & Green, C. A. (2013). Human factors issues associated with limited ability autonomous driving systems: drivers' allocation of visual attention to the forward roadway. In *Proceedings of the Seventh International Driving Symposium on Human Factors in Driver Assessment, Training, and Vehicle Design,* Bolton Landing, NY, June 17–20, pp. 92–98.

Melcher, V., Rauh, S., Diederichs, F., Widlroither, H., & Bauer, W. (2015). Take-over requests for automated driving. In *6th International Conference on Applied Human Factors and Ergonomics*, Las Vegas, NV, July 27–30.

Mercedes-Benz. (2015). The *Mercedes-Benz F 015 Luxury in Motion Concept Car.* https://www.mercedes-benz.com/en/mercedes-benz/innovation/research-vehicle-f-015-luxury-in-motion/. Accessed 27 June 2016.

National Center for Statistics and Analysis. (2015). *Early estimate of motor vehicle traffic fatalities for the first quarter of 2015* (Crash Stats Brief Statistical Summary Report No. DOT HS 812 190). Washington, DC: National Highway Traffic Safety Administration.

NHTSA. (2008). *National Motor Vehicle Crash Causation Survey: Report to Congress* (Technical Report DOT-HS-811-059). Washington, DC: U.S. Department of Transportation, the National Highway Traffic Safety Administration (NHTSA).

Naughton, N. (2015, July 29). Average age of the U.S. fleet hits record 11.5 years, IHS says. *Automotive News.* http://www.autonews.com/article/20150729/RETAIL/150729861/.average-age-of-u.s.-fleet-hits-record-11.5-years-ihs-says. Accessed 27 June 2016.

Radlmayr, J., Gold, C., Lorenz, L., Farid, M., & Bengler, K. (2014). How traffic situations and non-driving related tasks affect the take-over quality in highly automated driving. In *Proceedings of the Human Factors and Ergonomics Society 58th Annual Meeting*, Chicago, IL, Oct. 27–31.

Reilhac, P., Millet, N., & Hottelart, K. (2015). Thinking intuitive driving automation. tractatus technologico-humanicus. In: G. Meyer & S. Beiker (Eds.), *Road Vehicle Automation 2*. Lecture Notes in Mobility. Springer International Publishing Switzerland.

Reilhac, P., Moizard, J., Kaiser, F., & Hottelart, K. (2016). Cockpit concept for conditional automated driving. In: *ATZ Magazine*, No 3, 2016, p. 42–46 (In press).

SAE. (2014). *SAE J3016: Taxonomy and Definitions for Terms Related to On-Road Automated Motor Vehicles*. Warrendale, PA: Society of Automotive Engineers.

Schoettle, B., & Sivak, M. (2014a). *A Survey of Public Opinion About Autonomous and Self-Driving Vehicles in the U.S., the U.K., and Australia* (UMTRI-2014-21). Ann Arbor, MI.: University of Michigan Transportation Research Institute.

Schoettle, B., & Sivak, M. (2014b). *Public Opinion About Self-Driving Vehicles in China, India, Japan, the U.S., the U.K., and Australia* (UMTRI-2104-30). Ann Arbor, MI.: University of Michigan Transportation Research Institute.

Schoettle, B., & Sivak, M. (2015). *Motorists' Preferences for Different Levels of Vehicle Automation (UMTRI-2015-22)*. Ann Arbor, MI.: University of Michigan Transportation Research Institute.

Schoettle, B., & Sivak, M. (2016). *Motorists' Preferences for Different Levels of Vehicle Automation: 2016 (UMTRI-2016-8)*. Ann Arbor, MI.: University of Michigan Transportation Research Institute.

Stinson, L. (2014). IDEO Imagines the Wild Future of Self-Driving Cars. *Wired*. http://www.wired.com/2014/11/ideo-imagines-self-driving-cars/. Accessed 27 June 2016.

Stokes, S. (2015). *Consumers expect to use mobile devices, read and eat in self-driving cars*. http://engineering.cmu.edu/media/press/2015/01_22_autonomous_vehicle_survey.html. Accessed 5 July 2016.

Sulzmann, F., Melcher, V., Diederichs, F., & Sayar, R. (2013). Modular dashboard for flexible in car HMI testing. In N. A. Stanton (Ed.), *Advances in human aspects of road and rail transportation (Advances in human factors and ergonomics series)* (pp. 97–106). Boca Raton, FL: CRC.

Tesla Motors. (2016, June 30). *A Tragic Loss* [Tesla Motors Blog]. https://www.teslamotors.com/blog/tragic-loss. Accessed 30 June 2015.

Treat, J. R., Tumbas, N. S., McDonald, S. T., Shinar, D., Hume, R. D., Mayer, R. E., Stanisfer, R. L. & Castellan, N. J. (1977). *Tri-Level Study of the Causes of Traffic Accidents* (Technical Report DOT-HS-034-3-535-77). Washington, D.C.: U.S. Department of Transportation, National Highway Traffic Safety Administration (NHTSA).

Volvo. (2016). *Concept 26*. http://www.volvocars.com/intl/About/Our-Innovation-Brands/IntelliSafe/IntelliSafe-Autopilot/C26. Accessed 27 June 2016.

Chapter 18
AutoPlay: Unfolding Motivational Affordances of Autonomous Driving

Sven Krome, Jussi Holopainen and Stefan Greuter

Abstract The *AutoPlay* prototypes have been designed to explore the implementation of non-driving activities into the context of a future autonomous driving situation. The conceptual design goal was to maintain a pleasurable situational awareness of the inactive driver by integrating the driving context as a meaningful input into the interaction system. In this chapter, we introduce the design of three experimental applications for autonomous driving and report on explorative user studies conducted to investigate the impact of the three *AutoPlay* prototypes: *AutoGym,* an in-car exertion game that translates car speed and traffic situations into an individual exercise program. *AutoJam,* a touch sensitive steering wheel cover to generate interactive music experiences in a creative interplay with car's driving dynamics. *AutoRoute,* a discovery application for future urban commuting in autonomous cars that enables an exploration of the city based on spontaneous routing and rerouting. Furthermore, we reflect on the outcome of the user studies and propose three motivational affordances of autonomous driving: *drivability*, *performability*, and *explorability*. Each of these concepts, help to understand the motivational possibilities of the autonomous driving situation and facilitates a meaningful alignment of interaction systems and the driving context. We discuss the underlying concepts of the three affordances by relating them to the experiences identified in the user studies. Subsequently the contribution of this chapter is twofold: (1) We introduce the *AutoPlay* prototypes as inspirational concepts for aligning non-driving activities with the autonomous driving context and (2) we propose three motivational affordances as design targets for the implementation of non-driving activities in order to initiate a broader discussion on the pleasures of autonomous driving beyond instrumental motives.

The original version of this chapter was revised: Table 18.2 has been removed. The erratum to this chapter is available at 10.1007/978-3-319-49448-7_19

S. Krome (✉) · J. Holopainen · S. Greuter
Center for Game Design Research, RMIT University, Melbourne, VIC, Australia
e-mail: sven.krome@rmit.edu.au

G. Meixner and C. Müller (eds.), *Automotive User Interfaces*,
Human–Computer Interaction Series, DOI 10.1007/978-3-319-49448-7_18

18.1 Introduction

Autonomous driving technology promises an eventual relief from the driving task. From a safety perspective, the benefits of automation technologies are indisputable. Advanced driver assistance systems such as Audi's *multicollision brake assist*[1] are already available and assisting the driver during emergency situations. But also in noncritical situations, automated driving technology is already available saving the driver from undesired situations such as frustrating traffic congestions, monotonous highway drives or the chore of parking the car (Bengler et al. 2014). In the near future, even more advanced driving automation will be available taking over the complete control on the major parts of the trip.

The safety critical aspects of autonomous driving systems require rigorous human factors and ergonomics testing to ensure that the technology is usable and safe. However, besides usability challenges, user experience factors will be of crucial importance for user acceptance. For example, experience research on driver assistance systems, such as adaptive cruise control, concluded that the relieve from driving can have a negative impact on feelings of control, competency, and autonomy (Eckoldt et al. 2012). For many people driving a car means much more than just going from one place to another. Owning and using a car is associated with individualism, power, independency, and of course an expression of status (Sheller 2004). Up until now, these hedonic factors (Diefenbach et al. 2014) of automated driving have sparked only very little interest in research and industry. In our research, we focus on a design of hedonic factors of autonomous driving by maintaining and facilitating situational awareness and pleasurable driving experiences. Led by a research-through-design approach (Zimmerman et al. 2007), we investigated the design challenge of translating the (often unconscious) pleasures of actively driving into meaningful and fulfilling non-driving activities of future autonomous driving.

In doing so, we designed and evaluated a series of in-car prototypes (called *AutoPlay* prototypes). These prototypes articulate our vision of autonomous driving and subsequently enable a reflection of motivational affordances of autonomous driving. Each *AutoPlay* prototype addresses one particular noninstrumental motivation of driving by a playful implementation of a non-driving activity with the dynamics of the driving situation. The *AutoPlay* prototypes are:

- *AutoGym*: an in-car exertion game that translates the driving context into a challenging fitness program.
- *AutoJam*: a touch sensitive, interactive steering wheel cover that motivates the player to learn and play music in collaboration with the car.
- *AutoRoute*: a navigation application for commuters that motivate spontaneous routing and a playful exploration of urban environments.

[1]https://www.audi-mediacenter.com/en/press-releases/the-new-audi-q7-sportiness-efficiency-premium-comfort-410.

In the following, we present the *AutoPlay* prototypes and report on the findings from the user studies. Furthermore, we reflect on the design process and findings and postulate three motivational affordances of autonomous driving: *drivability*, *performability*, and *explorability*. These affordances are (unconsciously) present in the context of traditional driving (as motivational possibilities for actions) but are disguised by the context of autonomous driving. Through the implementation with the driving context, the *AutoPlay* prototypes make these affordances a tangible part of the experience and transform the autonomous driving situation into a pleasurable activity in itself. The discovery and articulation of motivational affordances of autonomous driving are of particular importance for designing situated user experiences that go beyond prominence of instrumental motives of current autonomous driving research.

18.2 Background

18.2.1 Motivational Affordances

The concept of *affordance* was coined by the ecological psychologist Gibson (2014). According to Gibson, an affordance is a physical property of the environment that describes an action possibility between an organism and its environment. Donald Norman introduced the concept of affordance into design and particular HCI and distinguishes affordances into perceived and actual properties of an object (Norman 2013). A perceived affordance does not necessary be an actual affordance and vice versa, an actual affordance can be obscured. In other words, an affordance can be available in the environment independently if the actor perceives the affordance as such.

In recent years, researchers situated the notion of affordance with theories of motivation and behavior change. For instance, Zhang suggests ten design principles for motivational affordance (Zhang 2008). She understands motivational affordances, as perceived properties of an object that *determine whether and how it can support one's motivational needs*. An interaction system has high motivational affordance if its use fulfills a basic psychological need. Those basic psychological needs are propagated by self-determination theory and contain constructs like autonomy, competency, and self-actualization (Sheldon et al. 2001; Ryan et al. 2006).

In accordance with Zhang's suggestions, we propose motivational affordances of autonomous driving as properties of the driving situation that (are sometimes hidden) can transform the driving situation into a need fulfilling experience. In correspondence to traditional driving, that can be a satisfying and pleasurable activity in itself, the autonomous driving situation becomes a satisfying activity in itself. Through the implementation of non-driving activities, the *AutoPlay* prototypes made three motivational affordances tangible: (1) *drivability* that affords a feeling of control over the car, (2) *performability* that affords self-expression and (3) *explorability* that affords a feeling of independency and autonomy of the driver.

18.2.2 Design Assumptions

18.2.2.1 The Driving Scenario

The lack of highly automated driving technology required us to imagine an autonomous driving scenario. Based on literature and discussions with industry experts, we defined a set of design assumption and articulated a possible scenario of a future autonomous driving use case.

We assume that in the near future, at least as long as traditional cars are on our roads, the driver would benefit from a maintaining a comfortable situational awareness during the trip. The inactive driver should be encouraged to intentionally or unintentionally observe the surrounding traffic situation. Not only to intervene in case of an emergency but also to experience a feeling of orientation and progression. At least while the car is driving, the inactive driver would benefit if he or she can face the windscreen and look at the traffic in front.

For the design scenario, we set the layout of the car as a contemporary standard passenger car with front-facing seats and corresponding safety requirements such as seatbelt. We assume that the current front-seat passenger experience would resemble an autonomous driving experience of the near future. Correspondingly, all types of non-driving activities and entertainment devices that comply with the current safety requirement of the front-seat passenger would be a suitable for our autonomous driving setup. The driving scenario has been defined as an urban commute with a single adult passenger. The drive takes place during rush-hour traffic in an urban environment resulting in dense traffic with many stop-and-go phases.

SAE International (formerly Society of Automotive Engineers) distinguishes between six consecutive levels of automated driving. The first three levels (level 0–2) represent the lower end of driving automation. Driver assistant systems, such as adaptive cruise control, that require the driver to retain monitoring the environment and in consequence be fully responsible for the driving operation. Level 3 to 5 represent higher end of driving automation. The monitoring of the environment as well as the responsibility of the driving task is transferred to the system. Whereas in level 3, the human driver needs to be available as a fallback, in level 4 and 5 the system is able to drive (parts of the trip in level 4) fully autonomously. In the following, when we talk about *autonomous driving*, we refer to these higher levels of driving automation. All three *AutoPlay* prototypes would require at least SAE level 3 automation, in the case of *AutoJam*, and level 4 and 5 for *AutoGym* and *AutoRoute*.

18.2.2.2 Conceptual Design Challenges

Based on our definition of the autonomous driving situation, we articulated four conceptual design challenges. These challenges are based on literature review and correspond to the definition of the driving scenario. The concepts of the prototype should ensure the following qualities:

(1) **Promote a non-driving activity beyond entertainment or work**

Research on activities in public transportation shows a strong dominance of entertainment and work-related activities (Lyons et al. 2007). Even in the driving context work-related activities such as writing emails has been ranked highly (Alt et al. 2010). In other words: relieved from driving, the car will not only transform into an extension of the living room but also in an extension of the office. We believe, however, that the car is a very unique space that can provide meaningful activities beyond entertainment or work. The design challenge was therefore to explore activities and embed them into the design space of the car, so that they provide an additional benefit to the user such as health (*AutoGym*), creativity (*AutoJam*), or exploration (*AutoRoute*).

(2) **The player should be alert and motivated to observe the traffic**

As argued in our speculation on autonomous driving, we believe that safety and well-being of autonomous driving will benefit from situational awareness of the inactive driver. Additionally, many car passengers report on motion sickness when reading a book or interacting with a computer. Hence, the inactive driver should maintain a heads-up in-car position (i.e., looking out of the car's windscreen) at least while the car is moving and the prototypes should encourage the user to engage with the world outside of the car.

(3) **The experience should reframe the traffic perception**

The driving scenario was defined as dense rush-hour traffic. Traffic and in particular when the car stops, was usually experienced as very frustrating. Similarly, commuting on the same route in congested traffic can be monotone and a daily hassle. The conceptual design of the prototypes should ensure that traffic is experience as less threatening. In preliminary research, we identified that the car's stop is experienced as a central conflict of driving in traffic (Krome et al. 2016). However, the stop also provides an opportunity for the user to engage and focus on an interaction. The challenge is to leverage the stop as an interactive opportunity and subsequently reframe the traffic experience.

(4) **Facilitating a relationship with the car and the dynamics of driving**

As research on adaptive cruise control shows, automation technology can have a strong impact on a feeling autonomy and competency (Eckoldt et al. 2012). We believe that there are many unconscious needs of driving that may be affected by autonomous driving technology. In order to explore the hidden pleasures of driving, the prototypes should establish a connection to the car and the dynamics of driving. The challenge was therefore a meaningful alignment of interaction system and car features to ensure that hedonic aspects of driving are addressed and established through an engaging relationship with the driving situation.

18.3 The AutoPlay Prototypes

In this section we will introduce three *AutoPlay* prototypes. Besides the conceptual design, we also present a brief digest of the findings from the prototypes' user studies based on the forthcoming publications (Krome et al. 2017a, b, c). For a better understanding of the design and study setup, we recommend the reader to watch the video documentation of the *AutoPlay* prototypes available online: https://vimeo.com/165761217.

18.3.1 AutoGym

Driving a car is arguably not the healthiest mode of transportation. The car cabin and the sitting position allows only very limited space for body movements which can result in sitting fatigue and even postural defects. One possible way to prevent negative effect of the driving situation could be in-car exercises. Inactivity of professional drivers indicates that exercises can have a beneficial impact of the general well-being and stress relieve (Taylor and Dorn 2006). Some airlines motivate their passengers to engage in in-flight exercises and some offices motivate the staff with exercise programs in order to prevent negative side effects of office work (Proper et al. 2003). The relief from driving provides new opportunities for in-car exercises and may transform autonomous driving into a healthy and fit activity.

AutoGym is an in-car fitness device that translates the traffic into an interactive and individualized exercise program. In its current iteration, *AutoGym* consists of a mini-exercise bike that is manually operated. In future iterations, the exertion interface could become any kind of exercises machines with adjustable resistance.

AutoGym represents itself as an exertion game that motivates exercises through game-like interaction. It requires the player to predict the traffic situation and choose the length of the workout segment based on this prediction. The resistance of the spinning wheel is connected to the speed so that the player is only able to beat the complete the selected workout segment as long as the car drives slowly enough.

18.3.1.1 Conceptual Design

The conceptual design of *AutoGym* was based on two central assumptions. First, a physical experience of the car's driving activity would establish a feeling of control and orientation. Second, using the car's speed as a playable input in the game would (positively) reframe the perception of traffic. Based on these two assumptions, was designed *AutoGym* by an iterative and context-based design process (Krome et al. 2015).

Figure 18.1 shows the in-car setup of the *AutoGym* with the manually operated mini-exercise bike placed on the lap. The resistance of the mini-exercise bike is connected to the cars speed: the faster the car, the higher the resistance. When the car has stopped, the resistance is low enough to comfortably spin the wheel. The resistance increases to the maximum at a speed of 60 km/h. Maximum resistance makes it very difficult to spin the wheel. The game interface of *AutoGym* is based on a tablet-pc mounted in front of the player.

18.3.1.2 Exertion Game and Interface Design

The goal of playing *AutoGym* is to complete an exercise program by selecting time segments based on a prediction of current car speed. Figure 18.2 shows the *AutoGym's* interface. The circle represents the whole exercise program. To complete the exercise program, the player needs to select one of the colorized segments of the circle. The longer the segment, the more turns have to be completed on the exercise bike. Each successful turn is rewarded by a visual progression of the segment as well as audio feedback.

For each segment the player has only a limited amount of time. The bigger the segment, the more turns are required for completion. The time for each segment is represented by a progress bar running in the inner boarder of the circle. If the player

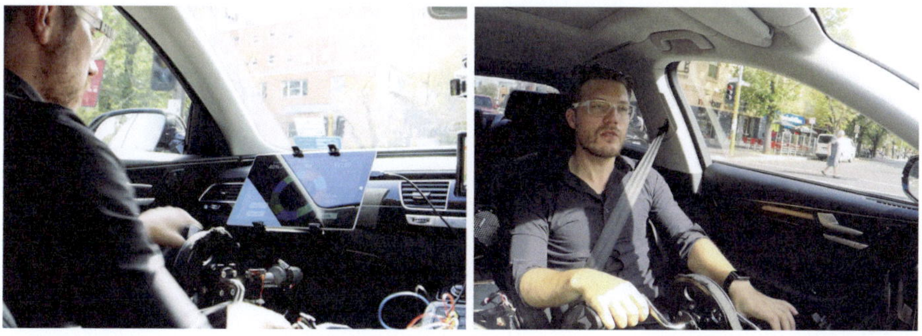

Fig. 18.1 *AutoGym* play test in the car with manually operated spinning bike and tablet-pc for the game interactions

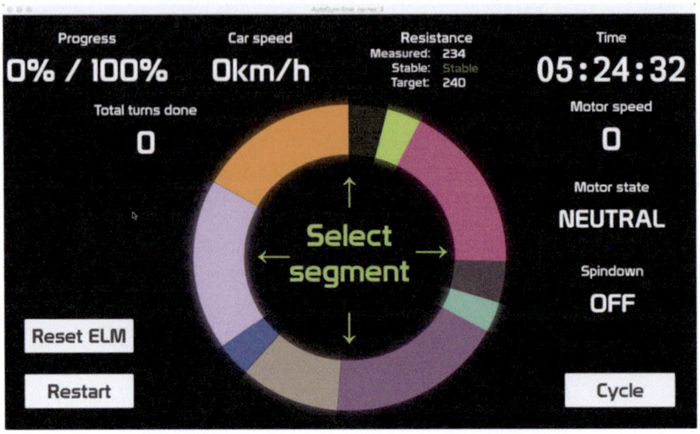

Fig. 18.2 *AutoGym* graphical user interface. The *circle* represents the whole exercise program that has to be completed by selecting the segments for workout

was not able to complete the required turns of the segment before the time runs out, the player failed this segment. The segment defaults and the player has to work on it again when the traffic is more suitable for this length.

Since the resistance of the exercise bike is connected to the speed, the player has to estimate how long the car is standing or drives slowly enough to be able to complete the segment in time.

The interaction loop of *AutoGym* from the player's perspective consists of three complementary elements: (1) The anticipation and prediction of the traffic situation, (2) The translation of the prediction into a time segments, and (3) the completion of the time segment by spinning the wheel often enough before the countdown expires.

The game finishes as soon as the player has successfully completed all segments of the ring. After completion the player is presented with performance statistics of his or her performance such as overall turns, failed segments, and total time.

18.3.1.3 Technical Implementation

Figure 18.3 shows the *AutoGym* exertion interface. The spinning wheel is based on a mini-exercise bike that can be used by hand or by foot. In order to adjust the resistance based on the car's speed and display the progress on a visual interface, we modified the exercise bike with a high-torque electrical motor and several sensors to register resistance and turns. Sensors and electrical motor are connected to the game interface via an Arduino microcontroller.

Fig. 18.3 *AutoGym* Technical Setup. The resistance is modified by a high-torque electrical motor. The motor is controlled by an *Arduino* microcontroller that translates the speed of the car into resistance

18.3.1.4 Evaluation

In order to explore the impact of *AutoGym* on the player experience, we conducted a simulator study based with a quantitative and qualitative evaluation with 28 participants. Goal of the study has been to explore the motivational impact of an exertion game in the context of traffic and identify how the game affects the autonomous driving experience.

Study Procedure

After an introduction and consent, the participants were briefed on the driving scenario, the *AutoGym* prototype and the purpose of the study. Before and after the play test, the participants were asked to complete a short questionnaire with 15 questions plus several demographic questions. The questionnaire consisted of five scales from (Sheldon et al. 2001): self-determination questionnaire: autonomy-scale, competency-scale, stimulation-scale, security-scare, and self-esteem scale. After the pre-questionnaire, the scenario was introduced as a regular rush-hour commute in an autonomous car. Stuck in traffic you are able to play an exertion game that keeps you fit and aware of the traffic situation. The test procedure has been concluded by a semi-structured interview focusing on experience topics.

Simulator Setup

We tested *AutoGym* in an improvised driving simulator. The traffic scenario was based on a traffic video recorded on an evening commute during rush-hour traffic in Melbourne. The video footage of the commuting clip was played on a 46" TV screen. Additional to the video recording, we also recorded the speed of the car captured through the car's OBD-II interface. The *AutoGym* simulator was playing the video in sync with the speed data that was controlling the application, i.e., the resistance of the exercise bike. Figure 18.4 shows a picture of the simulator setup.

18.3.1.5 Summary of Findings

The transcribed interviews were analyzed using a thematic analysis approach (Braun and Clarke 2006). The analysis resulted in three implications for designing exertion games in the autonomous driving context that are based on identified interpretations of how *AutoGym* motivates exertion.

(1) **Implementing exertion as a driving substitute**

Driving is a physical challenge of controlling the car. In order to implement exertion as a driving substitute, the interaction system has to substitute actual control with a feeling of control. *AutoGym* realized this by overcoming the physical dynamics of the car. The car's speed is the physical challenge to obstruct control over the car. The challenge to "beat the car", as a participant reported, framed the perception of the car as a competitor that has to be overcome. The rewarding

Fig. 18.4 AutoGym simulator setup

moment is to impose the own power over the car's power. The higher the speed the more challenging the game and the higher the feeling of exerting control over the situation. The stop phases of the car were used as time for recovery.

By defeating the physical dynamic of driving, the player experience control as self-efficacy. The player is in control of the driving task by mastering the power and speed of the car. Besides sportive and fitness motives, the pleasure of exertion is facilitated by the unusual context of the exercise program and particular by the exposition to the physical power of the car.

In a future of autonomous driving, the exertion motivation would benefit if the player's performance had a direct impact on the driving style. This implementation requires a direct translation of a car's power into an exertion interface. The game is a *play* of power, additional game mechanics such as the prediction mechanic are only necessary as long as their support the core experience of beating the car.

(2) **Implementation of exertion though car identification**

This implementation is also based on a physical experience of the car's power but frames it not as a challenge but as sympathy for the car. The car is perceived as a companion with whom you have to share the workload of progressing through traffic. Exertion task is interpreted as a medium to build a physical connection to the car and to progress in traffic.

The goal is to not to beat the power of the car but to feel the progression of traffic through the exercise program. In contrast to (1), the assumption is that neither traffic nor exercises are fun. However, it is pleasurable to experience traffic progression and gain orientation and control by feeling what the car is doing. The player does not overcome the car's power. The player is in a physically mediated dialog with the cars driving dynamics. The exertion facilitates an understanding and allows the player to progress.

(3) **Implementation of exertion as a contribution**

In strong contrast to the two other implementations, this one favors the prediction of traffic as the core experience of the game. Exercise and the feeling for the car (the core experiences of the other two implementations) are only a motivational affordance to make a correct prediction. In fact the goal was described as avoiding exertion through the precise anticipation of the driving situation.

Traffic is the playground and focus of the player's engagement: it has to be understood in order to progress in the game which is the primary goal. The interaction resembles a dialog that framed the car as a collaborator to progress the game by sharing the workload.

This implementation establishes a feeling of control by understanding traffic. A correct prediction is associated with a feeling of success and control over the traffic. By understanding the driving situation, traffic lost its character as an uncontrollable, disastrous event and turns into a playable environment.

This collaborative interpretation is based on an exchange of exertion work. This makes it interesting for the context of electrical cars. It promotes the idea of charging. Especially the spinning wheel was associated with charging the car, which adds another reason for exercising. Crucial for this implementation is a strict enforcement of the stop-and-go duality and the rules of the game.

18.3.2 AutoJam

Since the early beginning of in-car entertainment, music has played a central role. As long as the driving task required operational driving control, listening to music was one of the few accepted entertainment alternatives. The car cabin and the fixed seating position, make the car perfectly suitable for listening to a powerful sound system. In connection with the dynamics of driving, music can establish a unique emotional experience of driving (Bull 2004). Subsequently, it exists a huge after-market for car-based sound systems with advanced configuration resembling the power and clarity of a concert system.

With autonomous driving, sound and music experiences can be much more interactive. However, the potential sound and music is mostly investigated as an information display to establish trust, control or situational awareness such as (Beattie et al. 2014). Despite the close connection between of music and driving, research on interactive music experiences that involve creativity and exploration, such as known from digital music games and arcade machines, is very limited. In order to explore the connection of creative music games and autonomous driving, we design *AutoJam* as an in-car music experience that integrate music listening and gaming.

18.3.2.1 Conceptual Design

In an autonomous car an inactive steering wheel (i.e., a steering wheel not used for driving) can be a prominent and tempted interface for many types of interactions. Drumming on the steering wheel is a typical activity when listening to music. *AutoJam* is based on a touch sensitive steering wheel cover (as seen in Fig. 18.5) that enables interactive music experiences.

The general idea of *AutoJam* is that the player can chose his or her favorite song and become a part of the band. Thereby the song is dissected in three segments: A rhythm segment, in which the player learns the basic drum rhythms of the song (drum mode). A melodic segment, in which the player can improvise with an instrument of the song (freeplay-mode) and a progression segment in which the original song progresses and the player can listen to upcoming musical challenges (progress-mode). Each of those three segments of the song is connected and triggered by three defined stages of a driving situation: idle phase, when the car has stopped, acceleration phase, when car drives an consistent speed and cruise phase, when the car drives a consistent speed (Table 18.1).

Fig. 18.5 In-car game design session based on a second steering wheel (*left* image). *AutoJam* prototype in development: each drum or synth sound is assigned to one of the three touch pads with LED indicator (*right* image)

Table 18.1 *AutoJam* Modes: The three game modes are initiated by the driving phases

Mode	Driving phases	Description	Scoring
Drum	Idle phase (0–5 km/h)	Repeat a one bar drum loop	Yes
Freeplay	Acceleration/deceleration (5–60 km/h)	Improvise an instrument with car speed as input of the notes register	No
Progress	Cruise (const. speed for 15 s)	Song progresses	No

18.3.2.2 Game Design

To develop a proof-of-concept prototype of *AutoJam* we chose the Herbie Hancock's song *Chameleon* as the musical content. From the song, we extracted nine simplified drum rhythms and composed three instruments for the improvization-phase; a synth and two organs. Only when the car was standing, the player can progress in the drum mode and level up. Every time player has mastered three drum levels, he or she unlocks a new instrument that then becomes available during freeplay-mode.

Freeplay-mode is initiated when the car drives with inconsistent speed. In freeplay-mode, the player becomes the instrumentalist of the band and can improvise a solo with the unlocked instrument. Each of the three pads on the steering wheel triggers one note based on the songs harmonic scale. However, the register, i.e., the pitch of the note is based on the speed of the car. The faster the car, the higher the register of the played note. We predicted that the player's creativity would be increased by the necessity to adapt the improvization with the speed of the car.

Finally, if the car drove at a constant speed for more then 15 s, the game triggered progression-mode. This mode did not require any interactions from the player. The original song progressed and the player can listen upcoming challenges.

18.3.2.3 Implementation and Car Setup

AutoJam was played on three conductive pads implemented on the steering wheel cover as seen in Fig. 18.5. Besides the LED of the drum pads that visually indicate the drum patterns, *AutoJam* does not have any visual interface. All instructions, game progressions and levels are audio based through music or verbal commands. The game and car speed (taken via the OBD-II interface) are processed on an *Arduino Touchboard* that is sending MIDI data to an external sound processor (Ableton Live) that contains the samples of the song.

To enable an in situ user study, we mounted a second steering wheel with the *AutoJam* prototype on the passenger side of the test vehicle as seen in Fig. 18.6. Even though the steering wheel could not be turned it provided a realistic impression of a driving environment.

18.3.2.4 Evaluation

In order to explore how *AutoJam* impacts on the experience of traffic, the car and the game, we conducted an in situ user study with 14 participants. The study process was based on a play test of *AutoJam* in real traffic followed by a semi-structured in-car interview. All test drives have been taken place in Melbourne city traffic. The traffic situation ranged between busy and very busy depending on the time of testing. The play tests drives were between 20–60 min; all 14 participants were able to finish at least five of the nine drum levels so they were able to unlock at least the second improvization instrument. However, only seven participants were able to complete the whole game.

The semi-structured interviews focused on an evaluation of the player experience of *AutoJam*. The interviews were structured by 10 open questions that inquire on general feeling, the motivations to play and the perception of traffic and the

Fig. 18.6 *AutoJam* Setup. During the user study, the prototype was mounted on the stationary steering wheel on the dashboard of the passenger side

autonomous driving situation. Furthermore, we inquired on a subjective evaluation of the situational awareness of the participant as well as the experience of improvising in collaboration with the car. The video recordings of the play tests and the in-car interviews have been coded and analyzed by a thematic analysis approach (Braun and Clarke 2006). Besides shared experiences, we focused in the analysis on the unexpected aspect of the *AutoJam* experience.

18.3.2.5 Summary of Findings

The analysis of the interviews provided us with a detailed resolution of the *AutoJam* user experience. Particularly, the creative performances in the context of the driving situation enabled a new perspective, how creative play can frame driving and traffic into a space of self-expression and improvization.

Building Blocks of In-Car Performances

In the following, we reflect on the findings of the semi-structured interviews by discussing the dynamics of the in-car performance. We identified four crucial building blocks of in-car performances: (1) the stage, (2) the play, (3) the progression, and (4) the roles/actors. The following provides a brief overview how the dynamics of the user experience establishes and impacts the building blocks of the in-car performance.

(1) **The Stage: a private/public play**

Cars are public as well as private spaces; especially when standing in traffic and being exposed to other road users. On one hand, the encapsulation of the car established a feeling of being protected from the outside. On the other hand, the car is a genuine vehicle for (self-) presentation. The car is both, the stage and the backstage as outlined by (Goffman 1975). This combination often established a tension between the car as a private space and self-representation.

Our interviews confirm the tension. The eye-catching setup of the *AutoJam* motivated great performances but also resulted in awkward feelings of being exposed to the other people's glances, in other words, being exposed to an audience. According to (Goffman 1974), the frame is the definition of the situation and guiding principle, which organize the behavior and the subjective involvement of the actors. Situations where the actors have different frames may lead to awkwardness. For example, starting to sing an opera in a supermarket queue results in the actors (i.e., the singer and the other people around) framing the situation differently, usually leading to mutual embarrassment. Appropriate cues, such as a costume and other stage props, can help, however, to frame the situation as a show. In *AutoJam* setup the car is the costume and the stage props of the player that need to support and communicate the frame of creative play to prevent awkwardness.

We assume that awkwardness resulted entirely from *AutoJam's* eye-catching test setup and not from the type of interactions. Tapping the steering wheel is an interaction that may not break the frame of the situation. Nevertheless, we propose that future in-car activities will require a framing of the situation by communicating the play to other traffic participants. In case of AutoJam, a clear definition of the situation would transport the creative result to the outside. Examples could be simple as a LED strip or a windscreen display that would serve as visualization (such as equalizer bars) or the casting of the music to the outside world.

(2) **The Creative Play: improvization and situational awareness**

Freeplay-mode, i.e., the improvizations in the context of driving speed, is the most crucial creative component of AutoJam. The improvization of freeplay-mode was designed to enable a heads-up experience. The assumption was that the connection of music production and speed would motivate the player to creatively interact with traffic while being aware of the car speed and the surrounding traffic.

The user study shows that the motivation to engage in improvization swings between an exploration of the relationship between speed and pitch of the note, the accentuation of background music, and the translation of the surrounding into musical expression. We propose that this motivational dynamic is a very productive way to promote creativity in driving situations. However, it also strongly influences the direction of situational awareness. Creative in-car activities benefit from motivating the player to engage in a translation of the driving task into a creative product. In creative in-car games, the similarities between driving a car and playing an instrument can be exploited by phases of structure and improvization.

(3) **The dramatic Progression: the tension of stop-and-go**

AutoJam has been designed to enable a consistent feeling of progression while in traffic. The player is experiencing progression on three levels. When the car has stopped, the player can progress in the game and unlock new instruments. When the car drives with inconsistent speed, the player can jam with the unlocked instrument and finally when the car reaches a cruise speed, the song progresses and does the driving. Even though this cruise mode was only triggered a few times during the test drives, the participants reported that the alignment of game and traffic progression resulted in a consistent feeling of progressing. The alignment proofed to be a successful implementation to facilitate a feeling of progression.

However, *AutoJam* translated the tensions of stop-and-go traffic into a challenge that affords an ongoing switch between a gameful activity and a playful or creative activity. For some participants, this dialectic of instructions and improvizations turned out as excessively demanding as it would break the flow of the game. In contrast to this, musician reported that this switching of mindsets reminds them of playing in a band with phases of structures and looser parts.

One musical solution to relieve the tension between those two contrasting game modes could be to an integration of music created in one mode as content of the other mode. Another way could be the integration of game elements into freeplay-mode. Recent research on the intersection of play and games such as (Sicart 2014; Holopainen and Stain 2015; Stenros 2015) shows promising results to facilitate creative activities by gameful mechanics.

(4) **The Role of the car: play maker and instrument**

As cars transform into robots the in-car activity may shift from operational control to a more collaborative driving activity. With *AutoJam*, we explored how a playful integration of the driving context can transform creative music production into a dialog with the car or the driving situation. As the user study suggests, the experience of driving toggled between an accentuation of driving, a collaboration, and a feeling of being controlled by the car. Whereas the last interpretation (the car as a playmaker) resulted from the tensions discussed in (3), we assume that the other experiences resulted from the interpretation of speed as a feature of driving. The collaboration was perceived as an addition to driving rather than a dialog with the car. In other words, the car transformed into a medium for experiencing the driving situation from a musical perspective (with the positive side effect of a consistent feeling of progression). The findings from the interviews back this interpretation by the strong prominence of the driving context reported by the players. The connection to the car's speed accentuated the experience of driving and progressing through traffic. This perception is also manifested in the participant's desire to incorporating the expected length of the trip into the game. The song progresses with the trip; whereas the car controls the play and is played at the same time.

18.3.3 AutoRoute

Everyday millions of people commute to work by car. Car-based commuting is often experienced as a daily hassle. However, commuting also provides an important function in separating work and leisure time (Koslowsky and Kluger 1995). Moreover, the routinized driving task often provides a chance for self-reflection and relaxation. Relieved from the driving task the commuting experience in autonomous car may change fundamentally. Commuting time may be transform into extra work hours or activities that are in contrast with experiencing commuting as a transition and routine activity.

With *AutoRoute* we designed a playful navigation app that frames commuting as a fun, explorative, and self-fulfilling activity by unfolding the full potential of autonomous driving. Using *AutoRoute* enables the commuter to embody the commuting situation by navigating the self-driving car through a reflection of preferences and point-of-interests (POI). *AutoRoute* is designed for urban commuters by working as a two-stage process: in the morning, on the way to work, the commuter can predefine the route back home by selecting interesting tags and routing corresponding POIs. In the evening, on the way back home, the commuter can review the predefined route and spontaneously reroute the car to new POIs.

18.3.3.1 Conceptual Design

The concept of *AutoRoute* is based on three design goals that were identified by a contextual inquiry with Melbourne-based car commuters (Krome et al. 2016) and a review of commuter studies in cars and public transport.

Three design goals were identified. (1) The application should address commuting as an active part of the day by integrating the morning commute and the evening commute into a meaning activity. (2) The design should leverage the frustrating parts of the driving situation (in particular the car's standstill) in order to reframe them as an integral context of the applications core-interaction. (3) The potentials of car transportation should be a tangible experience such as a leverage of independency, spontaneity, or a feeling of privacy.

Based on those three goals, we designed *AutoRoute* as an in-car exploration/navigation application that is neatly connected to the car. Based on a contextual reflection of preferences, the commuter is able to spontaneously route and un-route POIs on the way. To address the emotional differences between the morning commute (i.e., the trip to work) and the evening commute (i.e., the return trip), *AutoRoute* has a slightly different mode for the morning and the evening commute.

Morning-Mode

AutoRoute morning-mode provides the user the opportunity to predefine the evening route by liking tags (descriptive keywords/phrases related to surrounding POI) and routing suggested POI. In this mode the app is completely exploratory. Tags and POIs can be selected without any limitations. In our proof-of-concept prototype the tags were vocalized by voice API but in future iteration we envision it as more subtle augmented reality visualization. We assume that in the morning, the commuter needs to take the fastest way to work and does not want any too exciting interactions. However, the commuter might be interested in a relaxed and subtle planning activity that does not require a particular attention.

Evening Mode

In the evening, when the commuter steps into the car to return home, *AutoRoute* asked the commuter to determine a time-budget for the return trip. Based on this time-budget (in the proof-of-concept the commuter was only able to choose between 20, 40 or 60 min), AutoRoute instantly calculates the best route depending on the POIs that were selected in the morning-mode.

If the commuter has routed too many POIs for the selected time-budget, he or she is asked to review the routed POIs and un-route unwanted POI. Therefore each POI indicates the amount of time required for its detour. If the commuter's time-budget is larger than the sum of the selected POI, the commuter can spontaneously route new POIs on the go.

The routing and un-routing of POIs happens while the autonomous car is driving. The car drives exactly the route that is currently selected. The spontaneous routing and un-routing of POI by the commuter is immediate reflected, i.e., the car would immediately adapt to the new route.

18.3.3.2 Interaction Design

For the proof-of-concept prototype of *AutoRoute*, we implemented the concept as a tablet application that is mounted on the dashboard of the front-seat passenger.

The central activities in both modes are the selection of tags by liking them and the routing or un-routing of POIs depending on the selected tags. In order to connect those two activities with the driving context, we toggled between the two activities based on the most fundamental driving parts: stop-and-go.

When the car is driving, the commuter can like up to five tags by touching the prominent *like* button as seen in Fig. 18.7. A liked tag is displayed in the top row of the screen. If all five spaces are full and the user likes another tag, he or she looses all selected tags. We implemented this restricting feature in order to motivate the user to deliberately choose tags and base the selection on expected traffic conditions. The selected tags are the basis for the POIs suggested in the stop phases.

In the stop-phase, such as waiting in traffic or a traffic light, the commuter can swipe through POIs as seen in Fig. 18.8. The POIs are ranked and presented to the player based on the collection of tags. If the commuter swiped to POI to the right, the POI is *routed*, i.e., added to the route. A routing of a POI may result in a change of the abstract route visualization and the expected arrival time at home.

Before starting the evening trip, the commuter is asked to choose a time-budget for the trip. This feature restricts how many POIs the route can contain. If the sum of the POIs' time-budgets exceeds the selected time-budget for the return trip, the user has to un-route selected POIs before he or she is able to route new POIs into the route.

18.3.3.3 Study Design

In order to explore the impact of *AutoRoute* on the user experience of commuting in a future autonomous car, we invited 10 Melbourne car-based commuters to test the application in the field during a usual commuting day. The work place of all participants was in Melbourne inner city (CBD and inner city suburbs). The participants' homes were located in suburban Melbourne. Their usual commuting time was between 20 and 50 min one-way depending on the suburb.

In the morning the participants were picked up according to their usual commuting schedule and were briefed on the study and the functionality of the AutoRoute prototype. The prototype was running on a tablet-pc that was mounted on the front-seat passenger dashboard. During the morning trip, the participants were able to freely route as many POIs they liked for the return trip. The commuters were informed that during the evening trip they would be able to un-route all selected POIs as desired. The participants were particularly encouraged to inquire about everything

Fig. 18.7 *AutoRoute* Driving Mode: When the car is driving, the player can like up to five tags that are vocalized during driving phases. A liked tag is added to the blank spaces in the *top row*. The tags are based on descriptions from POI in vicinity. In the bottom of the screen is as an abstract visualization of the current route. In driving mode only the progression of the route can change. The route itself can only be modified in stop-mode

unclear regarding the functionality of the app. This ensured that the participants were able to use *AutoRoute* without any assistance during the evening commute.

In the evening, the commuter was picked up from work by our improvised autonomous car service as seen in Fig. 18.9.

Our version of Wizard-of-Oz (WoZ) autonomous car setup isolates the driver from the passenger via a cardboard divider similar to (Baltodano et al. 2015). The cardboard was fitted into the car so that the driver can to see all three back mirrors but the passenger is not able to see the driving activity when sitting straight. With this setup we wanted to simulate an autonomous driving experience. However, the participants reported that they were still aware of the human driver. Nevertheless, the WoZ setup created a feeling of privacy and it prevented the participants to speak with the driver.

While the morning trip was based on the participant's usual route, the evening commute took a detour of about 20–15 min through the city of Melbourne as seen in Fig. 18.10. All POIs that the participants could route were located on this detour.

Fig. 18.8 *AutoRoute* Stop-Mode: If the car stops the route visualization moves up and presents the routing-display containing a stack of POIs. Each POI is presented with picture, name and a selection of descriptive tags. During the evening mode the POIs also contain the time required for the detour. This screenshot is during an un-routing session as indicated in *red*. Swiping the POI to the *left* would delete it from the route

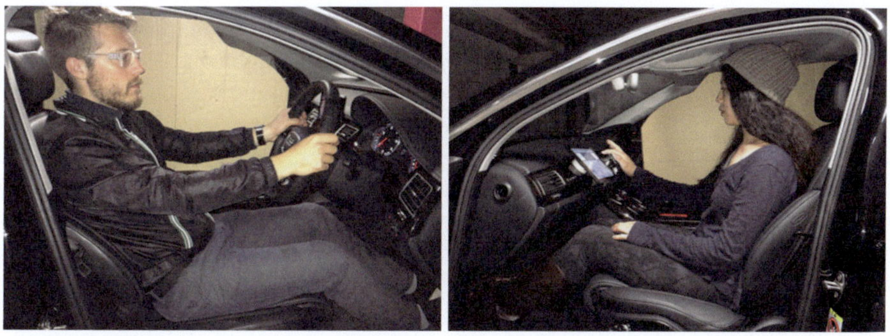

Fig. 18.9 *AutoRoute* Test Setup. Realized as Wizard-of-Oz autonomous car improvised by a cardboard divider. *Left picture* the driver. *Right picture* Setup for the study participant with AutoRoute app running on tablet-pc mounted on the passenger's dashboard

18.3.3.4 Summary of Findings

After the commuting experience with *AutoRoute*, we conducted a semi-structure in-car interview with our participants. All participant valued *AutoRoute* as an

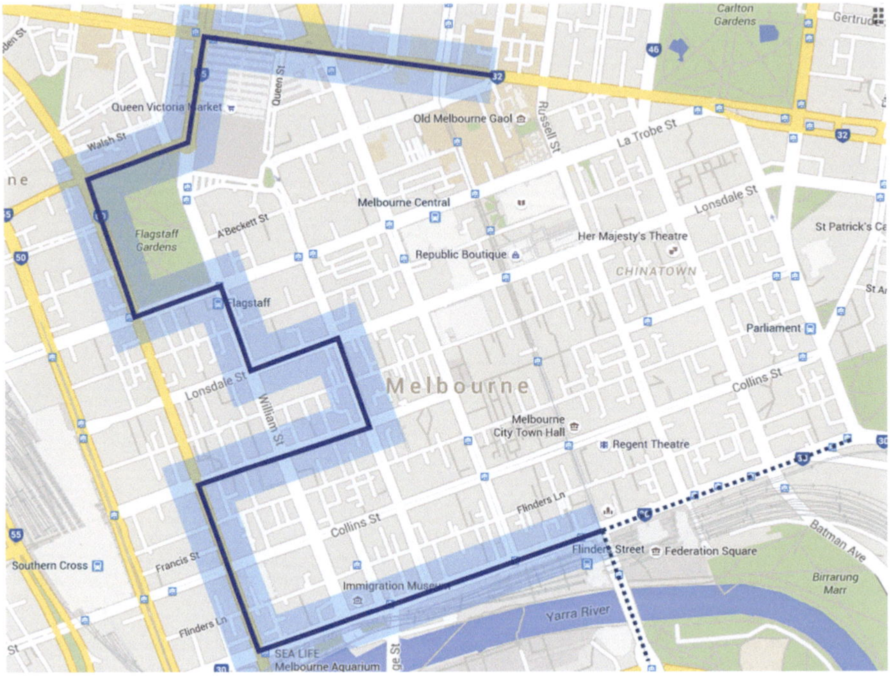

Fig. 18.10 Test route of the user studies. All POIs were located along the *blue* route. Even though all participants were traveling along the same route, they believed it was an individual route because of the AutoRoute interactions

overwhelmingly positive experience highlighting the unique approach of city exploration in everyday life. Most interestingly, even though *AutoRoute* has been designed to reframe and break the routine of given commuting patterns, the participants attributed *AutoRoute* as an experience of autonomy and independency as well as a new way of driving control. We propose that these attributes resulted from leveraging the relief from the driving task as a way to operate the car through reflection of preferences and real-world objects. *AutoRoute* integrates exploration as a feature of autonomous driving. From player interviews, we identified three design implications for promoting exploration in the context of autonomous driving.

(1) **Self-reflection as a motivator for exploration can be facilitated by an exposition to unfiltered realities of augmented information.**

Participants reported that *AutoRoute* provided the experience to navigate the city through an augmented layer of information. Hereby, the exploration consisted of a dialog between what the city has to offer (i.e., the navigation of the car) and the exposition to unfiltered tags that provoke curiosity and reasoning. As we have seen, navigating by preferences requires an understanding of what you want, i.e., a moment of self-reflection. A crucial design sensitivity for future explorative navigation systems will be the management of the inner dialog of the users with the chosen augmented layer of information related to the current location. Our findings imply that the car has a genuine potential to be a medium for exploration of new places. The features of the car, in particular the mobility and encapsulation, make it a perfect instrument for being exposed to augmented location-based realities that may be disturbing in other contexts. In other words, the evaluation of *AutoRoute* shows that the car has a great potential to facilitate an exploration of nonfiltered information.

(2) **Awareness of the commuting routine and playful exploration needs to be framed as a utility.**

AutoRoute connected the work commute with the home commute by making the home commute a topic while on the trip to work. The user study indicates that the commuting experience was fundamentally reframed by becoming a proactive, self-reflective activity. For designers of apps for the commuting context, it will be decisive to balance the commute's routine-nature with the inspiration to go new ways (literally and metaphorically). As a proof-of-concept *AutoRoute*, at least in its current state, does not provide the sensitivity for being an everyday activity. However, the user feedback clearly indicates that a more utility-based iteration such as weekday assignment or to-do list features could greatly benefit from the contextual implementation of *AutoRoute*. This implies that it is not enough to just make the user aware of the commuting activity. It is important that awareness is embedded into useful features for the everyday.

(3) **The shift to journey control requires a clear time frame to enable spontaneous decisions.**

AutoRoute was described as increasing freedom, autonomy, and control in the context of autonomous driving. The findings suggest that these feelings resulted from an alignment of *AutoRoute's* interactions with the driving context. Foremost, there is a meaningful connection of the core-interaction with the context of driving in rush-hour traffic: being exposed to new tags while driving and routing corresponding POIs when the car has stopped. Second, *AutoRoute* built on one of the strongest differentiators of car-based commuting in comparison to public transportation: the feeling of independence to potentially go everywhere and stop or return instantly. This spontaneity was additionally enforced through maintaining a journey control, i.e., defining the length of the trip by selecting a time-budget. The experience interviews implied that the connection of a contextual structure of suggesting preferences together with the leverage of instant and spontaneous rerouting in a consequence-free setting (i.e., the time-budget) resulted in a feeling of freedom and control over the situation. We assume that the user experience of autonomous driving may improve when we redefine the role of the car as a medium that facilitates a dialog between the city and the commuter's preferences.

18.4 Motivational Affordances

Even though the three prototypes share a very similar driving scenario, routine trips in a fully autonomous vehicle, the user studies show that they trigger fundamentally different transportation experience. Reflecting on the conceptual design of the three prototypes in the context of the results from the user studies, we identified a strong connection between the experience of the player and the motivational affordances of driving. Driving a car can be motivated by much more then just instrumental reasons of going efficiently from one place to another. Also affective and symbolic motives such as competency, thrill, status, and a feeling of independency are important values of driving (Steg 2005). These hedonic values of driving seem to be obscured by driving automation promoting a decoupling of driver and car. By implementing non-driving activities into the context of an autonomous driving, the *AutoPlay* prototypes helped us to translate some of the hedonic values of driving into the new context of the autonomous car. In the following, we present a generalized conception of motivational affordances that transformed autonomous driving into pleasurable interactive experience with a purpose in itself.

"Drivability" as a play of control

In traditional, human driven cars *drivability* is arguably one of the most important motivational affordances. The operation of driving required skill and a direct connection between the car and driver in order to enable a feeling of control over the situation. *Drivability* of a car affords a feeling of mastery and control over the situation. In the context of autonomous driving *drivability* as a motivational affordance require a new interactive basis other than the operation of the vehicle to generate a feeling of control.

In autonomous driving research, control and trust as an UX challenge has gained a lot of attention. So far most approaches try to translate cues of the driving situation in audible, visual, or tactile stimuli with the goal to afford a feeling of orientation and control such as (Beattie et al. 2013; Koo et al. 2015). In contrast to this *AutoGym* promotes *drivability* in an interactive way. The player can actively exert control over the car mediated by playful competition with the power of the car or the ambiguity of traffic.

The results from the *AutoGym* user study show that depending on game design and physical condition of the player, a feeling of control over the situation can be generated in several ways: as an understanding of traffic (car as collaborator), as playing with the cars behavior (car as companion) or as an overcoming of the power of the car (car as competitor). Even though all interpretations could possibly be designed without exertion as base activity, the traffic context turned out to be a very stimulating frame for exercises.

"Performability" as a play of progression and self-expression

Car ownership has always been motivated by symbolic values; be it as status symbol and lifestyle object. But also driving a car can be seen as a performance in itself; as an embodiment of the car design. In that way driving resembles a performance not only to demonstrate the car values and power but also to express personality. Moreover, traffic is a genuine social situation even though road users often ignore each other when standing side by side. With autonomous driving new non-driving activities become possible. These activities may challenge the private/public context of traffic situations and as long as it is not possible to transform the car in a complete private space, there will be a space to translate self-expression from car design and driving styles into an interactive performance. As *performability* we understand the interaction possibility that affords self-expression and creative play within the situation. The goal of *performability* is to reframe the situation into a new experience. In the case of *AutoJam*, the frustration of stop-and-go traffic was reframed into a creative experience of progressing through traffic. We believe that the interactive possibilities of autonomous driving can provide a large set of non-driving activities that will actively help to reframe the driving situations into various pleasurable experiences. There are already several research projects that have aimed on the roads a social place (Juhlin 2010) and in future cars with more interactive possibilities of the driver and an advanced car-to-x communication, driving would benefit from a new social frame for games, social media, or creative interactions.

"Explorability" as a play of self-reflection and surroundings

Driving and car ownership has always been associated with independency and the freedom to spontaneously change route, stop or explore and discovery new places and locations. Even though car trips are in general routine-based or destination-focused, the possibility to make an instant and spontaneous change (of route or destination) is literally in the hands of the driver. In the context of traditional cars this innate quality of driving affords a feeling of independency and has a close relation of being in control, i.e., drivability.

In a future of fully autonomous driving, in which the car passenger simply needs to enter the destination to get there, independency and subsequently the feeling of autonomy, seem to become obsolete qualities. However, as the experience of *AutoRoute* demonstrates, the alignment of preferences with surrounding points-of-interests of the trip enables a feeling of autonomy and independency as a function of self-reflection and exploration. This *explorability* redefines navigating with autonomous car as a spontaneous activity.

AutoRoute implements *explorability* through a reflection of own preferences in relation to things the world has to offer. By embedding *explorability* into an everyday frame such as commuting, together with the implantation of a timeframe for the activity, *AutoRoute* provided a sandbox for exploration. The interactive setup of *AutoRoute* promoted spontaneous interactions (addressing the car) by making options available that are based on the driving context. This implementation as a driving feature transferred self-reflection into an experience of navigating the car and most importantly it affords a feeling of independency and autonomy.

18.5 Conclusion

In this chapter, we presented concepts and designs of three *AutoPlay* prototypes. The prototypes have been designed as inspirational artifacts that demonstrate a possible implementations and integrations of non-driving activities into the context of autonomous driving. In order to gain insight about the impact on player and driving experience, we conducted a series of user studies with the prototypes. The results show that our implementation of non-driving activities is experienced as a pleasurable enhancement of the driving context and reframes frustrating traffic into a meaningful context for exercises, music creation, or exploration.

Furthermore, the user studies indicate that each of the three prototypes accentuates one basic psychological need in a pleasurable way: *AutoGym* through a play of control and self-efficacy, *AutoJam* as self-actualization and mastery, and *AutoRoute* as independency and autonomy. We reflect on those results by proposing three motivational affordances of autonomous driving: *drivability*, *performability*, and *explorability*. These three affordances became tangible through the interaction with the prototypes and represent inherent motivational possibilities of traditional cars that seem to be disguised by the autonomous driving context. The affordances act as starting points for discussing meaningful alignment of non-driving activities in the design space of autonomous driving. We believe that it is essential for the user experience of future autonomous driving to consider these affordances as lenses for fulfilling driving experiences.

References

Alt, F., Kern, D., Schulte, F., et al. (2010). Enabling micro-entertainment in vehicles based on context information. In *Proceedings of the 2nd International Conference on Automotive User Interfaces and Interactive Vehicular Applications* (pp. 117–124). ACM.

Baltodano, S., Sibi, S., Martelaro, N., et al. (2015). The RRADS platform: A real road autonomous driving simulator. In *Proceedings of the 7th International Conference on Automotive User Interfaces and Interactive Vehicular Applications* (pp. 281–288). ACM.

Beattie, D., Baillie, L., Halvey, M., McCall, R. (2013). Maintaining a sense of control in autonomous vehicles via auditory feedback. In *Proceedings of PQS Conference, Austria.*

Beattie, D., Baillie, L., Halvey, M., McCall, R. (2014). What's around the corner? enhancing driver awareness in autonomous vehicles via in-vehicle spatial auditory displays. In *NordiCHI'14* (p. 10). ACM.

Bengler, K., Dietmayer, K., Farber, B., et al. (2014). Three decades of driver assistance systems: Review and future perspectives. *IEEE Intelligent Transportation Systems Magazine, 6*, 6–22.

Braun, V., & Clarke, V. (2006). Using thematic analysis in psychology. *Qualitative Research in Psychology, 3*, 77–101.

Bull, M. (2004). Automobility and the power of sound. *Theory, Culture & Society, 21*, 243–259. doi:10.1177/0263276404046069.

Diefenbach, S., Kolb, N., & Hassenzahl, M. (2014). The "Hedonic" in Human-Computer Interaction—History, Contributions, and Future Research Directions. In *Proceedings of the DIS 2014*, pp. 305–314. doi:10.1145/2598510.2598549.

Eckoldt, K., Knobel, M., Hassenzahl, M., & Schumann, J. (2012). An experiential perspective on advanced driver assistance systems. *Information Technology, 54*, 165–171. doi:10.1524/itit.2012.0678.

Gibson, J. J. (2014). *The ecological approach to visual perception: Classic edition*. New York: Psychology Press.

Goffman, E. (1975). The presentation of self in everyday life.

Goffman, E. (1974). *Frame analysis: An essay on the organization of experience*. Cambridge: Harvard University Press.

Holopainen, J., & Stain, M. (2015) Dissecting playfulness for practical design. *Gameful World Approaches*, Issues, Appl 419.

Juhlin, O. (2010). *Social media on the road: The future of car based computing*. London: Springer.

Koo, J., Kwac, J., Ju, W., et al. (2015). Why did my car just do that? Explaining semi-autonomous driving actions to improve driver understanding, trust, and performance. *International Journal on Interactive Design and Manufacturing, 9*, 269–275. doi:10.1007/s12008-014-0227-2.

Koslowsky, M., & Kluger, A. N. (1995). *Commuting stress: Causes, effects and methods of coping*. New York: Springer.

Krome, S. (2017a). AutoGym: An exergame for autonomous driving. Manuscript submitted for publication.

Krome, S. (2017b). Autojam: Exploring interactive music experiences in stop-and-go traffic. Manuscript submitted for publication.

Krome, S. (2017c). AutoRoute: Designing pleasurable commuting experiences for future autonomuos driving. Manuscript submitted for publication.

Krome, S., Goddard, W., Greuter, S., et al. (2015). A context-based design process for future use cases of autonomous driving: Prototyping AutoGym. In *AutomotiveUI* 2015.

Krome, S., Walz, S. P., Greuter, S. (2016). Contextual inquiry of future commuting in autonomous cars. In *Proceedings of the 2016 CHI Conference Extended Abstracts on Human Factors in Computing Systems* (pp. 3122–3128). ACM.

Lyons, G., Jain, J., & Holley, D. (2007). The use of travel time by rail passengers in Great Britain. *Transportation Research Part A: Policy and Practice, 41*, 107–120. doi:10.1016/j.tra.2006.05.012.

Norman, D. A. (2013). *The design of everyday things: Revised and expanded edition*. Basic Books.

Proper, K. I., Koning, M., Van der Beek, A. J., et al. (2003). The effectiveness of worksite physical activity programs on physical activity, physical fitness, and health. *Clinical Journal of Sport Medicine, 13*, 106–117.

Ryan, R. M., Rigby, C. S., & Przybylski, A. (2006). The motivational pull of video games: A self-determination theory approach. *Motivation and Emotion, 30*, 344–360. doi:10.1007/s11031-006-9051-8.

Sheldon, K. M., Elliot, A. J., Kim, Y., & Kasser, T. (2001). What is satisfying about satisfying events ? *Testing 10 Candidate Psychological Needs, 80*, 325–339. doi:10.1037//O022-3514.80.2.325.

Sheller, M. (2004). Automotive emotions: Feeling the car. *Theory, Culture & Society, 21*, 221–242. doi:10.1177/0263276404046068.

Sicart, M. (2014). *Play matters*. New York: MIT Press.

Steg, L. (2005). Car use: lust and must. Instrumental, symbolic and affective motives for car use. *Transportation Research Part A: Policy and Practice, 39*, 147–162. doi:10.1016/j.tra.2004.07.001.

Stenros, J. (2015). Playfulness, play, and games: A constructionist ludology approach.

Taylor, A. H. A., & Dorn, L. (2006). Stress, fatigue, health, and risk of road traffic accidents among professional drivers: The contribution of physical inactivity. *Annual Review of Public Health, 27*, 371–391. doi:10.1146/annurev.publhealth.27.021405.102117.

Zhang, P. (2008). Motivational affordances: Reasons for ICT Design and Use. *Communications of the ACM, 51*, 145–147. doi:10.1145/1400214.1400244.

Zimmerman, J., Forlizzi, J., & Evenson, S. (2007). Research through design as a method for interaction design research in HCI. In *Proceedings of the SIGCHI Conference on Human Factors in Computing Systems—CHI '07 493*. doi:10.1145/1240624.1240704.

Erratum to: AutoPlay: Unfolding Motivational Affordances of Autonomous Driving

Sven Krome, Jussi Holopainen and Stefan Greuter

Erratum to:
Chapter 18 in: G. Meixner and C. Müller (eds.), *Automotive User Interfaces*, Human–Computer Interaction Series, DOI 10.1007/978-3-319-49448-7_18

The original version of the book was inadvertently published without removing Table 18.2 in Chap. 18. The erratum chapter and the book have been updated with the change.

The updated original online version for this chapter can be found at http://dx.doi.org/10.1007/978-3-319-49448-7_18

G. Meixner and C. Müller (eds.), *Automotive User Interfaces*,
Human–Computer Interaction Series, DOI 10.1007/978-3-319-49448-7_19

MIX
Papier aus verantwortungsvollen Quellen
Paper from responsible sources
FSC® C105338

If you have any concerns about our products,
you can contact us on
ProductSafety@springernature.com

In case Publisher is established outside the EU,
the EU authorized representative is:
Springer Nature Customer Service Center GmbH
Europaplatz 3, 69115 Heidelberg, Germany

Printed by Libri Plureos GmbH
in Hamburg, Germany